Higher Dimensional Categories

Categories

——— From Double to Multiple Categories ———

Higher Dimensional Categories

-------- From Double to Multiple Categories --------

Marco Grandis

Università di Genova, Italy

World Scientific

NEW JERSEY · LONDON · SINGAPORE · BEIJING · SHANGHAI · HONG KONG · TAIPEI · CHENNAI · TOKYO

Published by

World Scientific Publishing Co. Pte. Ltd.

5 Toh Tuck Link, Singapore 596224

USA office: 27 Warren Street, Suite 401-402, Hackensack, NJ 07601

UK office: 57 Shelton Street, Covent Garden, London WC2H 9HE

British Library Cataloguing-in-Publication Data

A catalogue record for this book is available from the British Library.

HIGHER DIMENSIONAL CATEGORIES
From Double to Multiple Categories

ISBN 978-981-120-510-1

For any available supplementary material, please visit
https://www.worldscientific.com/worldscibooks/10.1142/11406#t=suppl

To

Bob Paré

Preface

The study of higher dimensional categories has mostly been developed in the *globular* form of 2-categories, n-categories, ω-categories and their weak versions. Here we study a different form: double categories, n-tuple categories and multiple categories, with their weak and lax versions.

We want to show the advantages of this form for the theory of adjunctions and limits. Furthermore, this form is much simpler in higher dimension, starting with dimension three where weak 3-categories (also called tricategories) are already quite complicated, much more than weak or lax triple categories.

This book can be used as a textbook for graduate and postgraduate studies, and as a basis for self-study and research. Notions are presented in a 'concrete' way, with examples and exercises; the latter are endowed with a solution or hints. Part I, devoted to double categories, starts at basic category theory and is kept at a relatively simple level. Part II, on multiple categories, can be used independently by a reader acquainted with 2-dimensional categories.

Contents

Introduction

0.1 Strict and weak double categories

The first part of this study of higher dimensional categories presents the 2-dimensional case of double categories.

Double categories, in the strict sense, were introduced by C. Ehresmann, in 1962 [Eh1]. Weak double categories, also called pseudo double categories, began to be studied much later, in our series [GP1]–[GP4] (with R. Paré, as many other joint papers cited below). Other references are given below, and in Chapter 3.

We outline below some advantages of weak double categories, with respect to the better-known (and more particular) structure of bicategories, in the domain of adjunctions and limits. This requires some elementary notions of ordinary category theory, and the beginning of 2-dimensional categorical structures.

Essentially, we shall see that in a general adjunction of dimension 2, the left adjoint is colax and the right one is lax: they cannot be composed, but can be viewed as a vertical and a horizontal arrow in a suitable double category, where their adjunction lives.

0.2 A problem with adjunctions

Working for simplicity in the category **Ab** of abelian groups, let us start from the exponential law for a fixed abelian group A. It is a well-known adjunction, between two endofunctors F, G coming from the symmetric monoidal closed structure of **Ab**

$$F \colon \mathbf{Ab} \rightleftarrows \mathbf{Ab} \colon G, \qquad F(X) = X \otimes A, \quad G(Y) = \mathrm{Hom}(A, Y), \qquad (0.1)$$

$$\eta \colon 1 \to GF, \qquad \eta X \colon X \to \mathrm{Hom}(A, X \otimes A), \quad (\eta X(x))(a) = x \otimes a,$$

$$\varepsilon \colon FG \to 1, \qquad \varepsilon Y \colon \mathrm{Hom}(A, Y) \otimes A \to Y, \quad (\varepsilon Y)(h \otimes a) = h(a).$$

1

The left adjoint F preserves all colimits (and is right exact) while the right adjoint G preserves all limits (and is left exact).

The reader probably knows that the category **Ab** can be embedded in a larger category Rel**Ab**, of abelian groups and relations, extensively used in Homological Algebra (and reviewed in Section 2.1). A relation $u\colon A \nrightarrow B$ of abelian groups is a subgroup of the cartesian product $A \times B$, and will often be denoted by a dot-marked arrow. A homomorphism $A \to B$ is identified with the relation formed by its graph in $A \times B$.

This category of relations has an elementary structure of (locally ordered) 2-category, where a 2-cell $u \leqslant u'$ between two relations $u, u' \colon A \nrightarrow B$ amounts to an inclusion of subgroups of $A \times B$.

We shall see (in Section A4 of Appendix A) that the functors F, G can be extended to the 2-category of relations, obtaining – respectively – a *colax* functor and a *lax* functor Rel**Ab** $\to$ Rel**Ab**

$$F' = \mathrm{Rel}(F) \quad (colax), \qquad\qquad G' = \mathrm{Rel}(G) \quad (lax). \qquad (0.2)$$

This means that the composition of relations is only preserved up to comparisons, which here are just inequalities

$$F'(vu) \leqslant F'(v).F'(u), \qquad G'(v).G'(u) \leqslant G'(vu). \qquad (0.3)$$

Composing F' and G' would destroy their comparisons, and *extending the previous adjunction makes problems.*

(Let us note that F' is a functor if and only if F is exact, which means that A is torsion free. Similarly G' is a functor if and only if G is exact, which means that A is a free abelian group.)

0.3 The extension

This problem can be overcome if we 'amalgamate' **Ab** and Rel**Ab** in a two-dimensional structure, the double category $\mathbb{R}$el**Ab** of abelian groups, homomorphisms and relations (note the different notation for $\mathbb{R}$el), which has:

- the same *objects* as these two categories (namely the abelian groups),
- *horizontal arrows* as in **Ab**, with its identities and composition,
- *vertical arrows* as in Rel**Ab**, with its identities and composition,
- double cells $\alpha \colon (u \, {}^f_g \, v)$ as below

$$
\begin{array}{ccc}
A & \xrightarrow{\ f\ } & A' \\
u \downarrow & \leqslant & \downarrow v \\
B & \xrightarrow[\ g\ ]{} & B'
\end{array}
\qquad\qquad gu \leqslant vf. \qquad (0.4)
$$

(This structure is *flat*, i.e. each double cell is determined by its boundary. This has the advantage of simplicity and the drawback of not being 'representative' of general double categories, in the same way as the simple 2-categorical structure of Rel**Ab** given by the ordering $u \leqslant v$ is not representative of general 2-categories.)

Double cells have a horizontal composition $\alpha \mid \beta$, determined by composition in **Ab**, and a vertical composition $\frac{\alpha}{\gamma}$, determined by composition in Rel**Ab**. Both laws are categorical, and related by the interchange property. The 2-category Rel**Ab** is 'vertically' embedded in the double category, as the substructure whose horizontal arrows are identities.

Now the functors F, G can be extended to the double categories of relations, obtaining – respectively – a colax and a lax *double* functor (which also extend the previous F', G')

$$
\begin{aligned}
F'' &= \mathbb{Rel}(F) \colon \mathbb{Rel}\mathbf{Ab} \to \mathbb{Rel}\mathbf{Ab} & (colax), \\
G'' &= \mathbb{Rel}(G) \colon \mathbb{Rel}\mathbf{Ab} \to \mathbb{Rel}\mathbf{Ab} & (lax).
\end{aligned}
\tag{0.5}
$$

This means that the horizontal composition (of homomorphisms) is preserved, while the vertical composition (of relations) is preserved up to comparisons, as above, in (0.3). We still cannot (reasonably) compose lax and colax double functors, because their comparisons have conflicting directions.

Yet the whole adjunction can be extended, within the theory of double categories: the unit η and counit ε of the extended adjunction are double cells (with $\mathbb{X} = \mathbb{A} = \mathbb{Rel}\mathbf{Ab}$)

$$
\tag{0.6}
$$

which – leaving apart obvious set-theoretical problems of size – live in a crucial (strict) double category $\mathbb{Dbl}$ of weak double categories, with lax double functors as horizontal arrows, colax double functors as vertical arrows and suitably defined double cells (see Section 4.2). The triangular equations take the form

$$
\frac{\eta}{\varepsilon} = 1_{F''}, \qquad \varepsilon \mid \eta = e_{G''},
$$

where the double cell $1_{F''}$ is a horizontal identity, and $e_{G''}$ a vertical one.

Double adjunctions compose, by pasting units and counits in $\mathbb{Dbl}$. For $F \dashv G$ and $H \dashv K$, the unit and the counit of the composed adjunction are obtained as follows

$$
\begin{array}{ccc}
\mathbb{X} == \mathbb{X} == \mathbb{X} & \qquad & \mathbb{B} \xrightarrow{K} A \xrightarrow{G} \mathbb{X} \\
F\downarrow \quad 1 \quad F\downarrow \quad \eta \quad \| & & \| \quad e \quad \| \quad \varepsilon \quad \downarrow F \\
A == A \ -G\!\!\to \mathbb{X} & & \mathbb{B} \ -K\!\!\to A == A \\
H\downarrow \quad \eta' \quad \| \quad e \quad \| & & \| \quad \varepsilon' \quad H\downarrow \quad 1 \quad \downarrow H \\
\mathbb{B} \xrightarrow{K} A \xrightarrow{G} \mathbb{X} & & \mathbb{B} == \mathbb{B} == \mathbb{B}
\end{array}
\tag{0.7}
$$

Moving to a different subject, it is also interesting to note that the double category $\mathbb{R}\mathrm{el}\mathbf{Ab}$ has all (horizontal) double limits and colimits (see Chapter 5), while the 2-category $\mathrm{Rel}\mathbf{Ab}$ even lacks products and a terminal object.

As an elementary example, every family of relations $u_i \colon A_i \nrightarrow B_i$ $(i \in I)$ has an obvious cartesian product

$$
u \colon \textstyle\prod_i A_i \nrightarrow \prod_i B_i,
$$

$$
u = \{((a_i), (b_i)) \mid (a_i, b_i) \in u_i, \text{ for all } i \in I\}.
\tag{0.8}
$$

Its projections π_i are double cells

$$
\begin{array}{ccc}
A \xrightarrow{\ p_i\ } A_i & \qquad & A = \textstyle\prod_i A_i, \\
u\downarrow \quad \leqslant \quad \downarrow u_i & & \\
B \xrightarrow[\ q_i\ ]{} B_i & & B = \textstyle\prod_i B_i,
\end{array}
\tag{0.9}
$$

and satisfy the adequate universal property with respect to the horizontal composition of double cells. The reader will note that the vertical arrows are playing a role of higher dimensional objects, with morphisms $\pi_i \colon u \to u_i$ given by double cells.

As we shall see in Chapter 4, the existence of functorial double products in $\mathbb{R}\mathrm{el}\mathbf{Ab}$ essentially means that:

- the ordinary category of horizontal arrows (i.e. $\mathbf{Ab}$) has products,

- the ordinary category of vertical arrows and double cells (with horizontal composition) has products,

- the two procedures agree with respect to vertical domain, codomain and identities.

Loosely speaking, the sense of what we are doing by replacing a 2-category of relations with the corresponding double category can be outlined as follows: arrows which are *too relaxed* (like relations, spans, profunctors, ...) or *too strict* (like adjoint pairs) to have good universal properties can be studied in a (possibly weak) double category, correlating them with more ordinary 'main' arrows: those 'which preserve the structure'. We use thus the main direction to study the secondary one.

0.4 Other examples

For a reader acquainted with the bicategories SpanSet and CospSet, of spans or cospans of sets, we can outline another example of an adjunction which only makes sense within weak double categories. It will be dealt with in 4.5.6.

First, we amalgamate the category **Set** with the bicategory SpanSet, forming a weak double category SpanSet.

Objects, horizontal arrows and their composition are as in **Set**. A vertical arrow $u\colon X \nrightarrow Y$ is a span $X \leftarrow U \rightarrow Y$ of sets. A cell $\alpha\colon (u\,{}^{f}_{g}\,v)$ is a natural transformation $u \rightarrow v$ of these diagrams and amounts to the commutative right diagram below (in **Set**), with a middle arrow $m\alpha\colon U \rightarrow V$ completing the boundary (note that this structure is not flat)

$$
\begin{array}{ccc}
X & \xrightarrow{\ f\ } & X' \\
u\downarrow & \alpha & \downarrow v \\
Y & \xrightarrow[g]{} & Y'
\end{array}
\qquad
\begin{array}{c}
X \xrightarrow{\ f\ } X' \\[2pt]
{}^{u'}\nearrow \qquad \nearrow{}^{v'} \\
U \xrightarrow{m\alpha} V \\[2pt]
{}^{u''}\searrow \qquad \searrow{}^{v''} \\
Y \xrightarrow[g]{} Y' \quad \text{SpanSet}
\end{array}
\tag{0.10}
$$

The horizontal composition $\alpha|\beta$ of α with a second cell $\beta\colon v \rightarrow w$ is a composition of natural transformations. The vertical composition of spans is computed with pullbacks, as in the bicategory SpanSet, and keeps its comparisons (as double cells with trivial horizontal arrows); this composition is extended to double cells, in the obvious way.

Similarly we amalgamate the category **Set** with the bicategory CospSet, forming a weak double category CospSet, with the following double cells

$$
\begin{array}{ccc}
X & \xrightarrow{\ f\ } & X' \\
u\downarrow & \alpha & \downarrow v \\
Y & \xrightarrow[g]{} & Y'
\end{array}
\qquad
\begin{array}{c}
X \xrightarrow{\ f\ } X' \\[2pt]
{}^{u'}\searrow \qquad \searrow{}^{v'} \\
U \xrightarrow{m\alpha} V \\[2pt]
{}^{u''}\nearrow \qquad \nearrow{}^{v''} \\
Y \xrightarrow[g]{} Y' \quad \text{CospSet}
\end{array}
\tag{0.11}
$$

Now, there is a pair of 'double functors' which are the identity on objects and horizontal arrows

$$
F\colon \text{SpanSet} \rightleftarrows \text{CospSet} \colon G.
\tag{0.12}
$$

F acts on spans by pushout, and is colax; G acts on cospans by pullback, and is lax. Again we do not want to compose them, but we do have a colax-lax adjunction, whose unit and counit are double cells in $\mathbb{D}$bl, as in (0.6).

If we come back to the bicategories SpanSet and CospSet, we can view them as 'vertical' weak double categories, with trivial horizontal arrows. The previous adjunction can be restricted, because its unit and counit are trivial on the objects, but still lives in the double category $\mathbb{D}$bl, and cannot be interpreted in an n-category, strict or weak: any globular structure, of any dimension, can only have one sort of arrows.

Also here the weak double categories $\mathbb{S}$pan**Set** and $\mathbb{C}$osp**Set** have all double limits and colimits, while the bicategories Span**Set** and Cosp**Set** lack most limits and colimits.

0.5 Higher dimensional categories

The rest of this introduction is more technical and addressed to readers with some knowledge of higher category theory.

Weak double categories can be extended to *weak multiple categories*, keeping one strict direction, now called the *transversal direction*, or direction 0, and allowing infinitely many weak ones, called the *geometric directions*, indexed by an integer $i > 0$. The 2-dimensional truncation of a weak multiple category is a weak double category, with a strict direction $i = 0$ and one geometric direction $i = 1$.

Let us recall that the (more studied) *globular* form of 2-categories, n-categories and ω-categories is based on a (possibly truncated) *globular set*. This is a system X of sets X_n and mappings, called *faces* (∂^α, with $\alpha = \pm$) and *degeneracies* (e)

$$X_0 \underset{e}{\overset{\partial^\alpha}{\leftleftarrows}} X_1 \underset{e}{\overset{\partial^\alpha}{\leftleftarrows}} X_2 \ \ldots \qquad X_{n-1} \underset{e}{\overset{\partial^\alpha}{\leftleftarrows}} X_n \ \ldots \qquad (0.13)$$

that satisfy the globular relations: $\partial^\alpha \partial^- = \partial^\alpha \partial^+$, $\partial^\alpha e = \mathrm{id}$.

Here we are interested in a different, more general setting, that was introduced by Charles Ehresmann before the globular one and studied with Andrée C. Ehresmann [Eh2, BaE, EE]: the *multiple* form of double categories, n-tuple categories and multiple categories, based on a *multiple set*.

The latter is a system X of sets $X_{i_1, i_2, \ldots, i_n}$, indexed by multi-indices $\mathbf{i} = \{i_1, \ldots, i_n\} \subset \mathbb{N}$. It has faces and degeneracies, for $i_1 < \ldots < i_j < \ldots < i_n$ and $\alpha = \pm$

$$\begin{aligned} \partial_{i_j}^\alpha : X_{i_1, \ldots, i_n} &\longrightarrow X_{i_1, \ldots, \hat{i}_j, \ldots, i_n}, \\ e_{i_j} : X_{i_1, \ldots, \hat{i}_j, \ldots, i_n} &\longrightarrow X_{i_1, \ldots, i_n}, \end{aligned} \qquad (0.14)$$

that satisfy the *multiple relations* (see 6.2.2). A multiple category has categorical compositions in each direction $i \geqslant 0$

$$x +_i y \qquad (x, y \in X_{\mathbf{i}}, \ i \in \mathbf{i}, \ \partial_i^+ x = \partial_i^- y), \qquad (0.15)$$

and units $e_i(x)$ (for $x \in X_{\mathbf{i}}$ and $i \notin \mathbf{i}$).

This structure has been generalised to *weak* and *lax* multiple categories in [GP6]–[GP10]. The transversal direction $i = 0$ still has a categorical composition, but the geometric directions $i > 0$ are only assumed to have a weakly categorical composition, up to invertible transversal cells (see Section 6.4). In the partially lax case of a *chiral* multiple category (see 6.4.8) the geometric composition laws in directions $i < j$ have a directed interchange χ_{ij} which is not assumed to be invertible.

An important 3-dimensional example is the chiral triple category $\mathsf{SC}(\mathbf{Set})$ $= \mathsf{S}_1\mathsf{C}_1(\mathbf{Set})$ *of spans and cospans of sets* (in Section 6.6), where:

- 0-directed arrows are mappings, 1-directed arrows are spans of sets, 2-directed arrows are cospans,

- 12-cells are spans of cospans, or – equivalently – cospans of spans,

- 012-cells are natural transformations of the latter.

More generally, for a category $\mathbf{C}$ with pullbacks and pushouts, we have a chiral n-tuple category $\mathsf{S}_p\mathsf{C}_q(\mathbf{C})$ for $p, q > 0$ and $n = p + q + 1$: its geometric i-directed arrows are spans of $\mathbf{C}$ for $0 < i \leqslant p$ and cospans of $\mathbf{C}$ for $p < i \leqslant p + q$. There are various infinite-dimensional extensions, like the 'unbounded' chiral multiple category $\mathsf{S}_{-\infty}\mathsf{C}_\infty(\mathbf{C})$, with indices in $\mathbb{Z}$ (see 6.6.4).

Other examples can be found in Section 6.4: for instance the weak multiple category $\mathsf{Span}(\mathbf{C})$ of cubical spans over a category $\mathbf{C}$ with pullbacks, or the weak multiple category $\mathsf{Cosp}(\mathbf{C})$ of cubical cospans over a category $\mathbf{C}$ with pushouts (see 6.4.6).

Let us remark that the weak forms of multiple categories are *much simpler* than the globular ones, because here all the weak composition laws are associative, unitary and interchangeable up to *invertible* cells in the strict 0-indexed direction; the latter are strictly coherent. This aspect has been discussed in [GP5], where we showed how the 'simple' comparisons of a weak triple category (of cubical type) produce – via some associated cells – the 'complicated' ones of a tricategory.

Cubical categories can be viewed as a particular case of multiple categories, based on the geometry of *cubical sets*, well-known from Algebraic Topology; see 6.2.3, 6.3.1 and Appendix B.

0.6 Outline

Each chapter has its own introduction; here we only give a brief synopsis.

The first chapter is a review of the theory of ordinary (1-dimensional) categories. Chapter 2 briefly introduces 2-dimensional categorical structures, from ordered categories to 2-categories and bicategories (the weak version of 2-categories).

Chapter 3 deals with the general 2-dimensional categorical structures, namely strict and weak double categories, together with their 'functors' and 'transformations'. Various examples are given.

Chapters 4 and 5 are about adjunctions and limits for weak double categories, and are based on [GP1, GP2].

In Chapter 6 we define weak and lax multiple categories; our main form, partially lax, is called a *chiral multiple category*. Various examples are given. We also consider *intercategories*, a laxer form of multiple category introduced in [GP6, GP7] that comprises, besides weak and chiral multiple categories, other 3-dimensional structures studied in the literature, like duoidal categories, Gray categories, Verity double bicategories and monoidal double categories.

Chapter 7, about adjunctions in chiral multiple categories, is essentially based on [GP11]. Chapter 8 deals with monads and algebras in the same context, and is mostly new. The main non-standard results deal with the idempotent case. Multiple limits are only considered here in particular cases; the reader is referred to [GP10], for a general study.

Appendix A is about applications of weak double categories in Homological Algebra (from [G8]), and Algebraic Topology (from [G2]–[G4]).

Appendix B is about symmetries in cubical sets and weak cubical categories, from [G2, G3, G5]. It also gives a detailed construction of Cosp(**C**) as a weak cubical category, in Section B4.

Finally, Appendix C contains solutions or hints for the exercises of the text. However, the solution can follow the exercise when it is important in itself or to be used later on.

0.7 Literature

Higher dimensional category theory began with (strict) double and multiple categories, introduced and studied by C. Ehresmann and A.C. Ehresmann [Eh1, Eh2, Eh3, Eh4, BaE, EE], since 1962.

Higher category theory in globular form is studied in many papers and books; we only cite: Bénabou [Be1] for bicategories; Gordon, Power and Street [GPS] for tricategories; Leinster [Le] for weak ω-categories.

Infinite dimensional weak and lax multiple categories have been introduced in [GP8]–[GP10]; some of their variations (possibly of cubical type) have also been treated in the following papers (among others):

- Gray categories: [Gra1],

- Verity double bicategories: [Ve],

- weak double categories: [GP1, GP2, GP3, GP4, GP11, GP13],

- strict cubical categories: [AlBS],

- weak and lax cubical categories: [G2]–[G6], [G10],

- duoidal (or 2-monoidal) categories: [AM, BkS, St5],

- monoidal double categories: [Shu],

- weak triple categories and 3-dimensional intercategories: [GP6, GP7],

- links between the cubical and the globular setting, in the strict case: [BroM, AlBS],

- the same in the weak 3-dimensional case: [GP5].

0.8 Notation and conventions

The symbol $\subset$ denotes *weak* inclusion. For an n-tuple $(i_1, ..., i_n)$, the symbol $(i_1, ..., \hat{i}_j, ..., i_n)$ means that the term i_j has been taken out.

As usual, the symbols $\mathbb{N}, \mathbb{Z}, \mathbb{R}, \mathbb{C}$ denote the sets of natural, integral, real or complex numbers; $\mathbb{N}^*$ is the subset of the positive integers.

Categories, 2-categories and bicategories are generally denoted as $\mathbf{A}, \mathbf{B}$...; double categories as $\mathbb{A}, \mathbb{B}$...; multiple categories (of dimension three or higher) as A, B...

A section, subsection or part marked with * deals with some topic out of the main line of this book, and is generally addressed to readers having some knowledge of the subject.

0.9 Acknowledgements

The core of this book is based on a series of joint works with Robert Paré, on double and multiple categories [GP1]–[GP14], the result of a long and very pleasant collaboration which began in 1993. It would have been natural – and a pleasure to me – to write a joint book on this subject. After proposing him this project, for some years, I had to desist. I can only retaliate by dedicating this book to him.

The author is grateful to Andrée C. Ehresmann for a lasting support; many of the papers mentioned above were published in 'Cahiers'.

The research at the basis of this book was supported by Università di Genova, MIUR, GNSAGA-INDAM, CNR (Italy) and NSERC (Canada).

Part I
From categories to double categories

1

A review of basic category theory

This review of the basic notions of category theory is also meant to fix our terminology and notation for this domain – including ordered sets in 1.1.6 and Section 1.4. Other points are deferred to Chapter 2, when dealing with 'globular' 2-dimensional aspects, or to Appendix A, for abelian categories and their generalisations.

This chapter is elementary and reasonably self-contained, but rather compressed for a beginner in category theory. A reader of this book will probably have some knowledge and practice of this domain; with a good knowledge one can skip this chapter altogether and only check the terminology when necessary. Some proofs are referred to books devoted to this field, like those of Mac Lane [M4], Borceux [Bo1, Bo2], Adámek, Herrlich and Strecker [AHS], or the author [G11].

The foundational setting we use is based on standard set theory, assuming the existence of Grothendieck universes. This aspect, presented in 1.1.3, will be mostly left as understood.

1.1 Categories

Categories were introduced by Eilenberg and Mac Lane [EiM] in 1945, together with the other basic terms of category theory.

1.1.1 Some examples

Loosely speaking, a category $\mathbf{C}$ consists of *objects* and *morphisms* together with a (partial) *composition law*: given two 'consecutive' morphisms $f\colon X \to Y$ and $g\colon Y \to Z$ we have a composed morphism $gf\colon X \to Z$. This partial operation is associative (whenever composition is legitimate) and every object X has a unit, written as 1_X or $\mathrm{id}X$, which acts as an identity for legitimate compositions.

The prime example is the category **Set** *of sets* (and mappings), where:

- an object is a set,

- the morphisms $f\colon X \to Y$ between two given sets X and Y are the (set-theoretical) mappings from X to Y,

- the composition law is the usual composition of mappings.

The following categories of structured sets and structure-preserving mappings (with the usual composition) will often be used and analysed:

- the category **Top** *of topological spaces* (and continuous mappings),

- the category **Ab** *of abelian groups* (and their homomorphisms),

- the category **Gp** *of groups* (and homomorphisms),

- the category **Mon** *of monoids*, i.e. unitary semigroups (and homomorphisms),

- the category **Rng** *of rings*, understood to be associative and unitary (and homomorphisms),

- the category **CRng** *of commutative rings* (and homomorphisms),

- the category $R\,\mathbf{Mod}$ *of left modules* on a fixed unitary ring R (and homomorphisms),

- the category **Ord** of *ordered sets* (and monotone mappings),

- the category p**Ord** of *preordered sets* (and monotone mappings),

- the category **Set.** *of pointed sets* (and pointed mappings),

- the category **Top.** *of pointed topological spaces* (and pointed continuous mappings),

- the category **Ban** of *Banach spaces and continuous linear mappings*,

- the category $\mathbf{Ban}_1$ of *Banach spaces and linear weak contractions* (with norm $\leqslant 1$).

A homomorphism of a 'unitary' algebraic structure, like a monoid or a unitary ring, is always assumed to preserve units.

For **Set.** we recall that a *pointed set* is a pair (X, x_0) consisting of a set X and a base-element $x_0 \in X$, while a *pointed mapping* $f\colon (X, x_0) \to (Y, y_0)$ is a mapping $f\colon X \to Y$ such that $f(x_0) = y_0$. Similarly, a *pointed topological space* (X, x_0) is a space with a base-point, and a pointed map $f\colon (X, x_0) \to (Y, y_0)$ is a continuous mapping from X to Y such that $f(x_0) = y_0$. The reader may know that the category **Top.** is important in Algebraic Topology: for instance, the fundamental group $\pi_1(X, x_0)$ is defined for a pointed topological space.

For the category **Ban** (and $\mathbf{Ban}_1$) it is understood that we have chosen *either* the real *or* the complex field; when using both one can write $\mathbb{R}\mathbf{Ban}$ and $\mathbb{C}\mathbf{Ban}$.

When a category is named after its objects alone (e.g. the 'category of groups'), this means that the morphisms are understood to be the obvious ones (in this case the homomorphisms of groups), with the obvious composition law. Different categories with the same objects are given different names, like $\mathbf{Ban}$ and $\mathbf{Ban}_1$.

1.1.2 Definition

A *category* $\mathbf{C}$ consists of the following data:

(a) a set $\mathrm{Ob}\mathbf{C}$, whose elements are called *objects* of $\mathbf{C}$,

(b) for every pair X, Y of objects, a set $\mathbf{C}(X, Y)$ (called a *hom-set*) whose elements are called *morphisms* (or *maps*, or *arrows*) of $\mathbf{C}$ from X to Y and denoted as $f \colon X \to Y$,

(c) for every triple X, Y, Z of objects of $\mathbf{C}$, a mapping of *composition*

$$\mathbf{C}(X, Y) \times \mathbf{C}(Y, Z) \to \mathbf{C}(X, Z), \qquad (f, g) \mapsto gf,$$

where gf is also written as $g.f$.

These data must satisfy the following axioms.

(i) *Associativity.* Given three consecutive arrows, $f \colon X \to Y$, $g \colon Y \to Z$ and $h \colon Z \to W$, one has: $h(gf) = (hg)f$.

(ii) *Identities.* Given an object X, there exists an endomap $e \colon X \to X$ which acts as an identity whenever composition makes sense; in other words if $f \colon X' \to X$ and $g \colon X \to X''$, one has: $ef = f$ and $ge = g$. One shows, in the usual way, that e is determined by X; it is called the *identity* of X and written as 1_X or $\mathrm{id}X$.

We generally assume that the following condition is also satisfied.

(iii) *Separation.* For X, X', Y, Y' objects of $\mathbf{C}$, if $\mathbf{C}(X, Y) \cap \mathbf{C}(X', Y') \neq \emptyset$ then $X = X'$ and $Y = Y'$.

In other words, every map $f \colon X \to Y$ has a well-determined *domain* $\mathrm{Dom}\, f = X$ and *codomain* $\mathrm{Cod}\, f = Y$. Concretely, when constructing a category, one can forget about condition (iii), since one can always satisfy it by *redefining* a morphism $\hat{f} \colon X \to Y$ as a triple $(X, Y; f)$ where f is a morphism from X to Y in the original sense (possibly not satisfying the Separation axiom).

$\mathrm{Mor}\mathbf{C}$ denotes the set of all the morphisms of $\mathbf{C}$, which is the disjoint union of all hom-sets. Two morphisms f, g are said to be *parallel* when they have the same domain and the same codomain.

If **C** is a category, the *opposite* (or *dual*) category, written as $\mathbf{C}^{\mathrm{op}}$, has the same objects as **C**, reversed arrows and reversed composition $g * f$

$$\mathbf{C}^{\mathrm{op}}(X, Y) = \mathbf{C}(Y, X), \qquad g * f = fg, \quad \mathrm{id}^{\mathrm{op}} X = \mathrm{id} X. \qquad (1.1)$$

Every topic of category theory has a dual instance, which comes from the opposite category (or categories). A dual notion is generally distinguished by the prefix 'co-'.

A set X can be viewed as a *discrete* category: its objects are the elements of X, and the only arrows are their (formal) identities; here $X^{\mathrm{op}} = X$.

As usual in category theory, the term *graph* will be used to denote a simplified notion, with objects (or vertices) and morphisms (or arrows) $f \colon x \to y$, but no assigned composition nor identities. (This is called a *directed multigraph* in graph theory.) A *morphism of graphs* preserves objects, arrows, domain and codomain. A *reflexive graph* is an intermediate notion, where every object x has an assigned endomorphism $\mathrm{id} x \colon x \to x$.

Every category has an *underlying reflexive graph*, and an *underlying graph*.

1.1.3 Small and large categories

We will not insist on set-theoretical foundations. Yet some care is necessary, to avoid speaking of 'the set of all sets', or requiring of a category properties of completeness that are 'too large for its size' (as we shall see at the end of 1.3.6).

We assume the existence of a (Grothendieck) *universe* $\mathcal{U}$, which is fixed throughout. Its axioms, listed in 1.1.9, say that we can perform inside it the usual operations of set theory. Its elements are called *small* sets (or $\mathcal{U}$-small sets, if necessary).

A category is understood to have objects and arrows belonging to this universe, and is said to be *small* if its set of morphisms belongs to $\mathcal{U}$, *large* if it does not (and is just a subset of $\mathcal{U}$). Note that in a small category the set of objects also belongs to $\mathcal{U}$ (since it is a subset of $\mathcal{U}$ in bijective correspondence with an element of $\mathcal{U}$, namely the set of all identities). More generally, a category **C** is said to *have small hom-sets* if all its sets $\mathbf{C}(X, Y)$ are small; in this case **C** is small if and only if Ob**C** is.

The 'usual' categories of structured sets are large categories with small hom-sets, like the category **Set** of small sets, and all the examples listed in 1.1.1. In such cases we speak – as usual – of the 'category of sets', and so on, *leaving understood the term 'small' as referred to these structured sets*.

*While developing the theory, one often needs a *hierarchy* of universes.

For instance, **Cat** will denote the category of small categories and functors, introduced in 1.2.1. In order to view the (large) categories **Set**, **Top**, **Ab**, etc. as objects of a similar structure we should assume the existence of a second universe $\mathcal{V}$, with $\mathcal{U} \in \mathcal{V}$, and use the category $\mathbf{Cat}_{\mathcal{V}}$ of $\mathcal{V}$-small categories. In a more complex situation one may need a longer chain of universes. Most of the time *these points will be left as understood.**

1.1.4 Isomorphisms, monomorphisms and epimorphisms

In a category $\mathbf{C}$ a morphism $f \colon X \to Y$ is said to be *invertible*, or an *isomorphism*, if it has an inverse, i.e. a morphism $g \colon Y \to X$ such that $gf = 1_X$ and $fg = 1_Y$. The latter is uniquely determined by f; it is called the *inverse* of f and written as f^{-1}. In the categories listed in 1.1.1 this definition gives the usual isomorphisms of the various structures – called 'homeomorphisms' in the case of topological spaces.

The *isomorphism relation* $X \cong Y$ between objects of $\mathbf{C}$ (meaning that there exists an isomorphism $X \to Y$) is an equivalence relation.

A morphism $f \colon X \to Y$ is said to be a *monomorphism*, or *mono*, or *monic*, if it satisfies the following cancellation property: for every pair of maps $u, v \colon X' \to X$ such that $fu = fv$ one has $u = v$. Dually, the morphism $f \colon X \to Y$ is said to be an *epimorphism*, or *epi*, or *epic*, if it satisfies the dual cancellation property: for every pair of maps $u, v \colon Y \to Y'$ such that $uf = vf$ one has $u = v$.

An arrow $\rightarrowtail$ always denotes a monomorphism, while $\twoheadrightarrow$ stands for an epimorphism. (Subobjects and quotients will be dealt with below.)

Every isomorphism is mono and epi. A category is said to be *balanced* if the converse holds: every morphism which is mono and epi is invertible. For instance, **Set** (resp. **Ab**) is a balanced category: it is easy to verify that monos, epis and isos coincide with the injective, surjective and bijective mappings (resp. homomorphisms). **Top** is not balanced: its monos, epis and isos coincide with the injective maps, or surjective maps, or homeomorphisms. **Mon** is not balanced, as proved by the inclusion $\mathbb{N} \to \mathbb{Z}$ (of additive monoids), which is mono and epi. In fact epimorphisms in **Mon** have no elementary characterisation, and 'regular epimorphisms' are more important (see 1.3.2); on the other hand, in the 'usual categories of structured sets' the monomorphisms 'generally' coincide with the injective morphisms (see 1.2.9 for a precise result in this sense).

A *groupoid* is a category where every map is invertible; it is interesting to recall that this structure was introduced before categories, by H. Brandt in 1927 [Bra]. *The fundamental groupoid of a space X is an important structure, that contains all the fundamental groups $\pi_1(X, x)$, for $x \in X$.**

Suppose now that we have, in a category $\mathbf{C}$, two maps $m\colon A \to X$ and $p\colon X \to A$ such that $pm = \mathrm{id}A$. It follows that m is a monomorphism (called a *section*, or a *split monomorphism*), while p is an epimorphism (called a *retraction*, or a *split epimorphism*); A is said to be a *retract* of X.

In **Set** a retract of a set $X \neq \emptyset$ is any non-empty subset – a statement equivalent to the axiom of choice. In **Ab** retracts coincide with direct summands. There is no elementary characterisation of retracts in **Top** (see 1.2.3).

Let $f\colon X \to Y$ and $g\colon Y \to Z$ be consecutive maps in a category. The following facts are well known, and easy to prove (with (a*) dual to (a), and so on):

(a) if f and g are both mono, gf is also mono; if gf is mono, f is also,

(a*) if f and g are both epi, gf is also epi; if gf is epi, g is also,

(b) if f and g are both split mono, gf is also; if gf is a split mono, f is also,

(b*) if f and g are both split epi, gf is also; if gf is a split epi, g is also,

(c) if f is a split mono and an epi, then it is invertible,

(c*) if f is a split epi and a mono, then it is invertible.

A family of morphisms $f_i\colon X \to Y_i$ $(i \in I)$ with the same domain is said to be *jointly mono* if for every pair of maps $u, v\colon X' \to X$ such that $f_i u = f_i v$ (for all indices i) one has $u = v$. Dually a family $f_i\colon X_i \to Y$ is *jointly epi* if for all $u, v\colon Y \to Y'$ such that $uf_i = vf_i$ (for all i) one has $u = v$.

1.1.5 Subcategories, quotients and products of categories

(a) Let $\mathbf{C}$ be a category. A *subcategory* $\mathbf{D}$ is defined by assigning:

- a subset $\mathrm{Ob}\mathbf{D} \subset \mathrm{Ob}\mathbf{C}$, whose elements are called *objects of* $\mathbf{D}$,

- for every pair of objects X, Y of $\mathbf{D}$, a subset $\mathbf{D}(X, Y) \subset \mathbf{C}(X, Y)$, whose elements are called *morphisms of* $\mathbf{D}$ *from X to Y*,

so that the following conditions hold:

(i) for every pair of consecutive morphisms of $\mathbf{D}$, their composite in $\mathbf{C}$ belongs to $\mathbf{D}$,

(ii) for every object of $\mathbf{D}$, its identity in $\mathbf{C}$ belongs to $\mathbf{D}$.

$\mathbf{D}$, equipped with the induced composition law, is then a category.

One says that $\mathbf{D}$ is a *full* subcategory of $\mathbf{C}$ if, for every pair of objects X, Y of $\mathbf{D}$, we have $\mathbf{D}(X, Y) = \mathbf{C}(X, Y)$, so that $\mathbf{D}$ is determined by assigning its subset of objects. One says that $\mathbf{D}$ is a *wide* subcategory of $\mathbf{C}$

if it has the same objects, so that $\mathbf{D}$ is determined by assigning its subset of morphisms (closed under composition and identities). For instance, $\mathbf{Ab}$ is a full subcategory of $\mathbf{Gp}$ while $\mathbf{Ban}_1$ is a wide subcategory of $\mathbf{Ban}$. The only full and wide subcategory of a category is the total one.

(b) A *congruence* $R = (R_{XY})$ in a category $\mathbf{C}$ consists of a family of equivalence relations R_{XY} in each set of morphisms $\mathbf{C}(X,Y)$, that is consistent with composition:

(iii) if $f\, R_{XY}\, f'$ and $g\, R_{YZ}\, g'$, then $gf\, R_{XZ}\, g'f'$.

The *quotient category* $\mathbf{D} = \mathbf{C}/R$ has the same objects of $\mathbf{C}$ and $\mathbf{D}(X,Y)$ $= \mathbf{C}(X,Y)/R_{XY}$; in other words, a morphism $[f]\colon X \to Y$ in $\mathbf{D}$ is an equivalence class of morphisms $X \to Y$ in $\mathbf{C}$. The composition is induced by that of $\mathbf{C}$, which is legitimate because of condition (iii): $[g].[f] = [gf]$.

For instance, the homotopy relation $f \simeq f'$ in $\mathbf{Top}$ is a congruence of categories (examined in Section A5). The quotient category ho$\mathbf{Top}$ = $\mathbf{Top}/\simeq$ is called the *homotopy category of topological spaces*, and is important in Algebraic Topology. Plainly, a continuous mapping $f\colon X \to Y$ is a homotopy equivalence (i.e. there exists a continuous mapping $g\colon Y \to X$ such that $gf \simeq \text{id}X$ and $fg \simeq \text{id}Y$) if and only if its homotopy class $[f]$ is an isomorphism of ho$\mathbf{Top}$.

(c) If $\mathbf{C}$ and $\mathbf{D}$ are categories, one defines the *product category* $\mathbf{C} \times \mathbf{D}$. An object is a pair (X,Y) where X is in $\mathbf{C}$ and Y in $\mathbf{D}$. A morphism is a pair of morphisms

$$(f,g)\colon (X,Y) \to (X',Y'), \qquad (f \in \mathbf{C}(X,X'),\ g \in \mathbf{D}(Y,Y')), \qquad (1.2)$$

and its composition with $(f',g')\colon (X',Y') \to (X'',Y'')$ is component-wise: $(f',g').(f,g) = (f'f,g'g)$.

Similarly one defines the product $\Pi\,\mathbf{C}_i$ of a family of categories indexed by a small set.

1.1.6 Preorders and categories

We use the following terminology for orderings. A *preorder* relation $x \prec x'$ is reflexive and transitive. An *order* relation, generally written as $x \leqslant x'$, is also anti-symmetric: if $x \leqslant x' \leqslant x$ then $x = x'$.

The category of *ordered sets and increasing mappings* (the order preserving ones, also called *monotone*) will be written as $\mathbf{Ord}$, while we write as p$\mathbf{Ord}$ the category of *preordered sets and monotone mappings*.

An order relation is said to be *total* if for all x, x' we have $x \leqslant x'$ or $x' \leqslant x$. (An ordered set is often called a '*partially* ordered set', abbreviated to 'poset', to mean that its order is not assumed to be total. Accordingly, the reader can find the notation **Pos** instead of **Ord**.)

A preordered set X has an associated equivalence relation $x \sim x'$ defined by the conjunction: $x \prec x'$ and $x' \prec x$. The quotient $X/\sim$ has an induced order: $[x] \leqslant [x']$ if $x \prec x'$.

If X is a preordered set, X^{op} is the opposite one (with reversed preorder). If $a \in X$, the symbols $\downarrow a$ and $\uparrow a$ denote the downward and upward closed subsets of X generated by a

$$\downarrow a = \{x \in X \,|\, x \prec a\}, \qquad \uparrow a = \{x \in X \,|\, a \prec x\}. \tag{1.3}$$

In a preordered set, the greatest lower bound (resp. least upper bound) of a subset A is written as $\inf A$ or $\wedge A$ (resp. $\sup A$ or $\vee A$). The same notation is used for an indexed family (x_i).

A preordered set X will often be viewed as a category, where the objects are the elements of X and the set $X(x, x')$ contains precisely one (formal) arrow if $x \prec x'$ (which can be written as $(x, x') \colon x \to x'$), and no arrow otherwise. Composition and units are (necessarily)

$$(x', x'').(x, x') = (x, x''), \qquad \mathrm{id}x = (x, x).$$

All diagrams in these categories commute. Two elements x, x' are isomorphic objects if and only if $x \sim x'$.

In particular, each ordinal defines a category, written as $\mathbf{0}, \mathbf{1}, \mathbf{2}, \dots$ Thus, $\mathbf{0}$ is the empty category; $\mathbf{1}$ is the *singleton category*, i.e. the discrete category on one object 0; $\mathbf{2}$ is the *arrow category*, with two objects and precisely one non-identity arrow, $0 \to 1$.

We are also interested in the *formal-span* category $\vee$ and the *formal-cospan* category $\wedge = \vee^{\mathrm{op}}$, which are also defined by (partially) ordered sets

$$\vee \colon \; 0 \leftarrow \iota \to 1 \qquad\qquad \wedge \colon \; 0 \to \iota \leftarrow 1. \tag{1.4}$$

Every hom-set $\mathbf{Ord}(X, Y)$ is canonically ordered by the *pointwise order* relation, defined as follows for $f, g \colon X \to Y$

$$f \leqslant g \;\; \text{if for all } x \in X \text{ we have } f(x) \leqslant g(x) \text{ in } Y. \tag{1.5}$$

1.1.7 Subobject and quotients

Let A be an object of the category $\mathbf{C}$. A subobject of A is defined as an equivalence class of monomorphisms, or better as a *selected* representative of such a class.

(The first definition, as an equivalence class, is common in category theory, but the second is what is normally used in concrete categories of structured sets. Moreover, when working with the restriction of morphisms to subobjects, one is forced to shift to the second setting – even in the abstract theory.)

More precisely, given two monos m, n *with values in* A, we say that $m \prec n$ if there is a (uniquely determined) morphism u such that $m = nu$. We say that m and n *are equivalent*, or $m \sim n$, if $m \prec n \prec m$, which amounts to the existence of a (unique) isomorphism u such that $m = nu$.

In every class of equivalent monos with codomain A, precisely one is *selected* and called a *subobject* of A; in the class of isomorphisms we always choose the identity 1_A. The subobjects of A in $\mathbf{C}$ form the (possibly large) *ordered* set $\mathrm{Sub}(A)$, with maximum 1_A; here, the induced *order* $m \prec n$ is also written as $m \leqslant n$. (Equalisers and kernels will always be *chosen* as subobjects, see 1.3.1 and 1.8.2.)

Set, **Ab**, **Gp** have a canonical choice of subobjects, based on the inclusion mappings $m \colon X \subset A$ of subsets and subgroups; for **Top** we are more interested in 'regular subobjects', see 1.3.1.

Epimorphisms with a fixed domain A are dealt with in a dual way. Their preorder and equivalence relation are also written as $p \prec q$ (meaning that p factorises through q) and $p \sim q$. A *quotient* of A is a *selected* representative of an equivalence class of epimorphisms with domain A; they form the ordered set $\mathrm{Quo}(A)$, with maximum 1_A; again the induced order is also written as $p \leqslant q$. (Coequalisers and cokernels will always be chosen as quotients.)

The category $\mathbf{C}$ is said to be *well powered* (resp. *well copowered*) if all its sets $\mathrm{Sub}(A)$ (resp. $\mathrm{Quo}(A)$) are small, as is often the case with categories of (small) structured sets.

Duality of categories turns subobjects of $\mathbf{C}$ into quotients of $\mathbf{C}^{\mathrm{op}}$, preserving their order.

1.1.8 A digression on mathematical structures and categories

When studying a mathematical structure with the help of category theory, it is crucial to choose the 'right' kind of structure and the 'right' kind of morphisms, so that the result is sufficiently general and 'natural' to have good properties with respect to the goals of our study – even if we are interested in more particular situations.

The following remarks use some elementary notions that will be reviewed later, but are generally well known.

(a) A first point to be kept in mind is that the isomorphisms of the category (i.e. its invertible arrows) should indeed preserve the structure we are interested in, or we risk of studying something different from our purpose.

As a trivial example, the category **T** of topological spaces and *all* mappings between them has little to do with Topology: an isomorphism of **T** is any bijection between topological spaces. Indeed **T** is *equivalent to the category of sets* (as we shall see in 1.2.5), and is a modified way of looking at the latter.

Less trivially, the category **M** of metric spaces and continuous mappings misses crucial properties of metric spaces, since its invertible morphisms need not preserve completeness: e.g. the real line is homeomorphic to any non-empty open interval. In fact **M** is equivalent to the category of *metrisable topological spaces* and continuous mappings, and can be replaced with the latter. A 'reasonable' category of metric spaces should be based on *Lipschitz* maps, or – more particularly – on weak contractions, so that its isomorphisms (bi-Lipschitz or isometric bijections, respectively) do preserve metric properties, like being complete or bounded.

(b) Other points will become clearer below. For instance, the category **Top** of topological spaces and continuous mappings is a classical framework for studying topology. Among its good properties there is the fact that all categorical limits and colimits (studied in Section 1.3) exist, and are computed as in **Set**, then equipped with a suitable topology determined by the structural maps. (As we shall see, this is a consequence of the fact that the forgetful functor **Top** → **Set** has a left and a right adjoint, corresponding to discrete and chaotic topologies.) Hausdorff spaces are certainly important, but it is 'often' better to view them *in* **Top**, as their category is less well behaved: colimits exist, but are not computed as in **Set**.

*(c) Many category theorists would agree with Mac Lane, saying that even **Top** is not sufficiently good (cf. [M4], Section VII.8), because it is not a cartesian closed category (see 2.2.4), and prefer – for instance – the category of compactly generated spaces; however, researchers interested in Homotopy Theory and Algebraic Topology might be satisfied with the fact that the standard interval $[0, 1]$ (with its cartesian powers) is exponentiable in **Top**.*

(d) It is a common feature of Mathematics to look for the natural framework where certain properties should be studied: for instance, many properties of ordinary polynomials with real coefficients can be examined in general polynomial rings, or in more general algebras. Besides yielding more general results, the natural framework gives a deeper comprehension of what we are studying.

Category theory makes a further step in this sense. For instance, categories of modules are certainly important, but – since Buchsbaum's *Appendix* [Bu2] and Grothendieck's paper [Gt] – a consistent part of Homological Algebra finds its natural framework in *abelian categories* and their generalisations (see Appendix A). Similarly, the categories of structured sets can be viewed as (particular) *concrete categories* (see 1.2.3), and categories of algebras as *monadic categories* (see 1.6.5).

In other words, when studying a certain 'class' of categories, we can look for a structural definition including this class and abstracting its crucial properties, instead of some more or less general way of constructing the important examples we want to study.

(e) Artificial exclusions 'most of the time' give categories of little interest and poor properties, like the category of non-abelian groups, or non-empty semigroups. The latter case needs some further comment.

The reader may know that, in Universal Algebra, a 'variety of algebras' includes all the algebraic structures of a given signature, satisfying a given set of equational axioms (under universal quantifiers): e.g. all groups, or all rings; but not all fields, because multiplicative inverses for non-zero elements cannot be given by a 'general' unary operation satisfying some universal equations.

Here a *variety of algebras* will mean a category of objects defined in this way, with their homomorphisms. *We do not follow the convention that the underlying set should be non empty*, as commonly assumed in Universal Algebra (cf. Grätzer [Gr1]) – with a few exceptions like Cohn's book [Coh]. This convention has unlucky consequences for any theory without constants (or zeroary operations), like that of semigroups: for instance, two subalgebras of an algebra need not have a meet (as subalgebras).

For a reader with some knowledge of categorical limits and colimits, dealt with in Section 1.3, we can add that a variety of algebras (in the present sense) has all limits and colimits, while the category of non-empty semigroups lacks pullbacks and an initial object, precisely because we have artificially taken out some solutions. Other comments about free semigroups and monadicity can be found in 1.6.6(e).

(f) Finally, the solution of a universal problem, like categorical products and sums, completely depends on the category where we are considering it, as the following examples – and many others – make evident.

- The categorical sum $\mathbb{Z} + \mathbb{Z}$ in **Ab** is the direct sum $\mathbb{Z} \oplus \mathbb{Z}$, i.e. the free abelian group on two generators. In **Gp** we get the free product $\mathbb{Z} * \mathbb{Z}$, which is the free group on two generators.

- The categorical product $\{*\} \times \{*\}$ in **Set** is the singleton. We shall see that in the category **RelSet** of sets and relations this product has two elements (Exercise 2.1.4(c)), while in the category $\mathcal{S}$ of sets and partial mappings it has three elements (by (2.10)).

1.1.9 *Grothendieck universes*

For the interested reader, we recall the definition of a *universe* as given in [M4], Section I.6. It is a set $\mathcal{U}$ satisfying the following (redundant) properties:

(i) $x \in u \in \mathcal{U}$ implies $x \in \mathcal{U}$,

(ii) $u, v \in \mathcal{U}$ implies that the sets $\{u, v\}$, (u, v), $u \times v$ belong to $\mathcal{U}$,

(iii) $x \in \mathcal{U}$ implies that $\mathcal{P}x$ and $\bigcup x$ belong to $\mathcal{U}$,

(iv) the set $\mathbb{N}$ of finite ordinals belongs to $\mathcal{U}$,

(v) if $f \colon x \to y$ is a surjective mapping with $x \in \mathcal{U}$ and $y \subset \mathcal{U}$, then $y \in \mathcal{U}$.

Here $\mathcal{P}x$ is the set of subsets of x and $\bigcup x = \{y \mid y \in z \text{ for some } z \in x\}$.

1.2 Functors and natural transformations

Well-behaved mappings $F \colon \mathbf{C} \to \mathbf{D}$ between categories are called 'functors'. Given two *parallel* functors $F, G \colon \mathbf{C} \to \mathbf{D}$ (between the same categories) there can be 'second-order arrows' $\varphi \colon F \to G$, called 'natural transformations' (and functorial isomorphisms when they are invertible).

'Category', 'functor' and 'natural transformation' are the three basic terms of category theory, since the very beginning of the theory in [EiM]. It is interesting to note that only the last term is taken from the common language: one can say that Eilenberg and Mac Lane introduced categories and functors because they wanted to formalise the natural transformations that they were encountering in algebra and algebraic topology (as remarked in [M4], at the end of Section I.6). Much in the same way as a general theory of *continuity* (a familiar term for a familiar notion) requires the introduction of *topological spaces* (a theoretical term for a more abstract notion).

A reader acquainted with basic homotopy theory can take advantage of a formal parallelism, where *spaces* correspond to *categories*, *continuous mappings* to *functors*, *homotopies* of mappings to *invertible natural transformations*, and *homotopy equivalence* of spaces to *equivalence* of categories.

This analogy is even deeper in the domain of Directed Algebraic Topology, where *directed homotopies* need not be reversible and correspond to *natural transformations*; see [G7].

1.2.1 Functors

A (covariant) *functor* $F\colon \mathbf{C} \to \mathbf{D}$ consists of the following data:

(a) a mapping $F_0\colon \mathrm{Ob}\mathbf{C} \to \mathrm{Ob}\mathbf{D}$, whose action is generally written as $X \mapsto F(X)$,

(b) for every pair of objects X, X' in $\mathbf{C}$, a mapping

$$F_{XX'}\colon \mathbf{C}(X, X') \to \mathbf{D}(F(X), F(X')),$$

whose action is generally written as $f \mapsto F(f)$.

Composition and identities must be preserved. In other words:

(i) if f, g are consecutive maps of $\mathbf{C}$ then $F(gf) = F(g).F(f)$,

(ii) if X is an object of $\mathbf{C}$ then $F(\mathrm{id}X) = \mathrm{id}(F(X))$.

Given a second functor $G\colon \mathbf{D} \to \mathbf{E}$, one defines in the obvious way the *composed functor* $GF\colon \mathbf{C} \to \mathbf{E}$. This composition is associative and has identities: the *identity functor* of each category

$$\mathrm{id}\mathbf{C}\colon \mathbf{C} \to \mathbf{C}, \qquad X \mapsto X, \quad f \mapsto f. \tag{1.6}$$

An *isomorphism of categories* is a functor $F\colon \mathbf{C} \to \mathbf{D}$ which is invertible, i.e. it admits an *inverse* $G\colon \mathbf{D} \to \mathbf{C}$; this means a functor such that $GF = \mathrm{id}\mathbf{C}$ and $FG = \mathrm{id}\mathbf{D}$. Obviously, the functor F is an isomorphism if and only if all the mappings F_0 and $F_{XX'}$ considered above are bijective. Being isomorphic categories is an equivalence relation, written as $\mathbf{C} \cong \mathbf{D}$.

Categories linked by an obvious isomorphism are often perceived as 'the same thing'. For instance, $\mathbf{Ab}$ is isomorphic to the category $\mathbb{Z}\,\mathbf{Mod}$ of modules on the ring of integers; the various equivalent ways of defining topological spaces give rise to isomorphic categories that are nearly never distinguished.

A functor between two preordered sets, viewed as categories (see 1.1.6), is the same as a (weakly) increasing function, i.e. a preorder-preserving mapping.

A *contravariant functor* $F\colon \mathbf{C} \dashrightarrow \mathbf{D}$ can be defined as a covariant functor $\mathbf{C}^{\mathrm{op}} \to \mathbf{D}$.

A *functor on two variables* is an ordinary functor $F\colon \mathbf{C}\times\mathbf{D} \to \mathbf{E}$ defined on the product of two categories. Fixing an object X_0 in $\mathbf{C}$ we have a functor $F(X_0, -)\colon \mathbf{D} \to \mathbf{E}$; and symmetrically.

Cat will denote the category of small categories and their functors. (Its 2-dimensional structure, including the natural transformations, will be examined in 1.2.4 and Section 2.3.)

A functor $\mathbf{1} \to \mathbf{C}$ amounts to an object of $\mathbf{C}$, while a functor $\mathbf{2} \to \mathbf{C}$ 'is' a morphism of $\mathbf{C}$. A functor $\mathbf{2} \times \mathbf{2} \to \mathbf{C}$ is a commutative square in $\mathbf{C}$. A functor $\vee \to \mathbf{C}$ (see 1.1.6) is a *span* $X \leftarrow U \to Y$ (*from X to Y*); a functor $\wedge \to \mathbf{C}$ is a *cospan* $X \to U \leftarrow Y$, *from X to Y.*

More generally, a *diagram* in $\mathbf{C}$ can be formally defined as a functor $\mathbf{S} \to \mathbf{C}$ defined on a small category, the 'shape' of the diagram. Every diagram defined on a preordered set is commutative.

1.2.2 Forgetful and structural functors

(a) Forgetting structure, or part of it, yields various examples of functors between categories of structured sets, like the following obvious instances

$$\mathbf{Top} \to \mathbf{Set}, \qquad \mathbf{Rng} \to \mathbf{Ab} \to \mathbf{Set}. \qquad (1.7)$$

These are called *forgetful functors*, and often denoted by the letter U, which refers to the *underlying* set, or *underlying* abelian group, and so on.

(b) A subcategory $\mathbf{D}$ of $\mathbf{C}$ yields an *inclusion* functor $\mathbf{D} \to \mathbf{C}$, which we also write as $\mathbf{D} \subset \mathbf{C}$. For instance, $\mathbf{Ab} \subset \mathbf{Gp}$ and $\mathbf{Ban}_1 \subset \mathbf{Ban}$. These functors *forget properties*, rather than structure.

(c) A congruence R in a category $\mathbf{C}$ yields a *projection functor* $P \colon \mathbf{C} \to \mathbf{C}/R$, which is the identity on objects and sends a morphism f to its equivalence class $[f]$. For instance we have the projection functor $\mathbf{Top} \to \mathrm{ho}\mathbf{Top} = \mathbf{Top}/\simeq$.

(d) A product category $\mathbf{C} = \prod \mathbf{C}_i$ has *projection functors* $P_i \colon \mathbf{C} \to \mathbf{C}_i$.

(e) If $\mathbf{C}$ has small hom-sets (see 1.1.3), there is a functor of morphisms, or *hom-functor*:

$$\mathrm{Mor} \colon \mathbf{C}^{\mathrm{op}} \times \mathbf{C} \to \mathbf{Set},$$

$$(X, Y) \mapsto \mathbf{C}(X, Y), \qquad (f, g) \mapsto g. - .f, \qquad (1.8)$$

$$g. - .f \colon \mathbf{C}(X, Y) \to \mathbf{C}(X', Y'), \qquad u \mapsto guf,$$

for $f \colon X' \to X$ and $g \colon Y \to Y'$ in $\mathbf{C}$.

Fixing one of these variables one gets a 'representable functor' (covariant or contravariant on $\mathbf{C}$), see 1.2.7.

1.2.3 Faithful and full functors

For a functor $F \colon \mathbf{C} \to \mathbf{D}$ let us consider again the mappings (of sets, possibly large):

$$F_{XX'} \colon \mathbf{C}(X, X') \to \mathbf{D}(F(X), F(X')), \qquad f \mapsto F(f). \tag{1.9}$$

F is said to be *faithful* if all these mappings are injective (for X, X' in $\mathbf{C}$). F is said to be *full* if all of them are surjective.

An isomorphism of categories is always full and faithful (and bijective on objects). The inclusion functor $\mathbf{D} \to \mathbf{C}$ of a subcategory is always faithful; it is full if and only if $\mathbf{D}$ is a full subcategory of $\mathbf{C}$. A projection functor $P \colon \mathbf{C} \to \mathbf{C}/R$ is always full (and bijective on objects).

There are some obvious preservation and reflection properties of functors:

(a) every functor preserves commutative diagrams, isomorphisms, retracts, split monos and split epis,

(b) a faithful functor *reflects* monos and epis (i.e. if $F(f)$ is mono or epi, then f is also) and commutative diagrams,

(c) a full and faithful functor reflects isomorphisms, split monos and split epis.

Applying point (a), a common way of proving that a topological subspace $A \subset X$ is not a retract (in **Top**) is to find a functor $F \colon \mathbf{Top} \to \mathbf{Ab}$ such that the associated homomorphism $F(A) \to F(X)$ is not a split mono in **Ab**. In this way, any homology functor H_n can be used to prove that the n-sphere $\mathbb{S}^n$ is not a retract of the euclidean space $\mathbb{R}^{n+1}$.

As a formal alternative to the notion of 'category of structured sets' (either vague or formalised in different, complicated ways) a *concrete category* is defined as a category $\mathbf{A}$ *equipped* with a faithful functor $U \colon \mathbf{A} \to \mathbf{Set}$, called its *forgetful functor*. As a consequence U *reflects monos and epis*, but need not preserve them. Concrete categories are extensively studied in [AHS].

Even $\mathbf{Set}^{\mathrm{op}}$ can be made concrete over $\mathbf{Set}$ (by the contravariant functor of subsets, see 1.4.5). But it is interesting to note that the homotopy category ho**Top** *cannot* be made concrete, as was proved by P. Freyd [Fr2, Fr3].

More generally, a category $\mathbf{A}$ equipped with a faithful functor $U \colon \mathbf{A} \to \mathbf{C}$ is said to be *concrete over* $\mathbf{C}$.

1.2.4 *Natural transformations*

Given two functors $F, G\colon \mathbf{C} \to \mathbf{D}$ between the same categories, a *natural transformation* $\varphi\colon F \to G$ (or $\varphi\colon F \to G\colon \mathbf{C} \to \mathbf{D}$) consists of the following data:

- for each object X of $\mathbf{C}$ a morphism $\varphi X\colon FX \to GX$ in $\mathbf{D}$ (called the *component* of φ on X and also written as φ_X),

so that, for every arrow $f\colon X \to X'$ in $\mathbf{C}$, we have a commutative square in $\mathbf{D}$ (*naturality condition* of φ on f)

$$
\begin{array}{ccc}
FX & \xrightarrow{\ \varphi X\ } & GX \\
{\scriptstyle Ff}\big\downarrow & & \big\downarrow{\scriptstyle Gf} \\
FX' & \xrightarrow[\ \varphi X'\ ]{} & GX'
\end{array}
\qquad\qquad \varphi X'.F(f) = G(f).\varphi X = \varphi(f), \qquad (1.10)
$$

whose diagonal will be written as $\varphi(f)$, when useful (see (2.25)).

In particular, *the identity of a functor* $F\colon \mathbf{C} \to \mathbf{D}$ is the natural transformation $\mathrm{id}F\colon F \to F$, with components $(\mathrm{id}F)X = \mathrm{id}(FX)$.

For instance if $F\colon R\,\mathbf{Mod} \to R\,\mathbf{Mod}$ is the identity functor, every scalar λ in the centre of the (unitary) ring R gives a natural transformation $\lambda\colon F \to F$ whose component $\lambda A\colon A \to A$ on the left module A is the multiplication by λ. It is easy to prove that every natural transformation $F \to F$ is of this type (working on its component on R, as a module). This bijective correspondence shows that the category $R\,\mathbf{Mod}$ *determines* the centre of the ring R, in a structural way, and leads to considering the relation of Morita equivalence of rings (see 1.2.6).

A natural transformation $\varphi\colon F \to G$ often comes out of a 'canonical choice' of a family $(\varphi X\colon FX \to GX)_X$ of morphisms, *but these two aspects should not be confused.* There are canonical choices of such families that are not natural, and natural transformations deduced from the axiom of choice.

Natural transformations have a *vertical composition*, written as $\psi\varphi$ (or $\psi.\varphi$)

$$
\mathbf{C} \;\;\begin{array}{c} \xrightarrow{\ \ F\ \ } \\[-2pt] {\scriptstyle\downarrow\varphi} \\[-4pt] \xrightarrow{\ \ G\ \ } \\[-2pt] {\scriptstyle\downarrow\psi} \\[-4pt] \xrightarrow[\ \ H\ \ ]{} \end{array}\;\; \mathbf{D}
\qquad\qquad (\psi\varphi)(X) = \psi X.\varphi X\colon FX \to HX. \qquad (1.11)
$$

There is also a *whisker composition*, or *reduced horizontal composition*, of natural transformations with functors, written as $K\varphi H$ (or $K \circ \varphi \circ H$, when useful to distinguish compositions)

$$\mathbf{C'} \xrightarrow{\ H\ } \mathbf{C} \underset{G}{\overset{F}{\rightrightarrows}}{\scriptstyle\Downarrow\varphi} \mathbf{D} \xrightarrow{\ K\ } \mathbf{D'} \qquad (1.12)$$

$$K\varphi H \colon KFH \to KGH, \qquad (K\varphi H)(X') = K(\varphi(HX')).$$

An *isomorphism of functors*, or *natural isomorphism*, or *functorial isomorphism*, is a natural transformation $\varphi \colon F \to G$ which is invertible with respect to vertical composition. It is easy to see that this happens if and only if all the components φX are invertible in $\mathbf{D}$. The inverse is written as $\varphi^{-1} \colon G \to F$. Isomorphism of (parallel) functors is an equivalence relation, written as $F \cong G$. (The old term 'natural equivalence', for a functorial isomorphism, can be confusing and will not be used.)

This 2-dimensional structure of $\mathbf{Cat}$, where natural transformations play the role of 2-dimensional arrows between functors, will be further analysed in Section 2.3 (including the *full* horizontal composition of natural transformations, which extends the reduced one).

Replacing the category $\mathbf{C}$ with a graph Γ (see 1.1.2), one can consider a *natural transformation* $\varphi \colon F \to G \colon \Gamma \to \mathbf{D}$ between two morphisms of graphs defined on Γ, *with values in a category*.

1.2.5 Equivalence of categories

Isomorphisms of categories have been recalled in 1.2.1. More generally, an *equivalence of categories* is a functor $F \colon \mathbf{C} \to \mathbf{D}$ which is invertible up to functorial isomorphism, i.e. there exists a functor $G \colon \mathbf{D} \to \mathbf{C}$ such that $GF \cong \mathrm{id}\mathbf{C}$ and $FG \cong \mathrm{id}\mathbf{D}$. The functor G can be called a *quasi inverse* of F.

An *adjoint equivalence of categories* is a coherent version of this notion; namely it is a four-tuple $(F, G, \eta, \varepsilon)$ where

- $F \colon \mathbf{C} \to \mathbf{D}$ and $G \colon \mathbf{D} \to \mathbf{C}$ are functors,

- $\eta \colon \mathrm{id}\mathbf{C} \to GF$ and $\varepsilon \colon FG \to \mathrm{id}\mathbf{D}$ are isomorphisms of functors,

- $F\eta = (\varepsilon F)^{-1} \colon F \to FGF, \qquad \eta G = (G\varepsilon)^{-1} \colon G \to GFG.$

(The direction of η and ε is written above as in the general case of an adjunction $F \dashv G$, where these transformations need not be invertible and their direction is substantial: see Section 1.5.)

The following conditions on a functor $F \colon \mathbf{C} \to \mathbf{D}$ are equivalent, forming a very useful *characterisation of the equivalence of categories*:

(i) F is an equivalence of categories,

(ii) F can be completed to an adjoint equivalence of categories $(F, G, \eta, \varepsilon)$,

(iii) F is faithful, full and *essentially surjective on objects*.

The last property means that: for every object Y of $\mathbf{D}$ there exists some object X in $\mathbf{C}$ such that $F(X)$ is isomorphic to Y in $\mathbf{D}$. The proof of the equivalence of these three conditions requires the axiom of choice: see [M4], Section V.4, or [G11], Theorem 1.5.8.

One says that the categories $\mathbf{C}, \mathbf{D}$ are *equivalent*, written as $\mathbf{C} \simeq \mathbf{D}$, if there exists an equivalence of categories between them (or, equivalently, an *adjoint* equivalence of categories). This is indeed an equivalence relation, as can be easily proved (directly or from the previous characterisation).

1.2.6 Exercises and complements (Equivalences and skeletons)

Equivalences of categories play an important role in the theory of categories, as highlighted by the following points. We recall that solutions and hints for the exercises can be found in Appendix C, excluding the obvious ones and some of the exercises marked with a *.

(a) Prove that the category of finite sets (and mappings between them) is equivalent to its full subcategory of finite cardinals, which is small and therefore cannot be isomorphic to the former.

(b) Prove that a category $\mathbf{C}$ is equivalent to the singleton category $\mathbf{1}$ if and only if it is non-empty and *indiscrete*, i.e. each hom-set $\mathbf{C}(X, Y)$ has precisely one element.

(c) Prove that $\mathbf{Top}$ is a subcategory of a category equivalent to $\mathbf{Set}$.

(d) Extending point (c), prove that any category $\mathbf{A}$ concrete over $\mathbf{C}$ (by a faithful functor $U \colon \mathbf{A} \to \mathbf{C}$) can be embedded as a subcategory of a suitable category equivalent to $\mathbf{C}$.

*(e) The reader may know that two rings are said to be *Morita equivalent* if their categories of modules are equivalent. This is an important notion in ring theory, that becomes trivial in the domain of commutative rings: in fact, if two rings are Morita equivalent one can easily prove that their centres are isomorphic (using what we have seen in 1.2.4); but – quite interestingly – commutative rings can be Morita equivalent to non-commutative ones, like their rings of square matrices. Thus, studying left modules on any matrix ring $M_n(\mathbb{R})$ is equivalent to studying real vector spaces.

(f) A category is said to be *skeletal* if it has no pair of distinct isomorphic objects. Prove that every category has a *skeleton*, i.e. an equivalent skeletal category. The latter can be obtained by *choosing* precisely one object in every class of isomorphic objects.

(g) We have described above a skeleton of the category of finite sets that can be constructed without any choice, even though we do need the axiom of choice to prove that the inclusion of this skeleton has a quasi inverse. In a different way, a preordered set X has a natural skeleton *formed by a quotient*, the associated ordered set $X/\sim$ (see 1.1.6); again we need the axiom of choice to prove that the projection $X \to X/\sim$ has a quasi inverse.

(h) Prove that two categories are equivalent if and only if they have isomorphic skeletons. Loosely speaking, this says that an equivalence of categories amounts to multiplying or deleting isomorphic copies of objects, even though there may be no canonical way of doing this.

(i) *Modifying functors.* Starting from a functor $F \colon \mathbf{C} \to \mathbf{D}$ and an arbitrary family of isomorphisms $\varphi_X \colon F(X) \to G(X)$ in $\mathbf{D}$ (for X in $\mathbf{C}$), prove that there is precisely one way of extending these data to a functor $G \colon \mathbf{C} \to \mathbf{D}$ and a natural transformation $\varphi \colon F \to G$ (obviously invertible).

1.2.7 Functor categories and categories of presheaves

Let $\mathbf{S}$ be a small category. For any category $\mathbf{C}$ we write as $\mathbf{C}^{\mathbf{S}}$, or $\mathbf{Cat}(\mathbf{S}, \mathbf{C})$, the category whose objects are the functors $F \colon \mathbf{S} \to \mathbf{C}$ and whose morphisms are the natural transformations $\varphi \colon F \to G \colon \mathbf{S} \to \mathbf{C}$, with vertical composition; it is called a *functor category*. (Writing $\mathbf{Cat}(\mathbf{S}, \mathbf{C})$ is an abuse of notation, unless $\mathbf{C}$ too is small.) More generally one can consider the category $\mathbf{C}^{\Gamma}$ of graph-morphisms $F \colon \Gamma \to \mathbf{C}$, where Γ is a small graph: see 1.2.4.

In particular the arrow category $\mathbf{2}$ (with one non-identity arrow $0 \to 1$) gives the *category of morphisms* $\mathbf{C}^{\mathbf{2}}$ of $\mathbf{C}$, where a map $(f_0, f_1) \colon u \to v$ is a commutative square of $\mathbf{C}$; these are composed *by pasting squares*, as in the right diagram below

$$
\begin{array}{ccc}
A_0 & \xrightarrow{f_0} & B_0 \\
u\downarrow & & \downarrow v \\
A_1 & \xrightarrow[f_1]{} & B_1
\end{array}
\qquad
\begin{array}{ccccc}
A_0 & \xrightarrow{f_0} & B_0 & \xrightarrow{g_0} & C_0 \\
u\downarrow & & \downarrow v & & \downarrow w \\
A_1 & \xrightarrow[f_1]{} & B_1 & \xrightarrow[g_1]{} & C_1
\end{array}
\qquad (1.13)
$$

A natural transformation $\varphi \colon F \to G \colon \mathbf{C} \to \mathbf{D}$ can be viewed as a functor $\mathbf{C} \times \mathbf{2} \to \mathbf{D}$ or equivalently as a functor $\mathbf{C} \to \mathbf{D}^{\mathbf{2}}$. (This accounts for the notation $\varphi(f)$ in (1.10).)

We now assume that the category $\mathbf{C}$ has small hom-sets. (Note that this is also true of $\mathbf{C}^{\mathbf{S}}$.)

A functor $F \colon \mathbf{C} \to \mathbf{Set}$ is said to be *representable* if it is isomorphic to a functor $\mathbf{C}(X_0, -) \colon \mathbf{C} \to \mathbf{Set}$, for some object X_0 in $\mathbf{C}$. This object is said *to represent F*, *via* an isomorphism $\mathbf{C}(X_0, -) \to F$. In such a case the Yoneda Lemma describes the natural transformations $F \to G$, for every functor $G\colon \mathbf{C} \to \mathbf{Set}$, and proves that the object X_0 is determined by F, up to isomorphism (see 1.2.8). It is easy to verify that a representable functor $F\colon \mathbf{C} \to \mathbf{Set}$ *preserves monomorphisms*.

A functor $\mathbf{S}^{\mathrm{op}} \to \mathbf{C}$, defined on the opposite category $\mathbf{S}^{\mathrm{op}}$, is also called a *presheaf* of $\mathbf{C}$ on the (small) category $\mathbf{S}$. They form the presheaf category $\mathrm{Psh}(\mathbf{S}, \mathbf{C}) = \mathbf{C}^{S^{\mathrm{op}}}$, and $\mathbf{S}$ is called the *site* of the latter.

$\mathbf{S}$ is canonically embedded in $\mathrm{Psh}(\mathbf{S}, \mathbf{C})$, by the *Yoneda embedding*

$$Y\colon \mathbf{S} \to \mathbf{C}^{\mathbf{S}^{\mathrm{op}}}, \qquad Y(i) = \mathbf{S}(-, i)\colon \mathbf{S}^{\mathrm{op}} \to \mathbf{Set}, \qquad (1.14)$$

which sends every object i to the corresponding representable presheaf $Y(i)$.

Taking as $\mathbf{S}$ the category $\underline{\Delta}$ of finite positive ordinals (and increasing maps), one gets the category $\mathrm{Smp}\mathbf{C} = \mathbf{C}^{\underline{\Delta}^{\mathrm{op}}}$ of *simplicial objects* in $\mathbf{C}$, and – in particular – the well-known category $\mathrm{Smp}\mathbf{Set}$ of *simplicial sets*. Here the Yoneda embedding sends the positive ordinal $\{0, ..., n\}$ to the simplicial set Δ^n, freely generated by one simplex of dimension n. Multiple sets will be studied in Chapter 6, cubical sets in Appendix B.

A reader interested in categories of sheaves is referred to [MaM, Bo3].

1.2.8 Yoneda Lemma

(a) Let $F, G\colon \mathbf{C} \to \mathbf{Set}$ be two functors, with $F = \mathbf{C}(X_0, -)$. The canonical mapping

$$y\colon \mathrm{Nat}(F, G) \to G(X_0), \qquad y(\varphi) = (\varphi X_0)(\mathrm{id}X_0), \qquad (1.15)$$

from the set of natural transformations $\varphi\colon F \to G$ to the set $G(X_0)$ is a bijection.

(b) The functor $G\colon \mathbf{C} \to \mathbf{Set}$ is represented by the object X_0 of $\mathbf{C}$ if and only if there exists some $x_0 \in G(X_0)$ such that:

(*) for every X in $\mathbf{C}$ and every $x \in G(X)$ there is a unique morphism $f\colon X_0 \to X$ such that $G(f)(x_0) = x$.

This morphism f is an isomorphism if and only if G is also represented by X, via $x \in G(X)$.

Proof The crucial point is that, for every natural transformation $\varphi\colon F \to G$ and every morphism $f\colon X_0 \to X$, the naturality of φ on f gives

$$(\varphi X)(f) = G(f)(y(\varphi)) \in G(X). \qquad (1.16)$$

(a) One constructs the mapping

$$y' : G(X_0) \to \text{Nat}(F, G),$$
$$(y'(x))_X : \mathbf{C}(X_0, X) \to GX, \qquad (f : X_0 \to X) \mapsto (Gf)(x), \qquad (1.17)$$

and verifies that the mappings y, y' are inverse to each other, using (1.16).

(b) Giving a natural transformation $\varphi : F \to G$ is equivalent to giving an element $x_0 \in G(X_0)$, linked by the previous bijection: $x_0 = y(\varphi)$ and $\varphi = y'(x_0)$. Moreover φ is an isomorphism if and only if each component $\varphi X : FX \to GX$ is a bijection of sets, which amounts to condition (*), taking into account formula (1.16).   □

1.2.9 Universal arrows

There is a general way of formalising 'universal properties', based on a functor $U : \mathbf{A} \to \mathbf{X}$ and an object X of $\mathbf{X}$.

A *universal arrow from the object X to the functor U* is a pair $(A, \eta : X \to UA)$ consisting of an object A of $\mathbf{A}$ and an arrow η of $\mathbf{X}$, which is universal, in the sense that every similar pair $(B, f : X \to UB)$ factorises uniquely through (A, η): namely, there exists a unique map $g : A \to B$ in $\mathbf{A}$ such that the following triangle commutes in $\mathbf{X}$

$$
\begin{array}{ccc}
X & \xrightarrow{\;\eta\;} & UA \\
 & \searrow{\scriptstyle f} & \downarrow{\scriptstyle Ug} \\
 & & UB
\end{array}
\qquad Ug.\eta = f. \qquad (1.18)
$$

The pair (A, η) is then determined up to a unique isomorphism $A \to A'$ of $\mathbf{A}$.

Dually, a *universal arrow from the functor U to the object X* is a pair $(A, \varepsilon : UA \to X)$ consisting of an object A of $\mathbf{A}$ and an arrow ε of $\mathbf{X}$ such that every similar pair $(B, f : UB \to X)$ factorises uniquely through (A, ε): there exists a unique $g : B \to A$ in $\mathbf{A}$ such that the following triangle commutes in $\mathbf{X}$

$$
\begin{array}{ccc}
UA & \xrightarrow{\;\varepsilon\;} & X \\
\uparrow{\scriptstyle Ug} & \nearrow{\scriptstyle f} & \\
UB & &
\end{array}
\qquad \varepsilon.Ug = f. \qquad (1.19)
$$

A reader who is not familiar with this notion might begin by constructing the universal arrow from a set X to the forgetful functor $R\,\mathbf{Mod} \to \mathbf{Set}$, using the free module $A = \bigoplus_{x \in X} R$ generated by X. Or the universal arrow from a set X to the forgetful functor $\mathbf{Mon} \to \mathbf{Set}$, using the free monoid

$A = \sum_{n \in \mathbb{N}} X^n$ of *finite words* on the alphabet X. Or, quite easily, from a set X to the forgetful functor $U \colon \mathbf{Set}_{\bullet} \to \mathbf{Set}$.

More generally, if $U \colon \mathbf{A} \to \mathbf{Set}$ is a concrete category (see 1.2.3), a universal arrow $(A, \eta \colon X \to UA)$ from the set X to the forgetful functor U gives – by definition – the *free* $\mathbf{A}$-*object over* X (which may exist or not, of course). One can extend this terminology to any functor $U \colon \mathbf{A} \to \mathbf{X}$ which is *viewed*, in a given context, as a 'forgetful functor', even if it is not faithful.

In particular a functor $U \colon \mathbf{A} \to \mathbf{Set}$ is representable (see 1.2.7) if and only if the singleton $\{*\}$ has a free object A in $\mathbf{A}$, so that

$$U \cong \mathbf{Set}(\{*\}, U(-)) \cong \mathbf{A}(A, -).$$

This property often holds in a concrete category, and ensures that U *preserves monomorphisms* (and reflects them, if it is faithful).

Universal arrows compose: for a composed functor $UV \colon \mathbf{B} \to \mathbf{A} \to \mathbf{X}$, given

- a universal arrow $(A, \eta \colon X \to UA)$ from an object X to U,

- a universal arrow $(B, \zeta \colon A \to VB)$ from the previous object A to V,

it is easy to verify that we have a universal arrow from X to UV constructed as follows

$$(B, U\zeta.\eta \colon X \to UA \to UV(B)). \tag{1.20}$$

As an example one can think of the composed forgetful functor $\mathbf{Rng} \to \mathbf{Mon} \to \mathbf{Set}$. The free ring on the set X can be constructed in two steps: first the free monoid A on X (by finite words on this alphabet, or 'non-commutative monomials', where $xyx \neq x^2 y$), and then the free ring on this monoid (by $\mathbb{Z}$-linear combinations of these monomials); the result is the ring of 'non-commutative polynomials' with variables in X and integral coefficients (like $2 + xyx - 3x^2 y$). A different (less effective) construction of the 'same' ring can be obtained from the factorisation $\mathbf{Rng} \to \mathbf{Ab} \to \mathbf{Set}$. Ordinary polynomials can be similarly obtained working with categories of commutative rings or commutative R-algebras.

Universal arrows for 'two-dimensional categories' are considered in 2.3.5.

1.3 Limits and colimits

The general notion of limit of a functor includes cartesian products and equalisers (see 1.3.1), pullbacks (see 1.3.4) and the classical 'projective limits'. Dually, colimits comprise sums, coequalisers (see 1.3.2), pushouts (see 1.3.4) and the classical 'inductive limits' (see [M4], Chapter IX).

The construction of limits from products and equalisers is dealt with in 1.3.6; their description as universal arrows or by representable functors in 1.3.8. Their relationship with adjoints will be seen in 1.5.4(d), (e), with terminal objects of comma categories in 1.5.7(b).

1.3.1 *Products and equalisers*

The simplest case of a limit, in a category $\mathbf{C}$, is the *product* of a family $(X_i)_{i \in I}$ of objects, indexed by a *small set I*.

This is defined as an object X equipped with a family of morphisms $p_i \colon X \to X_i$ $(i \in I)$, called (cartesian) *projections*, which satisfies the following universal property

$$Y \xdashrightarrow{f} X$$
$$f_i \searrow \quad \downarrow p_i$$
$$X_i$$

(1.21)

(i) for every object Y and every family of morphisms $f_i \colon Y \to X_i$ there exists a unique morphism $f \colon Y \to X$ such that, for all $i \in I$, $p_i f = f$.

The map f is often written as (f_i), by its *components*.

The product of a family need not exist. If it does, it is determined up to a unique *coherent* isomorphism, in the sense that if Y is also a product of the family $(X_i)_{i \in I}$ with projections $q_i \colon Y \to X_i$, then the unique morphism $f \colon X \to Y$ which commutes with all projections (i.e. $q_i f = p_i$, for all indices i) is invertible. This follows easily from the fact that there is also a unique morphism $g \colon Y \to X$ such that $p_i g = q_i$; moreover $gf = \mathrm{id}X$ (because $p_i(gf) = p_i(\mathrm{id}X)$, for all i) and $fg = \mathrm{id}Y$.

Therefore one speaks of *the* product of the family (X_i), denoted as $\prod_i X_i$, or $X_1 \times \ldots \times X_n$ in a finite case.

We say that a category $\mathbf{C}$ *has products* (resp. *finite products*) if every family of objects indexed by a *small* set (resp. by a finite set) has a product in $\mathbf{C}$.

In particular the product of the empty family of objects $\emptyset \to \mathrm{Ob}\mathbf{C}$ means an object X such that for every object Y there is a unique morphism $Y \to X$. The solution, if it exists, is called the *terminal* object of $\mathbf{C}$ and can be written as $\top$.

In **Set**, **Top**, **Ab** and **Ord** all products exist and are the usual cartesian ones; the terminal object is the singleton (with the appropriate structure). In the category X associated to a preordered set, the categorical product of a family of points $x_i \in X$ amounts to the greatest lower bound inf x_i, while the terminal object amounts to the greatest element of X; notice that they are determined up to the equivalence relation associated to our preorder, and uniquely determined if X is an ordered set.

It is easy to prove that a category has finite products if and only if it has binary products $X_1 \times X_2$ and a terminal object. (Note that the *unary product* of an object X always exists and is X; this is a trivial example of an *absolute limit*, preserved by any functor.)

Products are a basic instance of a much more general concept recalled below, the limit of a functor. Another basic instance is the *equaliser* of a pair $f, g \colon X \to Y$ of parallel maps of $\mathbf{C}$; this is (an object E with) a map $m \colon E \to X$ such that $fm = gm$ and the following universal property holds:

$$E \xrightarrow{\ m\ } X \underset{g}{\overset{f}{\rightrightarrows}} Y \qquad\qquad m = \mathrm{eq}(f, g),$$

$$\text{(1.22)}$$

(ii) every map $h \colon Z \to X$ such that $fh = gh$ factorises uniquely through m (i.e. there exists a unique map $w \colon Z \to E$ such that $mw = h$).

The equaliser morphism is necessarily a monomorphism, and is called a *regular monomorphism*. It is determined up to the equivalence relation of monos with values in X (see 1.1.7). *We will always choose the strictly unique subobject* that satisfies the given property (if it exists, of course).

A *regular subobject* is, by definition, an equaliser (of some pair of maps). In **Set** (resp. **Top**, **Ab**), the equaliser of two parallel maps $f, g \colon X \to Y$ is the embedding in X of the whole subset (resp. subspace, subgroup) of X on which they coincide. It follows that in **Set** and **Ab** all subobjects are regular (and can be identified with subsets or subgroups, respectively). In **Top** the regular subobjects amount to *inclusion of subspaces*, while every injective continuous mapping is a mono; here the 'general subobjects' are less important than the regular ones. In the category X associated to a preordered set two parallel maps $x \to x'$ always coincide and their equaliser is the identity 1_x.

The reader may also be interested to know that all subgroups are regular subobjects in **Gp** (a non-trivial fact, see [AHS], Exercise 7H), while a subsemigroup need not be a regular subobject ([AHS], Exercise 7I); in the categories of algebraic structures the general subobjects are more important than the regular ones.

Products in **Cat** have been considered in 1.1.5(c); the terminal object is the singleton category **1**. The equaliser of two functors $F, G \colon \mathbf{C} \to \mathbf{D}$ is the subcategory of the items of $\mathbf{C}$ on which they coincide.

1.3.2 Sums and coequalisers

As already mentioned in Section 1.1, every notion of category theory has a dual notion.

The *sum*, or *coproduct*, of a family $(X_i)_{i \in I}$ of objects of **C** is dual to their product. Explicitly, it is an object X equipped with a family of morphisms $u_i \colon X_i \to X$ $(i \in I)$, called *injections*, which satisfies the following universal property:

$$X \dashrightarrow^{f} Y \qquad u_i \uparrow \quad \nearrow_{f_i} \qquad X_i \tag{1.23}$$

(i) for every object Y and every family of morphisms $f_i \colon X_i \to Y$, there exists a unique morphism $f \colon X \to Y$ such that, for all $i \in I$, $f u_i = f_i$.

The map f will be written as $[f_i]$, by its co-components.

Again, if the sum of the family (X_i) exists, it is determined up to a unique coherent isomorphism, and denoted as $\sum_i X_i$, or $X_1 + ... + X_n$ in a finite case. The sum of the empty family is the *initial* object $\bot$: this means that every object X has precisely one map $\bot \to X$.

Sums in **Set** and **Top** are realised as disjoint unions, for instance as follows

$$\sum_i X_i = \bigcup_i X_i \times \{i\}, \tag{1.24}$$

(with open subsets given by arbitrary unions of open subsets of the summands, in the second case); the initial object is the empty set or space.

In **Ab** and R**Mod** categorical sums are realised as 'direct sums' and the initial object is the zero module. In the category X associated to a preordered set the categorical sum of a family of points $x_i \in X$ amounts to the least upper bound $\sup x_i$, while the initial object amounts to the least element of X. The initial object in **Rng** and **CRng** is the ring of integers. *In **Gp** categorical sums are classically known as 'free products'; in **CRng** they are realised as tensor products over $\mathbb{Z}$; in **Rng** the construction is more complex.*

The *coequaliser* of a pair $f, g \colon X \to Y$ of parallel maps of **C** is a map $p \colon Y \to C$ such that $pf = pg$ and:

$$X \underset{g}{\overset{f}{\rightrightarrows}} Y \xrightarrow{p} C \qquad p = \mathrm{coeq}(f, g), \qquad h \downarrow \quad \swarrow_{w} \qquad Z \tag{1.25}$$

(ii) every map $h\colon Y \to Z$ such that $hf = hg$ factorises uniquely through p (i.e. there exists a unique map $w\colon C \to Z$ such that $wp = h$).

A coequaliser morphism is necessarily an epimorphism, and is called a *regular epimorphism*. Sums and coequalisers are particular instances of the *co*limit of a functor (see 1.3.3).

A *regular quotient* is defined as a coequaliser (of some pair of maps); its choice is determined by the choice of quotients (see 1.1.7). A regular quotient in **Set** or **Ab** is the same as an ordinary quotient. In **Top** a regular quotient amounts to the projection on a quotient space, while every surjective continuous mapping is an epi. Regular quotients are the important notion in **Top**, *as well as* in each variety of algebras.

In **Cat** a sum $\sum \mathbf{C}_i$ of categories is their obvious disjoint union and the initial object is the empty category **0**. *The coequaliser of two functors $F, G\colon \mathbf{C} \to \mathbf{D}$ is the quotient of **D** modulo the *generalised congruence* generated by this pair; the latter, as defined in [BBP], also involves equivalent objects. One can avoid giving a 'construction' of the coequaliser category (necessarily complicated) and just prove its existence by the Adjoint Functor Theorem, that will be recalled in 1.5.4.*

A *zero object* of a category, often written as 0, is both initial and terminal. This exists in $R\mathbf{Mod}$, **Gp**, **Set.**, **Top.**, ... but not in **Set**, **Top**, **Cat**, **Rng**... (see 1.8.1). A category with zero object is said to be *pointed*.

1.3.3 Limits and colimits

Let **I** be a small category and $X\colon \mathbf{I} \to \mathbf{C}$ a functor, written in 'index notation':

$$X\colon \mathbf{I} \to \mathbf{C}, \qquad i \mapsto X_i, \quad a \mapsto (X_a\colon X_i \to X_j), \qquad (1.26)$$

for $i \in I = \mathrm{Ob}\,\mathbf{I}$ and $a\colon i \to j$ in **I**.

A *cone* of X is an object A of **C** with a family of maps $(f_i\colon A \to X_i)_{i \in I}$ in **C** such that the following triangles commute, for $a\colon i \to j$ in **I**

$$A \xrightarrow{\ f_i\ } X_i \qquad\qquad X_a.f_i = f_j.$$
$$\searrow_{f_j} \quad \downarrow_{X_a} \qquad\qquad\qquad (1.27)$$
$$X_j$$

The *limit* of $X\colon \mathbf{I} \to \mathbf{C}$ is a universal cone $(L, (u_i\colon L \to X_i)_{i \in I})$. This means a cone of X such that every cone $(A, (f_i\colon A \to X_i)_{i \in I})$ factorises uniquely through the former; in other words, there is a unique map $f\colon A \to L$ such that, for all $i \in I$, $u_i f = f_i$. The solution need not exist; when it does, it is determined up to a unique coherent isomorphism and the limit object L is denoted as $\mathrm{Lim}\,(X)$.

The uniqueness part of the universal property amounts to saying that *the family* $u_i\colon L \to X_i$ $(i \in I)$ *is jointly mono* (see 1.1.4). A cone of X that satisfies the existence part of the universal property is called a *weak limit*; of course such a cone is constrained in a weak way.

Dually, a *cocone* $(A, (f_i\colon X_i \to A)_{i \in I})$ of X satisfies the condition $f_j.X_a = f_i$ for every $a\colon i \to j$ in $\mathbf{I}$. The *colimit* of the functor X is a universal cocone $(L', (u_i\colon X_i \to L')_{i \in I})$: the universal property says now that for every cocone $(A, (f_i))$ of X there exists a unique map $f\colon L' \to A$ such that $fu_i = f_i$ (for all i). The colimit object is denoted as $\mathrm{Colim}\,(X)$. The uniqueness part of the universal property means that the family $u_i\colon X_i \to L'$ is jointly epi. A cocone that satisfies the existence part of the universal property is called a *weak colimit*.

As an interesting exercise, the reader can prove that the limit of any functor $\mathbf{I} \to \mathbf{Set}$ exists, and can be constructed as the set of its cones with vertex at the singleton set $\{*\}$.

1.3.4 Particular cases, pullbacks and pushouts

(a) The product $\prod X_i$ of a family $(X_i)_{i \in I}$ of objects of $\mathbf{C}$ is the limit of the corresponding functor $X\colon \mathbf{I} \to \mathbf{C}$, defined on the discrete category whose objects are the elements $i \in I$ (and whose morphisms reduce to the formal identities of the objects). The sum $\sum X_i$ is the colimit of this functor X. In particular the terminal (resp. initial) object of $\mathbf{C}$ is the limit (resp. colimit) of the empty functor $\mathbf{0} \to \mathbf{C}$.

We recall that the family of cartesian projections (p_i) of a product is *jointly mono*. These projections are 'often' epi but *not necessarily*: in $\mathbf{Set}$ this fails whenever some of the factors X_i are empty and others are not. But the reader can easily show that, if $\mathbf{C}$ has a zero object (see 1.3.2), all cartesian projections are split epi.

(b) The equaliser in $\mathbf{C}$ of a pair of parallel morphisms $f, g\colon X_0 \to X_1$ is the limit of the obvious functor defined on the category $0 \rightrightarrows 1$. The coequaliser is the colimit of this functor.

(c) The *pullback* of a pair of morphisms $f\colon X_0 \to X \leftarrow X_1 : g$ with the same codomain is the limit of the corresponding functor (defined on the formal-cospan category $0 \to \iota \leftarrow 1$). This amounts to the usual definition: an object A equipped with a span $u_i\colon A \to X_i$ $(i = 0, 1)$ which forms a commutative square with f and g, in a universal way:

$$(1.28)$$

that is, $fu_0 = gu_1$, and for every span (B, v_0, v_1) such that $fv_0 = gv_1$ there exists a unique map $w \colon B \to A$ such that $u_0 w = v_0$, $u_1 w = v_1$. The span (u_0, u_1) is jointly mono. The pullback-object A is also called a *fibred product* over X and written as $X_0 \times_X X_1$, or $X_0 \overset{\bullet}{\times} X_1$. A pullback diagram is often marked as above.

In the ordered set X the pullback of the diagram $x_0 \to x \leftarrow x_1$ is the meet $x_0 \wedge x_1$. Saying that X has pullbacks means that every *upper-bounded* pair of elements has a meet.

In **Set** (resp. **Top**, **Ab**) the pullback-object can be realised as a subset (resp. subspace, subgroup) of the product $X_0 \times X_1$:

$$A = \{(x_0, x_1) \in X_0 \times X_1 \mid f(x_0) = g(x_1)\}.$$

Generalising this construction, it is easy to prove that a category that has binary products and equalisers also has pullbacks: A is constructed as the equaliser of the maps $fp_0, gp_1 \colon X_0 \times X_1 \to X$. On the other hand, the product $X_0 \times X_1$ amounts to the pullback of X_0 and X_1 over the terminal object, when the latter exists.

If $f = g$, the pullback of the diagram $(f, f) = (X_0 \to X \leftarrow X_0)$ is called the *kernel pair* of f. In **Set**, it can be realised as

$$\{(x, x') \in X_0 \times X_0 \mid f(x) = f(x')\},$$

and amounts to the equivalence relation associated to f. (But note that, in any category, the equaliser of f and f is the identity of its domain.)

(d) Dually, the *pushout* of a span $(f, g) = (X_0 \leftarrow X \to X_1)$ is the colimit of the corresponding functor defined on the formal-span category $0 \leftarrow \iota \to 1$. This amounts to an object A equipped with a cospan $u_i \colon X_i \to A$ $(i = 0, 1)$ which forms a commutative square with f and g, in a universal way:

$$(1.29)$$

that is, $u_0 f = u_1 g$, and for every triple (B, v_0, v_1) such that $v_0 f = v_1 g$ there exists a unique map $w \colon A \to B$ such that $w u_0 = v_0$, $w u_1 = v_1$.

A category that has binary sums and coequalisers also has pushouts: A is the coequaliser of two maps $X \to X_i \to X_0 + X_1$.

The pushout-object A is also called a *pasting* over X and written as $X_0 +_X X_1$. In **Set** this means a quotient of the sum $X_0 + X_1$ modulo the equivalence relation generated by identifying all pairs $f(x)$, $g(x)$ for $x \in X$. In **Top** a pasting $X_0 +_X X_1$ is constructed as in **Set** and equipped with the quotient topology of the sum – a useful way of constructing constructing or describing spaces (see Exercise 1.3.5(j)).

The well known Seifert–van Kampen Theorem exhibits the fundamental group of a space as a pushout of groups. R. Brown's version generalises this result, using fundamental groupoids [Bro1, Bro2].

1.3.5 *Exercises and complements* (Pullbacks and pushouts)

The following facts are important and will often be used.

(a) *Characterising monos.* Prove that the following conditions on a morphism $f \colon X \to Y$ are equivalent

(i) f is mono,
(ii) the left square below is a pullback (i.e. the kernel pair of f is $(1, 1)$),
(iii) if the right square below is a pullback, then $h = k$,
(iv) if the right square below is a pullback, then $h = k$ is invertible

$$
\begin{array}{ccc}
X & \xrightarrow{\;f\;} & Y \\
\big\uparrow{\scriptstyle 1} & & \big\uparrow{\scriptstyle f} \\
X & \xrightarrow[\;1\;]{} & X
\end{array}
\qquad\qquad
\begin{array}{ccc}
\bullet & \xrightarrow{\;f\;} & \bullet \\
\big\uparrow{\scriptstyle h} & & \big\uparrow{\scriptstyle f} \\
\bullet & \xrightarrow[\;k\;]{} & \bullet
\end{array}
\qquad (1.30)
$$

Note that these characterisations do not require the existence of pullbacks in our category.

(b) *Preimages.* If the following square is a pullback and n is mono, so is m

$$
\begin{array}{ccc}
X & \xrightarrow{\;f\;} & Y \\
\big\uparrow{\scriptstyle m} & & \big\uparrow{\scriptstyle n} \\
\bullet & \xrightarrow{\;\;\;} & \bullet
\end{array}
\qquad (1.31)
$$

The latter is determined up to equivalence of monomorphisms (see 1.1.7): it is called the *preimage* of n along f and written as $f^*(n)$. When we work with subobjects, $f^*(n)$ can be determined by their choice. In particular it will be useful to follow the following convention:

(*) *The unit constraint for pullbacks.* The pullback of an identity morphism along any morphism (always exists and) is chosen to be an identity: $f^*(1_Y) = 1_X$.

(c) *Symmetry.* If (P, u, v) is a pullback of (f, g), then (P, v, u) is – obviously – a pullback of (g, f). However, in a category with pullbacks, one can*not* assume the existence of a strictly symmetric choice of them.

(d) *Pasting property.* If the two squares below are pullbacks, so is their 'pasting', i.e. the outer rectangle

$$
\begin{array}{ccccc}
\bullet & \longrightarrow & \bullet & \longrightarrow & \bullet \\
\big\uparrow & & f\big\uparrow & & \big\uparrow \\
\bullet & \longrightarrow & \bullet & \underset{g}{\longrightarrow} & \bullet
\end{array}
\qquad (1.32)
$$

(e) *Depasting property.* If in the commutative diagram above the outer rectangle is a pullback and the pair (f, g) is jointly mono then the left square is a pullback.

The hypothesis on the pair (f, g) is automatically satisfied in two cases, frequently used: when the right square is a pullback or one of the morphisms f, g is mono.

(f) Pushouts have dual properties; we shall generally follow *the unit constraint for pushouts.*

(g) In **Set**, if $X = X_0 \cup X_1$, then X is the pushout of X_0 and X_1 over $X_0 \cap X_1$. Extend this fact to **Top**, under convenient hypotheses on the subspaces $X_i \subset X$.

*(h) In an abelian category pullbacks and pushouts are characterised by exactness properties of an associated sequence of morphisms: see A3.9.

*(i) Prove that a category with pullbacks and terminal object has equalisers, and therefore all finite limits.

*(j) In **Top**, the euclidean sphere $\mathbb{S}^2$ can be presented as a pushout, the cokernel pair of the embedding $\mathbb{S}^1 \to \mathbb{D}^2$ of the circle as the boundary of the compact disc; concretely, we are pasting two discs along their boundary. The reader will see (or already know) that this fact can be extended to any dimension, starting with $\mathbb{S}^1$ as the pasting of two compact intervals over $\mathbb{S}^0$. The homology of all spheres is computed, inductively, on this basis.

1.3.6 Complete categories and the preservation of limits

A category **C** is said to be **I**-*complete* if it has a limit for every functor $\mathbf{I} \to \mathbf{C}$. It is said to be *complete* (resp. *finitely complete*) if this holds for every small (resp. finite) category **I**. The smallness requirement is explained in a remark below.

One says that a functor $F: \mathbf{C} \to \mathbf{D}$ *preserves the limit*

$$(L, (u_i: L \to X_i)_{i \in I})$$

of a functor $X: \mathbf{I} \to \mathbf{C}$ if the cone $(FL, (Fu_i: FL \to FX_i)_{i \in I})$ is the limit of the composed functor $FX: \mathbf{I} \to \mathbf{D}$. One says that F *preserves limits* if it preserves all the limits which exist in $\mathbf{C}$. Analogously for the preservation of products, equalisers, etc.

A representable functor preserves all (the existing) limits. A functor which preserves pullbacks also preserves monomorphisms, by 1.3.5(a).

One proves, by a constructive argument, that *a category is complete (resp. finitely complete) if and only if it has equalisers and products (resp. finite products)*. Moreover, if $\mathbf{C}$ is complete (resp. finitely complete), a functor $F: \mathbf{C} \to \mathbf{D}$ preserves all limits (resp. all finite limits) if and only if it preserves equalisers and products (resp. finite products).

As a sketch of the proof, given a functor $X: \mathbf{I} \to \mathbf{C}$ defined over a small category and written in index notation as in 1.3.3, we consider the products $\prod_i X_i$ (for $i \in \mathrm{Ob}\mathbf{I}$) and $\prod_a X_{j(a)}$ (for $a: i(a) \to j(a)$ in $\mathrm{Mor}\mathbf{I}$), with projections (p_i) and (q_a), respectively. A cone of X amounts to a map $f: A \to \prod_i X_i$ that equalises the following two maps u, v

$$u, v: \prod_i X_i \to \prod_a X_{j(a)},$$

$$q_a u = p_{j(a)}, \quad q_a v = X_a . p_{i(a)} \qquad (\text{for } a: i(a) \to j(a) \text{ in } \mathbf{I}), \tag{1.33}$$

and the limit of X in $\mathbf{C}$ is obtained as the equaliser of u, v. (See [M4], Section V.2, Theorem 1.)

Dual results hold for colimits and *cocompleteness*. For instance, *a category is cocomplete (resp. finitely cocomplete) if and only if it has coequalisers and sums (resp. finite sums)*.

The categories **Set**, **Top**, **Ab** are complete and cocomplete; the forgetful functor **Top** $\to$ **Set** preserves limits and colimits, while **Ab** $\to$ **Set** only preserves limits.

One can easily see that, if the category $\mathbf{C}$ is **I**-complete, so is the functor category $\mathbf{C^S}$ (for any small category **S**), with **I**-limits computed pointwise on each object of **S**. The same holds for colimits. The proof is straightforward, and can be written down as an exercise; but a more formal and far quicker proof will be given in 1.5.4(f), using adjunctions.

The category associated to a preordered set X is complete if and only if the latter has all inf; since this fact is (well known to be) equivalent to the existence of all sup, X is complete if and only if it is cocomplete. In the ordered case this means that X is a complete lattice.

*More precisely one speaks of a *small-complete* category. Of course we cannot expect **Set** to have products indexed by a large set, or limits for functors $\mathbf{I} \to \mathbf{Set}$ defined over a large category. In other words, the basis $\mathbf{I}$ should be of a 'smaller size' than the category $\mathbf{C}$ where we construct limits, or we force $\mathbf{C}$ to be just a preordered set (where products are infima, and are not influenced by the size of the set of indices).

The question is settled by a neat result of P. Freyd, which is also easy to prove: a small category $\mathbf{C}$ has the limit of every functor defined on a small category (if and) only if it is a preordered set with all infima (see [M4], Section V.2, Proposition 3 or [G11], 2.2.5(f)). This is precisely why, *in an arbitrary category*, the existence of all (small) limits does not imply the existence of colimits, as is the case for preordered sets. In fact, the initial object of $\mathbf{C}$ is the colimit of the (small) functor $\mathbf{0} \to \mathbf{C}$; it can also be viewed as the limit of the identity functor $\mathbf{C} \to \mathbf{C}$, but the latter is not a small functor, generally.*

1.3.7 The creation of limits

There is another interesting property, related to preservation of limits; again, it does not assume their existence.

One says that a functor $F \colon \mathbf{C} \to \mathbf{D}$ *creates limits* for a functor $X \colon \mathbf{I} \to \mathbf{C}$ if:

(i) for every limit cone $(L', (v_i \colon L' \to FX_i)_{i \in I})$ of the composed functor $FX \colon \mathbf{I} \to \mathbf{D}$, there is precisely one cone $(L, (u_i \colon L \to X_i)_{i \in I})$ of X in $\mathbf{C}$ taken by F to $(L', (v_i))$,

(ii) the latter is a limit of X in $\mathbf{C}$ (preserved by F).

For instance the forgetful functor $U \colon \mathbf{Ab} \to \mathbf{Set}$ creates all small limits (which do exist). Indeed, a limit of abelian groups can be constructed by taking the limit L' of the underlying sets and putting on it the unique structure that makes the mappings v_i into homomorphisms. The same happens for all varieties of algebras. Note that these forgetful functors do not create *colimits*: they do not even preserve them, generally.

On the other hand the forgetful functor $\mathbf{Top} \to \mathbf{Set}$ preserves all limits but does not create them: for instance the product topology is the coarsest topology on the product set that makes all cartesian projections continuous, but is not the only one having this outcome – generally.

The relations between creating, preserving and reflecting limits are examined in [G11], Exercises 2.2.9.

1.3.8 Limits as universal arrows or representative objects

Consider the category $\mathbf{C}^{\mathbf{I}}$ of functors $\mathbf{I} \to \mathbf{C}$ and their natural transformations, for a small category $\mathbf{I}$ (see 1.2.7). The *diagonal functor*

$$\Delta: \mathbf{C} \to \mathbf{C}^{\mathbf{I}}, \qquad (\Delta A)_i = A, \quad (\Delta A)_a = \mathrm{id}A, \qquad (1.34)$$

sends an object A to the constant functor at A, and a morphism $f: A \to B$ to the natural transformation $\Delta f: \Delta A \to \Delta B: \mathbf{I} \to \mathbf{C}$ whose components are constant at f.

Let a functor $X: \mathbf{I} \to \mathbf{C}$ be given. A natural transformation $f: \Delta A \to X$ is the same as a cone of X of vertex A.

The limit of X in $\mathbf{C}$ is the same as a universal arrow $(L, \varepsilon: \Delta L \to X)$ from the functor Δ to the object X of $\mathbf{C}^{\mathbf{I}}$ (see 1.2.9). Dually, the colimit of X in $\mathbf{C}$ is the same as a universal arrow $(L', \eta: X \to \Delta L')$ from the object X of $\mathbf{C}^{\mathbf{I}}$ to the functor Δ.

As another characterisation, assuming that $\mathbf{C}$ has small hom-sets, consider the contravariant functor G *of sets of cones* of X

$$G: \mathbf{C}^{\mathrm{op}} \to \mathbf{Set}, \qquad G(A) = \mathbf{C}^{\mathbf{I}}(\Delta A, X). \qquad (1.35)$$

Then, by the Yoneda Lemma 1.2.8(b), X has a limit (L, ε) if and only if the functor G is representable, with representative object $X_0 = L$, via the element $x_0 = \varepsilon \in G(L)$.

Dually, the existence of a colimit (L', η) of X is equivalent to saying that the covariant functor $G'(A) = \mathbf{C}^{\mathbf{I}}(X, \Delta A)$ is representable, with representative object L', via the element $\eta \in G'(L')$.

1.3.9 Coends

A reader unacquainted with this tool of category theory can skip it until it is needed.

A functor $F: \mathbf{X}^{\mathrm{op}} \times \mathbf{X} \to \mathbf{C}$ is given. Its *coend* is an object C of $\mathbf{C}$ equipped with a family of morphisms $u_X: F(X, X) \to C$ indexed by the objects of $\mathbf{X}$. This family must be:

(i) *coherent*, in the sense that for every morphism $f: X \to Y$ in $\mathbf{X}$ we have $u_X.F(f, 1) = u_Y.F(1, f): F(Y, X) \to C$,

(ii) *universal*, in the sense that each coherent family $(v_X: F(X, X) \to C')_X$ factorises uniquely through the former: there exists precisely one morphism $c: C \to C'$ such that $c.u_X = v_X$, for all X.

The coend of F is written as $\int^X F(X, X)$.

If $\mathbf{X}$ is small and $\mathbf{C}$ is cocomplete, the coend of F exists and can be obtained as the coequalizer of the two morphisms

$$u, v \colon (\textstyle\sum_{f \colon X \to Y} F(Y, X)) \rightrightarrows (\textstyle\sum_X F(X, X)),$$

where the the first sum is indexed by all the morphisms of **X**, and the co-component u_f (resp. v_f) on the morphism $f \colon X \to Y$ comes from $F(f, 1) \colon F(Y, X) \to F(X, X)$ (resp. from $F(1, f) \colon F(Y, X) \to F(Y, Y)$).

*There is a classical example from Algebraic Topology. Let S be a simplicial set, i.e. a functor $S \colon \underline{\Delta}^{\mathrm{op}} \to \mathbf{Set}$ defined on the opposite of the category $\underline{\Delta}$ of finite positive ordinals (see 1.2.7). The *geometric realisation* of S can be viewed as the coend $\int^{\underline{n}} DS(\underline{n}) \times \Delta^n$, where $D \colon \mathbf{Set} \to \mathbf{Top}$ equips any set with the discrete topology and $\Delta \colon \underline{\Delta} \to \mathbf{Top}$ realises the positive ordinal $\underline{n} = \{0, ..., n\}$ of $\underline{\Delta}$ as the standard simplex $\Delta^n \subset \mathbb{R}^{n+1}$.*

The dual notion, called an *end*, is related to limits.

1.4 Lattices and Galois connections

We recall now some basic facts about the theory of lattices; the interested reader is referred to the classical texts of Birkhoff and Grätzer [Bi, Gr2]. We also review Galois connections between ordered sets.

1.4.1 Lattices

Classically, a lattice is defined as a (partially) ordered set X such that every pair x, x' of elements has a *join* $x \vee x'$ (the least element of X greater than both) and a *meet* $x \wedge x'$ (the greatest element of X smaller than both).

Here we follow a slightly different terminology, usual in category theory: *lattice* will always mean an ordered set with *finite* joins and meets (called a 'bounded lattice' in Lattice Theory). Thus, besides binary joins and meets, we are also assuming the existence of the least element $0 = \vee \emptyset$ (the empty join) and the greatest element $1 = \wedge \emptyset$ (the empty meet). These bounds are equal in the one-point lattice $\{*\}$, and only there.

Consistently with this terminology, a *lattice homomorphism* has to preserve finite joins and meets; a *sublattice* of a lattice X is closed under these operations (and has the same bounds as X). The category of lattices and homomorphisms will be written as **Lth**.

Occasionally we speak of a *quasi lattice* when we only assume the existence of binary joins and meets; a homomorphism of quasi lattices only has to preserve them. A *quasi sublattice* Y of a quasi lattice X is closed under binary joins and meets in X; when X is a lattice, Y may have different bounds, or lack one of them, or both.

For instance, if X is a lattice and $a \in X$, the downward and upward closed subsets $\downarrow a$, $\uparrow a$ of X generated by a (see 1.1.6) are quasi sublattices of X, and lattices in their own right.

1.4.2 Distributive and modular lattices

A lattice is said to be *distributive* if the meet operation distributes over the join operation, or equivalently if the join distributes over the meet. In fact, if we assume that meets distribute over joins, we have:

$$(x \vee y) \wedge (x \vee z) = ((x \vee y) \wedge x) \vee ((x \vee y) \wedge z)$$
$$= x \vee (x \wedge z) \vee (y \wedge z) = x \vee (y \wedge z).$$

A *boolean algebra* is a distributive (bounded) lattice where every element x has a *complement* x', defined and determined by the properties: $x \wedge x' = 0$, $x \vee x' = 1$.

The subsets of a set X form the classical boolean algebra $\mathcal{P}X$. The lattice $\mathrm{Sub}A$ of subgroups of an abelian group (or submodules of a module) is not distributive, generally; but one can easily check that it always satisfies a weaker, restricted form of distributivity, called modularity.

Namely, a lattice is said to be *modular* if it satisfies the following selfdual property (for all elements x, y, z)

(*) if $x \leqslant z$ then $(x \vee y) \wedge z = x \vee (y \wedge z)$.

The category of modular (resp. distributive) lattices and their homomorphisms will be written as **Mlh** (resp. **Dlh**).

By Birkhoff's representation theorem ([Bi] III.5, Theorem 5) the free distributive lattice on n generators is finite and isomorphic to a lattice of subsets. The reader may also be interested to know that the free modular lattice on three elements is finite and (obviously!) not distributive (see [Bi], III.6, Fig. 10), while four generators already give an infinite free modular lattice (see the final Remark in [Bi], III.6).

1.4.3 Galois connections

A covariant Galois connection is an adjunction between (partially) ordered sets, viewed as categories; this simple notion can serve as an introduction to general adjunctions, dealt with in the next section.

Given a pair X, Y of ordered sets, a (covariant) *Galois connection* between them can be presented in the following ways, whose equivalence can be easily verified.

(i) We assign two increasing mappings $f\colon X \to Y$ and $g\colon Y \to X$ such that, for all $x \in X$ and $y \in Y$:

$$f(x) \leqslant y \text{ in } Y \quad \Leftrightarrow \quad x \leqslant g(y) \text{ in } X.$$

(ii) We assign an increasing mapping $g\colon Y \to X$ such that, for every $x \in X$, there exists in Y:

$$f(x) = \min\{y \in Y \mid x \leqslant g(y)\}.$$

(ii*) We assign an increasing mapping $f\colon X \to Y$ such that, for every $y \in Y$, there exists in X:

$$g(y) = \max\{x \in X \mid f(x) \leqslant y\}.$$

(iii) We assign two increasing mappings $f\colon X \to Y$ and $g\colon Y \to X$ such that $\mathrm{id}X \leqslant gf$ and $fg \leqslant \mathrm{id}Y$.

By these formulas g determines f (called its *left adjoint*) and f determines g (its *right adjoint*). One writes $f \dashv g$ (as in the general notation of adjoints in category theory). The relations $\mathrm{id}X \leqslant gf$ and $fg \leqslant \mathrm{id}Y$ imply that $f = fgf$ and $g = gfg$.

Of course an isomorphism of ordered sets is, at the same time, left and right adjoint to its inverse. More generally, an increasing mapping *may* have one or both adjoints, which can be viewed as 'best approximations' to an inverse, of different kinds.

For instance, the embedding of ordered sets $i\colon \mathbb{Z} \to \mathbb{R}$ has a well-known right adjoint, the integral-part function, or *floor* function

$$[-]\colon \mathbb{R} \to \mathbb{Z}, \qquad [x] = \max\{k \in \mathbb{Z} \mid k \leqslant x\}. \tag{1.36}$$

The left adjoint also exists: it is the *ceiling* function

$$\min\{k \in \mathbb{Z} \mid k \geqslant x\} = -[-x],$$

related (here) to the right adjoint by the anti-isomorphism $x \mapsto (-x)$ of the real and integral lines.

On the other hand the embedding $\mathbb{Q} \to \mathbb{R}$ of the rational numbers has neither a right nor a left adjoint: an irrational number has no 'best' rational approximation, lower or upper.

1.4.4 Properties

Let us come back to a general Galois connection $f \dashv g$ between ordered sets X, Y.

The mapping f *preserves all the existing joins*, while g *preserves all the existing meets*. In fact, if $x = \vee x_i$ in X then $f(x_i) \leqslant f(x)$ (for all indices

i). Supposing that $f(x_i) \leqslant y$ in Y (for all i), it follows that $x_i \leqslant g(y)$ (for all i); but then $x \leqslant g(y)$ and $f(x) \leqslant y$.

From the relations $f = fgf$ and $g = gfg$ it follows that:

(a) $gf = \mathrm{id} \iff f$ is injective $\iff f$ is a split mono $\iff g$ is surjective $\iff g$ is a split epi $\iff f$ reflects the order relation,

(a*) $fg = \mathrm{id} \iff f$ is surjective $\iff f$ is a split epi $\iff g$ is injective $\iff g$ is a split mono $\iff g$ reflects the order relation.

Moreover the connection restricts to an isomorphism (of ordered sets) between the sets of *closed elements* of X and Y

$$g(Y) = \{x \in X \mid x = gf(x)\}, \qquad f(X) = \{y \in Y \mid y = fg(y)\}. \qquad (1.37)$$

An adjunction $f \dashv g$ will also be written as an arrow $(f, g) \colon X \nrightarrow Y$, often dot-marked and conventionally directed as the left adjoint $f \colon X \to Y$. Such arrows have an obvious composition

$$(f', g').(f, g) = (f'f, gg') \qquad (\text{for } (f', g') \colon Y \nrightarrow Z), \qquad (1.38)$$

and form the category Adj**Ord** *of ordered sets and Galois connections*; it is a selfdual category (see 2.1.1).

Each hom-set Adj**Ord**(X, Y) is canonically ordered: for two adjunctions $(f, g), (f', g') \colon X \nrightarrow Y$ we let $(f, g) \leqslant (f', g')$ if the following equivalent conditions hold

$$f \leqslant f', \qquad\qquad g' \leqslant g, \qquad (1.39)$$

since $f \leqslant f'$ gives $g' \leqslant gfg' \leqslant gf'g' \leqslant g$.

The relationship between Adj**Ord** and the usual category **Ord** (on the same objects) will become clearer in Chapter 3, where we shall amalgamate them to form a double category $\mathbb{A}$dj**Ord**, with horizontal arrows in **Ord** and vertical arrows in Adj**Ord** (see 3.1.7). Other similar double categories of lattices will also be of interest (see 3.1.7 and Section A1).

1.4.5 Direct and inverse images of subsets

The transfer of subobjects along morphisms is an important feature, that we examine here in **Set** and will be developed in Appendix A in other frameworks related to Homological Algebra.

Every set A has an ordered set $\mathrm{Sub}A = \mathcal{P}A$ of subsets, which actually is a complete boolean algebra (as recalled in 1.4.2). A mapping $f \colon A \to B$ gives two increasing mappings, of *direct* and *inverse image*

$$f_* \colon \mathrm{Sub}A \rightleftarrows \mathrm{Sub}B : f^*,$$

$$f_*(X) = f(X), \qquad f^*(Y) = f^{-1}(Y) \qquad (X \subset A, \, Y \subset B). \qquad (1.40)$$

These mappings form a Galois connection $f_* \dashv f^*$, since

$$X \subset f^* f_*(X), \qquad f_* f^*(Y) = Y \cap f(A) \subset Y, \qquad (1.41)$$

which implies, in particular, that f_* preserves unions and f^* preserves intersections. (In fact f^* is a homomorphism of complete boolean algebras and also has a right adjoint, related to the universal quantifier, while f_* is related to the existential one; we shall not use these facts.)

All this defines a *transfer functor* for subobjects of **Set**

$$\text{Sub}: \mathbf{Set} \to \text{Adj}\mathbf{Ord}, \quad \text{Sub}(f) = (f_*, f^*) : \text{Sub}A \rightarrowtail \text{Sub}B, \qquad (1.42)$$

with values in the category of ordered sets and Galois connections defined above.

In particular we have a (faithful) *contravariant functor of subsets* $\mathcal{P}^*$: **Set** $\dashrightarrow$ **Set**, that takes the mapping $f: A \to B$ to $f^*: \mathcal{P}B \to \mathcal{P}A$.

1.4.6 The graph of a Galois connection

The *graph* of the adjunction $(f, g): X \rightarrowtail$ will be the set

$$G(f,g) = \{(x,y) \in X \times Y \mid f(x) \leqslant y\}$$
$$= \{(x,y) \in X \times Y \mid x \leqslant g(y)\}, \qquad (1.43)$$

with the order induced by the cartesian product $X \times Y$ (that is a categorical product *in* **Ord**, of course).

The given adjunction has a canonical factorisation in two adjunctions

$$X \underset{g'}{\overset{f'}{\rightleftarrows}} G(f,g) \underset{g''}{\overset{f''}{\rightleftarrows}} Y \qquad (1.44)$$

$$f'(x) = (x, f(x)), \quad g'(x,y) = x, \qquad f''(x,y) = y, \quad g''(y) = (g(y), y).$$

*This forms a *natural weak factorisation system* in Adj**Ord**, in the sense of [GT], and the factorisation is mono-epi. In fact (f', g') is a monomorphism of Adj**Ord** (f' is an injective mapping, because $g'f' = \text{id}$) while (f'', g'') is an epimorphism (f'' is surjective, because $f''g'' = \text{id}$).*

1.4.7 *Exercises and complements* (Chains of adjunctions)

Long *chains of Galois connections* between two ordered sets X, Y

$$... f_{-2} \dashv f_{-1} \dashv f_0 \dashv f_1 \dashv f_2 ... \qquad f_{2i}: X \rightleftarrows Y : f_{2i+1}, \qquad (1.45)$$

seem not to be frequent, 'in nature', leaving apart the *trivial case* given by an isomorphism $f = f_{2i}$ of ordered sets and its inverse $g = f_{2i+1}$ ($i \in \mathbb{Z}$).

Yet, a Galois connection $\mathbb{Z} \rightleftarrows \mathbb{Z}$ on the integral ordered line *always* produces an unbounded chain of adjunctions, and there are chains where all maps are different. Replacing $\mathbb{Z}$ with the ordered real line $\mathbb{R}$, any non trivial chain has length $\leqslant 3$.

The reader might like to investigate this situation and study other cases, based on different ordered sets. Many results can be found in [G11], Sections 1.7.7 and 5.3.6, and in the article [Boo].

1.5 Adjoint functors

Adjunctions, a crucial step in category theory, were introduced by Kan in 1958 [K3]. They extend Galois connections (defined above, in 1.4.3) from ordered sets to general categories.

1.5.1 Main definitions

An *adjunction* $F \dashv G$, with a functor $F\colon \mathbf{C} \to \mathbf{D}$ *left adjoint* to a functor $G\colon \mathbf{D} \to \mathbf{C}$, can be equivalently presented in four main forms.

(i) We assign two functors $F\colon \mathbf{C} \to \mathbf{D}$ and $G\colon \mathbf{D} \to \mathbf{C}$ together with a family of bijections

$$\varphi_{XY}\colon \mathbf{D}(FX, Y) \to \mathbf{C}(X, GY) \qquad (X \text{ in } \mathbf{C}, \ Y \text{ in } \mathbf{D}),$$

which is natural in X and Y. More formally, the family (φ_{XY}) is a functorial isomorphism

$$\varphi\colon \mathbf{D}(F(-), =) \to \mathbf{C}(-, G(=))\colon \mathbf{C}^{\mathrm{op}} \times \mathbf{D} \to \mathbf{Set}.$$

(ii) We assign a functor $G\colon \mathbf{D} \to \mathbf{C}$ and, for every object X in $\mathbf{C}$, a universal arrow (see 1.2.9)

$$(F_0 X, \eta X\colon X \to GF_0 X) \qquad \text{from the object } X \text{ to the functor } G.$$

(ii*) We assign a functor $F\colon \mathbf{C} \to \mathbf{D}$ and, for every object Y in $\mathbf{D}$, a universal arrow

$$(G_0 Y, \varepsilon Y\colon FG_0 Y \to Y) \qquad \text{from the functor } F \text{ to the object } Y.$$

(iii) We assign two functors $F\colon \mathbf{C} \to \mathbf{D}$ and $G\colon \mathbf{D} \to \mathbf{C}$, together with two natural transformations

$$\eta\colon \mathrm{id}\mathbf{C} \to GF \quad (\text{the } unit), \qquad \varepsilon\colon FG \to \mathrm{id}\mathbf{D} \quad (\text{the } counit),$$

which satisfy the *triangular equations*:

$$\varepsilon F.F\eta = \mathrm{id}F, \qquad\qquad G\varepsilon.\eta G = \mathrm{id}G \qquad\qquad (1.46)$$

$$F \xrightarrow{F\eta} FGF \xrightarrow{\varepsilon F} F \qquad\qquad G \xrightarrow{\eta G} GFG \xrightarrow{G\varepsilon} G$$
$$\underbrace{\phantom{F \xrightarrow{F\eta} FGF \xrightarrow{\varepsilon F} F}}_{\mathrm{id}F} \qquad\qquad \underbrace{\phantom{G \xrightarrow{\eta G} GFG \xrightarrow{G\varepsilon} G}}_{\mathrm{id}G}$$

A proof of the equivalence can be found in [M4], Section IV.1, Theorem 2, or in [G11], Theorem 3.1.5. Essentially:

- given (i) one defines

$$\eta X = \varphi_{X,FX}(1_{FX}) : X \to GFX,$$

$$\varepsilon Y = (\varphi_{GY,Y})^{-1}(1_{GY}) : FGY \to Y,$$

- given (ii) one defines $F(X) = F_0X$, the morphism $F(f : X \to X')$ by the universal property of $\eta X'$ and the morphism $\varphi_{XY}(g : FX \to Y)$ as $Gg.\eta X : X \to GY$,
- given (iii) one defines the mapping $\varphi_{XY} : \mathbf{D}(F(X), Y) \to \mathbf{C}(X, G(Y))$ as above, a backward mapping $\psi_{XY}(f : X \to GY) = \varepsilon Y.Ff$, and proves that they are inverse to each other by the triangular equations.

1.5.2 Comments and complements

(a) The previous forms have different features.

Form (i) is the classical definition of an adjunction, and is at the origin of the name, by analogy with adjoint maps of Hilbert spaces.

Form (ii) is used when we start from a given functor G and want to construct its left adjoint (possibly less easy to define). Form (ii*) is used in a dual way.

The 'algebraic' form (iii) is adequate to the formal theory of adjunctions, as it makes sense in an abstract 2-category (see 2.3.2).

(b) Let $\mathbf{S}$ be a small category. A functor $F : \mathbf{C} \to \mathbf{D}$ has a canonical 'extension' to functor categories on $\mathbf{S}$, by composing a functor $X : \mathbf{S} \to \mathbf{C}$ with F

$$F^{\mathbf{S}} : \mathbf{C}^{\mathbf{S}} \to \mathbf{D}^{\mathbf{S}}, \qquad F^{\mathbf{S}}(X) = FX : \mathbf{S} \to \mathbf{D}. \tag{1.47}$$

A natural transformation $\varphi : F \to G : \mathbf{C} \to \mathbf{D}$ can be similarly extended

$$\varphi^{\mathbf{S}} : F^{\mathbf{S}} \to G^{\mathbf{S}} : \mathbf{C}^{\mathbf{S}} \to \mathbf{D}^{\mathbf{S}},$$
$$\varphi^{\mathbf{S}}(X) = \varphi X : FX \to GX : \mathbf{S} \to \mathbf{D}. \tag{1.48}$$

These extensions preserve all compositions, of functors and natural transformations.

(c) *Functor categories and adjunctions.* It follows that an adjunction $(\eta, \varepsilon) :$ $F \dashv G$ has a canonical extension to an adjunction

$$(\eta^{\mathbf{S}}, \varepsilon^{\mathbf{S}}) : F^{\mathbf{S}} \dashv G^{\mathbf{S}}, \tag{1.49}$$

just using the form (iii) of Definition 1.5.1.

1.5.3 Exercises and complements

(a) An adjoint equivalence $(F, G, \eta, \varepsilon)$, defined in 1.2.5, amounts to an adjunction where the unit and counit are invertible, so that $F \dashv G \dashv F$.

(b) For a 'forgetful functor' $U: \mathbf{A} \to \mathbf{X}$ the existence of the left adjoint $F \dashv U$ means that every object X of $\mathbf{X}$ has a free object $(FX, \eta: X \to UFX)$ in $\mathbf{A}$ (as defined in 1.2.9).

Using what we have already seen, one can easily construct the left adjoint to the following forgetful functors (and many others):

$$\begin{array}{lll} \mathbf{Mon} \to \mathbf{Set}, & R\,\mathbf{Mod} \to \mathbf{Set}, & \mathbf{Ab} \to \mathbf{Set}, \\ \mathbf{Rng} \to \mathbf{Mon}, & \mathbf{Rng} \to \mathbf{Set}, & \mathbf{Set}_\bullet \to \mathbf{Set}. \end{array} \tag{1.50}$$

*One can prove that the forgetful functor $U: \mathbf{A} \to \mathbf{Set}$ of any variety of algebras (in the sense of 1.1.8(e)) always has a left adjoint $F: \mathbf{Set} \to \mathbf{A}$, the *free-algebra functor* for $\mathbf{A}$. See, for instance, [M4, Bo2, G11].*

(c) The forgetful functor $U: \mathbf{Top} \to \mathbf{Set}$ has both adjoints $D \dashv U \dashv C$. The cylinder-cocylinder adjunction in $\mathbf{Top}$ will be examined in A5.1.

(d) The forgetful functor $U: \mathbf{pOrd} \to \mathbf{Set}$ of the category of preordered sets has a chain of adjunctions $\pi_0 \dashv D \dashv U \dashv C$.

(e) The (non-faithful) functor $\mathrm{Ob}: \mathbf{Cat} \to \mathbf{Set}$ has a similar chain $\pi_0 \dashv D \dashv \mathrm{Ob} \dashv C$, that extends the previous one. A category $\mathbf{C}$ is said to be *connected* if $\pi_0(\mathbf{C})$ is the singleton, a property that can be easily characterised. (This excludes the empty category.) Every category is the sum of its connected components (i.e. its maximal connected subcategories).

(f) The embedding $U: \mathbf{pOrd} \to \mathbf{Cat}$ has a left adjoint $\mathrm{po}: \mathbf{Cat} \to \mathbf{pOrd}$, where $\mathrm{po}(\mathbf{S})$ is the set $\mathrm{Ob}\mathbf{S}$ with an obvious preorder. A diagram $X: \mathbf{S} \to \mathbf{C}$ is commutative if and only if it factorises through the canonical projection $\eta: \mathbf{S} \to \mathrm{po}(\mathbf{S})$.

(g) Prove that a left adjoint preserves (the existing) colimits; dually, a right adjoint preserves limits.

(h) Prove that the chain $D \dashv U \dashv C$ in (c) cannot be extended.

(i) Prove that no functor in the list (1.50) has a right adjoint.

1.5.4 Main properties of adjunctions

(a) *Uniqueness and existence.* Given a functor, its left adjoint is uniquely determined up to isomorphism: this follows from the uniqueness property of universal arrows, in 1.2.9.

A crucial theorem for proving the existence (under suitable hypothesis) is the Adjoint Functor Theorem of P. Freyd: see [M4], Section V.6, Theorem 2, or [G11], Section 3.5.

(b) *Composing adjunctions.* As a consequence of the composition of universal arrows, in (1.20), two consecutive adjunctions

$$F: \mathbf{C} \rightleftarrows \mathbf{D} : G, \qquad \eta: 1 \to GF, \quad \varepsilon: FG \to 1,$$
$$H: \mathbf{D} \rightleftarrows \mathbf{E} : K, \qquad \rho: 1 \to KH, \quad \sigma: HK \to 1, \tag{1.51}$$

give a composed adjunction from the first to the third category

$$HF: \mathbf{C} \rightleftarrows \mathbf{E} : GK,$$
$$G\rho F.\eta: 1 \to GF \to GK.HF, \tag{1.52}$$
$$\sigma.H\varepsilon K: HF.GK \to HK \to 1.$$

There is thus a category **AdjCat** of small categories and adjunctions, with morphisms written in one of the following forms, and *conventionally directed as the left adjoint*

$$(F, G, \eta, \varepsilon): \mathbf{C} \rightarrowtail \mathbf{D}, \qquad F \dashv G: \mathbf{C} \rightarrowtail \mathbf{D}.$$

We also write $(\eta, \varepsilon): F \dashv G$, but one should not view this as an arrow from F to G in some category.

(c) *Involution.* Duality of categories interchanges left and right adjoint, unit and counit. **AdjCat** is thus a category with involution (as defined below, in 2.1.1)

$$\mathbf{C} \mapsto \mathbf{C}^{\mathrm{op}},$$
$$((F, G, \eta, \varepsilon): \mathbf{C} \rightarrowtail \mathbf{D}) \;\mapsto\; ((G^{\mathrm{op}}, F^{\mathrm{op}}, \varepsilon^{\mathrm{op}}, \eta^{\mathrm{op}}): \mathbf{D}^{\mathrm{op}} \rightarrowtail \mathbf{C}^{\mathrm{op}}). \tag{1.53}$$

(d) *Preserving limits and colimits.* We have already seen that a left adjoint preserves colimits, while a right adjoint preserves limits (Exercise 1.5.3(g)).

For a covariant Galois connection $f \dashv g$, this amounts to saying that f preserves the existing joins and g the existing meets, as already shown in 1.4.4. We also know that an equivalence $F: \mathbf{C} \to \mathbf{D}$ can always be completed to an adjoint equivalence, with $G \dashv F \dashv G$; therefore F preserves limits and colimits.

(e) *Limit functor as a right adjoint.* The description of the limit of a functor $X: \mathbf{I} \to \mathbf{C}$ as a universal arrow $(LX, \varepsilon_X : \Delta(LX) \to X)$ from the diagonal functor $\Delta\colon \mathbf{C} \to \mathbf{C}^{\mathbf{I}}$ to the object X of $\mathbf{C}^{\mathbf{I}}$ (in 1.3.8) shows that the existence of all **I**-limits in **C** amounts to the existence of a right adjoint to Δ, the *limit functor* $L\colon \mathbf{C}^{\mathbf{I}} \to \mathbf{C}$, with counit $\varepsilon\colon \Delta L \to \mathrm{id}\mathbf{C}$. Dually, a *colimit functor* $L'\colon \mathbf{C}^{\mathbf{I}} \to \mathbf{C}$ is left adjoint to Δ, and the unit $\eta\colon \mathrm{id}\mathbf{C} \to \Delta L'$ gives the universal cocone $\eta_X\colon X \to \Delta(L'X)$ of X.

(f) *Limits in functor categories.* One can now give a brief, synthetic proof of a result already stated in 1.3.6: if **C** has all **I**-limits, the same holds in any functor category $\mathbf{C}^{\mathbf{S}}$, with limits computed pointwise on **S**. The same applies to colimits.

In fact, we are assuming that the diagonal functor $\Delta\colon \mathbf{C} \to \mathbf{C}^{\mathbf{I}}$ has a (say) right adjoint $L\colon \mathbf{C}^{\mathbf{I}} \to \mathbf{C}$, with counit $\varepsilon\colon \Delta L \to 1$. By 1.5.2(c), the extension $\Delta^{\mathbf{S}}\colon \mathbf{C}^{\mathbf{S}} \to (\mathbf{C}^{\mathbf{I}})^{\mathbf{S}}$ has right adjoint $L^{\mathbf{S}}\colon (\mathbf{C}^{\mathbf{I}})^{\mathbf{S}} \to \mathbf{C}^{\mathbf{S}}$, with counit $\varepsilon^{\mathbf{S}}$, and the canonical isomorphisms $(\mathbf{C}^{\mathbf{I}})^{\mathbf{S}} = \mathbf{C}^{\mathbf{I}\times\mathbf{S}} = (\mathbf{C}^{\mathbf{S}})^{\mathbf{I}}$ show that $L^{\mathbf{S}}$ is right adjoint to the diagonal $\mathbf{C}^{\mathbf{S}} \to (\mathbf{C}^{\mathbf{S}})^{\mathbf{I}}$, with counit $\varepsilon^{\mathbf{S}}$.

(g) *Faithful and full adjoints.* Suppose we have an adjunction $(\eta, \varepsilon)\colon F \dashv G$. As proved in [M4], Section IV.3, Theorem 1, or [G11], Theorem 3.2.5:

(i) G is faithful if and only if all the components εY are epi,

(ii) G is full if and only if all the components εY are split mono,

(iii) G is full and faithful if and only if the counit ε is invertible,

(i*) F is faithful if and only if all the components ηX are mono,

(ii*) F is full if and only if all the components ηX are split epi,

(iii*) F is full and faithful if and only if the unit η is invertible.

1.5.5 Reflective and coreflective subcategories

A subcategory $\mathbf{D} \subset \mathbf{C}$ is said to be *reflective* if the inclusion functor $U\colon \mathbf{D} \to \mathbf{C}$ has a left adjoint, and *coreflective* if U has a right adjoint.

For a *full reflective subcategory* the counit ε is invertible (by 1.5.4(g)). One can always choose the reflector $F \dashv U$ so that $FU = \mathrm{id}\mathbf{D}$ and $\varepsilon = \mathrm{id}$. (In fact, one can constrain the unit $\eta\colon 1 \to UF$ so that $\eta U = \mathrm{id}$, and construct the reflector F from the constrained unit. Then $U\varepsilon = \mathrm{id}$ and $\varepsilon = \mathrm{id}$.)

For instance **Ab** is reflective in **Gp**, with reflector $(-)^{\mathrm{ab}}\colon \mathbf{Gp} \to \mathbf{Ab}$ sending a group G to the abelianised group $G^{\mathrm{ab}} = G/[G,G]$. The unit $\eta G\colon G \to G^{\mathrm{ab}}$ is given by the canonical projection, and the counit $\varepsilon A\colon A \to A/[A,A]$ is invertible. (It is actually the identity, if the trivial quotient $A/[A,A]$ is realised as A, according to our general convention on quotients.)

In **Ab** the full subcategory t**Ab** formed by all torsion abelian groups is coreflective; the counit $\varepsilon A\colon tA \to A$ is the embedding of the torsion subgroup of an abelian group. On the other hand the full subcategory tf**Ab** formed by all torsion-free abelian groups is reflective in **Ab**, with unit the canonical projection $\eta A \colon A \to A/tA$.

In the real ordered line $\mathbb{R}$, the ordered subset $\mathbb{Z}$ is reflective and coreflective, while the subset of rational numbers is neither (see 1.4.3).

1.5.6 Comma categories and slice categories

We now introduce an important construction. For functors $F \colon \mathbf{X} \to \mathbf{Z}$ and $G \colon \mathbf{Y} \to \mathbf{Z}$ with the same codomain one constructs a *comma category* $F \downarrow G$ equipped with functors P, Q and a natural transformation π

$$F \downarrow G \overset{P}{\nearrow} \overset{\mathbf{X}}{\underset{Q}{\searrow}} \overset{F}{\searrow} \mathbf{Z} \qquad \pi \colon FP \to GQ. \tag{1.54}$$

(The original notation was (F, G), whence the name.) The objects of $F \downarrow G$ are the triples $(X, Y, z \colon FX \to GY)$ formed of an object of $\mathbf{X}$, an object of $\mathbf{Y}$ and a morphism of $\mathbf{Z}$. A morphism

$$(f, g) \colon (X, Y, z \colon FX \to GY) \to (X', Y', z' \colon FX' \to GY'),$$

comes from a pair of maps $f \colon X \to X'$, $g \colon Y \to Y'$ that form a commutative square in $\mathbf{Z}$, namely $z'.Ff = Gg.z$.

Composition and identities come from those of $\mathbf{X} \times \mathbf{Y}$

$$(f', g').(f, g) = (f'f, g'g), \qquad \mathrm{id}(X, Y, z) = (\mathrm{id}X, \mathrm{id}Y).$$

There is an obvious universal property, that makes the triple (P, Q, π) a sort of *directed 2-dimensional pullback*: for every category $\mathbf{C}$ equipped with similar data $P' \colon \mathbf{C} \to \mathbf{X}$, $Q' \colon \mathbf{C} \to \mathbf{Y}$ and $\pi' \colon FP' \to GQ'$, there is precisely one functor $W \colon \mathbf{C} \to F \downarrow G$ which commutes with the structural data

$$W(C) = (P'C, Q'C, \pi'C), \qquad W(c \colon C \to C') = (P'c, Q'c),$$
$$PW = P', \qquad QW = Q', \qquad \pi W = \pi'. \tag{1.55}$$

As a matter of notation, one writes: $F \downarrow \mathrm{id}\mathbf{Z}$ as $F \downarrow \mathbf{Z}$ and $\mathrm{id}\mathbf{Z} \downarrow G$ as $\mathbf{Z} \downarrow G$. Moreover on object Z_0 of the category $\mathbf{Z}$ can be viewed as a functor $Z_0 \colon \mathbf{1} \to \mathbf{Z}$; therefore the comma category $F \downarrow Z_0$ has objects $(X, z \colon FX \to Z_0)$, while $Z_0 \downarrow G$ has objects $(Y, z \colon Z_0 \to GY)$.

In particular we have the *slice categories*

$$Z_0 \downarrow \mathbf{Z} = \mathbf{Z} \backslash Z_0, \qquad \mathbf{Z} \downarrow Z_0 = \mathbf{Z}/Z_0, \tag{1.56}$$

of *objects* $(Z, z \colon Z_0 \to Z)$ *below* Z_0 and *objects* $(Z, z \colon Z \to Z_0)$ *above* Z_0, respectively.

Note that **Set.** and **Top.** can be identified with the slice categories **Set**\{*} and **Top**\{*} of objects under the singleton.

*Less trivially the category **Rng**′ of 'associative rings' (not assumed to have a unit) is equivalent to the category **Rng**/$\mathbb{Z}$ of *copointed unitary rings*, or rings over the initial object $\mathbb{Z}$, as the interested reader can prove.*

1.5.7 Comma categories, adjunctions and limits

(a) As an extension of the case of Galois connections in 1.4.6, the *graph* of the adjunction $(F, G, \eta, \varepsilon) \colon \mathbf{X} \rightsquigarrow \mathbf{Y}$ will be the comma category

$$\mathbf{G}(F, G) = F \downarrow \mathbf{Y}, \tag{1.57}$$

with objects $(X, Y, c \colon FX \to Y)$ and morphisms

$$(f, g) \colon (X, Y, c \colon FX \to Y) \to (X', Y', c' \colon FX' \to Y'), \quad c'.Ff = g.c.$$

The adjunction $F \dashv G$ has a *graph factorisation* in Adj**Cat**

$$\mathbf{X} \; \underset{G'}{\overset{F'}{\rightleftarrows}} \; \mathbf{G}(F, G) \; \underset{G''}{\overset{F''}{\rightleftarrows}} \; Y \tag{1.58}$$

$$F'(X) = (X, FX, 1_{FX}), \quad G'(X, Y, c) = X, \quad \eta'X = 1_X,$$
$$F''(X, Y, c) = Y, \quad G''(Y) = (GY, Y, \varepsilon Y \colon FG(Y) \to Y), \quad \varepsilon''Y = 1_Y.$$

As in 1.4.6, this forms a *natural weak factorisation system* in Adj**Cat**, in the sense of [GT], which is mono-epi (because $G'F'$ and $F''G''$ are identities).

One can replace $\mathbf{G}(F, G)$ with the comma category $\mathbf{G}'(F, G) = \mathbf{X} \downarrow G$, which is isomorphic to the former because of the adjunction (and coincides with it when $\mathbf{X}$ and $\mathbf{Y}$ are order categories). One can also consider a factorisation in three adjunctions, through the isomorphism $\mathbf{G}(F, G) \rightleftarrows \mathbf{G}'(F, G)$, as we shall do in 4.3.8.

(b) Limits and colimits in the category $\mathbf{C}$ can be viewed as terminal or initial objects in comma categories of the diagonal functor $\Delta \colon \mathbf{C} \to \mathbf{C}^{\mathbf{I}}$ (defined in 1.3.8). In fact a cone $(A, (f_i \colon A \to X_i))$ of $X \colon \mathbf{I} \to \mathbf{C}$ is the same as an object of the comma category $\Delta \downarrow X$ and the limit of X is the same as the terminal object of $\Delta \downarrow X$. Dually a cocone (or the colimit) $(A, (f_i \colon X_i \to A))$ of X is an object (or the initial object) of the comma category $X \downarrow \Delta$.

1.5.8 Exercises and complements

(a) As we know from a remark in 1.3.3, the limit of a functor $X \colon \mathbf{I} \to$ **Set** defined on a small category can be computed as the set of its cones with vertices at the singleton. This can now be quickly proved, using an adjunction.

(b) Study the adjoints of the diagonal functor $\Delta \colon \mathbf{C} \to \mathbf{C}^2$ of a category $\mathbf{C}$ into the product $\mathbf{C} \times \mathbf{C}$.

(c) In particular, for $\mathbf{C} = R \, \mathbf{Mod}$, we have a periodic chain of adjunctions ... $B \dashv \Delta \dashv B \dashv \Delta$ ..., where $B(X, Y) = X \oplus Y$ is the direct sum, also called a *biproduct*.

A general definition of biproduct in a pointed category will be given in 1.8.4.

(Let us note that, in **Ord**, a relation $f \dashv g \dashv f$ trivially forces f and g to be inverse to each other.)

(d) Prove that, for a given $n \in \mathbb{N}$, the endofunctor $F(X) = X^n$ of $R \, \mathbf{Mod}$ is adjoint to itself. Prove that, for any small set I, the endofunctor $G(Y) = Y^I$ has a left adjoint; extend this fact to a category $\mathbf{C}$, under suitable hypotheses.

(e) Chains of Galois connections between ordered sets X and Y, even unbounded, have been considered in 1.4.7. Each of them can be transformed into a chain of adjunctions between the categories $\mathbf{C}^X$ and $\mathbf{C}^Y$, for any category $\mathbf{C}$. Other periodic chains of adjunctions can be seen in [G11], Section 3.2.8.

(f) Prove that the category **Mon** of monoids is coreflective in **Gp**.

*(g) Prove that it is also reflective.

1.6 Monads and algebras

Monads and their algebras give a wide formalisation of the 'algebraic character' of a category over another, typically used over the category of sets but also of interest in many other cases.

We follow the classical terminology and notation of Mac Lane [M4], Chapter VI. Monads are also called 'triples' or 'dual standard constructions' in other texts, e.g. [BaB, Bc, Du].

1.6.1 Monads

A *monad* on the category $\mathbf{X}$ is a triple (T, η, μ) where $T \colon \mathbf{X} \to \mathbf{X}$ is an endofunctor, while $\eta \colon 1 \to T$ and $\mu \colon T^2 \to T$ are natural transformations (called the *unit* and *multiplication* of the monad) that make the following diagrams commute

$$
\begin{array}{ccc}
T \xrightarrow{\ \eta T\ } T^2 \xleftarrow{\ T\eta\ } T & \qquad & T^3 \xrightarrow{\ T\mu\ } T^2 \\
\searrow \ \downarrow{\mu} \ \swarrow & & \mu T \downarrow \qquad \downarrow \mu \\
T & & T^2 \xrightarrow{\ \mu\ } T
\end{array}
\tag{1.59}
$$

These axioms are called *unitarity* and *associativity*. (In fact they are a rewriting of the diagrammatic presentation of a monoid T in **Set**, as we shall see in 2.2.6(e).)

1.6.2 From adjunctions to monads

It is easy to verify that an adjunction

$$
F \colon \mathbf{X} \rightleftarrows \mathbf{A} \colon G, \qquad \eta \colon 1 \to GF, \quad \varepsilon \colon FG \to 1,
\tag{1.60}
$$

yields a monad (T, η, μ) on $\mathbf{X}$ (the domain of the left adjoint), where $T = GF \colon \mathbf{X} \to \mathbf{X}$, the transformation $\eta \colon 1 \to T$ is the unit of the adjunction and $\mu = G\varepsilon F \colon GF.GF \to GF$.

Typically, a variety of algebras $\mathbf{A}$ has an associated monad $T = UF \colon$ **Set** $\to$ **Set** produced by the forgetful functor $U \colon \mathbf{A} \to$ **Set** and its left adjoint $F \colon$ **Set** $\to \mathbf{A}$, the *free-algebra functor* (see 1.5.3(b)).

1.6.3 Eilenberg–Moore algebras

The other way round, from monads to adjunctions, there are two main constructions; we begin from the more important one.

Given a monad (T, η, μ) on $\mathbf{X}$ one defines the category $\mathbf{X}^T$ of *T-algebras*, or *Eilenberg–Moore algebras* for T: these are pairs $(X, a \colon TX \to X)$ consisting of an object X of $\mathbf{X}$ and a map a (called the *algebraic structure*) satisfying two coherence axioms:

$$
a.\eta X = 1_X, \qquad a.Ta = a.\mu X,
\tag{1.61}
$$

$$
\begin{array}{ccc}
X \xrightarrow{\ \eta X\ } TX & \qquad & T^2 X \xrightarrow{\ Ta\ } TX \\
\searrow \ \downarrow{a} & & \mu X \downarrow \qquad \downarrow a \\
 X & & TX \xrightarrow{\ a\ } X
\end{array}
$$

A *morphism* of T-algebras $f \colon (X, a) \to (Y, b)$ is a morphism $f \colon X \to Y$ of $\mathbf{X}$ which preserves the algebraic structures, in the sense that $f.a = b.Tf$. They compose as in $\mathbf{X}$.

$\mathbf{X}^T$ is thus a full subcategory of the comma category $T \downarrow \mathbf{X}$.

One can find the category $\mathbf{X}^T$ written as $\mathbf{Alg}(T)$, but we prefer to follow Mac Lane's notation. We shall reserve the symbol $\mathbf{Alg}(T)$ for the category of algebraic objects of an *idempotent* monad, which is isomorphic to $\mathbf{X}^T$ (see Section 1.7).

For an ordered set X, viewed as a category, a monad $T \colon X \to X$ is just an increasing function which is *inflationary* ($1_X \leqslant T$) and *idempotent* ($T^2 = T$); this is also called a *closure* operator on X. An algebra for T is a closed element: $x = Tx$.

Obvious examples come from the topological closure $\mathcal{P}S \to \mathcal{P}S$ of a topological space S (which, moreover, has to preserve finite unions). Idempotent monads on arbitrary categories will be studied in Section 1.7.

1.6.4 From monads to adjunctions and back

Given a monad (T, η, μ) on $\mathbf{X}$, we can construct an adjunction with the category $\mathbf{X}^T$ of T-algebras

$$F^T \colon \mathbf{X} \rightleftarrows \mathbf{X}^T \colon G^T,$$

$$\eta^T = \eta \colon 1 \to G^T F^T, \qquad \varepsilon^T \colon F^T G^T \to 1, \tag{1.62}$$

so that the associated monad coincides with the given one.

First we have an obvious (faithful) forgetful functor

$$G^T \colon \mathbf{X}^T \to \mathbf{X},$$

$$G^T(X, a) = X, \qquad G^T(f \colon (X, a) \to (Y, b)) = (f \colon X \to Y). \tag{1.63}$$

Backwards we have the functor giving the *free T-algebra* on an object X of $\mathbf{X}$

$$F^T \colon \mathbf{X} \to \mathbf{X}^T,$$

$$F^T(X) = (TX, \; \mu X \colon T^2 X \to TX), \tag{1.64}$$

$$F^T(f \colon X \to Y) = Tf \colon (TX, \mu X) \to (TY, \mu Y).$$

Now $G^T F^T = T \colon \mathbf{X} \to \mathbf{X}$. The unit of the adjunction is $\eta^T = \eta \colon 1 \to T$, while the counit is defined as follows (and the triangular equations of 1.5.1 are satisfied)

$$\varepsilon^T \colon F^T G^T \to 1, \qquad \varepsilon^T(X, a) = a \colon (TX, \mu X) \to (X, a). \tag{1.65}$$

It is now easy to see that the monad associated to this adjunction is the original one, because the new multiplication also coincides with the old one: $G^T \varepsilon^T F^T(X) = G^T \varepsilon^T(TX, \mu X) = \mu X$.

1.6.5 Monadicity

On the other hand, if we start from an adjunction $(F, G, \eta, \varepsilon) \colon \mathbf{X} \rightharpoonup \mathbf{A}$ as in 1.6.2, form the associated monad (T, η, μ) and then the associated adjunction $(F^T, G^T, \eta, \varepsilon^T) \colon \mathbf{X} \rightharpoonup \mathbf{X}^T$, we get a comparison between the two adjunctions which may be an isomorphism or not; if it is the case we think of $\mathbf{A}$ as a 'category of algebras' over $\mathbf{X}$.

In fact there is a *comparison functor*:

$$K \colon \mathbf{A} \to \mathbf{X}^T, \qquad K(A) = (GA, G\varepsilon A \colon GFGA \to GA),$$
$$K(f \colon A \to B) \;=\; Gf \colon (GA, G\varepsilon A) \to (GB, G\varepsilon B). \tag{1.66}$$

K links the two adjunctions (which share the unit), in the sense that, as one can easily verify

$$
\begin{array}{cc}
\mathbf{X} \underset{G}{\overset{F}{\rightleftarrows}} \mathbf{A} & KF = F^T, \qquad G^T K = G, \\[2pt]
\Big\| \quad\;\; \Big\downarrow K & \\[2pt]
\mathbf{X} \underset{G^T}{\overset{F^T}{\rightleftarrows}} \mathbf{X}^T & \varepsilon^T K = K\varepsilon \qquad (\eta^T = \eta).
\end{array}
\tag{1.67}
$$

Now a functor $G \colon \mathbf{A} \to \mathbf{X}$ is said to be *monadic* (or *algebraic*), or to make $\mathbf{A}$ *monadic over* $\mathbf{X}$, if it has a left adjoint $F \colon \mathbf{X} \to \mathbf{A}$ and moreover the comparison functor $K \colon \mathbf{A} \to \mathbf{X}^T$ defined above is an *isomorphism* of categories (see 1.6.6(d)).

There are various 'monadicity theorems' (also called 'tripleability theorems') that give sufficient (or necessary and sufficient) conditions for a functor to be monadic: the interested reader can see [M4], Section VI.8, [Bo2], Section 4.4, and [Bc, Du].

1.6.6 Exercises, comments and complements

(a) Present **Set.** as a category of T-algebras over **Set**. Similarly for the category of semigroups.

*(b) One can prove that the forgetful functor $U \colon \mathbf{A} \to \mathbf{Set}$ of any variety of algebras is monadic (see [M4], Section VI.8). But the present *formalisation of the algebraic character of a category* is much wider: for instance the category of compact Hausdorff spaces is monadic over **Set** (see [M4], Section VI.9), and this monadicity depends on the closure operator of such spaces, a sort of 'infinitary operation'.

(c) (Limits of algebras). Let (T, η, μ) be a monad on a complete category $\mathbf{X}$. Prove that $\mathbf{X}^T$ is also complete, with limits *created* by the forgetful functor $G^T \colon \mathbf{X}^T \to \mathbf{X}$ (see 1.3.7).

(d) We are following Mac Lane ([M4], Section VI.3) in defining monadicity in a strong sense, *up to isomorphism of categories*, because concrete examples, based on structured sets, generally fall in this case. On the other hand [Bo2], in Section 4.4, and various other texts only ask that the comparison K be an *equivalence* of categories.

The difference can be appreciated at the light of this example. As we have already seen, the category **Set.** of pointed sets is a variety of algebras, produced by a single zeroary operation under no axioms, and is thus monadic over **Set**, via its forgetful functor U. The equivalent category $\mathcal{S}$ of sets and partial mappings (see 2.1.6) inherits a composed functor $\mathcal{S} \to \textbf{Set.} \to \textbf{Set}$, and the associated comparison is now an equivalence of categories. We prefer to view $\mathcal{S}$ as 'weakly algebraic' over **Set**, rather than as 'algebraic'.

(e) We also note that the free semigroup on the empty set is empty: the category of 'non-empty semigroups', discussed in 1.1.8(e), lacks such a free object and is not monadic over **Set**.

(f) Prove that the covariant endofunctor $\mathcal{P}$ of subsets forms a monad

$$\mathcal{P}\colon \textbf{Set} \to \textbf{Set},$$

$$X \mapsto \mathcal{P}X, \qquad (f\colon X \to Y) \mapsto (f_*\colon \mathcal{P}X \to \mathcal{P}Y),$$

$$\eta\colon 1 \to \mathcal{P}, \qquad \eta_X(x) = \{x\}, \tag{1.68}$$

$$\mu\colon \mathcal{P}^2 \to \mathcal{P}, \qquad \mu_X(\mathcal{A}) = \bigcup_{A \in \mathcal{A}} A \qquad (\mathcal{A} \subset \mathcal{P}X).$$

(g) Prove that its category of algebras is isomorphic to the category **C** of complete lattices and mappings that preserve arbitrary joins. Note that we are using an infinitary operation (arbitrary joins), as in the case recalled in (b).

(h) The last two exercises can be reorganised in an alternative way: we start from the forgetful functor $U\colon \textbf{C} \to \textbf{Set}$, and construct its left adjoint $F\colon \textbf{Set} \to \textbf{C}$ so that the associated monad (UF, η, μ) is the previous one; then we prove that U is monadic.

1.6.7 Kleisli algebras and their adjunction

The second construction of an adjunction from a monad (T, η, μ) on **X** is based on *Kleisli algebras* for T and their category $\textbf{X}_T$.

An object of $\textbf{X}_T$ is an object of **X**, *viewed as the basis of a free T-algebra*. A morphism $f^\sharp\colon X \to Y$ of $\textbf{X}_T$ is 'represented by' an arbitrary morphism $f\colon X \to TY$ of **X** (which tells us as $f^\sharp$ acts on the basis).

By definition, the composite $g^\sharp f^\sharp$ of $f^\sharp$ with $g^\sharp\colon Y \to Z$ is represented by the morphism

$$\mu Z.Tg.f \colon X \to TY \to T^2 Z \to TZ, \tag{1.69}$$

while the identity $\mathrm{id}X \colon X \rightarrowtail X$ is represented by the morphism $\eta X \colon X \to TX$. The axioms of unitarity and associativity for $\mathbf{X}_T$ are easily proved.

The forgetful functor of Kleisli algebras and its left adjoint are now:

$$G_T \colon \mathbf{X}_T \to \mathbf{X}, \qquad G_T(X) = TX,$$
$$G_T(f^\sharp \colon X \rightarrowtail Y) = \mu Y.Tf \colon TX \to TY,$$
$$F_T \colon \mathbf{X} \to \mathbf{X}_T, \qquad F_T(X) = X, \tag{1.70}$$
$$F_T(f \colon X \to Y) \ = \ (\eta Y.f)^\sharp \colon X \rightarrowtail Y.$$

Again $G_T F_T = T$. The unit and counit of the adjunction $F_T \dashv G_T$ are

$$\eta_T = \eta \colon 1 \to T = G_T F_T,$$
$$\varepsilon_T(X) \ = \ (1_{TX})^\sharp \colon F_T G_T(X) = TX \rightarrowtail X. \tag{1.71}$$

Also here the monad associated to this adjunction is the original one, since the new multiplication coincides with the previous one:

$$(G_T.\varepsilon_T.F_T)(X) = G_T((1_{TX})^\sharp) = \mu X.$$

If we start from an adjunction $(F, G, \eta, \varepsilon) \colon \mathbf{X} \rightarrowtail \mathbf{A}$, as in 1.6.2, form the associated monad (T, η, μ) and then the Kleisli adjunction

$$(F_T, G_T, \eta, \varepsilon_T) \colon \mathbf{X} \rightarrowtail \mathbf{X}_T,$$

there is – again – a comparison functor which links the two adjunctions

$$H \colon \mathbf{X}_T \to \mathbf{A}, \qquad H(X) = FX,$$
$$H(f^\sharp \colon X \rightarrowtail Y) = \varepsilon FY.Ff \colon FX \to FY, \tag{1.72}$$
$$HF_T = F, \quad GH = G_T, \quad H\varepsilon_T = \varepsilon H.$$

1.6.8 Exercises and complements

(a) For the monad $T = FU$ associated to the forgetful functor $U \colon \mathbf{Ab} \to \mathbf{Set}$, a morphism $f^\sharp \colon X \rightarrowtail Y$ of the Kleisli category $\mathbf{Set}_T$ can be viewed as a $\mathbb{Z}$-*weighted mapping between sets*. Compute the composition of these weighted mappings.

(b) (Kleisli algebras as free T-algebras). A monad (T, η, μ) gives a canonical functor $L \colon \mathbf{X}_T \to \mathbf{X}^T$

$$L(X) = (TX, \mu X), \qquad L(f^\sharp \colon X \rightarrowtail Y) \ = \ \mu Y.Tf \colon LX \to LY. \tag{1.73}$$

Prove that $\mathbf{X}_T$ is equivalent to the full subcategory of $\mathbf{X}^T$ containing its *free algebras*, namely (by definition) each algebra isomorphic to an algebra $(TX, \mu X)$.

(c) We shall see in Exercise 2.1.4(d) that the category RelSet of sets and relations can be presented as a category of Kleisli algebras over **Set**.

1.6.9 Comonads and coalgebras

Dually a *comonad* on the category **A** is a triple (S, ε, δ) formed of an endofunctor $S\colon \mathbf{A} \to \mathbf{A}$ with natural transformations $\varepsilon\colon S \to 1$ (*counit*) and $\delta\colon S \to S^2$ (*comultiplication*), which make the following diagrams commute

$$(1.74)$$

The category ${}^S\mathbf{A}$ of *S-coalgebras* $(A, c\colon A \to SA)$ is also defined by duality.

An adjunction $(F, G, \eta, \varepsilon)\colon \mathbf{X} \rightharpoonup \mathbf{A}$, as in 1.6.2, gives a comonad (S, ε, δ) on **A** (the domain of the right adjoint), where $S = FG\colon \mathbf{A} \to \mathbf{A}$, ε is the counit of the adjunction and $\delta = F\eta G\colon FG \to FG.FG$.

1.7 Idempotent monads and adjunctions

The theory of idempotent monads and idempotent adjunctions is well known within category theory, but it may be difficult to find it in a single text. The following exposition is based on [G11], Section 3.8. An adjunction is always 'equipped' with the associated monad (see 1.6.2).

We end by examining the strictly idempotent case, which will be useful in the infinite-dimensional extension.

1.7.1 Idempotent monads

A monad $T = (T, \eta, \mu)$ on the category **X** is said to be *idempotent* if the natural transformation $\mu\colon T^2 \to T$ is invertible. The inverse of μ is $T\eta = \eta T\colon T \to T^2$.

Equivalently one can define an idempotent monad on **X** as a pair (T, η) formed of a functor $T\colon \mathbf{X} \to \mathbf{X}$ and a natural transformation $\eta\colon 1 \to T$ such that $T\eta = \eta T\colon T \to T^2$ is invertible.

In this approach one defines $\mu = (T\eta)^{-1}\colon T^2 \to T$. The first axiom $\mu.\eta T = 1 = \mu.T\eta$ obviously holds; for the second it suffices to cancel $T^2\eta = T\eta T$ from the relation $(\mu.T\mu).T^2\eta = \mu = (\mu.\mu T).T\eta T$.

1.7.2 Theorem and Definition (Idempotent adjunctions)

Let $F \dashv G$ be an ordinary adjunction of categories (or more generally an adjunction in a 2-category, see 2.3.2)

$$F: \mathbf{X} \rightleftarrows \mathbf{A} : G, \qquad \eta: 1_{\mathbf{X}} \to GF, \quad \varepsilon: FG \to 1_{\mathbf{A}},$$
$$\varepsilon F.F\eta = 1_F, \qquad G\varepsilon.\eta G = 1_G. \tag{1.75}$$

The following conditions are equivalent:

(i) one of the four natural transformations which appear in the triangle equations (namely $F\eta$, εF, ηG and $G\varepsilon$) is invertible,

(ii) all of them are invertible, i.e. $F\eta.\varepsilon F = 1_{FGF}$ and $\eta G.G\varepsilon = 1_{GFG}$,

(iii) the associated monad $T = GF: \mathbf{X} \to \mathbf{X}$ is idempotent, i.e. its multiplication $\mu = G\varepsilon F: T^2 \to T$ is invertible (and then $T\eta = \eta T$ is its inverse),

(iv) $T\eta = GF\eta$ is invertible,

(v) $\eta T = \eta GF$ is invertible,

(iii*) the associated comonad $S = FG: \mathbf{A} \to \mathbf{A}$ is idempotent, i.e. its comultiplication $\delta = F\eta G: S \to S^2$ is invertible (and then $S\varepsilon = \varepsilon S$ is its inverse),

(iv*) $\varepsilon S = \varepsilon FG$ is invertible,

(v*) $S\varepsilon = FG\varepsilon$ is invertible.

When these conditions hold we say that the adjunction is idempotent. *The same is true of $G^{\mathrm{op}} \dashv F^{\mathrm{op}}$.*

Proof First, the triangle equations show that $F\eta$ is invertible if and only if εF is; the same holds for ηG and $G\varepsilon$. Moreover (iii) $\Leftrightarrow$ (iv) $\Leftrightarrow$ (v).

The main point is proving that (v) implies that $F\eta$ and εF are invertible, i.e. $F\eta.\varepsilon F = 1_{FGF}$. The natural transformation $F\eta.\varepsilon F$ forms the upper row of the following commutative diagram

$$\begin{array}{ccccc}
FGF & \xrightarrow{\ \varepsilon F\ } & F & \xrightarrow{\ F\eta\ } & FGF \\
{\scriptstyle FGF\eta}\downarrow & & \downarrow{\scriptstyle F\eta} & & \downarrow{\scriptstyle FGF\eta} \\
FGFGF & \xrightarrow[\varepsilon FGF]{} & FGF & \xrightarrow[F\eta GF]{} & FGFGF
\end{array} \tag{1.76}$$

The lower row $F\eta GF.\varepsilon FGF = (F\eta.\varepsilon F)GF$ is the identity, because $F\eta GF$ is invertible by hypothesis and the 'reversed composite' $(\varepsilon F.F\eta)GF$ is the identity, by (1.75). Therefore $FGF\eta.(F\eta.\varepsilon F) = FGF\eta$ and we get the conclusion by cancelling the transformation $FGF\eta$, which is invertible by hypothesis (in the equivalent form (iv)).

Adding an obvious implication we have:

- (ηG and $G\varepsilon$ are invertible) $\Rightarrow$ (v) $\Rightarrow$ ($F\eta$ and εF are invertible).

By duality all the properties of the statement are equivalent. $\square$

1.7.3 Theorem and Definition (Algebraic objects)

Let (T, η) be an idempotent monad on $\mathbf{X}$ as defined in 1.7.1: $T\eta = \eta T: T \to T^2$ is invertible and $\mu = (T\eta)^{-1}: T^2 \to T$.

(a) The following conditions on an object X of $\mathbf{X}$ are equivalent, and we say that X is an algebraic object (with respect to T) when they hold:

  (i) ηX is invertible,

  (ii) there exists a morphism $h: TX \to X$ such that $h.\eta X = 1_X$,

  (iii) there exists an (Eilenberg–Moore) T-algebra (X, h) over X,

  (iv) ηX is invertible and $(X, (\eta X)^{-1})$ is the unique T-algebra over X,

  (v) X is isomorphic to TX in $\mathbf{X}$.

In this case the algebra $\hat{X} = (X, (\eta X)^{-1})$ is isomorphic to the free T-algebra $(TX, \mu X)$ on X. (See the last point below.)

(b) An arrow of T-algebras $f: \hat{X} \to \hat{Y}$ is an ordinary $\mathbf{X}$-arrow $f: X \to Y$ where X and Y satisfy the equivalent conditions above. The forgetful functor $G^T: \mathbf{X}^T \to \mathbf{X}$ of the category of T-algebras is thus a full embedding

$$G^T(\hat{X}) = X, \qquad G^T(f: \hat{X} \to \hat{Y}) = (f: X \to Y), \qquad (1.77)$$

and makes $\mathbf{X}^T$ isomorphic to the full reflective subcategory $\mathbf{Alg}(T) \subset \mathbf{X}$ of algebraic objects defined above. The counit ε^T is invertible.

(c) Using this isomorphism the adjunction $F^T \dashv G^T$ associated to T can be rewritten in the following equivalent form, where G' is a full embedding with reflector F'

$$F': \mathbf{X} \rightleftarrows \mathbf{Alg}(T): G', \qquad \eta': 1 \to G'F' = T, \quad \varepsilon': F'G' \to 1,$$

$$F'(X) = TX, \qquad F'(f: X \to Y) = Tf: TX \to TY, \qquad (1.78)$$

$$\eta' = \eta, \qquad \varepsilon'(X) = (\eta X)^{-1}: F'G'(X) = TX \to X.$$

If T is the monad of an adjunction $F \dashv G: \mathbf{X} \to \mathbf{A}$, the comparison $K: \mathbf{A} \to \mathbf{X}^T$ can be rewritten as $K': \mathbf{A} \to \mathbf{Alg}(T)$, a codomain-restriction of $G: \mathbf{A} \to \mathbf{X}$, so that $G = G'K'$. If G is full, then K' is also (and K as well).

(d) The following composed functor, from $\mathbf{X}^T$ to the category $\mathbf{X}_T$ of Kleisli algebras

$$F_T G^T: \mathbf{X}^T \to \mathbf{X} \to \mathbf{X}_T, \qquad F_T G^T(X, (\eta X)^{-1}) = X, \qquad (1.79)$$

is an equivalence of categories (not an isomorphism, generally).

Proof (a) The implications (i) $\Rightarrow$ (ii) and (iv) $\Rightarrow$ (v) are obvious.

(ii) $\Rightarrow$ (iii) If $h.\eta X = 1_X$ then (X, h) is a T-algebra because the second axiom is here a consequence:

$$h.Th.T\eta X = h = h.\mu X.T\eta X.$$

(iii) $\Rightarrow$ (iv) The naturality of η on the morphism h and the property $T\eta = \eta T$ give

$$\eta X.h = Th.\eta TX = Th.T\eta X = T(h.\eta X) = 1_{TX},$$

so that ηX and h are inverses.

(v) $\Rightarrow$ (i) The existence of an isomorphism $i\colon X \to TX$ gives $Ti.\eta X = \eta TX.i$, whence ηX is invertible.

The last assertion follows from the isomorphism

$$\eta X\colon (X, (\eta X)^{-1}) \to (TX, \mu X).$$

(b) and (c). It is an obvious consequence of (a), applying the naturality of η.

(d) As in 1.6.7 we write as $f^{\sharp}\colon X \rightarrowtail Y$ the $\mathbf{X}_T$-arrow represented by the $\mathbf{X}$-morphism $f\colon X \to TY$; recall that $\mathrm{id}(X) = (\eta X)^{\sharp}$ and the composite of $f^{\sharp}$ with $g^{\sharp}\colon Y \rightarrowtail Z$ is represented by $\mu Z.Tg.f\colon X \to TZ$.

The functor $F_T G^T\colon \mathbf{X}^T \to \mathbf{X}_T$ sends the arrow $f\colon \hat{X} \to \hat{Y}$ to

$$(\eta Y.f)^{\sharp}\colon X \rightarrowtail Y.$$

Since ηY is invertible, this mapping $\mathbf{X}^T(\hat{X}, \hat{Y}) \to \mathbf{X}_T(X, Y)$ is bijective. It is now sufficient to prove that $F_T G^T$ is essentially surjective on objects.

Let us take an object X of $\mathbf{X}_T$, i.e. an arbitrary object of $\mathbf{X}$, and consider the T-algebra $(TX, \mu X)$; for an arbitrary monad this is the free T-algebra on X, with $F_T G^T(TX, \mu X) = TX$. Here X and TX are *isomorphic objects in* $\mathbf{X}_T$ (not in $\mathbf{X}$, generally), by the inverse morphisms

$$(T\eta X.\eta X)^{\sharp}\colon X \rightarrowtail TX, \qquad (1_X)^{\sharp}\colon TX \rightarrowtail X,$$

since

$$(\mu X.T1_X.T\eta X.\eta X)^{\sharp} = (\eta X)^{\sharp} = \mathrm{id}(X),$$
$$(\mu TX.T^2\eta X.T\eta X.1_X)^{\sharp} = (\mu TX.T\eta TX.T\eta X)^{\sharp} = (\eta TX)^{\sharp} = \mathrm{id}(TX).$$

$\square$

1.7.4 Idempotent comonads and coalgebras

An idempotent comonad (S, ε) on the category $\mathbf{A}$ has a dual characterisation, as an endofunctor $S \colon \mathbf{A} \to \mathbf{A}$ with a natural transformation $\varepsilon \colon S \to 1$ such that $S\varepsilon = \varepsilon S \colon S^2 \to S$ is invertible; the comultiplication $\delta \colon S \to S^2$ is the inverse of the latter.

Now an S-coalgebra amounts to a *coalgebraic object* A of $\mathbf{A}$, characterised by the following equivalent conditions:

(i) εA is invertible,

(ii) there exists a morphism $k \colon A \to SA$ such that $\varepsilon A.k = 1_A$,

(iii) there exists an S-coalgebra $(A, k \colon A \to SA)$ over A,

(iv) εA is invertible and $\hat{A} = (A, (\varepsilon A)^{-1})$ is the unique S-coalgebra over A,

(v) A is isomorphic to SA in $\mathbf{A}$.

In this case $\hat{A} = (A, (\varepsilon A)^{-1})$ is isomorphic to the cofree S-coalgebra $(SA, \delta A)$ on A. The category of S-coalgebras $^S\mathbf{A}$ is related to the full coreflective subcategory $\mathbf{Coalg}(S) \subset \mathbf{A}$ of coalgebraic objects, by the isomorphism

$$U \colon {}^S\mathbf{A} \to \mathbf{Coalg}(S), \qquad U(A, k) = A,$$
$$U(f \colon (A, k) \to (B, k')) = f \colon A \to B. \tag{1.80}$$

The adjunction associated to S

$$ {}^SF \colon {}^S\mathbf{A} \rightleftarrows \mathbf{A} \colon {}^SG,$$
$${}^S\eta \colon 1 \to {}^SG\, {}^SF, \quad {}^S\varepsilon = \varepsilon \colon {}^SF\, {}^SG \to 1, \tag{1.81}$$

is computed as follows:

$$ {}^SF(A, k) = A, \qquad {}^SF(f \colon (A, k) \to (B, k')) = f \colon A \to B,$$

$$ {}^SG(A) = (SA, \delta \colon SA \to S^2 A),$$

$$ {}^SG(f \colon A \to B) = Sf \colon {}^SG(A) \to {}^SG(B),$$

$$ {}^S\eta(A, k) = k \colon (A, k) \to (SA, \delta \colon SA \to S^2 A).$$

It can be rewritten in the following equivalent form, where F'' is a full embedding with coreflector G''

$$ F'' \colon \mathbf{Coalg}(S) \rightleftarrows \mathbf{A} \colon G'',$$

$$ \eta'' \colon 1 \to G''F'', \quad \varepsilon'' = \varepsilon \colon F''G'' \to 1,$$

$$ G''(A) = SA, \qquad G''(f \colon A \to B) = Sf \colon SA \to SB,$$

$$ \eta''(A) = (\varepsilon A)^{-1} \colon A \to SA, \qquad \varepsilon''(A) = \varepsilon A \colon SA \to A. \tag{1.82}$$

1.7.5 Full reflective subcategories

(a) Let **A** be a full reflective subcategory of **X** which is *replete* in **X**: any isomorphism $i\colon A \to X$ of **X** that involves an object of **A** belongs to **A**. In these hypotheses the embedding $G\colon \mathbf{A} \subset \mathbf{X}$ has a left adjoint $F \dashv G$, the adjunction is idempotent, **A** coincides with $\mathbf{Alg}(GF)$ and G is monadic.

In fact we already know that the counit $\varepsilon\colon FG \to 1$ is invertible (see 1.5.4(g)), which implies that the adjunction is idempotent. Moreover every $A \in \mathrm{Ob}\mathbf{A}$ is an algebraic object of **X** (because $\eta(GA)$ is invertible), and conversely an algebraic object $X \cong GF(X)$ belongs to **A**.

In particular we have seen (in 1.5.5) that **Ab** is full reflective in **Gp**, and obviously replete. The associated *abelianisation monad* is idempotent

$$T\colon \mathbf{Gp} \to \mathbf{Gp}, \qquad \eta X\colon X \to TX = X^{\mathrm{ab}} = X/[X,X], \qquad (1.83)$$

and the groups which are algebraic for T are precisely the abelian ones. Similar facts hold for many embeddings of algebraic varieties, like **Gp** $\subset$ **Mon** or **CRng** $\subset$ **Rng**; and also for the (full) embedding of Hausdorff spaces in **Top**.

(b) Dually the inclusion functor $U\colon \mathbf{A} \to \mathbf{X}$ of a full, replete and coreflective subcategory has a left adjoint $G\colon \mathbf{X} \to \mathbf{A}$ with invertible unit η. We have now an idempotent comonad $S = GU\colon \mathbf{A} \to \mathbf{A}$; the category of coalgebras $S\mathbf{A}$ is isomorphic to the full reflective subcategory $\mathbf{Coalg}(S) \subset \mathbf{A}$ of coalgebraic objects, namely the objects A of **A** such that $\varepsilon\mathbf{A}$ is invertible.

As an example we have seen in 1.5.5 the subcategory $t\mathbf{Ab} \subset \mathbf{Ab}$, with counit $\varepsilon A\colon tA \to A$ given by the embedding of the torsion subgroup of A (see 1.5.5); the abelian groups which are coalgebraic for the associated comonad $t\colon \mathbf{Ab} \to \mathbf{Ab}$ are precisely the torsion ones.

1.7.6 Exercises and complements

(a) Prove that any adjunction $X \rightharpoondown \mathbf{A}$ from an ordered set X to an arbitrary category is idempotent. By duality this also holds for every adjunction $\mathbf{A} \rightharpoondown X$.

(b) The idempotent monad of a full reflective subcategory has already been considered in 1.7.5. The reader can easily prove the following more general result.

We have an adjunction $F \dashv G$ where the functor G is full and faithful (whence $\varepsilon\colon FG \to 1$ is invertible and the adjunction is idempotent). Then the comparison $K\colon \mathbf{A} \to \mathbf{X}^T$ is an equivalence of categories, with quasi inverse $FG^T\colon \mathbf{X}^T \to \mathbf{A}$, and G is *monadic in the weak sense* mentioned in 1.6.6(d).

(c) Study the embedding **Rng** ⊂ **Rng′** (see 1.5.6) and the associated monad. Or, similarly, the embedding of **Mon** into the category of semigroups.

(d) Study the monads and comonads produced by the forgetful functor U: **Top** → **Set** with its adjoints $D \dashv U \dashv C$ (see Exercise 1.5.3(c)).

(e) For the abelianisation monad T, in (1.83), consider the equivalence $\mathbf{Gp}^T \to \mathbf{Gp}_T$ of 1.7.3(d).

1.7.7 Idempotent adjunctions, algebras and coalgebras

Let us start now from an idempotent adjunction $(F, G, \eta, \varepsilon)$: **X** ⇀ **A**, with associated idempotent monad (T, η) and associated idempotent comonad (S, ε), where $T = GF$: **X** → **X** and $S = FG$: **A** → **A**.

Rewriting the category of T-algebras as the full reflective subcategory **Alg**(T) ⊂ **X** (as in 1.7.3) and the category of S-coalgebras as the full coreflective subcategory **Coalg**(S) ⊂ **A** (as in 1.7.4), the given adjunction can be factorised as follows *up to isomorphism*

$$\mathbf{X} \underset{G'}{\overset{F'}{\rightleftarrows}} \mathbf{Alg}(T) \underset{G^\sharp}{\overset{F^\sharp}{\rightleftarrows}} \mathbf{Coalg}(S) \underset{G''}{\overset{F''}{\rightleftarrows}} \mathbf{A} \qquad (1.84)$$

(a) First, the reflective adjunction (η', ε'): $F' \dashv G'$ is described in (1.78). G' is the full embedding of algebraic objects, with reflector $F'(X) = TX$ and invertible counit $\varepsilon'(X) = (\eta X)^{-1}$. The comparison functor K is now computed as G

$$K: A \to \mathbf{Alg}(T), \qquad K(A) = GA, \quad K(f: A \to B) = Gf,$$
$$KF = F', \quad G'K = G, \quad K\varepsilon = \varepsilon'K. \qquad (1.85)$$

(b) In the third position, the coreflective adjunction (η'', ε''): $F'' \dashv G''$ is described in (1.82). F'' is the full embedding of coalgebraic objects, with coreflector $G''(A) = SA$ and invertible unit $\eta''(A) = (\varepsilon A)^{-1}$. The comparison functor H is computed as F

$$H: \mathbf{X} \to \mathbf{Coalg}(S), \qquad H(X) = FX, \quad H(f: X \to Y) = Ff,$$
$$HG = G'', \quad F''H = F, \quad H\eta = \eta'H. \qquad (1.86)$$

(c) In the central position, the given adjunction restricts to an adjunction between our full subcategories

$$(\eta^\sharp, \varepsilon^\sharp): F^\sharp \dashv G^\sharp: \mathbf{Alg}(T) \rightharpoonup \mathbf{Coalg}(S),$$
$$F^\sharp X = FX, \qquad G^\sharp A = GA, \qquad (1.87)$$

because, for every X in $\mathbf{X}$, FX is a coalgebraic object of $\mathbf{A}$ (with structure $F\eta X\colon FX \to SFX$), and dually. This is actually an *adjoint equivalence*, because $\eta^\sharp X = \eta X$ is invertible for every algebraic object X, and dually. It is even an isomorphism if T is the identity over all algebraic objects and S is the identity over all the coalgebraic ones.

(d) Composing the three adjunctions above we get $F''F^\sharp F' = FGF$, isomorphic to F (by $F\eta = (\varepsilon F)^{-1}$) and $G'G^\sharp G'' = GFG$, isomorphic to G (by $\eta G = (G\varepsilon)^{-1}$).

1.7.8 The strict case

An idempotent monad can be easily replaced with a strict one, up to isomorphism (as we prove in 1.7.9). *Concretely*, this is so obvious that we tend to make no difference between these notions. *Theoretically*, the distinction seems to be of little interest and pedantic. Yet the strict case will be useful in the infinite-dimensional extension of Chapter 8, where working in the general idempotent case would quickly become complicated.

The theoretical part is an obvious simplification of the previous one. We say that a monad $T = (T, \eta, \mu)$ on the category $\mathbf{X}$ is *strictly idempotent* if $\mu = 1_T$ (and therefore $T = T^2$); or equivalently $T\eta = 1_T$, or also $\eta T = 1_T$. Again, one can define a strictly idempotent monad on $\mathbf{X}$ as a pair (T, η) formed of a functor $T\colon \mathbf{X} \to \mathbf{X}$ and a natural transformation $\eta\colon 1 \to T$ such that $T\eta = 1_T = \eta T$ (to which we add $\mu = 1_T$).

An adjunction $(F, G, \eta, \varepsilon)\colon \mathbf{X} \rightharpoonup \mathbf{A}$, as in (1.75), will be said to be *left strict* if $\eta G = 1_G$, or equivalently $G\varepsilon = 1_G$ (which implies $GFG = G$). Then the adjunction is idempotent and the associated monad (T, η, μ) is strictly idempotent, because $\eta T = \eta GF = 1_{GF}$.

On the other hand, given a strictly idempotent monad (T, η) on $\mathbf{X}$, we let $\mathbf{Alg}_*(T)$ be the full subcategory of *strictly algebraic objects*, namely the objects X of $\mathbf{X}$ such that $TX = X$, or equivalently $\eta X = 1_X$ (from $TX = X$ it follows that $\eta X = \eta TX = 1_{TX} = 1_X$), or equivalently the objects of form TY. Let us note that $\mathbf{Alg}_*(T)$ is a full subcategory of $\mathbf{Alg}(T)$, equivalent to the latter (by 1.2.5) since any algebraic object X is isomorphic to TX, which is strictly algebraic.

The inclusion $G'\colon \mathbf{Alg}_*(T) \to \mathbf{X}$ has a left inverse F', which is a restriction of T on its codomain

$$F'\colon \mathbf{X} \to \mathbf{Alg}_*(T), \quad F'(X) = TX, \quad F'(f) = Tf \quad (F'G' = 1). \quad (1.88)$$

Moreover $G'F' = T \colon \mathbf{X} \to \mathbf{X}$ and we have an adjunction

$$F' \colon \mathbf{X} \rightleftarrows \mathbf{Alg}_*(T) \colon G',$$

$$\eta' = \eta \colon 1 \to G'F', \qquad \varepsilon' = 1 \colon F'G' \to 1, \tag{1.89}$$

which is left strict (since $\varepsilon' = 1$) and gives back the original monad (T, η).

The other way round, if we start from a left strict adjunction $(F, G, \eta, \varepsilon) \colon \mathbf{X} \rightarrowtail \mathbf{A}$, form the associated strictly idempotent monad (T, η) on $\mathbf{X}$ (with $T = GF$) and the associated adjunction (1.89), we have a comparison functor, which is a codomain-restriction of G (since $\eta G = 1$)

$$K \colon \mathbf{A} \to \mathbf{Alg}_*(T), \qquad K(A) = GA, \quad K(f) = Gf. \tag{1.90}$$

1.7.9 Proposition

An idempotent monad (T, η) on the category $\mathbf{X}$ has an associated strictly idempotent monad (S, η') on $\mathbf{X}$, with an isomorphism $\varphi \colon T \to S$ such that $\eta' = \varphi\eta \colon 1 \to S$.

The strictly algebraic objects of S are precisely the algebraic objects of T.

Proof We apply Exercise 1.2.6(i) to the functor $T \colon \mathbf{X} \to \mathbf{X}$ and the family $\varphi_X \colon TX \to SX$ defined as follows:

- if ηX is invertible, we let $S(X) = X$ and $\varphi_X = (\eta X)^{-1} \colon TX \to SX$,
- otherwise we let $S(X) = TX$ and $\varphi_X = 1_{TX}$.

We extend all this to a functor $S \colon \mathbf{X} \to \mathbf{X}$ and a functorial isomorphism $\varphi \colon T \to S$. Plainly $(S, \varphi\eta)$ is still an idempotent monad. It is strictly idempotent because $\varphi\eta S = 1_S$

$$(\varphi\eta S)_X = \varphi_X . \eta_X = 1_X = 1_{SX}, \qquad \text{if } \eta X \text{ is invertible,}$$

$$(\varphi\eta S)_X = \varphi_{TX} . \eta_{TX} = 1_{TX} = 1_{SX}, \qquad \text{otherwise.}$$

Now $SX = X$ if ηX is invertible; otherwise $SX = TX$ is not isomorphic to X (by 1.7.3) and cannot be X. (If one defines $S(X) = X$ only for the objects X of type TY, everything works as above except this last point – of course.) $\qquad \square$

1.8 Introducing abelian categories and exact functors

The beginning of category theory, in the 1950's, was focused on the study of abelian categories, which provided an extension of the categories of modules with two crucial advantages: the theory is self-dual and includes all categories of sheaves of modules on a given ring.

Here we briefly review the definition of abelian categories and exact functors. A more detailed study of exactness properties will be given in Appendix A.

1.8.1 Pointed categories

The notions of exactness considered here are based on kernels and cokernels.

To define the latter, we assume that we are working in a *pointed* category **E**. As already said in 1.3.2, this means that there is a *zero object* 0, that is both initial and terminal in **E**: for every object A there is precisely one morphism $0 \to A$ and precisely one morphism $A \to 0$. *The* zero object is determined up to isomorphism; as usual, one of them is chosen.

For instance **Ab**, $R\,\mathbf{Mod}$, **Gp**, **Set$_\bullet$**, **Top$_\bullet$**, **Rng$'$** are pointed categories, with zero object given by the singleton with the adequate structure.

Given two objects A, B in **E**, the composite $A \to 0 \to B$ is called the *zero morphism* from A to B, and written as $0_{AB}\colon A \to B$, or also as 0 when the context identifies it.

The morphism $0 \to A$ is plainly a monomorphism, also written as $0_A\colon 0 \rightarrowtail A$ and called the *zero subobject* of A; dually the morphism $A \to 0$ is necessarily an epimorphism, written as $0^A\colon A \twoheadrightarrow 0$ and called the *zero quotient* of A.

1.8.2 Kernels and cokernels

In the pointed category **E** the *kernel* of a morphism $f\colon A \to B$ is defined as a limit, namely the equaliser of f and $0_{AB}\colon A \to B$.

This means a morphism $\ker f\colon \mathrm{Ker}\, f \to A$ such that

(i) $f.(\ker f) = 0$, and for every map h such that $fh = 0$ there exists a unique morphism u such that $h = (\ker f)u$

$$\mathrm{Ker}\, f \xrightarrow{\ \ker f\ } A \xrightarrow{\ f\ } B \tag{1.91}$$

We shall choose the (regular) subobject that satisfies the given property, if it exists. (The existence of kernels does not require the existence of all equalisers, as many examples in A2.3 will show.)

More generally, a *normal* monomorphism is *a* kernel of some arrow f, i.e. a monomorphism equivalent to the *normal subobject* $\ker f$. Thus in **Ab** the natural 'kernel-object' is the usual subgroup $\mathrm{Ker}\, f$, and $\ker f$ is its embedding in the domain of f; however, if $u: K \to \mathrm{Ker}\, f$ is an isomorphism, also the composite $(\mathrm{ker}\, f)u: K \to A$ is *a* kernel of f.

Dually, the *cokernel* of f is the coequaliser of f and the zero morphism $A \to B$, namely a morphism $\mathrm{cok}\, f: B \to \mathrm{Cok}\, f$ such that

(i*) $(\mathrm{cok}\, f).f = 0$, and for every map h such that $hf = 0$ there exists a unique morphism u such that $h = u(\mathrm{cok}\, f)$

$$
A \xrightarrow{\ f\ } B \xrightarrow{\ \mathrm{cok}\, f\ } \mathrm{Cok}\, f \tag{1.92}
$$

A *normal epimorphism* is any cokernel of a morphism, and is always a regular epimorphism.

1.8.3 Exercises and complements

(a) Compute kernels and cokernels in $R\,\mathbf{Mod}$, $\mathbf{Gp}$, $\mathbf{Set_{\bullet}}$, $\mathbf{Top_{\bullet}}$ and $\mathbf{Rng}'$, characterising the normal subobjects and quotients.

(b) Let us note that in $R\,\mathbf{Mod}$ the image $f(A)$ *is determined as* $\mathrm{Ker}\,(\mathrm{cok}\, f)$, i.e. *the kernel-object of the morphism* $\mathrm{cok}\, f$, while $\mathrm{Cok}\,(\mathrm{ker}\, f) = A/\mathrm{Ker}\, f$ gives the (isomorphic) *coimage* of f. The other categories of point (a) do not behave in this 'simple' way, that will be used below in axiom (ab.2).

1.8.4 Biproducts

We have considered biproducts in $R\,\mathbf{Mod}$, in 1.5.8. In a pointed category, we define the *biproduct* of two objects A, B as an object C equipped with four maps u, v, p, q

$$
\tag{1.93}
$$

so that

- $pu = \mathrm{id}A$, $\quad qv = \mathrm{id}B$, $\qquad qu = 0$, $\quad pv = 0$,
- (C, p, q) is the product of A, B and (C, u, v) is their sum.

The object C is often written as $A \oplus B$; the maps p, q are called *projections* while u, v are called *injections*.

Similarly one defines the biproduct $\bigoplus_i A_i$ of any family of objects; the biproduct of the empty family is the zero object. A morphism $C \to \bigoplus_i A_i$

will often be written as (f_i), by its components $f_i \colon C \to A_i$, while a morphism $\bigoplus_i A_i \to D$ can be written as $[f_i]$, by its co-components $f_i \colon A_i \to D$ (as in our notation for products and coproducts, in 1.3.1 and 1.3.2).

All categories of modules and the category **Abm** of abelian monoids have finite biproducts. Other examples can be found in A3.6, including cases where *arbitrary* biproducts exist (like Rel**Set**). The relationship of finite biproducts with additive categories is deferred to Section A3.

1.8.5 Abelian categories

An *abelian category* **A** is a category satisfying the following three (redundant) self-dual axioms:

(ab.1) **A** is a pointed and every morphism $f \colon A \to B$ has a kernel and a cokernel,

(ab.2) in the *canonical factorisation* of a morphism f through its *coimage* and its *image*

$$
\begin{array}{ccccc}
\operatorname{Ker} f & \xrightarrow{\ k\ } & A & \xrightarrow{\ f\ } & B & \xrightarrow{\ c\ } & \operatorname{Cok} f \\
& & {\scriptstyle q}\big\downarrow & & \big\uparrow{\scriptstyle n} & & \\
& & \operatorname{Coim} f & \xrightarrow[\ g\ ]{} & \operatorname{Im} f & &
\end{array}
\tag{1.94}
$$

$$
\operatorname{Coim} f = \operatorname{Cok}(\ker f), \qquad\qquad \operatorname{Im} f = \operatorname{Ker}(\operatorname{cok} f),
$$
$$
q = \operatorname{coim} f = \operatorname{cok}(\ker f), \qquad n = \operatorname{im} f = \ker(\operatorname{cok} f),
$$

the unique morphism g such that $f = ngq$ is an *isomorphism*,

(ab.3) **A** has finite products and finite sums.

A finer analysis will be given in Sections A2 and A3. In particular, a category satisfying the first two axioms above is called Puppe-exact, or exact in the sense of Puppe–Mitchell. We shall see that such a category has finite products if and only if it has finite sums, which are then biproducts: then the category has finite limits and colimits, and a (unique) additive structure.

Every category $R\,\mathbf{Mod}$ is abelian. *Every category of sheaves of R-modules on a topological space is abelian. More generally, this holds for sheaves with values in any abelian category [Bo3].*

1.8.6 Exact functors

A functor between abelian categories is *left exact* (resp. *right exact*) if it preserves kernels (resp. cokernels), in the usual sense of preserving limits or colimits. It is *exact* if it is left and right exact.

A right adjoint functor between abelian categories is (obviously) left exact. We shall see in Appendix A that a left exact functor preserves all finite limits and the additive structure. Exact sequences of morphisms are also considered there.

2

Introducing two-dimensional category theory

Two-dimensional aspects have already appeared in various categories, like the order relation $f \leqslant g$ in **Ord** (in (1.5)) or – in a less elementary way – the second-order arrows $\varphi \colon F \to G$ in **Cat** (in 1.2.4). Here we begin to study 2-dimensional categorical structures, from ordered categories to bicategories.

The general 2-dimensional case, namely weak double categories, will be dealt with in Chapters 3 to 5.

2.1 Categories of relations and partial mappings

We review the construction of some ordered categories of relations and partial mappings, in concrete situations.

Relations and partial mappings will often (but not necessarily) be written as dot-marked arrows $X \dashrightarrow Y$, to distinguish them from ordinary morphisms. The categories of partial mappings will usually be denoted by calligraphic letters.

Relations will be extended to regular categories in Section 2.5, and to other generalisations of abelian categories in Appendix A. Double categories based on relations and partial maps will be studied in Chapters 3 and 5.

2.1.1 Categories and involutions

We begin by fixing our terminology for involutive endofunctors, starting from two examples related to ordered sets.

(i) The category **Ord**, of ordered sets and increasing mappings, has an involutive *covariant* endofunctor:

$$X \mapsto X^{\mathrm{op}}, \qquad (f \colon X \to Y) \mapsto (f^{\mathrm{op}} \colon X^{\mathrm{op}} \to Y^{\mathrm{op}}), \tag{2.1}$$

where X^{op} has reversed order and f^{op} is the 'same mapping' as f.

(ii) The category Adj**Ord**, of ordered sets and Galois connections (introduced in 1.4.4), has an involutive *contravariant* endofunctor:

$$X \mapsto X^{\mathrm{op}}, \qquad (u\colon X \rightarrowtail Y) \mapsto (u^{\mathrm{op}}\colon Y^{\mathrm{op}} \rightarrowtail X^{\mathrm{op}}), \qquad (2.2)$$

where $u^{\mathrm{op}} = (g^{\mathrm{op}}, f^{\mathrm{op}})$, if $u = (f, g)$ with $f \dashv g$.

This situation will become clearer in the domain of double categories: the two procedures (i) and (ii) – which have the same action on the objects – will be combined to give a *vertical involution* of the double category 𝔸dj**Ord**, horizontally covariant (on increasing mappings) and vertically contravariant (on adjunctions); see 3.2.2.

We shall use the following terms.

(a) A *category with reversor* will be a category **C** equipped with an involutive *covariant* endofunctor $R\colon \mathbf{C} \to \mathbf{C}$, often denoted as in (2.1), and said to be trivial when $R = \mathrm{id}\mathbf{C}$.

Obvious non-trivial examples are **Ord** and **Cat**. The name of 'reversor' is taken from Directed Algebraic Topology [G7], where the categories of 'directed spaces' in some sense (like preordered topological spaces, 'd-spaces', and small categories) have such an endofunctor, that turns a directed structure into the opposite one. In the trivial case, when $R = \mathrm{id}\mathbf{C}$, we go back to classical, non-directed 'spaces'.

(b) A *category with involution* will be a category **C** equipped with an involutive *contravariant* endofunctor $J\colon \mathbf{C} \dashrightarrow \mathbf{C}$. It follows that **C** is selfdual, i.e. isomorphic to $\mathbf{C}^{\mathrm{op}}$.

Obvious examples are Adj**Ord** and Adj**Cat**.

(c) More particularly, an *involutive category* will be a category **C** equipped with a (contravariant) involution *which is the identity on objects*. In other words every morphism $f\colon X \to Y$ has an associated *opposite morphism* $f^{\sharp}\colon Y \to X$ so that

$$(gf)^{\sharp} = f^{\sharp}g^{\sharp}, \qquad\qquad (1_X)^{\sharp} = 1_X. \qquad (2.3)$$

Classical examples are the categories of relations reviewed below.

The involution is said to be *regular* (in the sense of von Neumann) if $ff^{\sharp}f = f$ (for all maps f), as happens with relations of groups. In this case all the semigroups of endomorphisms $\mathbf{C}(X, X)$ are regular, in the well-known sense of semigroup theory. This has nothing to do with the notion of 'regular category', that will be recalled in Section 2.5.

2.1.2 Ordered categories

An *ordered category* will be a category **C** equipped with an order relation $f \leqslant g$ between *parallel morphisms*, which is consistent with composition: for $f, g \colon X \to Y$ and $f', g' \colon Y \to Z$

(i) if $f \leqslant g$ and $f' \leqslant g'$ then $f'f \leqslant g'g$.

(This notion is related to 'enrichment', see 2.4.2; a different 'internal' notion is recalled in 2.4.5.)

A functor $F \colon \mathbf{C} \to \mathbf{D}$ between ordered categories is *increasing*, or *monotone*, if it preserves the order relation. Any subcategory of an ordered category inherits an ordering.

For instance the category **Ord** of ordered sets is canonically ordered by the *pointwise order* $f \leqslant g$, defined in (1.5). This ordering is inherited by the subcategories **Lth**, **Mlh** and **Dlh** of lattices, modular lattices and distributive lattices (see 1.4.1, 1.4.2).

Also Adj**Ord** has a canonical order, defined in (1.39).

2.1.3 Relations of sets

The category Rel**Set** of *sets and relations* (or *correspondences*) is well known.

A relation $u \colon X \rightarrowtail Y$ is a subset of the cartesian product $X \times Y$, and will often be denoted by a dot-marked arrow. It can be viewed as a 'partially defined, multi-valued mapping', that sends an element $x \in X$ to the subset $\{y \in Y \mid (x, y) \in u\}$ of Y. (This aspect is analysed below, in 2.1.4(d).)

The identity relations and the composite of u with $v \colon Y \rightarrowtail Z$ are

$$\mathrm{id}X = \{(x, x) \mid x \in X\},$$
$$vu = \{(x, z) \in X \times Z \mid \exists \, y \in Y \colon (x, y) \in u \text{ and } (y, z) \in v\}. \tag{2.4}$$

It is easy to verify that Rel**Set** is a category. It is actually an *involutive ordered category*, in the following sense. First it is an involutive category as defined in 2.1.1: the *opposite relation* $u^{\sharp} \colon Y \rightarrowtail X$ is obtained by reversing pairs. Second, it is an ordered category, in the sense of 2.1.2: given two parallel relations $u, u' \colon X \rightarrowtail Y$ we say that $u \leqslant u'$ if $u \subset u'$ as subsets of $X \times Y$. Third, these structures are consistent: if $u \leqslant v$ then $u^{\sharp} \leqslant v^{\sharp}$ (and conversely, as a consequence).

The category **Set** is embedded in the category of relations Rel**Set**, identifying a mapping $f \colon X \to Y$ with its graph $\{(x, y) \in X \times Y \mid y = f(x)\}$; the condition $f \leqslant g$ for parallel mappings amounts to $f = g$.

2.1.4 Exercises and complements

(a) A relation $f\colon X \rightarrowtail Y$ of sets is a mapping if and only if it is everywhere defined and single-valued: for every $x \in X$ there exists precisely one $y \in Y$ such that (x, y) belongs to (the graph of) f. Prove that these conditions are characterised as follows

$$f^\sharp f \geqslant \operatorname{id} X, \qquad f f^\sharp \leqslant \operatorname{id} Y. \tag{2.5}$$

*(This means that $f \dashv f^\sharp$ *in* the ordered category RelSet, viewed as a 2-category: see 2.3.2.)*

(b) Prove that the involution of RelSet is not regular, in the sense of 2.1.1.

(c) Prove that RelSet has arbitrary products and sums, that are biproducts (see 1.8.4).

(d) RelSet can be viewed as a category of Kleisli algebras over **Set**, for the monad $\mathcal{P}$ of Exercise 1.6.6(f).

(e) For the same monad, compute the canonical functor $L\colon \mathbf{Set}_{\mathcal{P}} \to \mathbf{Set}^{\mathcal{P}}$ of Exercise 1.6.8(b).

(f) Show that the embedding $\mathbf{Set} \to \mathrm{RelSet}$ has a right adjoint.

2.1.5 Relations of abelian groups and modules

One proceeds in a similar way for abelian groups and their relations, also called *additive relations* or *correspondences*. The use of these categories in Homological Algebra goes back, at least, to Mac Lane and Hilton [M1, M2, Hi].

A *relation of abelian groups* $u\colon X \rightarrowtail Y$ is now a subgroup of the cartesian product $X \times Y$. Composition, order and involution are defined as above, and give the involutive ordered category Rel**Ab**. **Ab** is embedded in the former as the category of everywhere defined, single-valued relations.

The involution of Rel**Ab** is regular, i.e.

$$u u^\sharp u = u, \qquad \text{for all } u \text{ in RelAb}. \tag{2.6}$$

In fact the inclusion $u \leqslant u u^\sharp u$ is obvious (and also holds for relations of sets). The other way round, if (x, y), (x', y) and (x', y') are in u, we have

$$(x, y') = (x, y) - (x', y) + (x', y') \in u. \tag{2.7}$$

Let us note that Rel**Ab** lacks products (and sums), including the terminal object.

More generally for any ring R we have the involutive ordered category $\mathrm{Rel}(R\,\mathbf{Mod})$ of left R-modules, where a relation is a submodule of the cartesian product.

We also have the involutive ordered category Rel**Gp** of groups, where a relation is a subgroup of the cartesian product. The involution is still regular, with the same proof as in (2.7), where the commutativity of addition is not used.

In all these cases the original morphisms are characterised among relations by the conditions (2.5). The other points in 2.1.4 cannot be extended to the categories of relations listed here.

2.1.6 Partial mappings

A *partial mapping* of sets $f: X \nrightarrow Y$ is a mapping Def $f \to Y$ defined on a subset of X, its *definition* subset. Equivalently, it is a single-valued relation (characterised by $f f^{\sharp} \leqslant \mathrm{id} Y$).

Their composition is obvious (as in the well-known case of partially defined real functions $\mathbb{R} \nrightarrow \mathbb{R}$):

$$\mathrm{Def}\,(gf) = f^{-1}(\mathrm{Def}\,g), \quad (gf)(x) = g(f(x)) \quad (x \in \mathrm{Def}\,(gf)). \qquad (2.8)$$

We have thus the category $\mathcal{S}$ *of sets and partial mappings*, a wide subcategory of Rel**Set** that contains **Set**. In the induced order of $\mathcal{S}$, $f' \leqslant f$ means that f' is a restriction of f to a smaller subset of X.

The category $\mathcal{S}$ *should not be viewed as essentially different* from the usual categories of structured sets and 'total' mappings, since *it is equivalent* to the category **Set.** of pointed sets, via the functor

$$R: \mathbf{Set.} \to \mathcal{S}, \qquad (X, x_0) \mapsto X \setminus \{x_0\}, \qquad (2.9)$$

that sends a pointed mapping $f: (X, x_0) \to (Y, y_0)$ to its restriction

$$R(f): X \setminus \{x_0\} \nrightarrow Y \setminus \{y_0\}, \qquad \mathrm{Def}\,(R(f)) = f^{-1}(Y \setminus \{y_0\}).$$

In fact this functor is plainly full, faithful and surjective on objects (see 1.2.5). A quasi inverse adjoint $S: \mathcal{S} \to \mathbf{Set.}$ is obtained by choosing, for every set X, a pointed set $S(X) = (X \cup \{x_0\}, x_0)$ where $x_0 \notin X$, and defining S on partial mappings in the obvious (and unique) way that gives $RS = 1$. (Set theory can give a canonical – if 'confusing' – choice, letting $x_0 = X$, since $X \notin X$.)

The equivalence proves that $\mathcal{S}$ has all limits and colimits. Limits are easily constructed 'along the equivalence': for instance the product in $\mathcal{S}$ of a family (X_i) of sets can be computed as

$$R(\textstyle\prod_i S(X_i)) = (\textstyle\prod_i (X_i \cup \{x_i\})) \setminus \{(x_i)\}, \qquad (2.10)$$

first adding a base point $x_i \notin X_i$ to each factor, then taking their cartesian product in **Set.** and deleting the single base point (x_i) of the product.

(Thus 1×1 in $\mathcal{S}$ has *three* elements, and 2×3 has *eleven*.) Sums in $\mathcal{S}$ can also be constructed in this way, but they simply amount to disjoint unions of sets (as in Set and Rel(Set)).

It is easy to see that $\mathcal{S}$ and **Set.** are not isomorphic categories: the first has *one* initial object, the empty set, while **Set.** has a large set of them, all the singletons.

2.1.7 Partial bijections

More particularly a *partial bijection* (or partial isomorphism) of sets f: $X \rightarrowtail Y$ is a bijection between a subset Def f of X and a subset Val f of Y, or equivalently a single-valued, injective relation.

Partial bijections form a selfdual wide subcategory of $\mathcal{S}$ and RelSet, written as $\mathcal{I}$ in [G8], where it is used as a basic tool for graphic representations of spectral sequences. We shall see in A2.3 that it is a pointed category with good exactness properties, but lacks products and sums.

2.1.8 Exercises and complements (Partial mappings)

In various cases we may want to consider partial morphisms defined on sub-objects of a particular kind. For topological spaces, for instance, partial continuous mappings should be 'reasonably' *defined on subspaces*, i.e. regular subobjects, rather than arbitrary subobjects; but in fact the important structures of partial continuous mappings are even more restricted.

(a) Let us first consider the category $\mathcal{T}$ of topological spaces and partial continuous mappings, each defined on a subspace of its domain. The functor R: **Set.** $\to \mathcal{S}$ of (2.9) can be extended to a functor defined on the category **Top.** of pointed topological spaces

$$R \colon \textbf{Top.} \to \mathcal{T}, \qquad R(X, x_0) = X \setminus \{x_0\}, \qquad (2.11)$$

where $X \setminus \{x_0\}$ has the topology induced by X.

Prove that R is not an equivalence of categories.

*(b) While studying topological manifolds, it can be useful to use the category $\mathcal{C}$ of spaces and partial continuous mappings *defined on open subspaces*, together with its subcategory of spaces and partial homeomorphisms *between open subspaces* (see [G1]).

Prove that $\mathcal{C}$ is equivalent to a full subcategory $\textbf{Top.}' \subset \textbf{Top.}$ consisting of those objects (X, x_0) where the base point x_0 is closed and adherent to every other point (i.e. the only open subset containing x_0 is the total space X).

*(c) The quotient of the topological space $\mathbb{R}^n$ modulo the equivalence relation xRy:

$$\text{there exists some scalar } \lambda \neq 0 \text{ such that } x = \lambda y, \qquad (2.12)$$

is a topological space, which we consider as pointed at the singleton class $[0]$; the latter is closed and adherent to every other point. Taking out the base point, we get the real projective space $\mathbb{P}^{n-1}$, which has the advantage of being a topological manifold.

The reader is invited to note that the simplest way to define the category $\mathbb{R}\,\mathbf{Prj}_f$ of real projective spaces (of finite dimension) and linear projective mappings is to take the quotient of the (abelian) category $\mathbb{R}\,\mathbf{Mod}_f$ of finite-dimensional real vector spaces, modulo the congruence $f R g$ defined as in (2.12). (The quotient is not abelian, but Puppe-exact: see A2.3(f).)

This category can be made concrete over $\mathbf{Top}'_{\bullet}$ by sending a finite-dimensional real vector space X (with the unique consistent topology) to the topological space X/R, pointed as above. Obviously, $\mathbb{R}\,\mathbf{Prj}_f$ is also concrete over $\mathcal{C}$, taking out each base point: in this way the objects are represented by the usual projective manifolds, but the linear projective mappings are represented by partial maps.

*(d) Prove that the category of topological spaces and partial continuous mappings *defined on closed subspaces* is equivalent to the full subcategory $\mathbf{Top}''_{\bullet}$ of $\mathbf{Top}_{\bullet}$ formed by the objects (X, x_0) where the base point x_0 is open and dense (i.e. the only closed subspace containing x_0 is X).

2.2 Monoidal categories and exponentials

The 'weak' structures that we mention here and in the next section, like monoidal categories and bicategories, are only sketched. They are particular cases of weak double categories, that will be introduced in Chapter 3.

2.2.1 A classical example from multilinear algebra

Let us consider the category $R\,\mathbf{Mod}$ of modules on a *commutative* unitary ring R. Their tensor product and Hom-functor form a classical instance of a 'symmetric monoidal closed' structure, which we briefly recall now as an introduction to these structures. Further information on the present topic can be found in any text on multilinear algebra or homological algebra, for instance in [M2]; tensor products of bimodules over possibly non-commutative rings will be briefly examined in 3.4.5.

(a) The tensor product $A \otimes_R B$ of two modules A, B is equipped with an *R-bilinear* mapping $\varphi \colon A \times B \to A \otimes_R B$ (linear in each variable), and characterised up to isomorphism by the following universal property: every bilinear mapping $\psi \colon A \times B \to C$ factorises through φ by a unique homomorphism h

$$
\begin{array}{ccc}
A \times B \xrightarrow{\ \psi\ } C & \qquad & U(A) \times U(B) \xrightarrow{\ \psi\ } U(C) \\[2pt]
\varphi \downarrow \quad \nearrow h & & \varphi \downarrow \quad \nearrow Uh \\[2pt]
A \otimes_R B & & U(A \otimes_R B)
\end{array}
\qquad (2.13)
$$

(Note that this diagram lives in **Set**; to be more precise – or perhaps pedantic – one can rewrite it as at the right, where $U = |{-}| \colon R\,\mathbf{Mod} \to \mathbf{Set}$ is the forgetful functor.)

The existence of the solution is proved by constructing a quotient of the free R-module $F(|A| \times |B|)$, so to force the embedding of the basis $|A| \times |B| \to F(|A| \times |B|)$ to become an R-bilinear mapping. Its uniqueness up to isomorphism is granted from the start. Fixing a solution, one writes $a \otimes b = \varphi(a, b)$, for $a \in A$, $b \in B$.

We form thus a functor in two variables, with an obvious action on a pair of homomorphisms

$$
\begin{aligned}
- \otimes - \colon R\,\mathbf{Mod} \times R\,\mathbf{Mod} &\to R\,\mathbf{Mod}, \\
(A, B) \mapsto A \otimes_R B, \qquad\qquad (f, g) &\mapsto f \otimes g.
\end{aligned}
\qquad (2.14)
$$

This 'product' has well-known properties of unitarity, associativity and commutativity, up to canonical, natural isomorphisms. The unit object is R as a module on itself: $A \otimes_R R \cong A \cong R \otimes_R A$.

(b) The category $R\,\mathbf{Mod}$ has an 'internal' hom-functor in two variables, contravariant in the first

$$
\mathrm{Hom}_R \colon R\,\mathbf{Mod}^{\mathrm{op}} \times R\,\mathbf{Mod} \to R\,\mathbf{Mod}.
\qquad (2.15)
$$

Here $\mathrm{Hom}_R(A, B)$ is the set of homomorphisms $R\,\mathbf{Mod}(A, B)$ equipped with the pointwise sum and scalar product:

$$
(u + v)(a) = u(a) + v(a), \qquad (\lambda u)(a) = \lambda.u(a).
$$

On the morphisms we have the usual action: $\mathrm{Hom}_R(f, g) = g. - .f$.

(c) These two functors are linked by a family of adjunctions. For a fixed R-module A, there is a natural family of isomorphisms

$$
\begin{aligned}
\varphi_{XY} \colon \mathrm{Hom}_R(X \otimes_R A, Y) &\to \mathrm{Hom}_R(X, \mathrm{Hom}_R(A, Y)), \\
\varphi_{XY}(u)(x) &= u(x \otimes -) \colon A \to Y,
\end{aligned}
\qquad (2.16)
$$

for $u\colon X \otimes_R A \to Y$ and $x \in X$. The underlying bijections

$$\varphi_{XY}\colon R\mathbf{Mod}(X \otimes_R A, Y) \to R\mathbf{Mod}(X, \mathrm{Hom}_R(A, Y)),$$

give an adjunction $F \dashv G$ (depending on the parameter A)

$$F\colon R\mathbf{Mod} \rightleftarrows R\mathbf{Mod}\colon G,$$

$$\begin{array}{cc}
F(X) = X \otimes_R A, & F(f) = f \otimes \mathrm{id} A, \\
G(Y) = \mathrm{Hom}_R(A, Y), & G(g) = \mathrm{Hom}_R(A, g) = g.\,-\,.
\end{array} \qquad (2.17)$$

In particular, $F = -\otimes_R A$ preserves all colimits, while $G = \mathrm{Hom}_R(A, -)$ preserves limits.

With this structure $R\mathbf{Mod}$ is a symmetric monoidal closed category. The unit and counit of the adjunction *at A* are obvious, and the counit is called *evaluation*

$$\begin{array}{cc}
\eta_X\colon X \to \mathrm{Hom}_R(A, X \otimes_R A), & \eta_X(x)\colon a \mapsto x \otimes a, \\
\mathrm{ev}_Y\colon \mathrm{Hom}_R(A, Y) \otimes_R A \to Y, & \mathrm{ev}_Y(f \otimes a) = f(a).
\end{array} \qquad (2.18)$$

In particular, this holds for **Ab**, viewed as the category of $\mathbb{Z}$-modules. (Here one can simplify things, since the action of the scalars is determined by the additive structure, and bilinear just means additive in each variable.)

2.2.2 *Monoidal categories*

A *monoidal category* $(\mathbf{C}, \otimes, E, \kappa, \lambda, \rho)$ is a category equipped with a functor in two variables, often called a *tensor product*

$$\mathbf{C} \times \mathbf{C} \to \mathbf{C}, \qquad (A, B) \mapsto A \otimes B, \qquad (2.19)$$

and an object E, called the *unit*. This operation is assumed to be associative up to a functorial isomorphism κ of components

$$\kappa(A, B, C)\colon A \otimes (B \otimes C) \to (A \otimes B) \otimes C \qquad (\textit{associator}),$$

and the object E is assumed to be an identity, up to functorial isomorphisms

$$\lambda(A)\colon E \otimes A \to A, \qquad \rho(A)\colon A \otimes E \to A \qquad (\textit{unitors}).$$

Without entering into details (that will be analysed in the more general frameworks of bicategories and weak double categories), these isomorphisms must satify two coherence axioms (see 2.3.6) and – as a consequence – form a 'coherent system' (see 2.2.7 or [M4, Ke1, EiK, Ke2]). With an abuse of notation, this allows us to 'forget' them and write

$$A \otimes (B \otimes C) = (A \otimes B) \otimes C, \qquad E \otimes A = A = A \otimes E.$$

The case of a *strict* monoidal category, where these comparison isomor phisms are indeed identities, is less frequent but has important examples: see 2.4.2(d) and 2.4.3.

A *symmetric* monoidal category is further equipped with a symmetry isomorphism, coherent with the other ones:

$$s(A, B) \colon A \otimes B \to B \otimes A. \tag{2.20}$$

The latter cannot be omitted: note that $s(A, A) \colon A \otimes A \to A \otimes A$ is not the identity, in general.

A category $\mathbf{C}$ with finite products has a symmetric monoidal structure given by the categorical product; this structure is called *cartesian*. The category $R\mathbf{Mod}$ also has the (more important) tensor product reviewed above (for a commutative R). The *smash product* of $\mathbf{Set_{\bullet}}$ and $\mathbf{Top_{\bullet}}$ will be recalled in 6.8.6.

2.2.3 Exponentiable objects and internal homs

In a *symmetric* monoidal category $\mathbf{C}$ an object A is said to be *exponentiable* if the functor $- \otimes A \colon \mathbf{C} \to \mathbf{C}$ has a right adjoint, often written as $\mathrm{Hom}(A, -) \colon \mathbf{C} \to \mathbf{C}$ or $(-)^A$, and called an *internal hom*.

There is thus a family of bijections, natural in the variables X and Y

$$\varphi^A_{XY} \colon \mathbf{C}(X \otimes A, Y) \to \mathbf{C}(X, \mathrm{Hom}(A, Y)) \qquad (X, Y \text{ in } \mathbf{C}), \tag{2.21}$$

whose counit is called *evaluation* at A

$$\mathrm{ev}_Y \colon \mathrm{Hom}(A, Y) \otimes A \to Y. \tag{2.22}$$

Since adjunctions compose, all the tensor powers $A^{\otimes n} = A \otimes \ldots \otimes A$ are also exponentiable, with

$$\mathrm{Hom}(A^{\otimes n}, -) = (\mathrm{Hom}(A, -))^n. \tag{2.23}$$

A symmetric monoidal category is said to be *closed* if all its objects are exponentiable. In the non-symmetric case one should consider a left and a right hom-functor, as is the case with cubical sets (see Appendix B).

A category $\mathbf{C}$ with finite products is said to be *cartesian closed* if all the objects are exponentiable for the cartesian monoidal structure.

2.2.4 Exercises and complements

(a) **Set** is cartesian closed, with internal hom $Y^A = \textbf{Set}(A, Y)$. The natural bijection

$$\varphi_{XY} \colon \textbf{Set}(X \times A, Y) \to \textbf{Set}(X, Y^A)$$
$$(f \colon X \times A \to Y) \mapsto (g \colon X \to Y^A), \tag{2.24}$$

is expressed by the relation $f(x, a) = g(x)(a)$. It is also called the *exponential law* in **Set**, as it can be written in the form: $Y^{X \times A} \cong (Y^A)^X$. Compute unit and counit of this adjunction; the latter is called *evaluation*.

(b) The category **Ord** of ordered sets is cartesian closed, with internal hom **Ord**(A, Y) equipped with the pointwise order (see (1.5)).

(c) The category **Cat** of small categories is cartesian closed, with the internal hom $\textbf{Cat}(S, C) = C^S$ described in 1.2.7. For a fixed **S**, we have used in 1.5.2(b) an extension of the representable covariant functor $\textbf{Cat}(S, -) \colon \textbf{Cat} \to \textbf{Cat}$, acting on categories which are not assumed to be small.

The counit of the adjunction, the *evaluation* functor

$$\text{ev} \colon C^S \times S \to C, \qquad \text{ev}(F, X) = F(X), \quad \text{ev}(\varphi, f) = \varphi(f), \tag{2.25}$$

evaluates a natural transformation $\varphi \colon F \to G \colon S \to C$ on a morphism f of **S**, as in (1.10).

(d) The category **Ab** of abelian groups is not cartesian closed.

(e) Prove more generally that a cartesian closed category **C** with zero object is equivalent to the singleton category **1**.

(f) We have already recalled that $R\,\textbf{Mod}$ (for a commutative unitary ring R) is symmetric monoidal closed, with respect to the usual tensor product and hom-functor. This includes $\textbf{Ab} = \mathbb{Z}\,\textbf{Mod}$.

(g) The category **Abm** of abelian monoids (or semimodules on the semiring $\mathbb{N}$) is also a symmetric monoidal closed category, with tensor product and hom-functor essentially defined as in **Ab**. *(This can be extended to semimodules on a commutative semiring.)*

(h) **Top** is not cartesian closed, but it is well known (and not difficult to prove) that *every locally compact Hausdorff space A is exponentiable*: Y^A is the set of maps **Top**(A, Y) endowed with the *compact-open* topology (for an arbitrary space Y).

In particular the standard euclidean interval $[0, 1]$ is exponentiable, with all its cartesian powers. This is a crucial fact in homotopy theory, as we shall recall in A5.1.

*(As a partial converse, every exponentiable space A which is Hausdorff

must be locally compact, because the product $-\times A$ preserves coequalisers only in this case; see [Mic], Theorem 2.1 and footnote (5)*.

(i) The category **Rng** of rings is symmetric monoidal, with a tensor product $R \otimes S$ given by the tensor product of the underlying abelian groups, equipped with a natural multiplication. This can be extended to the category of algebras over a commutative ring.

(j) Restricting to the full subcategory **CRng** of commutative rings, $R \otimes S$ is a categorical sum and the monoidal structure is cocartesian. The tensor product $\mathbb{Z}[X] \otimes \mathbb{Z}[Y]$ of (commutative) polynomial rings on two sets of variables X, Y is easily computed.

The categorical sum in **Rng** is more complex. For the polynomial ring $R = \mathbb{Z}[x]$ on one variable, compute the categorical sum $R + R$ in **Rng** and **CRng**.

*(k) Every category of presheaves of sets (see 1.2.7) is cartesian closed: see [MaM].

2.2.5 Internal monoids

Let $\mathbf{X} = (\mathbf{X}, \otimes, E)$ be a monoidal category.

An *internal monoid* in $\mathbf{X}$ is a triple (M, e, m) consisting of an object M and two morphisms $e \colon E \to M$, $m \colon M^{\otimes 2} \to M$ of $\mathbf{X}$, called the *unit* and *multiplication*, which make the following diagrams commute

$$
\begin{array}{ccc}
M \xrightarrow{e\otimes M} M^{\otimes 2} \xleftarrow{M\otimes e} M & \qquad & M^{\otimes 3} \xrightarrow{M\otimes m} M^{\otimes 2} \\
\searrow \quad \downarrow m \quad \swarrow & & \quad {\scriptstyle m\otimes M}\downarrow \qquad\quad \downarrow m \\
M & & M^{\otimes 2} \xrightarrow{\ m\ } M
\end{array}
\qquad (2.26)
$$

These axioms are called *unitarity* and *associativity*.

2.2.6 Exercises and complements

(a) A monoid in **Set** (resp. **Top** or **Ord**), with respect to the cartesian product, is an ordinary monoid (resp. a topological or ordered monoid).

(b) A monoid in **Ab** (resp. $R\,\mathbf{Mod}$), with respect to the usual tensor product, is a $\mathbb{Z}$-algebra (resp. an R-algebra).

(c) In the cartesian case $\mathbf{X} = (\mathbf{X}, \times, \top)$, define an internal group G in $\mathbf{X}$ by adding to the structure of internal monoid an *inversion* morphism $i \colon G \to G$ satisfying suitable axioms.

(d) An internal group in **Top** is a topological group. But an ordered group is *not* an internal group in **Ord**.

(e) Verify that the category End(**C**) of endofunctors of the category **C** (with their natural transformations), has a strict, non-symmetric monoidal structure given by the composition of endofunctors. A monad on **C** is the same as a monoid in the category End(**C**).

*(f) An internal group in the category of differentiable manifolds is a Lie group.

2.2.7 Introducing Coherence Theorems

Mac Lane's Coherence Theorem of monoidal categories, the prototype of all coherence theorems in category theory, was first proved in 1961 [M3]. It can also be found in [M4], Section VII.2, in a more precise form.

Loosely speaking, the theorem says that, in a monoidal category $\mathbf{C} = (\mathbf{C}, \otimes, E, \kappa, \lambda, \rho)$, any 'natural diagram' made up by tensoring instances of the comparisons λ, ρ, κ, their inverses λ', ρ', κ' and identities, is commutative.

For instance, the following diagram always commute

$$
\begin{array}{ccc}
(A \otimes E) \otimes (B \otimes C) & \xrightarrow{\rho \otimes 1} & A \otimes (B \otimes C) \\
{\scriptstyle \kappa} \downarrow & & \downarrow {\scriptstyle \kappa} \\
((A \otimes E) \otimes B) \otimes C & \xrightarrow[(\rho \otimes 1) \otimes 1]{} & (A \otimes B) \otimes C
\end{array}
\tag{2.27}
$$

The tricky problem is a precise formulation of these 'well formed diagrams', because we have to get rid of possible coincidences between iterated tensor products, which might occur in **C**.

This can be made by replacing actual tensor products by 'formulas for them', as in Mac Lane's [M4]. The interested reader can see, in Section 2.6, the extension of this approach to bicategories, following Mac Lane – Paré [MaP].

2.3 From sesquicategories to 2-categories and bicategories

The category **Cat**, equipped with the natural transformations $\varphi \colon F \to G$ as second-order arrows (or 2-cells), is the prime example of a 2-category. We present it with the operations already examined in 1.2.4, the *vertical composition* and the *whisker composition* with functors, which determine the *horizontal composition* of 2-cells.

2.3.1 Sesquicategories

As a preliminary notion, a *sesquicategory* [St2] is a category $\mathbf{C}$ equipped with:

(a) for each pair of parallel morphisms $f, g \colon X \to Y$, a set of *2-cells*, or *homotopies*, $\mathbf{C}_2(f, g)$ whose elements are written as $\varphi \colon f \to g$ (or $\varphi \colon f \to g \colon X \to Y$), so that each map f has an *identity endocell* $\mathrm{id} f \colon f \to f$,

(b) a *concatenation*, or *vertical composition* of 2-cells, written as $\psi\varphi$ or $\psi.\varphi$

$$X \quad \substack{\xrightarrow{f} \\ \downarrow\varphi \\ \xrightarrow{} \\ \downarrow\psi \\ \xrightarrow{h}} \quad Y \qquad \psi\varphi \colon f \to h \colon X \to Y. \tag{2.28}$$

(c) a *whisker composition* for 2-cells and maps, or *reduced horizontal composition*, written as $k\varphi h$ (or $k\circ\varphi\circ h$, when useful to distinguish it)

$$X' \xrightarrow{h} X \quad \substack{\xrightarrow{f} \\ \downarrow\varphi \\ \xrightarrow{g}} \quad Y \xrightarrow{k} Y' \tag{2.29}$$

$$k\varphi h \colon kfh \to kgh \colon X' \to Y'.$$

These data must satisfy the following axioms (for *associativities, identities* and *distributivity of the whisker composition*):

$$\chi(\psi\varphi) = (\chi\psi)\varphi, \qquad k'(k\varphi h)h' = (k'k)\varphi(hh'),$$

$$\varphi.\mathrm{id} f = \varphi = \mathrm{id} g.\varphi, \quad 1_Y\circ\varphi\circ 1_X = \varphi, \quad k\circ\mathrm{id} f\circ h = \mathrm{id}(kfh), \tag{2.30}$$

$$k(\psi\varphi)h = (k\psi h).(k\varphi h).$$

(Equivalently, one can use two one-sided whisker compositions: $\varphi\circ h = 1_Y\circ\varphi\circ h$ and $k\circ\varphi = k\circ\varphi\circ 1_X$.) Note that each set $\mathbf{C}_2(f, g)$ is a category, under vertical composition.

*The category **Top** equipped with homotopies $\varphi \colon f \to g \colon X \to Y$ does *not* become a sesquicategory, because the vertical composition of homotopies is not associative and has no identities: we have a 2-dimensional categorical structure of a more complicated nature: see [G7] and its references. (One can form a sesquicategory by replacing standard homotopies with Moore homotopies, parametrised over compact euclidean intervals, but this approach is generally less interesting than the standard one.)

On the other hand it is easy to see that chain complexes, chain morphisms and homotopies, over an additive category $\mathbf{A}$, form a sesquicategory $\mathrm{Ch}_*(\mathbf{A})$.*

2.3.2 Definition

A *2-category* can be defined as a sesquicategory **C** which satisfies the following *reduced interchange property*:

$$X \; \overset{f}{\underset{g}{\rightrightarrows}} \; Y \; \overset{h}{\underset{k}{\rightrightarrows}} \; Z \qquad\qquad (\sigma g).(h\varphi) = (k\varphi).(\sigma f). \qquad (2.31)$$

(This approach is essentially followed in [EiK], Chapter 1.) Then one defines the *horizontal composition* of 2-cells φ, σ which are *horizontally consecutive*, as in diagram (2.31), using the previous equation:

$$\sigma \circ \varphi = (\sigma g).(h\varphi) = (k\varphi).(\sigma f) \colon hf \to kg \colon X \to Z. \qquad (2.32)$$

Note that the domain and codomain of φ for the new composition are objects of **C** instead of arrows; this can be expressed writing $\varphi \colon X \Rightarrow Y$.

We can now recover the usual definition of 2-categories [Be1, KeS], based on the vertical and horizontal composition of 2-cells: the horizontal composition is proved to be associative, to have identities $\mathrm{id}(1_X)$ (the identity cells of identity arrows) and to satisfy the *interchange laws* with the vertical composition:

$$(\tau.\sigma)\circ(\psi.\varphi) = (\tau\circ\psi).(\sigma\circ\varphi), \qquad\qquad 1_h \circ 1_f = 1_{hf}, \qquad (2.33)$$

$$X \; \overset{\downarrow\varphi}{\underset{\downarrow\psi}{\Rrightarrow}} \; Y \; \overset{\downarrow\sigma}{\underset{\downarrow\tau}{\Rrightarrow}} \; Z \qquad\qquad X \overset{f}{\longrightarrow} Y \overset{h}{\longrightarrow} Z$$

(The first is called *middle-four interchange*.) As a prime example of such a structure, **Cat** will also denote the *2-category* of small categories, their functors *and* their natural transformations. A set $\mathrm{Nat}(F, G)$ of natural transformations $F \to G$ can now be written as $\mathbf{Cat}_2(F, G)$.

There are sesquicategories where the reduced interchange property does not hold (and one does not define a full horizontal composition): for instance, the sesquicategory $\mathrm{Ch}_*(\mathbf{A})$ of chain complexes, recalled above.

An ordered category **C**, as defined in 2.1.2, is the same as a 2-category where each category $\mathbf{C}_2(f, g)$ is an ordered set; it is also called a *locally ordered 2-category*.

The categories of relations or partial morphisms belong to this case: see Section 2.1.

A *2-functor* $F \colon \mathbf{C} \to \mathbf{D}$ between 2-categories sends objects to objects, arrows to arrows and cells to cells, strictly preserving the whole structure: (co)domains, units and compositions. (Lax versions can be found in [Be1, KeS]; here they will be defined for weak double categories, in the next chapter.)

A 2-*natural transformation* $\varphi\colon F \to G\colon \mathbf{C} \to \mathbf{D}$ has components φX: $FX \to GX$ (for X in $\mathbf{C}$), as a natural transformation, but is also natural on every cell $\alpha\colon X \Rightarrow Y$, which means that $\varphi Y.F\alpha = G\alpha.\varphi X$.

A 2-*adjunction* $(F, G, \eta, \varepsilon)\colon \mathbf{C} \to \mathbf{D}$ is defined as in 1.5.1(iii); it involves 2-categories, 2-functors and 2-natural transformations.

2.3.3 Natural transformations and mates

Let us suppose that we have two adjunctions (between ordinary categories)

$$F\colon \mathbf{X} \rightleftarrows \mathbf{Y}\colon G, \qquad \eta\colon 1 \to GF, \qquad \varepsilon\colon FG \to 1,$$
$$F'\colon \mathbf{X}' \rightleftarrows \mathbf{Y}'\colon G', \qquad \eta'\colon 1 \to G'F', \qquad \varepsilon'\colon F'G' \to 1, \tag{2.34}$$

and two functors $H\colon \mathbf{X} \to \mathbf{X}'$, $K\colon \mathbf{Y} \to \mathbf{Y}'$.

As shown in [KeS], Section 2.2, there is a bijection between sets of natural transformations:

$$\mathrm{Nat}(F'H, KF) \to \mathrm{Nat}(HG, G'K), \qquad \varphi \leftrightarrow \psi,$$

$$\psi = G'K\varepsilon.G'\varphi G.\eta'HG\colon HG \to G'F'HG \to G'KFG \to G'K, \tag{2.35}$$

$$\varphi = \varepsilon'KF.F'\psi F.F'H\eta\colon F'H \to F'HGF \to F'G'KF \to KF,$$

The natural transformations φ, ψ are said to be *mates under the given data*; or *under the given adjunctions* when H and K are identities.

All this holds true for *internal* adjunctions in any 2-category (see 2.3.8).

2.3.4 The 2-category of adjoint functors

The category Adj**Cat** of small categories and adjunctions $(F, G, \eta, \varepsilon)\colon$ $\mathbf{C} \to \mathbf{D}$ has been introduced in 1.5.4. We show now that it has a natural 2-categorical structure, that extends the 2-categorical structure of Adj**Ord** (given by its order, see 2.3.2).

Given two adjunctions between the same categories

$$(F, G, \eta, \varepsilon)\colon \mathbf{C} \to \mathbf{D}, \qquad (F', G', \eta', \varepsilon')\colon \mathbf{C} \to \mathbf{D}, \tag{2.36}$$

a 2-cell of Adj**Cat**

$$(\varphi, \psi)\colon (F, G, \eta, \varepsilon) \to (F', G', \eta', \varepsilon')\colon \mathbf{C} \to \mathbf{D}, \tag{2.37}$$

is defined as a pair of mate natural transformations (under the adjunctions $F \dashv G$ and $F' \dashv G'$), so that each of them determines the other:

$$\varphi \colon F \to F' \colon \mathbf{C} \to \mathbf{D}, \qquad \psi \colon G' \to G \colon \mathbf{D} \to \mathbf{C},$$

$$\psi = (G\varepsilon'.G\varphi G'.\eta G' \colon G' \to GFG' \to GF'G' \to G), \qquad (2.38)$$

$$\varphi = (\varepsilon F'.F\psi F'.F\eta' \colon F \to FG'F' \to FGF' \to F').$$

The vertical composition of (φ, ψ) with a 2-cell

$$(\varphi', \psi') \colon (F', G') \to (F'', G'') \colon \mathbf{C} \nrightarrow \mathbf{D}$$

comes from the vertical composition of natural transformations

$$(\varphi', \psi').(\varphi, \psi) = (\varphi'\varphi, \psi\psi'). \qquad (2.39)$$

The whisker composition of (φ, ψ) with two adjunctions

$$(H, K, \rho, \sigma) \colon \mathbf{C}' \nrightarrow \mathbf{C}, \qquad (H', K', \rho', \sigma') \colon \mathbf{D} \nrightarrow \mathbf{D}',$$

is given by two whisker compositions of natural transformations

$$H'\varphi H \colon H'FH \to H'F'H \colon \mathbf{C}' \to \mathbf{D}',$$
$$K\psi K' \colon KG'K' \to KGK' \colon \mathbf{D}' \to \mathbf{C}'. \qquad (2.40)$$

(One verifies that $H'\varphi H$ and $K\psi K'$ are mate under $KFH \dashv KGH$.) Finally the (contravariant) involution of (1.53) can be extended (covariantly) to 2-cells

$$(-)^{\mathrm{op}} \colon \mathrm{Adj}\mathbf{Cat} \to \mathrm{Adj}\mathbf{Cat}, \qquad \mathbf{C} \mapsto \mathbf{C}^{\mathrm{op}},$$

$$((F, G, \eta, \varepsilon) \colon \mathbf{C} \nrightarrow \mathbf{D}) \mapsto ((G^{\mathrm{op}}, F^{\mathrm{op}}, \varepsilon^{\mathrm{op}}, \eta^{\mathrm{op}}) \colon \mathbf{D}^{\mathrm{op}} \nrightarrow \mathbf{C}^{\mathrm{op}}), \qquad (2.41)$$

$$((\varphi, \psi) \colon (F, G, \eta, \varepsilon) \to (F', G', \eta', \varepsilon')) \mapsto$$

$$((\psi^{\mathrm{op}}, \varphi^{\mathrm{op}}) \colon (G^{\mathrm{op}}, F^{\mathrm{op}}, \varepsilon^{\mathrm{op}}, \eta^{\mathrm{op}}) \to (G'^{\mathrm{op}}, F'^{\mathrm{op}}, \varepsilon'^{\mathrm{op}}, \eta'^{\mathrm{op}})).$$

In the same way (except for the involution) one can construct a 2-category $\mathrm{Adj}\mathbf{C}$ of *internal adjunctions* in an arbitrary 2-category $\mathbf{C}$.

2.3.5 *Two-dimensional universal arrows*

Let $U \colon \mathbf{A} \to \mathbf{X}$ be a 2-*functor* between 2-categories, as defined in 2.3.2. Universal arrows (see 1.2.9) have a strict and a weak extension to the 2-dimensional case.

(a) A 2-*universal arrow* from an object X of $\mathbf{X}$ to the 2-functor $U \colon \mathbf{A} \to \mathbf{X}$ is a pair $(A_0, h \colon X \to UA_0)$ which gives an isomorphism of categories (of arrows and cells, with vertical composition):

$$\mathbf{A}(A_0, A) \to \mathbf{X}(X, UA), \qquad g \mapsto Ug.h. \qquad (2.42)$$

This amounts to saying that the functor (2.42) is bijective on objects, full and faithful, i.e.

(i) for every A in $\mathbf{A}$ and every $f\colon X \to UA$ in $\mathbf{X}$ there exists a unique $g\colon A_0 \to A$ in $\mathbf{A}$ such that $f = Ug.h$,

(ii) for every pair of arrows $g, g'\colon A_0 \to A$ in $\mathbf{A}$ and every cell $\varphi\colon Ug.h \to Ug'.h$ in $\mathbf{X}$, there is a unique cell $\psi\colon g \to g'$ in $\mathbf{A}$ such that $\varphi = U\psi.h$.

(Equivalently, one can use a global universal property: for every cell $\varphi\colon f \to f'\colon X \to UA$ in $\mathbf{X}$, there is a unique cell $\psi\colon g \to g'$ in $\mathbf{A}$ such that $\varphi = U\psi.h$. This implies that $f = Ug.h$ and $f' = Ug'.h$.)

The solution of a 2-universal problem is *determined up to isomorphism*. Limits in 2-categories will be considered in Section 5.6.

*(b) More generally, a *biuniversal arrow* from X to $U\colon \mathbf{A} \to \mathbf{X}$ is a pair $(A_0, h\colon X \to UA_0)$ so that the functor (2.42) is an *equivalence* of categories. This can be rephrased saying that the functor (2.42) is essentially surjective on objects, full and faithful (by 1.2.5). In other words, we replace (i) with a weaker property (and keep (ii) as it is)

(i′) for every A in $\mathbf{A}$ and every $f\colon X \to UA$ in $\mathbf{X}$ there exists *some* $g\colon A_0 \to A$ in $\mathbf{A}$ such that $f \cong Ug.h$ (*isomorphic objects* in the category $\mathbf{X}(X, UA)$).

The solution of a biuniversal problem is *determined up to equivalence* (in a 2-category). In concrete situations one can often use the stronger property of *surjectivity* on objects, intermediate between (i) and (i′)

(i″) for every A in $\mathbf{A}$ and every $f\colon X \to UA$ in $\mathbf{X}$ there exists some $g\colon A_0 \to A$ in $\mathbf{A}$ such that $f = Ug.h$.

2.3.6 Bicategories

A bicategory is a weak version of a 2-category, introduced by Bénabou [Be1]. Here we briefly describe this structure; later, in Chapter 3, the more general structure of a weak double category will be studied in a detailed way.

Essentially, a bicategory is a weak version of a 2-category, where the horizontal composition is associative and has units up to (coherent) vertically invertible cells. It can also be viewed as a many-object generalisation of a weak monoidal category, and we shall adapt notation to this viewpoint, writing the weak composition as a tensor product, in diagrammatic order.

A *bicategory* $\mathbf{A}$ has objects $A, B, \dots$, arrows $u\colon A \nrightarrow B, \dots$ and 2-cells $\varphi\colon u \to v\colon A \nrightarrow B$. Fixing the objects A, B, these cells form a category $\mathbf{A}_2(A, B)$, with *vertical composition* $\psi\varphi\colon u \to w$ (for $\psi\colon v \to w$) and identities 1_u.

There is also a horizontal composition of arrows and 2-cells, written as

$$u \otimes u' : A \rightarrow C, \qquad \varphi \otimes \sigma : u \otimes u' \to v \otimes v' : A \rightarrow C, \qquad (2.43)$$

$$A \underset{v}{\overset{u}{\underset{\Downarrow\varphi}{\rightrightarrows}}} B \underset{v'}{\overset{u'}{\underset{\Downarrow\sigma}{\rightrightarrows}}} C$$

with identities $e_A : A \rightarrow A$ and $1_{e_A} : e_A \to e_A$. The compositions satisfy the *interchange laws* (as in (2.33))

$$\psi\varphi \otimes \tau\sigma = (\psi \otimes \tau)(\varphi \otimes \sigma), \qquad 1_u \otimes 1_{u'} = 1_{u \otimes u'}. \qquad (2.44)$$

The horizontal composition of arrows is categorical up to comparison *isocells* (vertically invertible)

- for an arrow $u : A \rightarrow B$ we have a *left unitor* $\lambda u : e_A \otimes u \to u$ and a *right unitor* $\rho u : u \otimes e_B \to u$, natural on a cell $\varphi : u \to v : A \rightarrow B$

$$(\lambda v)(e_f \otimes \varphi) = \varphi.(\lambda u) : e_A \otimes u \to v,$$
$$(\rho v)(\varphi \otimes e_g) = \varphi.(\rho u) : u \otimes e_B \to v. \qquad (2.45)$$

- for three consecutive arrows $u : A \rightarrow B$, $v : B \rightarrow C$ and $w : C \rightarrow D$ we have an *associator* $\kappa(u, v, w) : u \otimes (v \otimes w) \to (u \otimes v) \otimes w$, natural on a triple of cells $\varphi : u \to u'$, $\chi : v \to v'$, $\psi : w \to w'$

$$\kappa(u', v', w').(\varphi \otimes (\chi \otimes \psi)) = ((\varphi \otimes \chi) \otimes \psi).(\kappa(u, v, w)). \qquad (2.46)$$

The following diagrams of comparison cells must commute under vertical composition.

(a) *Coherence pentagon of the associator* κ, on four consecutive arrows x, y, z, t

$$
\begin{array}{ccc}
& (x \otimes y) \otimes (z \otimes t) & \\
{}^{\kappa}\nearrow & & \searrow^{\kappa} \\
x \otimes (y \otimes (z \otimes t)) & & ((x \otimes y) \otimes z) \otimes t \\
{}_{1\otimes\kappa}\searrow & & \nearrow_{\kappa\otimes 1} \\
x \otimes ((y \otimes z) \otimes t) & \xrightarrow{\ \kappa\ } & (x \otimes (y \otimes z)) \otimes t
\end{array}
\qquad (2.47)
$$

(b) *Coherence condition for the unitors*, for $x : A \rightarrow B$ and $y : B \rightarrow C$

$$
\begin{array}{ccc}
x \otimes (e_B \otimes y) & \xrightarrow{\ \kappa\ } & (x \otimes e_B) \otimes y \\
{}_{1\otimes\lambda}\searrow & & \swarrow_{\rho\otimes 1} \\
& x \otimes y &
\end{array}
\qquad (2.48)
$$

A monoidal (resp. strict monoidal) category $\mathbf{A}$ is the same as a bicategory (resp. 2-category) on one formal object $*$. The arrows $x\colon *\to *$ are the objects of $\mathbf{A}$, the cells $\varphi\colon x\to y\colon *\to *$ are the maps of $\mathbf{A}$, the vertical composition is that of $\mathbf{A}$, the horizontal one is tensor product.

2.3.7 Exercises and complements

(a) (*Spans*) For a category $\mathbf{C}$ with (a fixed choice) of pullbacks, the interested reader can construct now the bicategory Span$\mathbf{C}$ *of spans* of $\mathbf{C}$, which will be analysed in 3.4.1 in the extended form of a weak double category.

The objects of the bicategory Span$\mathbf{C}$ are those of $\mathbf{C}$. An arrow $u\colon X\to Y$ is a span $\vee\to\mathbf{C}$ *from X to Y* (see 1.2.1), i.e. a diagram $u=(u',u'')=(X\leftarrow U\to Y)$. An identity is a pair $e_X=(1_X,1_X)\colon X\to X$. One goes on defining 2-cells, the required compositions and comparisons.

(b) (*Cospans*) Dually, for a category $\mathbf{C}$ with (a fixed choice) of pushouts, one can construct a bicategory Cosp$\mathbf{C}$ *of cospans* of $\mathbf{C}$, where an arrow $u\colon X\to Y$ is a cospan $u\colon \wedge\to\mathbf{C}$ from X to Y, i.e. a diagram $X\to U\leftarrow Y$ in $\mathbf{C}$.

(c) Prove that the two mappings defined in (2.35) are inverse to each other.

*(d) The following result was proved by Kelly [Ke1] for monoidal categories; its extension to bicategories is straightforward. In a bicategory $\mathbf{C}$ we have the following commutative diagrams and identity, for $x\colon A\to B$ and $y\colon B\to C$

$$
e_A\otimes(x\otimes y)\xrightarrow{\ \kappa\ }(e_A\otimes x)\otimes y \qquad x\otimes(y\otimes e_C)\xrightarrow{\ \kappa\ }(x\otimes y)\otimes e_C
$$
$$
\lambda\searrow\quad\swarrow\lambda\otimes 1 \qquad\qquad 1\otimes\rho\searrow\quad\swarrow\rho
$$
$$
x\otimes y \qquad\qquad\qquad x\otimes y \qquad\qquad (2.49)
$$

$$
\lambda(e_A)=\rho(e_A)\colon e_A\otimes e_A\to e_A.
$$

2.3.8 Internal adjunctions and monads in a 2-category

Let $\mathbf{C}$ be a 2-category.

Extending what we have seen in $\mathbf{Cat}$, adjunctions, monads, equivalences and adjoint equivalences can be easily defined *inside* $\mathbf{C}$. Of course, for an adjunction one should use the 'algebraic' form (iii) of Definition 1.5.1.

Every adjunction $(F,G,\eta,\varepsilon)\colon X\to A$ in $\mathbf{C}$ has an associated monad $(X,T)=(X,T,\eta,\mu)$ on the object X, constructed in the usual way

$$
T=GF\colon X\to X, \qquad \eta\colon 1\to T, \qquad \mu=G\varepsilon F\colon T^2\to T. \qquad (2.50)
$$

*On the other hand, introducing the *object of algebras* for a monad, and the property of monadicity, is not obvious. We briefly review how this can be made, following Street's paper [St1]. This will only be used in 8.7.3.

Monads in $\mathbf{C}$ form a 2-category $\mathrm{Mnd}(\mathbf{C})$. An object is a monad $(X, T) = (X, T, \eta, \mu)$. A morphism $(F, \varphi) \colon (X, T) \to (Y, T')$ is a morphism $F \colon X \to Y$ of $\mathbf{C}$ with a cell $\varphi \colon TF \to FT'$ that 'commutes' with units and multiplications. A cell $\alpha \colon (F, \varphi) \to (G, \psi) \colon (X, T) \to (Y, T')$ is a cell $\alpha \colon F \to G \colon X \to Y$ of $\mathbf{C}$ which 'commutes' with φ and ψ.

The forgetful 2-functor

$$U \colon \mathrm{Mnd}(\mathbf{C}) \to \mathbf{C},$$
$$U(X, T) = X, \qquad U(F, \varphi) = F, \qquad U(\alpha) = \alpha, \tag{2.51}$$

has an obvious right 2-adjoint (see 2.3.2)

$$I \colon \mathbf{C} \to \mathrm{Mnd}(\mathbf{C}), \quad I(X) = (X, 1_X) \qquad (\varepsilon X \colon UI(X) = X). \tag{2.52}$$

One says that the 2-category $\mathbf{C}$ *admits the construction of algebras* if I has a right 2-adjoint, which will be written as

$$\mathrm{Alg} \colon \mathrm{Mnd}(\mathbf{C}) \to \mathbf{C},$$
$$\mathrm{Alg}(X, T) = X^T, \qquad \varepsilon^T X = (G^T, \gamma) \colon (X^T, 1) \to (X, T). \tag{2.53}$$

(When $\mathbf{C}$ is $\mathbf{Cat}$, X^T is the category of T-algebras $(x, a \colon Tx \to x)$ on the category X, $G^T \colon X^T \to X$ is the forgetful functor, and the natural transformation $\gamma \colon TG^T \to G^T$ has components $\gamma(x, a) = a \colon Tx \to x$.)

Assume that this holds. For every adjunction $(F, G, \eta, \varepsilon) \colon X \to A$ in $\mathbf{C}$ we have the associated monad (X, T) as in (2.50), and then its object of algebras $\mathrm{Alg}(X, T) = X^T$.

There is a comparison morphism, determined by the universal property of the counit ε^T

$$K \colon A \to X^T, \qquad \varepsilon^T.I(K) = (G, G\varepsilon) \colon (A, 1) \to (X, T). \tag{2.54}$$

The object A is said to be (strongly) *monadic* over X, via the morphism G, when K is an isomorphism.*

2.4 Enriched, internal and ordered categories

The classical references for these subjects are Eilenberg–Kelly [EiK], Kelly [Ke2] and Street [St3, St4].

2.4.1 Enriched categories

Let $\mathbf{V} = (\mathbf{V}, \otimes, E)$ be a symmetric monoidal category, whose comparisons will be left understood. A $\mathbf{V}$-*enriched category* $\mathbf{C}$, or $\mathbf{V}$-*category*, consists of:

(a) a set $\mathrm{Ob}\mathbf{C}$ of objects,

(b) for every pair $X, Y \in \mathrm{Ob}\mathbf{C}$, a $\mathbf{V}$-*object of morphisms* $\mathbf{C}(X, Y)$,

(c) for every $X \in \mathrm{Ob}\mathbf{C}$, a $\mathbf{V}$-*morphism of identity* $\mathrm{id}_X \colon E \to \mathbf{C}(X, X)$,

(d) for every triple $X, Y, Z \in \mathrm{Ob}\mathbf{C}$, a $\mathbf{V}$-morphism of *composition*

$$k \colon \mathbf{C}(X, Y) \otimes \mathbf{C}(Y, Z) \to \mathbf{C}(X, Z),$$

(whose action will be written as $(f, g) \mapsto gf$ *when* $\mathbf{V}$ *is a category of structured sets*).

These data are to satisfy three axioms of unitarity and associativity: the following diagrams must commute, for all $X, Y, Z, U \in \mathrm{Ob}\mathbf{C}$

$$\begin{array}{ccc}
E \otimes \mathbf{C}(X, Y) & \xrightarrow{\ \mathrm{id}\otimes 1\ } & \mathbf{C}(X, X) \otimes \mathbf{C}(X, Y) \\
& \searrow{\scriptstyle 1} & \downarrow{\scriptstyle k} \\
& & \mathbf{C}(X, Y)
\end{array} \qquad (2.55)$$

$$\begin{array}{ccc}
\mathbf{C}(X, Y) \otimes E & \xrightarrow{\ 1\otimes\mathrm{id}\ } & \mathbf{C}(X, Y) \otimes \mathbf{C}(Y, Y) \\
& \searrow{\scriptstyle 1} & \downarrow{\scriptstyle k} \\
& & \mathbf{C}(X, Y)
\end{array} \qquad (2.56)$$

$$\begin{array}{ccc}
\mathbf{C}(X, Y) \otimes \mathbf{C}(Y, Z) \otimes \mathbf{C}(Z, U) & \xrightarrow{\ 1\otimes k\ } & \mathbf{C}(X, Y) \otimes \mathbf{C}(Y, U) \\
{\scriptstyle k\otimes 1}\downarrow & & \downarrow{\scriptstyle k} \\
\mathbf{C}(X, Z) \otimes \mathbf{C}(Z, U) & \xrightarrow[\ k\ ]{} & \mathbf{C}(X, U)
\end{array} \qquad (2.57)$$

An enriched category $\mathbf{C}$ has an underlying category $U(\mathbf{C})$, constructed by means of the canonical forgetful functor $\mathbf{V}(E, -) \colon \mathbf{V} \to \mathbf{Set}$. The latter need not be faithful.

A $\mathbf{V}$-*functor* $F \colon \mathbf{C} \to \mathbf{D}$ between $\mathbf{V}$-categories consists of:

(e) a mapping $F_0 \colon \mathrm{Ob}\mathbf{C} \to \mathrm{Ob}\mathbf{D}$ whose action is written as $X \mapsto F(X)$,

(f) for every pair X, Y in $\mathrm{Ob}\mathbf{C}$, a $\mathbf{V}$-morphism

$$F_{XY} \colon \mathbf{C}(X, Y) \to \mathbf{D}(FX, FY)$$

(which will be written as $f \mapsto F(f)$ *when* $\mathbf{V}$ *is a category of structured sets*).

The following diagrams are assumed to commute, for all $X, Y, Z \in \mathrm{Ob}\mathbf{C}$

$$
\begin{array}{ccc}
E & \xrightarrow{\ \mathrm{id}\ } & \mathbf{C}(X,X) \\
 & {\scriptstyle \mathrm{id}}\searrow & \downarrow{\scriptstyle F} \\
 & & \mathbf{D}(FX, FX)
\end{array}
\qquad (2.58)
$$

$$
\begin{array}{ccc}
\mathbf{C}(X,Y) \otimes \mathbf{C}(Y,Z) & \xrightarrow{\ k\ } & \mathbf{C}(X,Z) \\
{\scriptstyle F \otimes F}\downarrow & & \downarrow{\scriptstyle F} \\
\mathbf{D}(FX, FY) \otimes \mathbf{D}(FY, FZ) & \xrightarrow[k]{} & \mathbf{D}(FX, FZ)
\end{array}
\qquad (2.59)
$$

A **V**-*transformation* $\varphi \colon F \to G \colon \mathbf{C} \to \mathbf{D}$ between **V**-functors consists of a family of **V**-morphisms

$$
\varphi X \colon E \to \mathbf{D}(FX, GX) \qquad (X \in \mathrm{Ob}\mathbf{C}),
$$

making the following diagrams commutative

$$
\begin{array}{ccccc}
E \otimes \mathbf{C}(X,Y) & \xrightarrow{\varphi X \otimes G} & \mathbf{D}(FX, GX) \otimes \mathbf{D}(GX, GY) & \xrightarrow{k} & \mathbf{D}(FX, GY) \\
\| & & & & \| \\
\mathbf{C}(X,Y) \otimes E & \xrightarrow[F \otimes \varphi Y]{} & \mathbf{D}(FX, FY) \otimes \mathbf{D}(FY, GY) & \xrightarrow[k]{} & \mathbf{D}(FX, GY)
\end{array}
$$

Appropriate compositions give the 2-category **VCat** of **V**-categories (see [Ke2]). There is a forgetful 2-functor

$$
U \colon \mathbf{VCat} \to \mathbf{Cat}. \qquad (2.60)
$$

2.4.2 *Examples and complements*

The following examples of enriched categories are based on the cartesian structure of the categories **Set**, **Cat**, **Ord**, **2**, and the symmetric monoidal structure of the categories **Ab** and $R\mathbf{Mod}$. In each case the enriched functors and their transformations are the obvious ones.

(a) A **Set**-enriched category is the same as an ordinary category with small hom-sets.

(b) A **Cat**-enriched category is the same as a 2-category with small hom-categories.

(c) An **Ord**-enriched category is an ordered category, as defined in 2.1.2, with small hom-sets. (A different notion of 'ordered category' will be recalled in 2.4.5.)

(d) A **2**-enriched category is a preordered set, possibly large.

(e) A *preadditive category* is a category $\mathbf{C}$ where every hom-set $\mathbf{C}(A, B)$ is equipped with a structure of abelian group, generally written as $f + g$, so that composition is additive in each variable

$$(f + g)h = fh + gh, \qquad k(f + g) = kf + kg. \qquad (2.61)$$

If $\mathbf{C}$ has small hom-sets, this amounts to saying that $\mathbf{C}$ is $\mathbf{Ab}$-enriched. A preadditive small category on one object is the same as a (unitary) ring.

An *additive functor* $F\colon \mathbf{C} \to \mathbf{D}$ between preadditive categories preserves the sum of morphisms.

(f) More generally we can consider R-*linear categories* and enrichment on $R\,\mathbf{Mod}$. For instance, the category $\mathbf{Ban}$ is linear on its scalar field – the real or the complex one.

2.4.3 Lawvere metric spaces

The strict monoidal category $\mathbb{R}_+$ was introduced by Lawvere [Law] to formalise generalised metric spaces as enriched categories. An object of $\mathbb{R}_+$ is an 'extended' positive real number $\lambda \in [0, \infty]$, an arrow $\lambda \to \mu$ is given by the order relation $\lambda \geqslant \mu$; the tensor product is given by the sum $\lambda + \lambda'$.

An object of the category $\mathbf{Mtr} = \mathbb{R}_+\mathbf{Cat}$, called a (Lawvere) *generalised metric space* X, or L-*metric space*, is thus a set X equipped with a *distance* d making it an $\mathbb{R}_+$-enriched category. This amounts to the following axioms

$$d(x, y) \in [0, \infty],$$
$$d(x, x) = 0, \qquad d(x, y) + d(y, z) \geqslant d(x, z). \qquad (2.62)$$

A morphism is a (weak) contraction $f\colon X \to X'$, satisfying $d(x, y) \geqslant d(fx, fy)$ for all $x, y \in X$.

All this leads to the notion of Cauchy-completeness of enriched categories, see [Law].

2.4.4 Internal categories

(a) Let us begin by remarking that a small category $\mathbf{C}$ can be described as a diagram in $\mathbf{Set}$ having the form of a 3-truncated simplicial set

$$C_0 \; \underset{e_0}{\overset{\partial_i}{\rightleftarrows}} \; C_1 \; \underset{e_i}{\overset{\partial_i}{\rightleftarrows}} \; C_2 \; \overset{\partial_i}{\underset{e_i}{\rightleftarrows}} \; C_3 \,. \qquad (2.63)$$

Here $C_0 = \mathrm{Ob}\mathbf{C}$ is the (small) set of objects of $\mathbf{C}$, $C_1 = \mathrm{Mor}\mathbf{C}$ is the set of morphisms, C_2 (resp. C_3) is the set of consecutive pairs (resp. triples) of morphisms. The faces ∂_i and the degeneracies e_i are defined as follows (letting t denote the triple (f, g, h)):

$$e_0(X) = \mathrm{id}_X, \qquad \partial_0(f) = \mathrm{Dom}\,(f), \qquad \partial_1(f) = \mathrm{Cod}\,(f),$$
$$e_0(f) = (f, 1), \qquad e_1(f) = (1, f).$$
$$\partial_0(f, g) = f, \qquad \partial_1(f, g) = gf, \qquad \partial_2(f, g) = g,$$
$$e_0(f, g) = (f, g, 1), \quad e_1(f, g) = (f, 1, g), \quad e_2(f, g) = (1, f, g),$$
$$\partial_0(t) = (f, g), \qquad \partial_1(t) = (f, hg), \qquad \partial_2(t) = (gf, h),$$
$$\partial_3(t) = (g, h).$$

The simplicial relations hold, so that (2.63) is indeed a 3-truncated simplicial set:

$$\partial_i \partial_j = \partial_{j-1} \partial_i \ \text{ for } i < j, \qquad e_j e_i = e_i e_{j-1} \ \text{ for } i < j,$$
$$\partial_i e_j = e_{j-1} \partial_i \ \text{ for } i < j, \qquad \partial_i e_j = e_j \partial_{i-1} \ \text{ for } i > j+1, \qquad (2.64)$$
$$\partial_i e_j = \mathrm{id} \quad \text{ for } \ i = j, j+1.$$

Finally, the following two squares are pullbacks (in **Set**)

$$
\begin{array}{ccc}
C_2 & \xrightarrow{\ \partial_0\ } & C_1 \\
{\scriptstyle \partial_2}\big\downarrow & \diagup & \big\downarrow{\scriptstyle \partial_1} \\
C_1 & \xrightarrow{\ \partial_0\ } & C_0
\end{array}
\qquad
\begin{array}{ccc}
C_3 & \xrightarrow{\ \partial_0\ } & C_2 \\
{\scriptstyle \partial_3}\big\downarrow & \diagup & \big\downarrow{\scriptstyle \partial_2} \\
C_2 & \xrightarrow{\ \partial_0\ } & C_1
\end{array}
\qquad (2.65)
$$

(b) Replacing **Set** with an arbitrary category **X**, an *internal category* **C** in **X**, or a *category object* in **C**, consists of a diagram (2.63) in **X** where $C_0 = \mathrm{Ob}\mathbf{C}$ is called the *object of objects* of **C**, $C_1 = \mathrm{Mor}\mathbf{C}$ is called the *object of morphisms* and C_2 is called the *object of consecutive pairs of morphisms*. The morphism $m = \partial_1 \colon C_2 \to C_1$ is called the *composition morphism* of **C**, or *partial multiplication*.

The following (redundant) axioms must be satisfied:

(i) the simplicial identities (2.64) hold, so that (2.63) is a 3-truncated simplicial object in **X**,

(ii) the square diagrams of (2.65) are pullbacks in **X**.

(The category **X** is not required to have all pullbacks.) In a category with finite products, an internal monoid with respect to the cartesian structure (as defined in 2.2.5) is the same as an internal category **C** where $\mathrm{Ob}\mathbf{C}$ is the terminal object.

(c) An *internal functor* $F \colon \mathbf{C} \to \mathbf{D}$ between internal categories in **X** is defined as a morphism of 3-simplicial objects. In other words we have four arrows

$$F_n \colon C_n \to D_n \qquad (n = 0, ..., 3), \qquad (2.66)$$

that commute with faces and degeneracies. The components F_2, F_3 are determined by the pullback condition, so that for $\mathbf{X} = \mathbf{Set}$ we simply have

$$F_2(f, g) = (F_1 f, F_1 g), \qquad F_3(f, g, h) = (F_1 f, F_1 g, F_1 h).$$

The composition of internal functors is obvious.

(d) An *internal transformation* $\varphi \colon F \to G \colon \mathbf{C} \to \mathbf{D}$ between internal functors in $\mathbf{X}$ is given (or represented) by an $\mathbf{X}$-morphism $\hat{\varphi}$ satisfying the following conditions

$$\hat{\varphi} \colon C_0 \to D_1, \qquad \partial_0.\hat{\varphi} = F_0, \qquad \partial_1.\hat{\varphi} = G_0,$$
$$m.\langle F_1, \hat{\varphi}\partial_1 \rangle = m.\langle \hat{\varphi}\partial_0, G_1 \rangle. \qquad (2.67)$$

For the last condition (of *naturality*), $m = \partial_1 \colon D_2 \to D_1$ is the partial multiplication of $\mathbf{D}$. The morphisms

$$\langle F_1, \hat{\varphi}\partial_1 \rangle \colon C_1 \to D_2, \qquad \langle \hat{\varphi}\partial_0, G_1 \rangle \colon C_1 \to D_2,$$

with values in the pullback D_2 are well defined because

$$\partial_1 F_1 = F_0 \partial_1 = \partial_0 \, \hat{\varphi} \, \partial_1, \qquad \partial_1 \, \hat{\varphi} \, \partial_0 = G_0 \partial_0 = \partial_0 G_1.$$

The interested reader can verify that there is a 2-category $\mathbf{Cat}(\mathbf{X})$ of internal categories, internal functors and internal transformations in $\mathbf{X}$. The vertical composition of internal transformations and their whisker composition with internal functors are defined as follows

$$(\psi\varphi)\hat{\ } = m.\langle \hat{\varphi}, \hat{\psi} \rangle \colon C_0 \to D_1,$$
$$(M\varphi L)\hat{\ } = M_1.\hat{\varphi}.L_0 \colon C_0' \to D_1', \qquad (2.68)$$

for $\psi \colon G \to H \colon \mathbf{C} \to \mathbf{D}$ and $L \colon \mathbf{C}' \to \mathbf{C}$, $M \colon \mathbf{D} \to \mathbf{D}'$.

2.4.5 Examples and complements

(a) We have already seen that an internal category in $\mathbf{Set}$ is a small category.

(b) An internal category in $\mathbf{Cat}$ is a small double category, studied in Chapter 3. Internal functors are the double functors; internal transformations are the horizontal ones (or the vertical ones, according to the way we are 'presenting' the double category, see 3.2.3).

*(c) We say that a category $\mathbf{C}$ is *internally ordered* when the sets $\mathrm{Ob}\mathbf{C}$ and

Mor**C** are equipped with consistent order relations (respected by domain, codomain, identity and composition). If **C** is small, this the same as an internal category in **Ord**. (An internally ordered category with a discrete order on the set of objects is the same as an ordered category in the enriched sense of 2.1.2.)

This notion was extensively used by C. Ehresmann for his theory of 'pseudogroups' in Differential Geometry. In fact, an *ordered category in the sense of Ehresmann* is an internally ordered category such that the order induced on each hom-set **C**(X, Y) is discrete (see [Eh5], p. 711).

The main example of such a structure is the category **Set** ordered by (weak) inclusion, of objects $(X \subset X')$ and morphisms: the mapping $f \colon X \to Y$ is *included* in $f' \colon X' \to Y'$ if $X \subset X'$, $Y \subset Y'$ and $f(x) = f'(x)$ for all $x \in X$. (When $X = X'$ and $Y = Y'$, this does imply $f = f'$.)

This structure rests heavily on set theory. One can replace it by the category $\mathcal{S}$ of sets and partial mappings, ordered in the present sense, where inclusion is only used *between subsets of given sets*. Similarly, the ordered category $\mathcal{C}$ of partial continuous mappings defined on open subspaces (see 2.1.8(b)) can be used in the theory of topological manifolds, and its differential analogues for differentiable manifolds. The interested reader is referred to [G1].

2.4.6 Exercises

(a) Compute the forgetful functor $\mathbb{R}_+ \to$ **Set** associated to the monoidal structure of $\mathbb{R}_+$, in 2.4.3.

(b) Compute the forgetful functor **Mtr** $\to$ **Cat** associated to the enriched structure of **Mtr**.

2.5 Relations for regular categories

Categories of relations will give interesting examples of double categories; the present extension covers all the cases we have considered in Section 2.1.

We begin by reviewing the ordered category Rel**Set**, as a *locally ordered 2-category*. Regular categories are briefly introduced, in 2.5.2. Then we extend the construction of Rel**Set** to a regular category **C**: this gives the ordered category of relations Rel**C**, where a relation $X \to Y$ is a subobject of the product $X \times Y$ (as in [Grt, Bo2]).

Marginally, we sketch a second approach to relations (followed for instance in [Me]), where a relation $X \to Y$ is a any monomorphism with values in $X \times Y$. In this way one gets a weaker structure, a *locally preordered bicategory*, which is written here as Rel$'$**C**.

2.5.1 Cartesian relations and locally ordered 2-categories

Let us recall that $\mathbf{R} = \text{Rel}\mathbf{Set}$ has been constructed in 2.1.3, as an involutive ordered category.

A relation $r\colon X \nrightarrow Y$ is a subset $r \subset X \times Y$, and will be called a *cartesian relation* when we want to distinguish it from the similar item of 2.5.7 (a monomorphism $r\colon R \rightarrowtail X \times Y$). Their composition has been recalled in (2.4). Given two parallel relations $r, s\colon X \nrightarrow Y$, the ordering $r \leqslant s$ means that $r \subset s$ as subsets of $X \times Y$; it agrees with composition.

In this way, each hom-set $\mathbf{R}(X, Y)$ is the ordered set $\mathcal{P}(X \times Y)$, and the relation $r\colon X \nrightarrow Y$ can be represented as in the diagram below, where the vertical map is the inclusion of a subset

$$X \xleftarrow{\ p'\ } X \times Y \xrightarrow{\ p''\ } Y \qquad \qquad \uparrow r \qquad \qquad (2.69)$$

We have also seen, in 2.3.2, that this ordered category is viewed as a locally ordered 2-category, turning the relationship $r \leqslant s$ into a 2-cell $r \to s$. The operations of vertical composition and whisker composition are determined by domains and codomains

$$X \xrightarrow[\substack{\downarrow \\ \downarrow \\ t}]{r} Y \qquad\qquad X' \xrightarrow{u} X \xrightarrow[s]{\substack{r \\ \downarrow}} Y \xrightarrow{v} Y' \qquad (2.70)$$

The vertical composition of $r \to s$ with $s \to t$ is thus the (unique) 2-cell $r \to t$, and the whisker composition above, at the right, is the (unique) 2-cell $vru \to vsu$. The coherence axioms of a 2-category are automatically satisfied, because of the uniqueness of a 2-cell between two given arrows.

2.5.2 Regular categories

Let $\mathbf{C}$ be a *regular category*. This means a category with finite limits, where the kernel pair $R \rightrightarrows A$ of every map $f\colon A \to B$ (see 1.3.4) has a coequaliser; moreover, the pullback of any regular epi (see 1.3.2) along any arrow must be a regular epi.

(Some texts give a more general definition, where the only limits assumed to exist are the kernel pairs and the pullback of any regular epi along any arrow [Ba, Bo2]; but this extension is presented in [Bo2] as "essentially a matter of personal taste".)

As a consequence every map f has a canonical factorisation $f = mp$

formed of a regular epimorphism p (the coequaliser of its kernel pair) and a monomorphism m (see [Bo2], Theorem 2.1.3, or [G11], Theorem 4.2.2).

A reader interested in regular categories is referred to [Bo2], Chapter 2, or [Ba, Grt]. It is easy to prove that **Set** and every variety of algebras are regular categories. Moreover, every abelian category is regular, as will also be seen in Exercise A3.9(d).

Extending Subsection 2.5.1, we want to construct an involutive ordered category $\mathbf{R} = \mathrm{Rel}(\mathbf{C})$ of relations on $\mathbf{C}$, in the 'cartesian form'.

For this goal, we assume that $\mathbf{C}$ is equipped with a choice of binary products and subobjects. (In the usual concrete categories these selections can be made without using the axiom of choice, by cartesian products and subsets in **Set**, via a forgetful functor $U \colon \mathbf{C} \to \mathbf{Set}$.)

Now every morphism $f \colon X \to Y$ has *precisely one* factorisation $f = mp$ where p is a regular epi and m is a *subobject* of Y, and we shall write

$$f = mp, \qquad m = \operatorname{im} f \colon \operatorname{Im} f \to Y. \tag{2.71}$$

2.5.3 Constructing the category of relations

For every pair of objects X, Y we let $\mathbf{R}(X, Y)$ be the ordered set of subobjects of the product $X \times Y$.

Given two consecutive relations

$$r = (r', r'') \colon R \rightarrowtail X \times Y, \qquad s = (s', s'') \colon S \rightarrowtail Y \times Z,$$

their composite $s.r = w$ is a subobject of $X \times Z$, determined by a construction based on a pullback P of (r'', s'), and the strict factorisation (2.71), where p is a regular epi and w is a subobject

$$P \xrightarrow{\ p\ } W \xrightarrow{\ w\ } X \times Z \tag{2.72}$$
$$\underset{(r'u', s''u'')}{\underbrace{\hphantom{P \xrightarrow{\ p\ } W \xrightarrow{\ w\ } X \times Z}}}$$

$$s.r = \operatorname{im}(r'u', s''u'') \colon W \rightarrowtail X \times Z. \tag{2.73}$$

The reader will note that any choice of the pullback gives the same subobject of $X \times Z$. The identity 1_X is the subobject associated to the diagonal monomorphism

$$1_X = \operatorname{im}((1, 1) \colon X \rightarrowtail X \times X) = (u, u) \colon \Delta_X \rightarrowtail X \times X, \tag{2.74}$$

where u is an isomorphism. The involution sends a subobject $r \colon R \rightarrowtail X \times Y$ to the opposite relation

$$r^{\sharp} = \operatorname{im} (ir \colon R \rightarrowtail X \times Y \to Y \times X), \qquad (2.75)$$

where $i = (p_2, p_1) \colon X \times Y \to Y \times X$ is the canonical symmetry of the cartesian product.

For two parallel relations $r, r' \colon X \nrightarrow Y$, the order relation

$$r \leqslant r' \qquad (2.76)$$

has already been defined, as the canonical order of the subobjects of $X \times Y$.

2.5.4 Theorem and Definition

These definitions produce an involutive ordered category $\mathbf{R} = \operatorname{Rel}(\mathbf{C})$, *called the category of (cartesian) relations over the regular category* $\mathbf{C}$.

The embedding $\mathbf{C} \to \operatorname{Rel}(\mathbf{C})$ *is the identity on the objects and sends the morphism* $f \colon X \to Y$ *to its graph, namely the subobject associated to the monomorphism* $(1, f) \colon X \to X \times Y$.

The category $\mathbf{R}$ *has small hom-sets if and only if every object of* $\mathbf{C}$ *has a small set of subobjects.*

Proof First, the associativity of the composition law, for three relations $r \colon X \nrightarrow Y$, $s \colon Y \nrightarrow Z$, $t \colon Z \nrightarrow U$, is proved by the following diagram

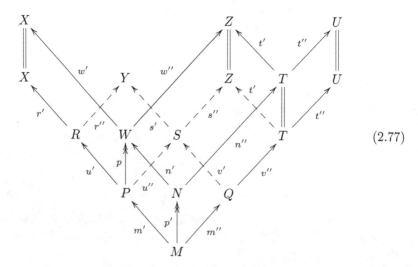

$$(2.77)$$

At the lower level we have a symmetric construction, based on three

pullbacks (P, u', u''), (Q, v', v'') and (M, m', m''). At the upper level we first compute

$$sr = \mathrm{im}\,(r'u', s''u'') = (w', w''): W \rightarrowtail X \times Z,$$

then the pullback (N, n', n'') of (w'', t'), so that $t(sr) = \mathrm{im}\,(N \to X \times U)$.

Now, the square $pm' = n'p'$ is a pullback, by Lemma 2.5.9 (below), and p is a regular epi, whence p' is also. But then $\mathrm{im}\,(N \to X \times U)$ coincides with $\mathrm{im}\,(M \to X \times U)$, and this subobject of $X \times U$ is determined by the symmetric construction of the lower level. By symmetry, $(ts)r$ too coincides with this subobject.

Second, the identity-relation $1_X : \Delta_X \rightarrowtail X \times X$ gives $r1_X = r$

$$
\begin{array}{ccccccc}
X & & & X & & & Y \\
 & \overset{u}{\nwarrow} & \overset{u}{\nearrow} & & \overset{r'}{\nwarrow} & \overset{r''}{\nearrow} & \\
 & & \Delta_X & & & R & \\
 & & \overset{v}{\nwarrow} & & \overset{1}{\nearrow} & & \\
 & & & R & & &
\end{array}
\tag{2.78}
$$

because u is an isomorphism and we can take R itself as the pullback of (u, r'); therefore the image of

$$(uv, r''.1) = (r', r''): R \rightarrowtail X \times Y$$

is precisely the relation r. Symmetrically, $1_Y s = s$.

The remaining points are obvious:

- the involution defined in (2.75) is consistent with the composition (in a contravariant way),

- the order defined in (2.76) is consistent with composition and involution,

- the embedding $\mathbf{C} \to \mathrm{Rel}(\mathbf{C})$ is a functor,

including the last statement about smallness. $\qquad\square$

2.5.5 Exercises and complements

(a) Extending 2.1.4(a), prove that every morphism f of $\mathbf{C}$ is left adjoint to $f^\sharp$ in the 2-category $\mathrm{Rel}(\mathbf{C})$, i.e. $1 \leqslant f^\sharp f$ and $ff^\sharp \leqslant 1$.

(b) It follows that the order of $\mathbf{R}(X, Y)$ restricted to the morphisms of $\mathbf{C}$ is discrete.

*(c) Conversely, any relation f satisfying the conditions above is a morphism of $\mathbf{C}$.

2.5.6 Exact squares and bicommutative squares

(a) $\mathbf{C}$ is always a regular category. Using the previous results, it is easy to see that a square of morphisms of $\mathbf{C}$

$$
\begin{array}{ccc}
X & \xrightarrow{f} & A \\
{\scriptstyle g}\downarrow & & \downarrow{\scriptstyle h} \\
B & \xrightarrow{k} & Y
\end{array}
\qquad (2.79)
$$

is commutative if and only if we have

$$
gf^{\sharp} \leqslant k^{\sharp}h \ \text{ in Rel}(\mathbf{C}) \qquad (\Leftrightarrow fg^{\sharp} \leqslant h^{\sharp}k).
$$

In fact

$$
hf = kg \ \Rightarrow \ gf^{\sharp} \leqslant k^{\sharp}kgf^{\sharp} = k^{\sharp}hff^{\sharp} \leqslant k^{\sharp}h,
$$
$$
gf^{\sharp} \leqslant k^{\sharp}h \ \Rightarrow \ kg \leqslant kgf^{\sharp}f \leqslant kk^{\sharp}hf \leqslant hf \ \Rightarrow \ kg = hf.
$$

We say that the square (2.79) is *v-exact* in $\mathbf{C}$ if it becomes *bicommutative* in Rel$(\mathbf{C})$, i.e. if it satisfies the stronger condition

(i) $gf^{\sharp} = k^{\sharp}h\colon A \rightarrowtail B$ $\qquad$ $(\Leftrightarrow fg^{\sharp} = h^{\sharp}k\colon B \rightarrowtail A),$

which means that the original square of $\mathbf{C}$ commutes in Rel$(\mathbf{C})$ when we 'reverse' two parallel edges.

Now, the relation $k^{\sharp}h\colon A \rightarrowtail B$ is computed by the pullback (P, f', g') of the cospan (h, k), and we conclude that property (i) is equivalent to each of the following ones:

(ii) the canonical morphism $p\colon X \to P$, determined by the following commutative diagram, is epi

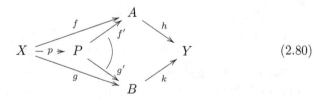

$$(2.80)$$

(iii) the span (f, g) and the pullback of (h, k) have the same commuting cospans (or the same pushout, provided it exists).

These conditions can be easily interpreted in **Set** and $R\,\mathbf{Mod}$, see 2.5.8.

2.5.7 Jointly monic relations

We sketch here a slightly different construction, that will only be used in Chapter 8. It gives a preordered structure, which we shall write as $\mathbf{R}' = \mathrm{Rel}'(\mathbf{C})$. It is not a category, but a *locally preordered bicategory*. The essential differences are:

- a hom-set $\mathbf{R}'(X, Y)$ is now a *preordered* set (and generally a large one),

- the composition of arrows is 'weakly associative' and has 'weak identities', *up* to the equivalence relation associated to the preorder.

A *jointly monic relation* $r = (r', r'') \colon X \nrightarrow Y$ is a jointly monic pair of morphisms of $\mathbf{C}$, as in the left diagram below, with an arbitrary domain R. It is also called a *jointly monic span*, from X to Y

$$
\begin{array}{ccc}
X & & Y \\
& \searrow^{r'} \quad \nearrow^{r''} & \\
& R &
\end{array}
\qquad
\begin{array}{ccc}
X \xleftarrow{\ s'\ } \bullet \xrightarrow{\ s''\ } Y \\
\searrow_{r'} \ \uparrow u \ \nearrow_{r''} \\
\bullet
\end{array}
\qquad (2.81)
$$

$\mathbf{R}'(X, Y)$ is the preordered set of such spans, with a 2-cell $r \prec s$ if there exists a (unique) mapping u that makes the right diagram above commute. Equivalently, $\mathbf{R}'(X, Y)$ is the set of all monomorphisms $r \colon \bullet \rightarrowtail X \times Y$, with their canonical preorder (see 1.1.7); the previous $\mathbf{R}(X, Y)$ can be identified with the associated ordered set $\mathbf{R}'(X, Y)/\!\sim$.

For a consecutive jointly monic relation $s \colon Y \nrightarrow Z$, the composition $w = sr$ is obtained much as in (2.72):

- *choosing* a pullback (P, u', u'') of the pair (r'', s'),

- then replacing the mapping $(r'u', s''u'') \colon P \to X \times Z$ with an associated jointly monic span, i.e. *any* monomorphism $w \colon W \rightarrowtail X \times Z$ of the factorisation regular epi-mono $(r'u', s''u'') = wp$ in $\mathbf{C}$.

Then one defines the vertical composition and the whisker composition of 2-cells. We only remark that the vertical composition is associated to a preorder, and is therefore strictly associative, with strict identities; on the other hand, the composition of morphisms is weakly associative and has weak identities, up to the equivalence relation associated to the preorder (as follows readily from the first approach). In this way, one gets a *locally preordered bicategory* $\mathbf{R}' = \mathrm{Rel}'(\mathbf{C})$.

The latter gives back the locally ordered 2-category $\mathbf{R}$, by identifying its arrows up to the equivalence relation associated to the preorder. Moreover, *defining $\mathbf{R}$ as this quotient gives an ordered category that is independent of any choice.*

2.5.8 Exercises and complements (Exact squares)

Let us come back to the commutative square $hf = kg$ of (2.79), in a regular category $\mathbf{C}$.

(a) In **Set** this square is v-exact if and only if it satisfies the following equivalent conditions:

(iv) for all $(a, b) \in A \times B$, we have $h(a) = k(b)$ if and only if there is some $x \in X$ such that $f(x) = a$, $g(x) = b$,

(v) the square is a weak pullback in **Set** (see 1.3.3),

but the equivalence of (v) depends on the axiom of choice.

(b) In $R\,\mathbf{Mod}$ and in **Gp** the characterisation (iv) still holds.

*(c) It is easy to verify that **Set** is also *coregular*, i.e. $\mathbf{Set}^{\mathrm{op}}$ is regular. This gives the involutive category $\mathrm{CorSet} = \mathrm{Rel}(\mathbf{Set}^{\mathrm{op}})$ of *corelations*, briefly examined in [G11], Section 1.8: they can be constructed as quotients of the sum of domain and codomain. The square $hf = kg$ is *v*-exact* in **Set** if it is bicommutative in CorSet, if and only if it satisfies a property dual to 2.5.6(ii): the canonical map from the pushout of (f, g) to Y is mono.

One can easily find in **Set** a v-exact square which is not v*-exact, and vice versa.

(d) We will see that, in any abelian category, v-exact and v*-exact squares coincide and are characterised by an exact sequence due to Hilton (in A3.9).

2.5.9 Lemma

In a category $\mathbf{C}$ we suppose that the following diagram is commutative and the two lateral squares are pullbacks (with vertices P, P'). Then the front square is a pullback

$$(2.82)$$

Proof (This lemma has been used above.) The pasting of the front and right squares coincides with the left square and is a pullback; but the right square is also, whence the front square is a pullback (by 1.3.5(e)). $\square$

2.6 *Coherence and strictification of bicategories

The Coherence Theorem of monoidal categories of Mac Lane [M3, M4] has been recalled in 2.2.7.

Dealing with the same problem, in the more general framework of bicategories, we follow the clear approach of Mac Lane and Paré [MaP], based on an equivalent bicategory $M(\mathbf{A})$ where 'tensor products determine their factors' and occasional coincidences cannot occur: a construction introduced by Laplaza [La].

In a bicategory $\mathbf{A} = (\mathbf{A}, \otimes, E, \kappa, \lambda, \rho)$ a composite $u \otimes v$ is written in the order of composition (as in 2.3.6). The inverses of the invertible 2-cells κ, λ, ρ are written as κ', λ', ρ'. Taking out all 2-cells we get a 'possibly non-associative, possibly non-unitary category', which we call a *magmoid*.

The exposition below is more elementary and detailed than the original one in [MaP], but the proof of the main theorem 2.6.6 is only referred to: an interested reader can see [MaP], and complete some points with [M3, M4].

2.6.1 Magmoids

A *magmoid* $A = (A_0, A_1, \partial^{\pm}, e, \otimes)$ will be a reflexive graph (see 1.1.2) with a composition law $u \otimes v$ of consecutive arrows $(\partial^+ u = \partial^- v)$

$$\partial^{\pm}: A_1 \underset{\longrightarrow}{\overset{\longrightarrow}{\rightleftarrows}} A_0 : e, \qquad \partial^- e = \mathrm{id} A_0 = \partial^+ e,$$
$$- \otimes -: A_1 \times_{A_0} A_1 \to A_1. \tag{2.83}$$

It is a category if and only if the composition is associative and admits all *units* e_x as partial units.

Mgd will denote the category of magmoids and their *functors* (which preserve the whole structure). The magmoid A has an underlying reflexive graph $|A|$, forgetting its composition.

The *free magmoid* $M(X)$ *on a reflexive graph* X has the same vertices, with arrows that can be expressed by (dichotomically) *branched strings* of consecutive arrows of X_1, like the following ones

u	uv	$(uv)(wt)$	$((uv)w)t$	(2.84)

The arrows of $M(X)$ are defined by recursion (together with their faces):

if $u\colon x \to y$ belongs to X_1 then $(u)\colon x \to y$ is a branched string; if $\underline{u}\colon x \to y$ and $\underline{v}\colon y \to z$ are (consecutive) branched strings then $(\underline{u})(\underline{v})\colon x \to z$ is also. Parentheses around a single term of X_1 are omitted, as in (2.84).

The composition of consecutive branched strings $\underline{u}, \underline{v}$ is by *branched concatenation* $\underline{u} \otimes \underline{v} = (\underline{u})(\underline{v})$. The unit of the object x is the string e_x.

The crucial fact is that any branched string of length greater than 1 is a concatenation $\underline{u} \otimes \underline{v}$ of two *determined* strings, so that we can reconstruct backwards its branching tree.

If A is a magmoid, the component of the counit of the adjunction $M \dashv |-|$ at A

$$p\colon M|A| \to A \qquad\qquad (evaluation\ functor), \qquad (2.85)$$

sends a branched string to an actual iterated composition in A. Thus $p((uv)(wt)) = (u \otimes v) \otimes (w \otimes t)$, if the composition in A is written by $\otimes$.

2.6.2 Arity and standard form

The free magmoid $M(X)$ on a reflexive graph X is *graded*, by the *arity* $\mathrm{ar}(\underline{u})$ of its arrows: this is the number of occurrences of *old arrows which are not units*

$$\mathrm{ar}(e_x) = 0, \qquad \mathrm{ar}(u) = 1 \qquad (u \in X_1 \setminus e(X_0)),$$
$$\mathrm{ar}(\underline{u} \otimes \underline{v}) = \mathrm{ar}(\underline{u}) + \mathrm{ar}(\underline{v}). \qquad\qquad (2.86)$$

Moreover, there is a graded idempotent endomorphism of reflexive graphs

$$\mathrm{St}\colon M(X) \to M(X) \qquad\qquad (standard\ form), \qquad (2.87)$$

that is the identity on all vertices. On a branched string, it takes all parentheses to the left and removes all occurrences of units, unless the string consists of one unit. For instance the branched word $((uv)e)(wt)$ has standard form $((uv)w)t$, provided that none of u, v, w, t is a unit.

This endomorphism is not a functor, but we have the following property

$$\mathrm{St}(\underline{u} \otimes \underline{v}) = \mathrm{St}(\mathrm{St}(\underline{u}) \otimes \mathrm{St}(\underline{v})). \qquad\qquad (2.88)$$

2.6.3 Expanding a bicategory

A bicategory $\mathbf{A}$ has an *underlying magmoid* $\mathrm{Mgd}\mathbf{A}$, formed of its objects and arrows, with their composition and units.

A functor of magmoids $p\colon M \to \mathrm{Mgd}\mathbf{A}$ can be extended to a strict functor of bicategories, full and faithful on 2-cells

$$p\colon \mathbf{B} \to \mathbf{A} \qquad\qquad (expansion\ of\ \mathbf{A}\ by\ the\ functor\ p), \qquad (2.89)$$

as follows. The underlying magmoid of **B** is M. For every pair arrows $u, u' \colon A \to B$ in **B**, a new 2-cell $\varphi \colon u \to u'$ is represented by an old cell $p\varphi \colon pu \to pu'$ (in **A**); their vertical composition and the identities 1_u are lifted from those of **A**.

As to the horizontal composition, we already have a composition $u \otimes v$ of arrows in the magmoid M (preserved by p), and we define the horizontal composition $\varphi \otimes \psi$ of new cells by lifting the old one

$$p(\varphi \otimes \psi) = p\varphi \otimes p\psi. \tag{2.90}$$

The new comparison isocells

$$\kappa(u, v, w) \colon u \otimes (v \otimes w) \to (u \otimes v) \otimes w,$$
$$\lambda u \colon 1 \otimes u \to u, \qquad \rho u \colon u \otimes 1 \to u, \tag{2.91}$$

are also represented by the old ones: $p\kappa(u, v, w) = \kappa(pu, pv, pw)$, etc.

2.6.4 The bicategory of branched words

Starting from the bicategory **A**, and using the previous points, we construct a bicategory $M(\mathbf{A})$ where *a composite $u \otimes v$ determines its factors u and v*.

The magmoid of $M(\mathbf{A})$ is the free magmoid $M = M(|\mathbf{A}|)$ generated by the underlying reflexive graph of **A**: it has the same objects, and the new arrows are *branched strings* of consecutive arrows of **A**, as in (2.84). This magmoid is graded by arity: the number of items in a string which are not units of **A**. The counit of the adjunction gives a functor of magmoids

$$p \colon M \to \mathbf{A} \qquad (\text{evaluation of branched strings}), \tag{2.92}$$

which, for instance, realises the branched string $\underline{u} = u(vw)$ as the arrow $p\underline{u} = u \otimes (v \otimes w)$ of **A**.

The expansion of the monoidal category **A** by this functor of magmoids gives a bicategory $M(\mathbf{A})$, called the *bicategory of branched words* of **A**. It is equipped with a strict functor of bicategories, full and faithful on cells

$$p \colon M(\mathbf{A}) \to \mathbf{A} \qquad (\text{evaluation}), \tag{2.93}$$

which will be proved to be an equivalence of bicategories (with a quasi-inverse pseudo functor).

Therefore, a cell $\varphi \colon u \to v$ of $M(\mathbf{A})$ is *represented* by a cell $p\varphi \colon pu \to pv$ of **A**. These cells are composed and tensored as their representatives in **A**.

The comparison isocells of $M(\mathbf{A})$ are defined as in (2.91). *Each of them can be given an arity*: the common arity of its two faces in the graded magmoid M underlying $M(\mathbf{A})$.

2.6.5 Canonical isocells

We form a sub-bicategory $C(\mathbf{A}) \subset M(\mathbf{A})$, defined as the smallest sub-bicategory containing all the objects and arrows of $M(\mathbf{A})$, all the comparison isocells (see (2.91)) and their inverses.

The cells of $C(\mathbf{A})$ are generated by the previous ones under vertical composition and tensor composition in $M(\mathbf{A})$. They are invertible, and will be called the *canonical isocells* of $M(\mathbf{A})$.

The bicategory $C(\mathbf{A})$ is graded by arity, as $M(\mathbf{A})$. The embedding $J \colon C(\mathbf{A}) \to M(\mathbf{A})$ gives a strict functor of bicategories $pJ \colon C(\mathbf{A}) \to \mathbf{A}$.

2.6.6 Theorem (Standard form and canonical isocells [MaP])

Let $\mathbf{A}$ be a bicategory and $M(\mathbf{A})$ its bicategory of branched words, with canonical isocells in $C(\mathbf{A})$.

For every branched word $\underline{u}$ in $M(\mathbf{A})$ there is a unique canonical isocell

$$\sigma\underline{u} \colon \underline{u} \to \mathrm{St}(\underline{u}). \tag{2.94}$$

Two words $\underline{u}, \underline{v}$ are connected in $C(\mathbf{A})$ if and only if they have the same standard form, and then there is a unique canonical isocell $\underline{u} \to \underline{v}$, namely $(\sigma\underline{v})^{-1}.\sigma\underline{u}$.

Proof See [MaP], Section 2. □

2.6.7 Corollary (Coherence Theorem for bicategories)

Let $\mathbf{A}$ be a bicategory. Every diagram of canonical isocells of $M(\mathbf{A})$ commutes.

Proof A straightforward consequence of the previous theorem. □

2.6.8 Proposition

Every bicategory $\mathbf{A}$ is equivalent to the bicategory $M(\mathbf{A})$, by means of the strict functor $p \colon M(\mathbf{A}) \to \mathbf{A}$ and a pseudo functor $i \colon \mathbf{A} \to M(\mathbf{A})$, with $pi = 1$ and $ip \cong 1$.

Proof The embedding $i \colon \mathbf{A} \to M(\mathbf{A})$ is obvious: we send any object to itself, any arrow and any cell to the corresponding unary branched string. It preserves the tensor product up to comparison isocells

$$\underline{i}(u,v) \colon (u,v) \to u \otimes v, \tag{2.95}$$

represented by the identity of the old composite $p(u, v) = p(u \otimes v) = u \otimes v$ in **A**. The coherence of these comparisons follows from the previous Coherence Theorem.

Finally $pi = 1$, while ip is isomorphic to the identity of $M(\mathbf{A})$, by the isomorphism $h\underline{u} \colon ip(\underline{u}) \to \underline{u}$ represented by the identity of $p\underline{u}$, for each branched string $\underline{u}$. □

2.6.9 Strictification Theorem

Every bicategory **A** *has an associated 2-category* $F(\mathbf{A})$, *equivalent to it by means of pseudo functors* $p \colon F(\mathbf{A}) \to \mathbf{A}$ *and* $i \colon \mathbf{A} \to F(\mathbf{A})$, *with* $pi = 1$ *and* $ip \cong 1$.

Every pseudo functor $S \colon \mathbf{A} \to \mathbf{B}$ *can be similarly replaced by a 2-functor* $S' \colon F(\mathbf{A}) \to F(\mathbf{B})$.

Proof We shall give a proof of the corresponding, more general statement for weak double categories: see 3.5.8. □

3

Double categories

Extending 2-categories (resp. bicategories), a strict (resp. weak) double category has horizontal arrows $f\colon X \to X'$, vertical arrows $u\colon X \dashrightarrow Y$ and double cells with a boundary as below

$$
\begin{array}{ccc}
X & \xrightarrow{\ f\ } & X' \\
{\scriptstyle u}\downarrow & \alpha & \downarrow{\scriptstyle v} \\
Y & \xrightarrow[\ g\ ]{} & Y'
\end{array}
\tag{3.1}
$$

which can be composed horizontally and vertically.

The main definitions about double categories, double functors and their transformations are given in Sections 3.1–3.2 (for the strict case) and 3.3–3.5 (for the weak one). Another important cleavage is presented in 3.2.5 and 3.2.6, between *transpositive* (strict) double categories, where the horizontal and vertical structure can be interchanged up to isomorphism, and *polarised* double categories (possibly weak) where such a symmetry does not exist.

As already mentioned in the general Introduction, in a polarised double category we use the main direction (written horizontally) to study the secondary one. Arrows which are *too relaxed* or *too strict* to have good universal properties are thus correlated with more ordinary arrows, typically the structure-preserving ones, through double cells. (The setting of '2-equipments', introduced by Carboni, Kelly, Verity and Wood [CaKVW], is a different approach to a similar goal.)

Most of our examples, in Sections 3.1 and 3.4, are polarised. Nevertheless, Section 4.2 will introduce a crucial transpositive structure, the double category $\mathbb{D}$bl of (small) weak double categories, with lax and colax double functors and suitable cells: double adjunctions will live inside this structure (forgetting size).

Weak double categories are a common generalisation of Ehresmann's double categories and Bénabou's bicategories. Besides the references mentioned in the general Introduction, weak double categories are also studied in various articles, like [BroM, BruMM, Da, DaP1, DaP2, DaPP1, DaPP2, Fi, FGK1, FGK2, Ga, Hr, Ko1, Ko2, Ni, Pa3, Pa4, Pa5].

Part of the theory of double categories might be deduced from the theory of internal categories (cf. Street [St3]) or indexed categories (see Paré–Schumacher [PaS]), but we want to give this subject an independent treatment, founded on basic category theory.

Double categories will generally be denoted as $\mathbb{A}, \mathbb{B}, \ldots$.

3.1 Introducing double categories

Historically, strict double categories were the first 2-dimensional categorical structure, introduced by C. Ehresmann [Eh1]–[Eh4] since 1962. After defining them, we present here various examples.

Relations and adjunctions form 'unusual categories', with few limits and colimits. Combining them with the corresponding 'usual categories' of mappings and functors, we form double categories which will be seen to have all horizontal double limits and colimits.

In all these cases the double structure gives a better understanding of the 2-category of its vertical arrows.

3.1.1 *Definition*

A (strict) *double category* $\mathbb{A}$ consists of the following structure.

(a) A set $\mathrm{Ob}\mathbb{A}$ of *objects* of $\mathbb{A}$.

(b) *Horizontal morphisms* $f\colon X \to Y$ between the previous objects; they form the category $\mathrm{Hor}_0\mathbb{A}$ *of the objects and horizontal maps* of $\mathbb{A}$, with composition written as gf and identities $1_X\colon X \to X$.

(c) *Vertical morphisms* $u\colon X \rightarrowtail Y$ (often denoted by a dot-marked arrow) between the same objects; they form the category $\mathrm{Ver}_0\mathbb{A}$ *of the objects and vertical maps* of $\mathbb{A}$, with composition written as $u \otimes v$, in diagrammatic order (or vu when convenient) and identities written as $e_X\colon X \rightarrowtail X$.

(d) (*Double*) *cells* with a *boundary* formed of two vertical arrows u, v and two horizontal arrows f, g

$$
\begin{array}{ccc}
X & \xrightarrow{\ f\ } & X' \\
u{\big\downarrow} & \alpha & {\big\downarrow}v \\
Y & \xrightarrow[\ g\ ]{} & Y'
\end{array}
\qquad\qquad
\alpha\colon (u \,{}_g^f\, v)\colon \left(\begin{smallmatrix} X & X' \\ Y & Y' \end{smallmatrix}\right). \tag{3.2}
$$

Writing $\alpha\colon (X \overset{X}{\underset{g}{\to}} v)$ or $\alpha\colon (e \overset{1}{\underset{g}{\to}} v)$ we mean that $f = 1_X$ and $u = e_X$. The cell α is also written as $\alpha\colon u \to v$ (with respect to its *horizontal* domain and codomain, which are vertical arrows) or as $\alpha\colon f \to g$ (with respect to its *vertical* domain and codomain, which are horizontal arrows).

We refer now to the following diagrams of cells, where the first is called a *consistent matrix* $\begin{pmatrix} \alpha & \beta \\ \gamma & \delta \end{pmatrix}$ of cells

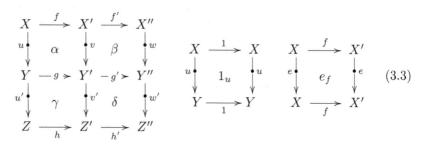

(e) Cells have a *horizontal composition*, consistent with the horizontal composition of arrows and written as $\alpha \,|\, \beta\colon (u \overset{f'f}{\underset{g'g}{\to}} w)$; this composition gives the category $\mathrm{Hor}_1\mathbb{A}$ *of vertical arrows and cells* $\alpha\colon u \to v$ of $\mathbb{A}$, with identities $1_u\colon (u \overset{1}{\underset{1}{\to}} u)$.

(f) Cells have also a *vertical composition*, consistent with the vertical composition of arrows and written as $\dfrac{\alpha}{\gamma}\colon (u \otimes u' \overset{f}{\underset{h}{\to}} v \otimes v')$, or $\alpha \otimes \gamma$; this composition gives the category $\mathrm{Ver}_1\mathbb{A}$ *of horizontal arrows and cells* $\alpha\colon f \to g$ of $\mathbb{A}$, with identities $e_f\colon (e \overset{f}{\underset{f}{\to}} e)$.

(g) The two compositions satisfy the *interchange laws* (for binary and zeroary compositions), which means that we have:

$$\frac{\alpha\,|\,\beta}{\gamma\,|\,\delta} = \frac{\alpha}{\gamma} \,\Big|\, \frac{\beta}{\delta}, \qquad \frac{1_u}{1_{u'}} = 1_{u\otimes u'},$$
$$e_f\,|\,e_{f'} = e_{f'f}, \qquad 1_{ex} = e_{1_X}. \tag{3.4}$$

The first condition says that a consistent matrix has a precise *pasting*; the last says that an object X has an *identity cell* $\square_X = 1_{ex} = e_{1_X}$. The expressions $\alpha\,|\,f'$ and $f\,|\,\beta$ will stand for $\alpha\,|\,e_{f'}$ and $e_f\,|\,\beta$, when this makes sense (i.e. when $X' = Y'$ and $v = e_{X'}$, in diagram (3.3)).

The pasting of double cells satisfies a general associativity property, established in [DaP1].

The dualities of double categories will be examined in the next section, but we already have in the examples below various instances of a vertical involution $\mathbb{A} \dashrightarrow \mathbb{A}$, contravariant in the vertical direction (see 3.2.2). We also use from now the obvious notion of double subcategory.

3.1.2 Relations of sets

The category RelSet of *sets and relations* has been reviewed in 2.1.3: a relation $u\colon X \rightarrowtail Y$ is a subset of the cartesian product $X \times Y$.

RelSet is an involutive ordered category, and contains Set as the subcategory of everywhere defined, single-valued relations, characterised as follows in RelSet

$$f^{\sharp}f \geqslant \mathrm{id}X, \qquad\qquad ff^{\sharp} \leqslant \mathrm{id}Y. \qquad\qquad (3.5)$$

We now amalgamate the ordinary categories Set and RelSet to form the double category $\mathbb{R}$elSet *of sets, mappings and relations* (notice the different notation of $\mathbb{R}$el). It has horizontal arrows in Set and vertical arrows in RelSet.

A double cell $(f, g)\colon u \to v$ is determined by its boundary, and exists when the following condition holds (the condition in parenthesis is equivalent, by (3.5))

$$
\begin{array}{ccc}
X & \xrightarrow{\ f\ } & X' \\
{\scriptstyle u}\big\downarrow & \leqslant & \big\downarrow{\scriptstyle v} \\
Y & \xrightarrow[\ g\ ]{} & Y'
\end{array}
\qquad\qquad gu \leqslant vf \quad (fu^{\sharp} \leqslant v^{\sharp}g). \qquad (3.6)
$$

We express this fact saying that $\mathbb{R}$elSet is a *flat* double category. Accordingly, the horizontal and vertical composition of double cells are determined by those of the arrows, and we only have to verify that the cells are closed under pasting in both directions, as verified below for the horizontal composition

$$
\begin{array}{ccccc}
X & \xrightarrow{\ f\ } & X' & \xrightarrow{\ f'\ } & X'' \\
{\scriptstyle u}\big\downarrow & \leqslant & {\scriptstyle v}\big\downarrow & \leqslant & \big\downarrow{\scriptstyle w} \\
Y & \xrightarrow[\ g\ ]{} & Y & \xrightarrow[\ g'\ ]{} & Y''
\end{array}
\qquad g'gu \leqslant g'vf \leqslant wf'f. \qquad (3.7)
$$

The axioms of associativity, unitarity and interchange are automatically satisfied.

$\mathbb{R}$elSet has an obvious vertical involution, that extends the involution $u \mapsto u^{\sharp}$ of relations and sends the cell $(f, g)\colon u \to v$ to the vertically opposite cell $(g, f)\colon u^{\sharp} \to v^{\sharp}$.

3.1.3 Other double categories of relations

(a) The involutive ordered category Rel**Ab** of *abelian groups and (additive) relations* has been reviewed in 2.1.5: a relation of abelian groups $u: X \to Y$ is a subgroup of the cartesian product $X \times Y$. **Ab** is embedded in the former as the category of everywhere defined, single-valued relations.

The double category $\mathbb{R}\text{el}\mathbf{Ab}$ *of abelian groups, homomorphisms and relations* has horizontal arrows in **Ab** and vertical arrows in Rel**Ab**. Double cells are defined as in 3.1.2 and give a flat structure with a regular vertical involution.

(b) More generally for any (unitary) ring R we have the involutive ordered category $\text{Rel}(R\,\mathbf{Mod})$ of left R-modules, and the flat double category $\mathbb{R}\text{el}(R\,\mathbf{Mod})$ of *modules, homomorphisms and relations*, with a regular vertical involution.

(c) More generally again, we can amalgamate any regular category **C** with $\text{Rel}(\mathbf{C})$ (see Section 2.5) and form the flat double category $\mathbb{R}\text{el}(\mathbf{C})$ *of objects, morphisms and relations* of **C**; also called the double category *of relations* of **C**, for short.

Other double categories of relations will be examined in Appendix A.

3.1.4 Double categories of quintets

(a) A 2-category **C** determines a double category $\mathbb{Q}\mathbf{C}$ *of quintets* of **C** (as named by C. Ehresmann), whose horizontal and vertical maps are the morphisms of **C**. Double cells are formed with 2-cells of **C**

$$
\begin{array}{ccc}
X & \xrightarrow{\;f\;} & X' \\
{\scriptstyle u}\downarrow & {\scriptstyle \alpha} & \downarrow{\scriptstyle v} \\
Y & \xrightarrow[\;g\;]{} & Y'
\end{array}
\qquad\qquad \alpha: vf \to gu: X \to Y'. \qquad\qquad (3.8)
$$

The horizontal and vertical composition of cells is obvious (see Exercise 3.1.8(a)).

If **C** is an ordinary category (viewed as a 2-category with trivial 2-cells), $\mathbb{Q}\mathbf{C}$ is the (flat) double category *of commutative squares* of **C**.

(b) Inside the double category $\mathbb{Q}\mathbf{Cat}$ of quintets of the 2-category **Cat**, we have the cell-wise full double subcategory $\mathbb{L}\mathbb{C}\mathbf{Cat}$ of small categories, with limit-preserving functors $f, g, \ldots$ as horizontal arrows and colimit-preserving functors $u, v, \ldots$ as vertical arrows.

(c) We write $\mathbb{L}\mathbb{R}\mathbf{Ac}$ for the cell-wise full double subcategory of $\mathbb{Q}\mathbf{Cat}$ formed by small abelian categories (Section 1.8), with left exact functors

$f, g, \ldots$ (that preserve finite limits) as horizontal arrows and right exact functors $u, v, \ldots$ as vertical arrows.

These double categories can be embedded in triple categories, adding 'strong arrows' in a new direction 0 (see 6.1.7): namely the functors which preserve limits and colimits in case (b), and exact functors (that preserve finite limits and colimits) in case (c).

3.1.5 Double categories of adjunctions

From a 2-category $\mathbf{C}$ we can construct another important prototype, the double category $\mathbb{A}\mathrm{dj}\mathbf{C}$ *of adjunctions in* $\mathbf{C}$, by a suitable amalgamation of $\mathbf{C}$ with the 2-category $\mathrm{Adj}\mathbf{C}$ of internal adjunctions in $\mathbf{C}$ (introduced in 2.3.4). (This double category can be also found in [KeS], Section 2.2, in a slightly different form.)

The objects are those of $\mathbf{C}$. In a general cell

$$
\begin{array}{ccc}
X & \xrightarrow{\ f\ } & X' \\
u \downarrow & \downarrow \alpha & \downarrow v \\
Y & \xrightarrow[\ g\ ]{} & Y'
\end{array}
\qquad (3.9)
$$

a horizontal arrow is a morphism of $\mathbf{C}$, while a vertical arrow is an internal adjunction in $\mathbf{C}$ directed as its left adjoint

$$u = (u_\bullet, u^\bullet, \eta, \varepsilon) \colon X \nrightarrow Y, \quad (u_\bullet \colon X \to Y) \dashv (u^\bullet \colon Y \to X),$$

$$\eta \colon 1_X \to u^\bullet u_\bullet, \qquad\qquad \varepsilon \colon u_\bullet u^\bullet \to 1_Y, \qquad (3.10)$$

$$\varepsilon u_\bullet . u_\bullet \eta = \mathrm{id}(u_\bullet), \qquad\qquad u^\bullet \varepsilon . \eta u^\bullet = \mathrm{id}(u^\bullet).$$

Finally the double cell $\alpha = (\alpha_\bullet, \alpha^\bullet)$ is a pair of mate 2-cells (see 2.3.3), each of them determining the other via the units and counits of the two adjunctions

$$\alpha_\bullet \colon v_\bullet f \to g u_\bullet, \qquad \alpha^\bullet \colon f u^\bullet \to v^\bullet g,$$

$$\alpha^\bullet = (f u^\bullet \to v^\bullet (v_\bullet f) u^\bullet \to v^\bullet (g u_\bullet) u^\bullet \to v^\bullet g), \qquad (3.11)$$

$$\alpha_\bullet = (v_\bullet f \to v_\bullet (f u^\bullet) u_\bullet \to v_\bullet (v^\bullet g) u_\bullet \to g u_\bullet).$$

The vertical composition of adjunctions is defined as in 1.5.4(b).

The horizontal and vertical compositions of cells is defined on the co-variant parts (marked with lower dots) as in the double category $\mathbb{Q}\mathbf{C}$ of quintets in$\mathbf{C}$. The axioms of double category are automatically satisfied. (Working with the contravariant parts gives an equivalent result, see Exercise 3.1.8(d).)

There is thus a forgetful double functor (preserving the whole structure)

$$(-)_\bullet \colon \mathsf{Adj}\mathbf{C} \to \mathbb{Q}\mathbf{C}$$
$$f \mapsto f, \qquad u \mapsto u_\bullet, \qquad \alpha \mapsto \alpha_\bullet. \tag{3.12}$$

We are also interested in the flat double subcategory $\mathsf{Adj}_0\,\mathbf{C} \subset \mathsf{Adj}\mathbf{C}$ *of adjunctions in* $\mathbf{C}$ *and bicommutative cells*. It has the same objects, horizontal arrows and vertical arrows, with double cells (3.9) reduced to the *bicommutative* ones, where both $\alpha_\bullet$ and $\alpha^\bullet$ are identities

$$v_\bullet f = g u_\bullet, \qquad\qquad f u^\bullet = v^\bullet g. \tag{3.13}$$

3.1.6 A double category of adjoint functors

An important case is the double category $\mathsf{Adj}\mathbf{Cat}$ of (small) *categories, functors and adjunctions*, where a vertical arrow $u = (u_\bullet, u^\bullet, \eta, \varepsilon) \colon X \rightarrowtail Y$ is an adjunction of small categories and a double cell $\alpha = (\alpha_\bullet, \alpha^\bullet)$ is a pair of mate natural transformations, as above.

In $\mathsf{Adj}\mathbf{Cat}$ there is a vertical involution, produced by opposite categories

$$u^{\mathrm{op}} = (u^\bullet, u_\bullet) \colon Y^{\mathrm{op}} \rightarrowtail X^{\mathrm{op}},$$
$$(\alpha^{\mathrm{op}})_\bullet = \alpha^{\bullet\,\mathrm{op}} \colon v^\bullet g \to f u^\bullet, \qquad (\alpha^{\mathrm{op}})^\bullet = \alpha_\bullet{}^{\mathrm{op}} \colon g u_\bullet \to v_\bullet f.$$

Let us note that the coherence of this involution with horizontal composition depends on a remark in 3.1.5, to be dealt with in Exercise 3.1.8(d).

3.1.7 Double categories of ordered sets

(a) The double category $\mathsf{Adj}\mathbf{Ord}$ of (small) *ordered sets, increasing functions and Galois connections* is a double subcategory of $\mathsf{Adj}\mathbf{Cat}$, closed under vertical involution. It amalgamates the usual category $\mathbf{Ord}$ of ordered sets and increasing mappings with the category $\mathsf{Adj}\mathbf{Ord}$ of Galois connections, described in 1.4.4.

$\mathsf{Adj}\mathbf{Ord}$ is flat, since a cell α as above (in (3.11)) exists if and only if

$v_\bullet f \leqslant gu_\bullet$, and is then determined by its boundary

$$
\begin{array}{ccc}
X & \xrightarrow{\ f\ } & X' \\
u\downarrow & \downarrow & \downarrow v \\
Y & \xrightarrow[\ g\ ]{} & Y'
\end{array}
\qquad v_\bullet f \leqslant gu_\bullet \qquad (\Leftrightarrow fu^\bullet \leqslant v^\bullet g). \qquad (3.15)
$$

(b) The double subcategory $\mathbb{A}\mathrm{dj}_0\mathbf{Ord}$, with the same objects and arrows, but *bicommutative cells*

$$
\begin{array}{ccc}
X & \xrightarrow{\ f\ } & X' \\
u\downarrow & = & \downarrow v \\
Y & \xrightarrow[\ g\ ]{} & Y'
\end{array}
\qquad v_\bullet f = gu_\bullet, \quad fu^\bullet = v^\bullet g, \qquad (3.16)
$$

will also be important: in $\mathbb{A}\mathrm{dj}_0\mathbf{Ord}$ we shall single out double subcategories of lattices, homomorphisms and adjunctions that are of interest in Homological Algebra, see A1.5.

3.1.8 Exercises and complements

(a) Define the horizontal and vertical composition of double cells in the double category $\mathbb{Q}\mathbf{C}$ of quintets of $\mathbf{C}$ (in 3.1.4).

(b) Examine the double category $\mathbb{A}\mathrm{dj}(\mathrm{Rel}\mathbf{Set})$, built over the 2-category $\mathrm{Rel}\mathbf{Set}$, and its relationship with the double category $\mathbb{R}\mathrm{el}\mathbf{Set}$.

*(c) Extend the previous point, replacing **Set** with a regular category.

*(d) For a 2-category $\mathbf{C}$, prove that the horizontal composition of double cells in $\mathbb{A}\mathrm{dj}\mathbf{C}$ (see 3.1.5) is consistent with mating. More explicitly, for a horizontal composite $\varphi = \alpha \mid \beta$, one should compute $\varphi_\bullet$ and prove that its mate $\varphi^\bullet$ coincides with the obvious pasting of $\alpha^\bullet$ and $\beta^\bullet$.

*(e) A reader interested in abelian categories or topos theory can consider the following structures, derived from $\mathbb{A}\mathrm{dj}\mathbf{Cat}$, and marginally used below.

$\mathbb{A}\mathrm{dj}\mathbf{Ac}$ will denote the cellwise-full double subcategory of $\mathbb{A}\mathrm{dj}\mathbf{Cat}$ formed by abelian categories, their exact functors and general adjunctions (where the left adjoint is right exact, and symmetrically).

$\mathbb{A}\mathrm{dj}\mathbf{Tp}$ will be the cellwise-full double subcategory of $\mathbb{A}\mathrm{dj}\mathbf{Cat}^\mathrm{v}$ (vertically reversed), consisting of toposes, their *logical morphisms* (functors which preserve the topos structure) and *geometric morphisms* (adjunctions whose left adjoint preserves finite limits). The latter are now directed according to the *right* adjoint, as convenient in topos theory.

3.1.9 Double categories of partial morphisms

(a) In the double category $\mathbb{R}\mathbf{elSet}$ of sets, mappings and relations we have a double subcategory $\mathbb{P}\mathrm{map}(\mathbf{Set})$ *of sets, mappings and partial mappings*: the vertical arrows belong now to the category $\mathcal{S}$ of partial mappings (see 2.1.6).

This can be easily extended to form a double category $\mathbb{P}\mathrm{map}(\mathbf{C})$ over a category $\mathbf{C}$ with pullbacks of monomorphisms along arbitrary morphisms (see 1.3.5(b)). For the sake of simplicity, we assume that $\mathbf{C}$ has a choice of subobjects, *stable under composition*; preimages of subobjects are (chosen to be) subobjects, and therefore uniquely determined.

A *partial map* $(u, h) \colon X \nrightarrow Y$ of $\mathbf{C}$ will be a span $X \leftarrowtail U \to Y$ where u is a subobject of X in $\mathbf{C}$. Their composition – by preimages – is obvious, and is strictly categorical

$$
X \leftarrowtail U \xrightarrow{\;h\;} Y \leftarrowtail V \xrightarrow{\;k\;} Z \qquad \qquad (3.17)
$$

with $h^*(V)$ and h'.

The double category $\mathbb{P}\mathrm{map}(\mathbf{C})$ *of partial morphisms* of $\mathbf{C}$ has double cells given by commutative diagrams

$$
\begin{array}{ccc}
X & \xrightarrow{\;f\;} & X' \\
\big\uparrow & & \big\uparrow \\
U & \longrightarrow & V \\
\big\downarrow & & \big\downarrow \\
Y & \xrightarrow[\;g\;]{} & Y'
\end{array}
\qquad\qquad (3.18)
$$

and obvious operations; it is flat. (When all pullbacks exist, general spans will produce a weak double category, see 3.4.1.)

(b) Similarly, if the category $\mathbf{C}$ has pullbacks of monos along monos (in $\mathbf{C}$), and a choice of subobjects stable under composition, there is a double category $\mathbb{P}\mathrm{iso}(\mathbf{C})$ *of partial isomorphisms*, constructed with the spans where the first map is a subobject and the second a monomorphism.

3.2 Double categories and their dualities

This section deals with with the basic structure of double categories, including their dualities, in the symmetry group of the square. This gives rise to the opposition between transpositive and polarised double categories, examined in 3.2.5 and 3.2.6.

3.2.1 Basic notions

(a) Let us recall that a double category is said to be *flat* if its cells are determined by their domains and codomains.

(b) The following notions are obvious:

- *products of double categories,*

- *double subcategory,*

- *totally full* double subcategory (determined by a subset of objects),

- *horizontally full* double subcategory (determined by a subcategory of vertical maps),

- *cellwise full* double subcategory (determined by a subcategory of horizontal maps and a subcategory of vertical maps with the same objects).

(c) A (double) *functor* $F: \mathbb{A} \to \mathbb{B}$ between double categories strictly preserves the whole structure: objects, horizontal and vertical arrows, double cells, domains and codomains, units and compositions. Double functors have an associative composition and identities. The lax and colax versions will be dealt with in Section 3.5. Horizontally full and faithful double functors are considered in 3.5.3.

(d) An *isomorphism* of double categories is an invertible double functor. It amounts to a double functor which is bijective on objects, horizontal arrows, vertical arrows and double cells.

(e) We write as **0** the empty double category, and as **1** the singleton double category. The latter has one object, say 0, with the required identities 1_0, e_0 and $\square_0$.

The *horizontal-arrow* double category **2** has a non-trivial horizontal arrow $0 \to 1$ and the required identities. Similarly, the *vertical-arrow* double category $\mathbf{2}^t$ has a non-trivial vertical arrow $0 \to 1$. (It is the transpose of **2**, as defined in 3.2.2.)

(f) A *double graph* has objects, arrows, cells, domains and codomains as above, without assigned compositions and identities. Morphisms of double graphs are obvious.

Let us note that, since there are no identity arrows, each term in the boundary $(u\ {}_g^f\ v)$ of a cell can be an object.

3.2.2 Dualities

The 8-element symmetry group of the square acts on a double category $\mathbb{A}$.

There is thus the *horizontal opposite* $\mathbb{A}^h$ (reversing the horizontal direction), the *vertical opposite* $\mathbb{A}^v$ (reversing the vertical direction) and the *transpose* $\mathbb{A}^t$ (interchanging the horizontal and vertical structures), under the (redundant) relations

$$\mathrm{hh} = \mathrm{vv} = \mathrm{tt} = 1, \qquad \mathrm{hv} = \mathrm{vh}, \qquad \mathrm{ht} = \mathrm{tv}, \qquad \mathrm{vt} = \mathrm{th}. \qquad (3.19)$$

The prefix 'co', as in *colimit, coequaliser* or *colax double functor*, will generally refer to horizontal duality; thus – for instance – coequalisers in **Set** will become coequalisers in the double category $\mathbb{R}el\mathbf{Set}$. Exceptionally this prefix can refer to the vertical direction, when the latter is viewed as the main one: this only happens in a few points, like 5.4.6, 5.4.7 and A2.7, where we consider *vertical limits* and *colimits*. (The prefix 'op' is not used in this book.)

We speak of a *horizontally contravariant*, or *vertically contravariant*, or *t-contravariant* double functor $\mathbb{A} \dashrightarrow \mathbb{B}$ to mean a double functor $\mathbb{A}^h \to \mathbb{B}$, or $\mathbb{A}^v \to \mathbb{B}$, or $\mathbb{A}^t \to \mathbb{B}$, respectively.

We say that a double category $\mathbb{A}$ is *horizontally selfdual*, or *vertically selfdual*, or *t-selfdual*, if it is isomorphic to $\mathbb{A}^h$, or $\mathbb{A}^v$, or $\mathbb{A}^t$, respectively. These isomorphisms are typically given by involutions.

A *vertical involution* is an isomorphism $V \colon \mathbb{A}^v \to \mathbb{A}$ whose inverse is $V^v \colon \mathbb{A} \to \mathbb{A}^v$. More simply, V can be considered as an involutive, vertically contravariant double functor $\mathbb{A} \dashrightarrow \mathbb{A}$. We have already seen various examples in the previous section, for all double categories of relations and some double categories of adjunctions. Transpositive involutions will be examined in 3.2.5.

3.2.3 Double categories as internal categories

A double category $\mathbb{A}$ can be viewed as a 3×3 array of sets connected by functions

$$\begin{array}{ccc}
A_{00} \rightrightarrows A_{01} \leftleftarrows A_{02} & \qquad & \mathrm{Hor}_0\mathbb{A} \\
\Updownarrow \quad\quad \Updownarrow \quad\quad \Updownarrow & & \partial^- \Updownarrow \partial^+ \\
A_{10} \rightrightarrows A_{11} \leftleftarrows A_{12} & & \mathrm{Hor}_1\mathbb{A} \\
\Vvert \quad\quad \Vvert \quad\quad \Vvert & & \Vvert \qquad\qquad (3.20) \\
A_{20} \rightrightarrows A_{21} \leftleftarrows A_{22} & & \mathrm{Hor}_2\mathbb{A}
\end{array}$$

$$\mathrm{Ver}_0\mathbb{A} \rightrightarrows \mathrm{Ver}_1\mathbb{A} \leftleftarrows \mathrm{Ver}_2\mathbb{A}$$

Each row forms a category $\mathrm{Hor}_n\mathbb{A}$; leaving apart – as usual – the problems of size, the right column presents $\mathbb{A}$ as a category object in **Cat** (see 2.4.4(b)). Explicitly

- in degree 0, $\mathrm{Hor}_0\mathbb{A}$ is the category of objects (in $A_{00} = \mathrm{Ob}\mathbb{A}$) and horizontal maps (in A_{01}) of $\mathbb{A}$,

- in degree 1, $\mathrm{Hor}_1\mathbb{A}$ is the category of vertical maps (in A_{10}) and cells $\alpha\colon u \to v$ (in A_{11}), with horizontal composition,

- $\mathrm{Hor}_2\mathbb{A}$ is the analogous category whose objects are the composable pairs of vertical maps (in A_{20}),

- vertical faces (domain and codomain) and degeneracy (or identity) give three functors

$$\partial^-, \partial^+ \colon \mathrm{Hor}_1\mathbb{A} \rightrightarrows \mathrm{Hor}_0\mathbb{A} : e. \qquad (3.21)$$

Similarly, each column is a category $\mathrm{Ver}_n\mathbb{A} = \mathrm{Hor}_n(\mathbb{A}^t)$ and the bottom row gives a second presentation of $\mathbb{A}$ as a category object in **Cat**. Giving priority to the horizontal composition, we consider the *first* presentation as the main one.

We say that $\mathbb{A}$ has *small horizontal hom-sets* if the categories $\mathrm{Hor}_0\mathbb{A}$ and $\mathrm{Hor}_1\mathbb{A}$ have small hom-sets. A double functor $F\colon \mathbb{A} \to \mathbb{B}$ determines three ordinary functors

$$F_n = \mathrm{Hor}_n F \colon \mathrm{Hor}_n\mathbb{A} \to \mathrm{Hor}_n\mathbb{B}. \qquad (3.22)$$

3.2.4 Double categories and 2-categories

A double category $\mathbb{A}$ has two associated 2-categories.

(a) The *horizontal 2-category* $\mathbf{Hor}\mathbb{A}$ of $\mathbb{A}$ is the category $\mathrm{Hor}_0\mathbb{A}$ equipped with 2-cells $\varphi\colon f \to g$ provided by the double cells $\varphi\colon (e\ {}_g^f\ e)$ of $\mathbb{A}$ whose vertical arrows are identities. These are called *globular cells*, and *globular isocells* when they are vertically invertible

$$X \overset{f}{\underset{g}{\rightrightarrows}} {\downarrow\varphi}\ X' \qquad (3.23)$$

A double category $\mathbb{A}$ whose vertical arrows are identities will be called a *horizontal double category*, and identified with the corresponding 2-category $\mathbf{A} = \mathbf{Hor}\mathbb{A}$. In particular we have the horizontal double category **2** (see 3.2.1).

(b) The *vertical 2-category* $\mathbf{Ver}\mathbb{A} = \mathbf{Hor}(\mathbb{A}^t)$ of $\mathbb{A}$ has 2-cells $\alpha\colon u \to v$ coming from the double cells $\alpha\colon (u\ {}_1^1\ v)$ of $\mathbb{A}$ whose horizontal arrows are identities

$$u \left(\begin{array}{c} X \\ \overset{\alpha}{\underset{\rightarrow}{}} \\ Y \end{array} \right) v \tag{3.24}$$

These are called *special cells*, and *special isocells* when they are horizontally invertible. In some cases it will be convenient to replace $\mathbf{Ver}\mathbb{A}$ with the 2-category $\mathbf{Ver}^*\mathbb{A}$ having reversed 2-cells $\alpha\colon u \to v$ (coming from the double cells $\alpha\colon (v \; {}^{1}_{1} \; u)$ of $\mathbb{A}$).

A double category $\mathbb{A}$ whose horizontal arrows are identities will be called a *vertical double category*, and viewed as a 2-category $\mathbf{Ver}\mathbb{A}$ written vertically. In particular we have the vertical double category $\mathbf{2}^{\mathrm{t}}$ (see 3.2.1).

(c) The other way round, we have already seen in 3.1.4 that a 2-category $\mathbf{C}$ determines the Ehresmann double category $\mathbb{Q}\mathbf{C}$ of quintets of $\mathbf{C}$, with double cells

$$\begin{array}{ccc} X & \overset{f}{\longrightarrow} & X' \\ u \downarrow & \overset{\nearrow}{\alpha} & \downarrow v \\ Y & \underset{g}{\longrightarrow} & Y' \end{array} \qquad \alpha\colon vf \to gu\colon X \to Y'. \tag{3.25}$$

Let us note that $\mathbf{Hor}(\mathbb{Q}\mathbf{C})$ and $\mathbf{Ver}^*(\mathbb{Q}\mathbf{C})$ both give back the 2-category $\mathbf{C}$.

(d) In a flat double category $\mathbb{A}$ the horizontal 2-category $\mathbf{Hor}\mathbb{A}$ is a preordered category, with $f \prec g$ when there exists a (unique) cell $(e \; {}^{f}_{g} \; e)$ in $\mathbb{A}$; the same holds in $\mathbf{Ver}\mathbb{A}$. We say that the double category $\mathbb{A}$ is *ultraflat* if it is flat and both $\mathbf{Hor}\mathbb{A}$ and $\mathbf{Ver}\mathbb{A}$ are *ordered* categories; in other words the second condition means that each vertically invertible cell $(e \; {}^{f}_{g} \; e)$ and each horizontally invertible cell $(u \; {}^{1}_{1} \; v)$ are identities.

For instance, $\mathbb{Q}\mathbf{Ord}$ is ultraflat, while the double category of quintets on the 2-category of *preordered* sets is just flat. The double category $\mathbb{R}\mathrm{el}\mathbf{C}$ (in 3.1.2, 3.1.3) is ultraflat. The same holds for $\mathbb{A}\mathrm{dj}\mathbf{Ord}$ (in 3.1.7): the 2-categories $\mathbf{Ord} = \mathbf{Hor}(\mathbb{A}\mathrm{dj}\mathbf{Ord})$ and $\mathbf{AdjOrd} = \mathbf{Ver}^*(\mathbb{A}\mathrm{dj}\mathbf{Ord})$ are locally ordered by the relation $f \leqslant g$ (as in (1.5)) and $u_{\bullet} \leqslant v_{\bullet}$ (as in (1.39)).

3.2.5 *Transpositive double categories*

As already seen in 3.2.2, in a t-selfdual double category the horizontal and vertical structure can be interchanged.

In particular a *transpositive* double category $\mathbb{A}$ is equipped with a *transposition isomorphism* $T = (-)^{\mathrm{t}}\colon \mathbb{A}^{\mathrm{t}} \to \mathbb{A}$ whose inverse is $T^{\mathrm{t}}\colon \mathbb{A} \to \mathbb{A}^{\mathrm{t}}$.

More simply, T can be considered as a *t-contravariant* involution $\mathbb{A} \dashrightarrow \mathbb{A}$ (as defined in 3.2.2).

With the notation of 3.2.3, this amounts to:

(a) an involution $(-)^t \colon A_{00} \to A_{00}$ on the set of objects,

(b) two inverse mappings $(-)^t \colon A_{01} \to A_{10}$ and $(-)^t \colon A_{10} \to A_{01}$ that interchange horizontal and vertical arrows,

(c) an involution $(-)^t \colon A_{11} \to A_{11}$ on the set of double cells,

so that faces, units and compositions are preserved, interchanging the horizontal and vertical issues.

For instance, the double category $\mathbb{A} = \mathbb{Q}\mathbf{Cat}$ of quintets of the 2-category $\mathbf{Cat}$ is transpositive, with transposition $\mathbb{A} \dashrightarrow \mathbb{A}$ given by the involution $(-)^{\mathrm{op}}$ of categories, functors and natural transformations. This acts as follows on the general double cell (3.25), inhabited by a natural transformation $\alpha \colon vf \to gu \colon A \to D$

$$
\begin{array}{ccc}
\begin{array}{ccc}
A & \xrightarrow{f} & B \\
u \downarrow & \swarrow \alpha & \downarrow v \\
C & \xrightarrow[g]{} & D
\end{array}
&
\mapsto
&
\begin{array}{ccc}
A^{\mathrm{op}} & \xrightarrow{u^{\mathrm{op}}} & C^{\mathrm{op}} \\
f^{\mathrm{op}} \downarrow & \swarrow \alpha^{\mathrm{op}} & \downarrow g^{\mathrm{op}} \\
B^{\mathrm{op}} & \xrightarrow[v^{\mathrm{op}}]{} & D^{\mathrm{op}}
\end{array}
\end{array}
\tag{3.26}
$$

If $\mathbf{C}$ is a category, the double category $\mathbb{Q}\mathbf{C}$ of commutative squares of $\mathbf{C}$ coincides with its transpose, and is thus transpositive in a strict sense.

The transpositive double category $\mathbb{D}\mathrm{bl}$, crucial for our analysis of double adjunctions, will be studied in Section 4.2.

3.2.6 Polarised double categories

We say that a double category $\mathbb{A}$ is *polarised* if it is not t-selfdual, i.e. not isomorphic to the transposed double category $\mathbb{A}^t$.

In such a double category it is often convenient to adopt a different terminology (ready to be extended to higher dimensional multiple categories, in Chapter 6): the horizontal direction is also called *transversal*, and viewed as the 'dynamic' direction, while the vertical direction is viewed as the 'static' or *geometric* one. Thus:

- objects and vertical arrows are viewed as *objects of degree 0 and 1*, respectively,

- horizontal arrows are viewed as *transversal 0-maps*, or *maps of degree 0*; they form, with the objects, the *transversal category* $\mathrm{tv}_*\mathbb{A} = \mathrm{Hor}_0\mathbb{A}$ *of degree 0*,

- double cells $\alpha \colon (u \overset{f}{\underset{g}{\;}} v)$ are viewed as *transversal 1-maps* $\alpha \colon u \to v$, or *maps of degree* 1; they form, with the 1-dimensional objects, the *transversal category* $\mathrm{tv}_1\mathbb{A} = \mathrm{Hor}_1\mathbb{A}$ *of degree* 1.

We have seen in Section 3.1 various examples of polarised double categories, based on relations or adjunctions. In all these cases a vertical arrow is something more complex than a mapping, its direction is conventional and the vertical composition looks as a sort of concatenation, or pasting, or tensor product. (All this will appear even more clearly in the *weak* double category $\mathbb{R}\mathrm{ng}$ of rings, homomorphisms and bimodules, in 3.4.5.)

3.2.7 Transformations and modifications

(a) A *horizontal transformation of double functors* $h\colon F \to G\colon \mathbb{X} \to \mathbb{A}$ can be defined as a double functor $h\colon \mathbb{X} \times \mathbf{2} \to \mathbb{A}$, with $F = h.(- \times 0)$ and $G = h.(- \times 1)$.

The transformation h amounts thus to giving two double functors F, G, with additional data:

- a horizontal arrow $hX\colon FX \to GX$ of $\mathbb{A}$ for every object X of $\mathbb{X}$,

- a double cell $hu\colon (Fu \, {}^{hX}_{hY} \, Gu)$ of $\mathbb{A}$ for every vertical map $u\colon X \nrightarrow Y$ of $\mathbb{X}$,

satisfying the following axioms of naturality and coherence:

$$F\xi \mid hv = hu \mid G\xi, \quad h(e_X) = e_{hX}, \quad h(u \otimes v) = hu \otimes hv, \qquad (3.27)$$

for every double cell $\xi\colon (u \, {}^{f}_{g} \, v)$, every object X and every pair u, v of consecutive vertical arrows, in $\mathbb{X}$.

(b) By transversal duality, a *vertical transformation* of double functors $r\colon F \nrightarrow G\colon \mathbb{X} \to \mathbb{A}$ comes from a double functor $r\colon \mathbb{X} \times \mathbf{2}^{\mathrm{t}} \to \mathbb{A}$. It assigns:

- a vertical arrow $rX\colon FX \nrightarrow GX$ of $\mathbb{A}$, to every object X of $\mathbb{X}$,

- a double cell $rf\colon (rX \, {}^{Ff}_{Gf} \, rY)$ of $\mathbb{A}$, to every horizontal map $f\colon X \to Y$ of $\mathbb{X}$,

satisfying the following axioms:

$$F\xi \otimes rg = rf \otimes G\xi, \quad r(1_X) = 1_{rX}, \quad r(gf) = rf \mid rg, \qquad (3.28)$$

for every double cell $\xi\colon (u \, {}^{f}_{g} \, v)$, every object X and every pair f, g of consecutive horizontal arrows, in $\mathbb{X}$.

(c) Given, in the diagram below, two horizontal transformations h, k and two vertical transformations r, s between four double functors $\mathbb{X} \to \mathbb{A}$, a *modification* $\mu\colon (r\,{}^{h}_{k}\,s)$

$$
\begin{array}{ccc}
F & \xrightarrow{\;h\;} & F' \\
{\scriptstyle r}\downarrow & \mu & \downarrow{\scriptstyle s} \\
G & \xrightarrow[\;k\;]{} & G'
\end{array}
\qquad\qquad (3.29)
$$

assigns to any object X of $\mathbb{X}$ a cell $\mu X\colon (rX\,{}^{hX}_{hY}\,sX)$ of $\mathbb{A}$ so that:

(mod.1) for every horizontal arrow $f\colon X \to Y$ in $\mathbb{X}$,

$$
\mu X \mid sf \;=\; rf \mid \mu Y, \qquad\qquad (3.30)
$$

$$
\begin{array}{ccccc}
FX \xrightarrow{\;hX\;} F'X \xrightarrow{\;F'f\;} F'Y & & FX \xrightarrow{\;Ff\;} FY \xrightarrow{\;hY\;} F'Y \\
{\scriptstyle rX}\downarrow \quad \mu X \quad \downarrow{\scriptstyle sX}\; sf \;\downarrow{\scriptstyle sY} & = & {\scriptstyle rX}\downarrow \quad rf \quad \downarrow{\scriptstyle rY}\; \mu Y \;\downarrow{\scriptstyle sY} \\
GX \xrightarrow[\;kX\;]{} G'X \xrightarrow[\;G'f\;]{} G'Y & & GX \xrightarrow[\;Gf\;]{} GY \xrightarrow[\;kY\;]{} G'Y
\end{array}
$$

(mod.2) for every vertical arrow $u\colon X \nrightarrow Y$ in $\mathbb{X}$,

$$
\mu X \otimes ku = hu \otimes \mu Y.
$$

If $\mathbb{X}$ is small, we have constructed the double category $\mathbb{A}^{\mathbb{X}}$ of double functors $\mathbb{X} \to \mathbb{A}$. The horizontal and vertical morphisms are given by the horizontal and vertical transformations, respectively; a double cell is a modification. The compositions of morphisms and cells are obvious.

3.2.8 Exercises and complements (Exponential)

(a) Define the double functor of *evaluation*, for double categories

$$
\mathrm{ev}\colon \mathbb{A}^{\mathbb{X}} \times \mathbb{X} \to \mathbb{A}, \qquad\qquad (3.31)
$$

and prove the universal property of the exponential: for every double functor $L\colon \mathbb{B}\times\mathbb{X} \to \mathbb{A}$ there is precisely one double functor $M\colon \mathbb{B} \to \mathbb{A}^{\mathbb{X}}$ such that $\mathrm{ev}.(M\times\mathbb{X}) = L$. In particular, the category of small double categories and double functors is cartesian closed, see 2.2.3.

(b) A horizontal transformation $h\colon \mathbb{X}\times\mathbf{2} \to \mathbb{A}$ can thus be viewed as a double functor $\mathbb{X} \to \mathbb{A}^{\mathbf{2}}$. Describe the objects, arrows and double cells of $\mathbb{A}^{\mathbf{2}}$.

(c) Describe the items of the double category $\mathbb{A}^{\mathbf{X}}$, where $\mathbf{X}$ is a small ordinary category (viewed as a horizontal double category).

3.3 Weak double categories

Weak double categories, also called pseudo double categories, were introduced in [GP1]. The strict case, defined in Section 3.1, is modified so that the horizontal composition is still strict but the vertical one is unitary and associative up to comparison double cells, invertible for horizontal composition. The terminology of the strict case is extended, as far as it applies.

3.3.1 Definition

A *weak double category* $\mathbb{A}$ consists of the following structure.

(a) A set $\mathrm{Ob}\mathbb{A}$ of *objects* of $\mathbb{A}$.

(b) *Horizontal morphisms* $f\colon X \to Y$ between the previous objects; they form the category $\mathrm{Hor}_0\mathbb{A}$ *of the objects and horizontal maps* of $\mathbb{A}$, with composition written as gf and identities $1_X\colon X \to X$.

(c) *Vertical morphisms* $u\colon X \nrightarrow Y$ between the same objects, with a composition written as $u \otimes v$ and identities $e_X\colon X \nrightarrow X$, whose properties are analysed below.

(d) (*Double*) *cells* $\alpha\colon (u \,{}^f_g\, v)$ as in (3.2), also written as $\alpha\colon u \to v$ in the horizontal direction or $\alpha\colon f \nrightarrow g$ in the vertical one.

We refer now to the diagrams of cells of (3.3).

(e) Cells have a *horizontal composition*, consistent with the horizontal composition of arrows and written as $\alpha \,|\, \beta\colon (u \,{}^{f'f}_{g'g}\, w)$; this composition gives the category $\mathrm{Hor}_1\mathbb{A}$ *of vertical arrows and cells* $\alpha\colon u \to v$ of $\mathbb{A}$, with identities $1_u\colon (u \,{}^1_1\, u)$. Special cells $\alpha\colon (u \,{}^1_1\, v)$ and special isocells are defined as in the strict case (in 3.2.4).

(f) Cells have also a *vertical composition*, consistent with the vertical composition of arrows and written as $\dfrac{\alpha}{\gamma}\colon (u \otimes u' \,{}^f_h\, v \otimes v')$, or $\alpha \otimes \gamma$; there are identities $e_f\colon (e \,{}^f_f\, e)$.

(g) The compositions satisfy the *interchange laws* (3.4). Again the identity cell of the object X is written as $\square_X = 1_{e_X} = e_{1_X}$.

(h) For $u\colon X \nrightarrow Y$ there are special isocells

$$\lambda u\colon e_X \otimes u \to u, \qquad \rho u\colon u \otimes e_Y \to u \qquad (\textit{left, right unitor}),$$

which are natural: for a cell $\alpha\colon (u \,{}^f_g\, v)$ we have

$$(e_f \otimes \alpha) \,|\, \lambda v = \lambda u \,|\, \alpha, \qquad (\alpha \otimes e_g) \,|\, \rho v = \rho u \,|\, \alpha, \qquad (3.32)$$

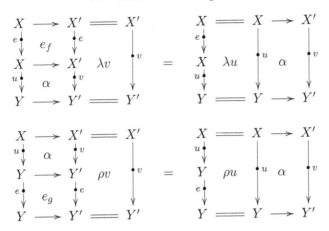

(i) For three consecutive vertical arrows u, v, w there is a special isocell

$$\kappa(u, v, w) \colon u \otimes (v \otimes w) \to (u \otimes v) \otimes w \qquad (associator),$$

which is natural: for three vertically consecutive double cells $\alpha \colon (u \; {}^f_g \; u')$, $\beta \colon (v \; {}^g_h \; v')$ and $\gamma \colon (w \; {}^h_k \; w')$, we have

$$(\alpha \otimes (\beta \otimes \gamma)) \mid \kappa(u', v', w') = \kappa(u, v, w) \mid ((\alpha \otimes \beta) \otimes \gamma), \qquad (3.33)$$

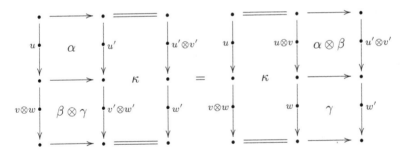

(j) Finally, as for bicategories (in 2.3.6), there are coherence conditions for the comparisons, which we write below in 3.3.2.

Again, vertical faces and vertical degeneracy give three functors

$$\partial^-, \partial^+ \colon \operatorname{Hor}_1(\mathbb{A}) \; \underset{\longrightarrow}{\overset{\longrightarrow}{\rightrightarrows}} \; \operatorname{Hor}_0(\mathbb{A}) \colon e. \qquad (3.34)$$

A weak double category has a *horizontal opposite* $\mathbb{A}^h$ (reversing the horizontal direction) and a *vertical opposite* $\mathbb{A}^v$ (reversing the vertical direction). The transpose $\mathbb{A}^t$ can only be considered in the strict case. Again, the prefix 'co' refers to horizontal duality, unless differently specified.

*In the light of 3.2.3, a small weak double category can be viewed as a *pseudo category object* in **Cat** (as defined in [Mar]).*

3.3.2 Coherence conditions

The following diagrams of comparison special isocells must commute under horizontal composition.

(a) *Coherence pentagon of the associator* κ, on four consecutive vertical arrows x, y, z, t

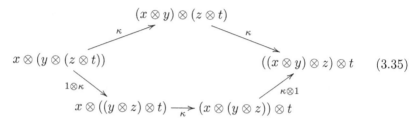

$$(3.35)$$

(b) *Coherence condition for the unitors*, for $x\colon X \rightarrow Y$ and $y\colon Y \rightarrow Z$

$$
\begin{array}{ccc}
x \otimes (e_Y \otimes y) & \xrightarrow{\ \kappa\ } & (x \otimes e_Y) \otimes y \\
& & \\
{}_{1\otimes\lambda} \searrow & & \swarrow {}_{\rho\otimes 1} \\
& x \otimes y &
\end{array}
\qquad (3.36)
$$

After assuming, in 3.3.1, that the associator and unitors are natural on *all* cells, the coherence conditions live in the structure **Ver**$\mathbb{A}$ of special cells, and amount to saying that this structure is a bicategory (see 2.3.6). Therefore the Coherence Theorem of bicategories applies, and says that all 'well-formed diagrams' of comparison cells commute. (A precise formulation was given in Section 2.6.)

In particular we have the following commutative diagrams and the following identity:

$$
\begin{array}{ccc}
e \otimes (x \otimes y) \xrightarrow{\ \kappa\ } (e \otimes x) \otimes y & \qquad & x \otimes (y \otimes e) \xrightarrow{\ \kappa\ } (x \otimes y) \otimes e \\
{}_{\lambda} \searrow \quad \swarrow {}_{\lambda\otimes 1} & & {}_{1\otimes\rho} \searrow \quad \swarrow {}_{\rho} \\
x \otimes y & & x \otimes y
\end{array}
\qquad (3.37)
$$

$$\lambda(e_X) = \rho(e_X)\colon e_X \otimes e_X \to e_X.$$

In the *flat* case, where double cells are determined by their domains and codomains, the coherence conditions are automatically satisfied.

3.3.3 Unitarity

A strict double category is the same as a weak one where all comparisons λ, ρ, κ are identities.

More generally we say that a weak double category $\mathbb{A}$ is *unitary* if its unitors λ, ρ are identities. The vertical composite of three arrows is then well-defined *whenever one of them is an identity*: in fact, in this case the associativity isocell $\kappa(u, v, w)$ is an identity, because of the coherence conditions (3.36) and (3.37).

As a consequence, the *vertical composite of three cells is well-defined whenever each of their two triples of vertical arrows falls in the previous situation*: both associativity isocells are identities (even if the unit vertical arrows are in different positions). We shall refer to this situation as a *normal* ternary (vertical) composition, of vertical arrows or cells.

As a much weaker condition we say that $\mathbb{A}$ is *preunitary* if the double cell

$$\lambda(e_X) = \rho(e_X) \colon e_X \otimes e_X \to e_X \qquad (3.38)$$

is an identity, for all objects X. Then also $\kappa(e_X, e_X, e_X)$ must be an identity.

Unitarity (or preunitarity) will often be assumed below, to avoid 'useless' complications. In our main examples the composition of the vertical arrows (typically spans, cospans, profunctors...) is settled by choice, and can be made unitary by obvious constraints on the latter.

3.3.4 Special cells

In the weak double category $\mathbb{A}$ we are particularly interested in the special cells and in the globular cells.

We already said that a cell $\alpha \colon u \to v$ is *special* if its horizontal arrows are identities

$$
\begin{array}{ccc}
X & \xrightarrow{1} & X \\
u \downarrow & \alpha & \downarrow v \\
Y & \xrightarrow{1} & Y
\end{array}
\qquad\qquad
\begin{array}{c}
X \\
u \left(\begin{array}{c} \alpha \\ \to \end{array} \right) v \\
Y
\end{array}
\qquad (3.39)
$$

Restricting $\mathbb{A}$ to these cells we obtain a *vertical weak double category* $\mathrm{Ver}\mathbb{A}$, which consists of the objects of $\mathbb{A}$, its horizontal identities, its vertical arrows and the special cells $\alpha \colon u \to v$. It is the same as a *bicategory written vertically*, with arrows and weak composition in the vertical direction and strict composition in the horizontal one, as in the right diagram above.

In particular, a monoidal category $\mathbf{C}$ can be viewed as a vertical weak double category $\mathbb{A}$ on one formal object, letting $\mathrm{Hor}_0\mathbb{A} = \mathbf{1}$ and $\mathrm{Hor}_1\mathbb{A} = \mathbf{C}$. The non-trivial items of $\mathbb{A}$ are thus the vertical arrows X (which are objects of $\mathbf{C}$) and the special double cells $f \colon X \to X'$ (which are morphisms of $\mathbf{C}$).

The vertical operations and their comparisons come from the monoidal structure $(\otimes, E, \lambda, \rho, \kappa)$ of **C**.

3.3.5 Globular cells

On the other hand, a cell $\varphi: f \twoheadrightarrow g$ is said to be *globular* if its vertical arrows are identities

$$
\begin{array}{ccc}
X & \xrightarrow{\ f\ } & X' \\
e\downarrow & \varphi & \downarrow e \\
X & \xrightarrow[\ g\ ]{} & X'
\end{array}
\qquad\qquad
X \underset{g}{\overset{f}{\underset{\Downarrow\varphi}{\rightrightarrows}}} X'
\qquad (3.40)
$$

If $\mathbb{A}$ is preunitary (see 3.3.3), it contains a *horizontal 2-category* **Hor**$\mathbb{A}$ formed by the category $\mathrm{Hor}_0\mathbb{A}$ enriched with 2-cells $\varphi: f \twoheadrightarrow g$ provided by the globular cells of $\mathbb{A}$, as in the right diagram above.

(Without assuming preunitarity, we should here modify the vertical composition of these cells, and **Hor**$\mathbb{A}$ could not be embedded in $\mathbb{A}$. This would leave us with two vertical compositions, which might result in misunderstandings.)

This cell φ is *vertically invertible*, or a *globular isocell*, if it has a vertical inverse in the 2-category **Hor**$\mathbb{A}$, written as $\varphi^{\check{}}: g \twoheadrightarrow f$

$$
\varphi \otimes \varphi^{\check{}} = e_f, \qquad\qquad \varphi^{\check{}} \otimes \varphi = e_g. \qquad (3.41)
$$

Then the arrows f, g are said to be *vertically isomorphic*. A *vertically invertible cell will always be a globular one.* (A general notion of 'vertically invertible cell' in a weak double category would be ill-founded, in the same way as to speak of 'invertible spans' or 'invertible profunctors', in a strict sense.)

A double category whose vertical arrows are identities will be called a *horizontal double category*. It is the same as a 2-category (horizontally written, as usual).

3.3.6 Exercises and complements

(a) Let X be a set equipped with two operations $x \mid y$ and $x \otimes y$ that have the same identity e and satisfy the middle-four interchange property (as in (2.33)):

$$
(x \mid y) \otimes (z \mid t) = (x \otimes z) \mid (y \otimes t). \qquad (3.42)
$$

Then these operations *coincide and are commutative.* (Note that the associativity of the given operations is not assumed.)

Classically, this simple algebraic result is used to prove that all homotopy groups of degree 2 or higher are commutative, using the concatenation of homotopy classes of singular cubes in two different directions.

(b) For a unitary weak double category $\mathbb{A}$, apply this result to the set of cells $\alpha\colon (e_X \,\substack{1\\1}\, e_X)$ whose boundary consists of identities of a fixed object X. We obtain thus a commutative monoid.

*(c) For a *weak* double category $\mathbb{A}$ define the exponential $\mathbb{A}^{\mathbf{X}}$, where $\mathbf{X}$ is a small ordinary category, extending what we have seen in 3.2.8(c) for a strict $\mathbb{A}$.

3.4 Examples of weak double categories

Spans on a category $\mathbf{C}$ with pullbacks form the well-known bicategory $\mathrm{Span}\mathbf{C}$ (already sketched in Exercise 2.3.7(a)), which has few limits and colimits. Combining these spans with the ordinary arrows of $\mathbf{C}$ we get a weak double category $\mathbb{S}\mathrm{pan}\mathbf{C}$, which will be proved to have all double limits and colimits.

Another important weak double category, written as $\mathbb{C}\mathrm{at}$, is obtained by amalgamating the category $\mathbf{Cat}$ with the bicategory of profunctors, also called distributors or bimodules (see Bénabou [Be2], Lawvere [Law]).

The enriched case gives profunctors between additive categories, or preordered sets, or generalised (Lawvere) metric spaces. Bimodules between rings also belong to this framework.

3.4.1 Spans

Let $\mathbf{C}$ be a category with (a fixed choice of) pullbacks. The weak double category $\mathbb{S}\mathrm{pan}\mathbf{C}$ *of spans* over $\mathbf{C}$ will play an important, 'representative' role in the theory of weak double categories (as we will see in Section 3.6 and Subsection 5.4.7).

Objects, horizontal arrows and their composition come from $\mathbf{C}$, so that $\mathrm{Hor}_0(\mathbb{S}\mathrm{pan}\mathbf{C}) = \mathbf{C}$.

A vertical arrow $u\colon X \nrightarrow Y$ is a span from X to Y. This is a diagram $X \leftarrow U \to Y$ in $\mathbf{C}$, or a functor $u = (u', u'')\colon \mathsf{V} \to \mathbf{C}$ defined on the formal-span category $0 \leftarrow \iota \to 1$. A vertical identity is a pair $e_X = (1_X, 1_X)$.

A cell $\alpha\colon (u \,\substack{f\\g}\, v)$ is a natural transformation $u \to v$ of such functors and amounts to the commutative left diagram below, with a middle arrow

$m\alpha\colon U \to V$ completing the boundary (note that this structure is not flat)

$$
\begin{array}{ccc}
X & \xrightarrow{\ f\ } & X' \\
u'\uparrow & & \uparrow v' \\
U & \xrightarrow{m\alpha} & V \\
u''\downarrow & & \downarrow v'' \\
Y & \xrightarrow[\ g\ ]{} & Y'
\end{array}
\qquad
\begin{array}{ccccc}
X & \longrightarrow & X' & \longrightarrow & X'' \\
u'\uparrow & & \uparrow v' & & \uparrow w' \\
U & \xrightarrow{m\alpha} & V & \xrightarrow{m\beta} & W \\
u''\downarrow & & \downarrow v'' & & \downarrow w'' \\
Y & \longrightarrow & Y' & \longrightarrow & Y''
\end{array}
\qquad (3.43)
$$

The horizontal composition $\alpha\,|\,\beta$ of α with a second cell $\beta\colon v \to w$ is a composition of natural transformations as in the right diagram above; it gives the category

$$\mathrm{Hor}_1(\mathbb{S}\mathrm{pan}\mathbf{C}) = \mathbf{Cat}(\vee, \mathbf{C}).$$

In particular, a special isocell $u \to v\colon X \rightarrowtail Y$ is given by an isomorphism $i\colon U \to V$ with $v'i = u'$, $v''i = u''$.

The vertical composition $u \otimes v$ of spans is computed with pullbacks in $\mathbf{C}$, by the fixed choice of pullbacks that we have assumed

$$
\begin{array}{ccccc}
X & & Y & & Z \\
& \nwarrow \quad \nearrow & & \nwarrow \quad \nearrow & \\
& U & & V & \qquad W = U\times_Y V. \qquad (3.44) \\
& & \nwarrow \quad \nearrow & & \\
& & W & &
\end{array}
$$

This is extended to double cells, in the obvious way. The comparisons λ, ρ, κ are determined by the universal property of pullbacks. Actually the unit constraint for pullbacks, specified in 1.3.5, makes $\mathbb{S}\mathrm{pan}\mathbf{C}$ *unitary*.

Reversing spans gives a *weak* vertical involution in $\mathbb{S}\mathrm{pan}\mathbf{C}$, which is a *pseudo* double functor (see 3.5.1): we have already seen in 1.3.5(c) that one *cannot* adopt a (strict) 'symmetry constraint' for pullbacks.

3.4.2 Cospans

Dually, for a category $\mathbf{C}$ with (a fixed choice of) pushouts there is a unitary weak double category $\mathbb{C}\mathrm{osp}\mathbf{C}$ *of cospans* over $\mathbf{C}$, that is horizontally dual to $\mathbb{S}\mathrm{pan}(\mathbf{C}^{\mathrm{op}})$.

We have now

$$\mathrm{Hor}_0(\mathbb{C}\mathrm{osp}\mathbf{C}) = \mathbf{C},$$

$$\mathrm{Hor}_1(\mathbb{C}\mathrm{osp}\mathbf{C}) = \mathbf{Cat}(\wedge, \mathbf{C}), \qquad (3.45)$$

where $\wedge = \vee^{\mathrm{op}}$ is the formal-cospan category $0 \to \iota \leftarrow 1$.

A vertical arrow $u = (u', u'')\colon \wedge \to \mathbf{C}$ is now a cospan, i.e. a diagram $X \to U \leftarrow Y$ in $\mathbf{C}$, and a cell $\alpha\colon u \to v$ is a natural transformation of such diagrams. Their vertical composition is computed with (the fixed choice of) pushouts in **C**.

3.4.3 Profunctors

Another important prototype is the weak double category $\mathbb{C}$at *of (small) categories, functors and profunctors*. In a general cell

$$
\begin{array}{ccc}
X & \xrightarrow{\ f\ } & X' \\
{\scriptstyle u}\Big\downarrow & \begin{array}{c}\alpha\\ \to\end{array} & \Big\downarrow{\scriptstyle v} \\
Y & \xrightarrow[\ g\ ]{} & Y'
\end{array}
\tag{3.46}
$$

an object is a category, a horizontal arrow is a functor, a vertical arrow $u\colon X \nrightarrow Y$ is a profunctor, i.e. a functor $X^{\mathrm{op}} \times Y \to \mathbf{Set}$. The cell 'is' a natural transformation

$$
\alpha\colon u \to v.(f^{\mathrm{op}} \times g)\colon X^{\mathrm{op}} \times Y \to \mathbf{Set}.
$$

The vertical composition of u with $u'\colon Y \nrightarrow Z$ is given by (choosing) a coend in **Set** (see 1.3.9)

$$
(u \otimes u')(x, z) = \int^{y} u(x, y) \times u'(y, z), \quad e_X(x, x') = X(x, x').
\tag{3.47}
$$

Its comparison cells for unitarity and associativity derive from the universal property of coends. The horizontal composition of double cells is obvious, the vertical one is computed by coends.

All this can be interpreted in a more elementary way.

In a profunctor $u\colon X \nrightarrow Y$ it is useful to view the elements $\lambda \in u(x, y)$ as *new arrows* $\lambda\colon x \dashrightarrow y$ from the objects of X to those of Y. Together with the disjoint union of the objects and arrows of X and Y, we form thus a new category $X +_u Y$ known as the *gluing*, or *collage*, of X and Y along u. (This category will be shown to be a double colimit, the *cotabulator of* u in $\mathbb{C}$at, in 3.7.4.)

The whisker composition between 'old' and 'new' arrows is determined by the action of u on the old ones

$$
\eta\lambda\xi = u(\xi, \eta)(\lambda) \qquad (\text{for } \xi\colon x' \to x \text{ in } X \text{ and } \eta\colon y \to y' \text{ in } Y).
\tag{3.48}
$$

The profunctor u amounts thus to a category $U = X +_u Y$ containing the disjoint union $X + Y$ and new arrows from objects of X to objects of Y (possibly none). More formally, U is a *category over* **2**, i.e. an object $U \to \mathbf{2}$ of $\mathbf{Cat}/\mathbf{2} = \mathbf{Cat} \downarrow \mathbf{2}$ (see (1.56)), whose fibres over 0 and 1 are X and Y, respectively.

For the vertical composite $u \otimes u' \colon X \nrightarrow Z$, an element of $(u \otimes u')(x, z)$ is an equivalence class $\mu \circ \lambda \colon x \dashrightarrow z$, for $\lambda \colon x \dashrightarrow y$ and $\mu \colon y \dashrightarrow z$; the equivalence relation is generated by

$$(\mu\eta) \circ \lambda \sim \mu \circ (\eta\lambda). \qquad (3.49)$$

Finally the cell α in (3.46) corresponds to a *functor over* **2** that restricts to f and g over 0 and 1

$$\alpha \colon X +_u Y \to X' +_v Y',$$
$$(\lambda \colon x \dashrightarrow y) \mapsto (\alpha(\lambda) \colon fx \dashrightarrow gy). \qquad (3.50)$$

The horizontal composition of cells is the composition of these functors, so that $\mathrm{Hor}_1 \mathbb{C}\mathrm{at} = \mathbf{Cat}/\mathbf{2}$. In a vertical composite $\alpha \otimes \beta \colon u \otimes u' \to v \otimes v'$ the equivalence class $\mu \circ \lambda \colon x \dashrightarrow z$ of $X +_{u \otimes u'} Z$ is sent to the equivalence class $\beta(\mu) \circ \alpha(\lambda) \colon \alpha(x) \dashrightarrow \beta(z)$.

Also here there is a weak vertical involution (see 3.5.1), produced by opposite categories

$$(3.51)$$

To simplify things we assume that $\mathbb{C}\mathrm{at}$ is unitary (by a constraint on the choices involved in vertical composition).

*It is also possible to replace the weak double category $\mathbb{C}\mathrm{at}$ with a *strict* one (up to special equivalence, see 3.5.5) by letting a profunctor $u \colon X \nrightarrow Y$ be defined as a functor $\hat{u}$ that preserves colimits

$$\hat{u} \colon \mathbf{Set}^X \to \mathbf{Set}^Y, \qquad \hat{u}(\Phi)(y) = \int^x u(x, y) \times \Phi(x), \qquad (3.52)$$

in the same way as a relation $r \colon X \nrightarrow Y$ of sets can be defined as a sup-preserving mapping $\hat{r} \colon 2^X \to 2^Y$ (see 2.1.4(e)). However we shall stay in the usual setting, where computing is simpler.*

3.4.4 *Enriched profunctors

Let a symmetric monoidal closed category **V** be given.

V-categories, **V**-functors and **V**-transformations have been reviewed in

2.4.1. Under suitable hypotheses on **V**, one can define a **V**-*profunctor* $u: X \rightarrow Y$ between **V**-categories, and form a bicategory with such morphisms, as shown in Street's paper [St4]. The prototype $\mathbb{C}$at considered above can thus be generalised, forming the weak double category **V**$\mathbb{C}$at of **V**-*categories*, **V**-*functors* and **V**-*profunctors*: leaving aside questions of size, $\mathbb{C}$at is obtained in this way from the cartesian closed category **V** = **Set**.

The general case is left to an interested reader. We work out directly some interesting cases, adapting what we have seen for the weak double category $\mathbb{C}$at. Again, all questions of size are left out.

3.4.5 Additive profunctors and bimodules

(a) The monoidal category **V** = **Ab** (with the usual tensor product) gives the weak double category **Ab**$\mathbb{C}$at of *preadditive categories, additive functors and additive profunctors* (see 2.4.2(e)).

A vertical arrow $u: X \rightarrow Y$ is now an additive functor $u: X^{\mathrm{op}} \times Y \to \mathbf{Ab}$. Equivalently, it can be presented as a preadditive category $U = X +_u Y$ in **Cat/2**, whose fibres over 0 and 1 are X and Y, respectively; we have thus new arrows $\lambda: x \dashrightarrow y$ (for x in X and y in Y) that form a preadditive category with the old ones.

The remaining points are described as in 3.4.3, replacing the cartesian closed category **Set** with the symmetric monoidal closed category **Ab**.

(b) Restricting **Ab**$\mathbb{C}$at to the totally full substructure of preadditive categories having one object we obtain the weak double category $\mathbb{R}$ng *of (unitary) rings, homomorphisms and bimodules*.

Now a vertical arrow $X: R \rightarrow S$ is an (R, S)-*bimodule*, i.e. a left-R and right-S module, with coherent multiplications by scalars

$$(rx)s = r(xs) \qquad \text{(for } r \in R, \ x \in X, \ s \in S).$$

Its composite with $Y: S \rightarrow T$ is the tensor product $X_S \otimes_S Y: R \rightarrow T$. This is the quotient of the abelian group $X \otimes Y$ that forces $xs \otimes y = x \otimes sy$ (for all x, y, s), with the obvious structure of (R, T)-bimodule: $r.(x \otimes y).t = rx \otimes yt$. The vertical identity e_R is R as a bimodule on itself.

A double cell $h: (X \begin{smallmatrix} f \\ g \end{smallmatrix} X')$ is an (R, S)-homomorphism $h: X \to X'$ (where the structure of X' comes from the ring-homomorphisms $f: R \to R'$ and $g: S \to S'$, which means that $r.x'.s = f(r).x'.g(s)$). Horizontal composition of cells is obvious and the vertical one is tensoring, namely $h \otimes k: X_S \otimes_S Y \to X'_S \otimes_S Y'$. The comparison isocells are obvious.

Interesting extensions of the weak double category $\mathbb{R}$ng, related to quantum field theory, are studied in [Pa6].

3.4.6 Preorder profunctors

The ordinal category $\mathbf{V} = \mathbf{2}$ (with cartesian product) gives the (strict) double category pOrd of *preordered sets, increasing functions and preorder profunctors*.

A vertical arrow $u\colon X \nrightarrow Y$ is an increasing mapping $u\colon X^{\mathrm{op}} \times Y \to \mathbf{2}$. It can be identified with the corresponding subset of $X \times Y$

$$\{(x,y) \in X \times Y \mid u(x,y) = 1\} \subset X \times Y, \tag{3.53}$$

as a relation $u\colon X \nrightarrow Y$ down-closed in X and up-closed in Y; this is also called an *order ideal* of $X^{\mathrm{op}} \times Y$. Preorder profunctors compose as relations.

A cell with boundary $(u\, {}_g^f\, v)$ exists if and only if $u \leqslant v(f^{\mathrm{op}} \times g)$, which means that $(x,y) \in u$ implies $(fx, gy) \in v$, and is equivalent to give a cell with the same boundary in RelSet.

There is thus a totally full embedding

$$D\colon \mathrm{RelSet} \to \mathrm{pOrd}, \tag{3.54}$$

that identifies sets with discrete preordered sets. Also pOrd is ultraflat (see 3.2.4(d)).

The profunctor $u\colon X \nrightarrow Y$ can also be viewed as a new preordered set, namely the collage $X +_u Y$, giving to the set $|X| + |Y|$ some preorder that coincides with the original preorders on X and Y, under the condition that no element of Y precedes an element of X. In this preorder, $x \prec y$ means that $u(x,y) = 1$.

3.4.7 Metric profunctors

The Lawvere strict monoidal category $\mathbb{R}_+ = [0, \infty]$, with arrows $\lambda \geqslant \mu$ and tensor product $\lambda + \mu$, has been recalled in 2.4.3. Extending the category $\mathbf{Mtr} = \mathbb{R}_+\mathbf{Cat}$ we have now the (strict) double category $\mathbb{M}\mathrm{tr} = \mathbb{R}_+\mathbb{C}\mathrm{at}$.

As we have already seen, an object is an L-metric space X with a distance d satisfying the axioms

$$d(x,y) \in [0, \infty],$$
$$d(x,x) = 0, \qquad d(x,y) + d(y,z) \geqslant d(x,z). \tag{3.55}$$

A horizontal arrow is a weak contraction $f\colon X \to X'$, satisfying $d(x,y) \geqslant d(fx, fy)$. A vertical arrow is a *metric profunctor* $u\colon X \nrightarrow Y$, represented by an $\mathbb{R}_+$-functor

$$u\colon X^{\mathrm{op}} \times Y \to \mathbb{R}_+,$$
$$u(x,y) + d(y,y') \geqslant u(x,y'), \qquad d(x,x') + u(x',y) \geqslant u(x,y). \tag{3.56}$$

Their composition is defined by a coend, strictly determined as a greatest lower bound in $[0, \infty]$

$$(u \otimes v)(x, z) = \inf_y (u(x, y) + v(y, z)). \tag{3.57}$$

Also here $u(x, y)$ can be viewed as the distance from a point of X to a point of Y, defining a new L-metric space, the collage $X +_u Y$ (with $d(y, x) = \infty$). Again $\mathbb{M}$tr is ultraflat, since a cell $u \to v.(f^{op} \times g)$ corresponds to the following inequalities

$$\begin{array}{ccc} X & \xrightarrow{\ f\ } & X' \\ u \downarrow & \geqslant & \downarrow v \\ Y & \xrightarrow[\ g\ ]{} & Y' \end{array} \qquad u(x, y) \geqslant v(fx, gy) \qquad (x \in X,\ y \in Y). \tag{3.58}$$

The embedding of strict monoidal categories

$$\mathbf{2} \to \mathbb{R}_+, \qquad 0 \mapsto \infty, \quad 1 \mapsto 0, \tag{3.59}$$

gives an embedding of double categories (that will be seen to be reflective and lax coreflective in 4.5.1)

$$M: \mathrm{p}\mathbb{O}\mathrm{rd} \to \mathbb{M}\mathrm{tr}, $$
$$d_{MX}(x, x') = 0 \quad (or\ \infty) \quad \Leftrightarrow \quad x \prec x' \quad (otherwise). \tag{3.60}$$

This embedding identifies $\mathrm{p}\mathbb{O}\mathrm{rd}$ with the double subcategory of $\mathbb{M}$tr consisting of those L-metric spaces whose distance takes values in $\{0, \infty\}$, their weak contractions and their profunctors with values in $\{0, \infty\}$

$$Mu: X \times Y \to \{0, \infty\}, \qquad Mu(x, y) = 0 \Leftrightarrow (x, y) \in u. \tag{3.61}$$

An L-metric space X can thus be viewed as a preordered set equipped with a finer information $d(x, x') \in [0, \infty]$, not reduced to the boolean values 0 (here meaning 'yes': $x \prec x'$) and ∞ (meaning 'no': one cannot go from x to x').

The embedding M is horizontally full. Taking into account the totally full embedding D of (3.54), we have a composed embedding

$$MD : \mathbb{R}\mathrm{el}\mathbf{Set} \to \mathrm{p}\mathbb{O}\mathrm{rd} \to \mathbb{M}\mathrm{tr}. \tag{3.62}$$

In this way $\mathbb{R}\mathrm{el}\mathbf{Set}$ is identified with a horizontally full double subcategory of $\mathbb{M}$tr, consisting of all *discrete* L-metric spaces (with $d(x, x') = 0$ if $x = x'$ and ∞ otherwise), together with their weak contractions and their profunctors with values in $\{0, \infty\}$.

3.4.8 Exercise

As a variation on the weak double category $\mathbb{R}$ng, in 3.4.5(b), one can study the restriction of $\mathbb{C}$at to the totally full substructure of categories having one object.

3.5 Lax functors and horizontal equivalence

We deal now with lax functors, their horizontal transformations and horizontal equivalences of weak double categories. (More general transformations will be seen in the next section.)

We end this section by proving a Strictification Theorem: every weak double category is 'specially equivalent' to a strict one.

3.5.1 Definition (Lax functors)

A *lax (double) functor* $F\colon \mathbb{X} \to \mathbb{A}$ between weak double categories amounts to assigning:

(a) two functors $F_n = \mathrm{Hor}_n F\colon \mathrm{Hor}_n\mathbb{X} \to \mathrm{Hor}_n\mathbb{A}$ for $n = 0,1$ (see 3.3.1), consistent with the vertical faces (domain and codomain)

$$
\begin{array}{ccc}
\mathrm{Hor}_1\mathbb{X} & \longrightarrow & \mathrm{Hor}_1\mathbb{A} \\
\partial^- \downarrow & & \downarrow \partial^- \\
\mathrm{Hor}_0\mathbb{X} & \longrightarrow & \mathrm{Hor}_0\mathbb{A}
\end{array}
\qquad
\begin{array}{ccc}
\mathrm{Hor}_1\mathbb{X} & \longrightarrow & \mathrm{Hor}_1\mathbb{A} \\
\partial^+ \downarrow & & \downarrow \partial^+ \\
\mathrm{Hor}_0\mathbb{X} & \longrightarrow & \mathrm{Hor}_0\mathbb{A}
\end{array}
$$

(b) for any object X in $\mathbb{X}$ a special cell, the *identity comparison*

$$
\underline{F}(X)\colon e_{FX} \to Fe_X\colon FX \nrightarrow FX,
$$

(c) for any vertical composite $u \otimes v\colon X \nrightarrow Y \nrightarrow Z$ in $\mathbb{X}$ a special cell, the *composition comparison*

$$
\underline{F}(u,v)\colon Fu \otimes Fv \to F(u \otimes v)\colon FX \nrightarrow FZ.
$$

The following axioms must be satisfied.

(ldf.1) (*Naturality of identity comparisons*) For a horizontal morphism $f\colon X \to X'$ in $\mathbb{X}$ we have

$$
e_{Ff}\,|\,\underline{F}X' = \underline{F}X\,|\,Fe_f,
$$

$$
\begin{array}{ccccc}
FX & \longrightarrow & FX' & = & FX' \\
e\downarrow & e_{Ff} & \downarrow e & \underline{F} & \downarrow Fe \\
FX & \longrightarrow & FX' & = & FX'
\end{array}
\quad = \quad
\begin{array}{ccccc}
FX & = & FX & \longrightarrow & FX' \\
e\downarrow & \underline{F} & \downarrow Fe & Fe_F & \downarrow Fe \\
FX & = & FX & \longrightarrow & FX'
\end{array}
$$

(ldf.2) (*Naturality of composition comparisons*) For a vertical composite of cells $\alpha \otimes \beta$ we have

$$(F\alpha \otimes F\beta) \,|\, \underline{F}(u', v') \;=\; \underline{F}(u, v) \,|\, F(\alpha \otimes \beta),$$

$$\begin{array}{ccccc} FX & \longrightarrow & FX' & = & FX' \\ {\scriptstyle Fu}\downarrow & F\alpha & \downarrow{\scriptstyle Fu'} & & \downarrow \\ FY & \longrightarrow & FY' & \underline{F} & \\ {\scriptstyle Fv}\downarrow & F\beta & \downarrow{\scriptstyle Fv'} & & \downarrow \\ FZ & \longrightarrow & FZ' & = & FZ' \end{array} \quad = \quad \begin{array}{ccccc} FX & = & FX & \longrightarrow & FX' \\ {\scriptstyle Fu}\downarrow & & \downarrow{\scriptstyle F(u\otimes v)} & & {\scriptstyle F(u'\otimes v')}\downarrow \\ FY & \underline{F} & & & \\ {\scriptstyle Fv}\downarrow & & \downarrow & F(\alpha \otimes \beta) & \downarrow \\ FZ & = & FZ & \longrightarrow & FZ' \end{array}$$

(ldf.3) (*Coherence with unitors*) For a vertical map $u \colon X \to\!\!\!\!\!\bullet\; Y$ the following diagrams of special cells are commutative:

$$\begin{array}{ccc} e(FX) \otimes Fu & \xrightarrow{\lambda(Fu)} & Fu \\ {\scriptstyle \underline{F}X\otimes 1}\downarrow & & \uparrow{\scriptstyle F(\lambda u)} \\ Fe_X \otimes Fu & \underset{\underline{F}(e,u)}{\longrightarrow} & F(e_X \otimes u) \end{array} \qquad \begin{array}{ccc} Fu \otimes e(FY) & \xrightarrow{\rho(Fu)} & Fu \\ {\scriptstyle 1\otimes \underline{F}Y}\downarrow & & \uparrow{\scriptstyle F(\rho u)} \\ Fu \otimes Fe_Y & \underset{\underline{F}(u,e)}{\longrightarrow} & F(x \otimes e_Y) \end{array}$$

(ldf.4) (*Coherence with associators*) For consecutive vertical arrows u, v, w in $\mathbb{X}$ the following diagram of special cells is commutative

$$\begin{array}{ccc} Fu \otimes (Fv \otimes Fw) & \xrightarrow{\kappa F} & (Fu \otimes Fv) \otimes Fw \\ {\scriptstyle 1\otimes \underline{F}}\downarrow & & \downarrow{\scriptstyle \underline{F}\otimes 1} \\ Fu \otimes F(v \otimes w) & & F(u \otimes v) \otimes Fw \\ {\scriptstyle \underline{F}}\downarrow & & \downarrow{\scriptstyle \underline{F}} \\ F(u \otimes (v \otimes w)) & \underset{F\kappa}{\longrightarrow} & F((u \otimes v) \otimes w) \end{array}$$

Lax double functors compose, in a categorical way: for $G \colon \mathbb{A} \to \mathbb{B}$ the composite GF has the following comparisons:

$$\begin{aligned} \underline{GF}(X) &= (\underline{G}(FX) \,|\, G\underline{F}(X)) \colon e_{GFX} \to GFe_X, \\ \underline{GF}(u, v) &= (\underline{G}(Fu, Fv) \,|\, G\underline{F}(u, v)) \colon GFu \otimes GFv \to GF(u \otimes v). \end{aligned} \tag{3.63}$$

A *colax* (double) functor $F \colon \mathbb{X} \to \mathbb{A}$ has comparison cells in the opposite direction

$$\underline{F}X \colon Fe_X \to e_{FX}, \qquad \underline{F}(u, v) \colon F(u \otimes v) \to Fu \otimes Fv, \tag{3.64}$$

under horizontally dual axioms (cdf.1–4).

A *pseudo (double) functor* is a lax functor whose comparisons are invertible; it is made colax by the inverse comparisons. More particularly, a *strict (double) functor* is a lax functor whose comparisons are identities; this amounts to strictly preserving the whole structure: faces, identities, compositions, unitors and associators.

For instance, if $\mathbf{C}$ is a category with pullbacks, the inclusion $\mathbb{P}\mathrm{map}(\mathbf{C}) \to \mathbb{S}\mathrm{pan}\mathbf{C}$ of the double category of partial maps in the weak double category of spans is a *pseudo* functor, with comparisons determined by the choice of pullbacks.

A pseudo functor with values in an ultraflat double category (see 3.2.4(d)) is necessarily strict.

As in the examples of Section 3.4, a *weak vertical involution* for the weak double category $\mathbb{A}$ is a pseudo functor $V \colon \mathbb{A}^{\mathrm{v}} \to \mathbb{A}$ with inverse V^{v}.

3.5.2 Unitary lax functors

The lax or colax functor F is said to be *unitary* if all its unit comparisons $\underline{F}(X)$ are identities. Then $F(e_X) = e_{FX}$, and axiom (ldf.3) proves that all comparisons $\underline{F}(e, u)$ and $\underline{F}(u, e)$ are identities. (The reader can verify that a pseudo unitary lax functor can always be made unitary.)

Unitarity of functors is essentially different from unitarity of weak double categories. The interest of the latter is 'practical' and a matter of taste: it simply allows us to simplify many points, at little cost. On the other hand, *the importance of unitarity for lax or colax functors is structural,* and will clearly appear when dealing with limits and their relationship with adjunctions (see 4.5.2(c), 5.1.1 and Section 5.5).

Basically, an object of $\mathbb{A}$ 'is' a strict functor $A \colon \mathbf{1} \to \mathbb{A}$ defined on the singleton double category; composing this functor with a *unitary* lax (or colax) functor $\mathbb{A} \to \mathbb{B}$ we get an object of $\mathbb{B}$, but composing it with a general lax functor we get a vertical monad in $\mathbb{B}$ (see Exercise 3.8.6(b)).

3.5.3 Full and faithful double functors

Let $F \colon \mathbb{X} \to \mathbb{A}$ be a lax or colax double functor. The main notions of fullness and faithfulness we are interested in are of a horizontal kind, related to the notions of equivalence that we shall see in 3.5.5 and Section 4.4. (Of course *in the transpositive case* the vertical analogues are of the same interest.)

(a) We say that F is *horizontally full* if both of the ordinary functors

$$F_0 \colon \mathrm{Hor}_0\mathbb{X} \to \mathrm{Hor}_0\mathbb{A}, \qquad F_1 \colon \mathrm{Hor}_1\mathbb{X} \to \mathrm{Hor}_1\mathbb{A}, \qquad (3.65)$$

are full.

If F is unitary it is easy to see that the second condition implies the first:

for a horizontal morphism $g \colon F(X) \to F(X')$ there is a cell $\alpha \colon e_X \to e_{X'}$ such that $F(\alpha) = e_g$ and both of its horizontal arrows $f_i \colon X \to X'$ satisfy $F(f_i) = g$.

(b) We say that F is *horizontally faithful* if both of the ordinary functors $\mathrm{Hor}_0 F$ and $\mathrm{Hor}_1 F$ are faithful. Again, if F is unitary the second condition implies the first.

(c) A weak double subcategory $\mathbb{X} \subset \mathbb{A}$ is horizontally full if the inclusion double functor is. $\mathbb{X}$ is then determined by any set of vertical arrows of $\mathbb{A}$ closed with respect to vertical identities (of domains and codomains) and vertical composition.

3.5.4 *Horizontal transformations*

A *horizontal transformation of lax double functors* $h \colon F \to G \colon \mathbb{X} \to \mathbb{A}$ can be defined as a lax double functor $h \colon \mathbb{X} \times \mathbf{2} \to \mathbb{A}$ with $F = h.(- \times 0)$ and $G = h.(- \times 1)$.

The transformation h amounts thus to giving two lax double functors F, G with additional data:

(a) a horizontal map $hX \colon FX \to GX$ in $\mathbb{A}$, for each object X in $\mathbb{X}$,

(b) a cell $hu \colon (Fu \, {}^{hX}_{hY} \, Gu)$ in $\mathbb{A}$, for each vertical map $u \colon X \nrightarrow Y$ in $\mathbb{X}$,

which must satisfy the following conditions.

(ht.1) (*Naturality*)

$$F\alpha \,|\, hv = hu \,|\, G\alpha, \quad \text{for a cell } \alpha \colon u \to v \text{ in } \mathbb{X}.$$

(ht.2) (*Coherence with vertical identities*)

$$\underline{F}X \,|\, h(e_X) = e_{hX} \,|\, \underline{G}X, \quad \text{for an object } X \text{ in } \mathbb{X}.$$

(ht.3) (*Coherence with vertical composition*)

$$\underline{F}(u,v) \,|\, hw = (hu \otimes hv) \,|\, \underline{G}(u,v), \quad \text{for } w = u \otimes v \text{ in } \mathbb{X}.$$

The naturality condition comprises its level of degree 0

$$hX'.Ff = Gf.hX, \quad \text{for } f \colon X \to X' \text{ in } \mathbb{X}. \tag{3.66}$$

These transformations have obvious vertical composition and whisker composition. First, for a horizontal transformation $k \colon G \to H \colon \mathbb{X} \to \mathbb{A}$, the transformation $kh \colon F \to H$ has components:

$$(kh)X = kX.hX \colon FX \to HX,$$
$$(kh)u = (hu \,|\, ku) \colon (Fu \, {}^{(kh)X}_{(kh)Y} \, Hu). \tag{3.67}$$

Secondly, for two lax functors $L: \mathbb{X}' \to \mathbb{X}$ and $M: \mathbb{A} \to \mathbb{A}'$, the transformation $MhL: MFL \to MGL$ has components:

$$(MhL)X' = M(h(LX')): MFLX' \to MGLX',$$
$$(MhL)u' = M(h(Lu')). \tag{3.68}$$

We have now the 2-category LxDbl of weak double categories, lax functors and horizontal transformations. Similarly we have the 2-categories CxDbl, PsDbl and StDbl for colax, pseudo and strict functors.

A horizontal transformation of lax double functors $h: F \to G$ is said to be:

- a *horizontal isomorphism of lax double functors* if it has an inverse h^{-1}: $G \to F$, which amounts to saying that all components hu (and hX) are horizontally invertible in $\mathbb{X}$ (then we write $F \cong G$),

- *special* if all cells $hu: Fu \to Gu$ are special, i.e. all morphisms hX are identities (equivalently, the natural transformation $\mathrm{Hor}_0(h)$ is an identity),

- a *special isomorphism of lax double functors* if both conditions above are satisfied, i.e. all cells $hu: Fu \to Gu$ are special isocells.

3.5.5 Horizontal equivalences of weak double categories

A *(horizontal) equivalence* of weak double categories will be a pseudo functor $F: \mathbb{X} \to \mathbb{A}$ that has a *quasi-inverse* $G: \mathbb{A} \to \mathbb{X}$, i.e. a pseudo functor with horizontal isomorphisms $h: GF \to \mathrm{id}\mathbb{X}$ and $k: FG \to \mathrm{id}\mathbb{A}$. (The adjoint case will be considered in 4.4.4.)

We speak of a *special equivalence* when both h and k can be chosen to be special isomorphisms, so that F and G give inverse isomorphisms between $\mathrm{Hor}_0\mathbb{X}$ and $\mathrm{Hor}_0\mathbb{A}$. In many examples of this kind, F and G even reduce to the identity on $\mathrm{Hor}_0(\mathbb{X}) = \mathrm{Hor}_0(\mathbb{A})$.

3.5.6 Relations of sets as profunctors and spans

The double category $\mathbb{R}\mathrm{el}\mathbf{Set}$ of sets, mappings and relations (see 3.1.2) can be embedded in various others.

(a) We have already seen, in 3.4.6 and 3.4.7, the canonical embeddings

$$\mathbb{R}\mathrm{el}\mathbf{Set} \to \mathrm{p}\mathbb{O}\mathrm{rd} \to \mathbb{M}\mathrm{tr}, \tag{3.69}$$

where the first is totally full, while the second is horizontally full.

(b) There is also a *lax* embedding S in the weak double category of spans, with a strict retraction

$$S: \mathbb{R}\mathrm{el}\mathbf{Set} \to \mathbb{S}\mathrm{pan}\mathbf{Set}, \quad R: \mathbb{S}\mathrm{pan}\mathbf{Set} \to \mathbb{R}\mathrm{el}\mathbf{Set} \quad (RS = 1). \tag{3.70}$$

Here R is the obvious double functor taking a span to the associated relation, while S is the lax double functor that takes a relation $u \subset X \times Y$ to the jointly monic span $Su = (X \leftarrow u \rightarrow Y)$ whose mappings are restrictions of cartesian projections; its comparison cell $(Su) \otimes (Sv) \rightarrow S(u \otimes v)$ goes from a span to the jointly monic span defining the same relation.

We shall see that the lax functor S is produced by a limit, the tabulator of a relation (in 3.7.1), and is right adjoint to R (in 4.5.3).

3.5.7 Exercises and complements (Spans as discrete profunctors)

(a) Prove that $\mathbb{S}\mathrm{pan}\mathbf{Set}$ is specially equivalent (in the sense of 3.5.5) to the totally full substructure of $\mathbb{C}\mathrm{at}$ determined by all discrete small categories.

(b) The discrete embedding $D: \mathbb{S}\mathrm{pan}\mathbf{Set} \rightarrow \mathbb{C}\mathrm{at}$ is a double functor. At level Hor_0, the functor $D_0: \mathbf{Set} \rightarrow \mathbf{Cat}$ has a left adjoint $\pi_0: \mathbf{Cat} \rightarrow \mathbf{Set}$ that associates to a category its set of connected components, and to a functor the induced mapping (Exercise 1.5.3(e)).

The theory of adjoints for double functors, developed in the next chapter, will give to the double functor D a *colax* left adjoint $\pi_0: \mathbb{C}\mathrm{at} \rightarrow \mathbb{S}\mathrm{pan}\mathbf{Set}$ (see Exercise 4.5.2(a)).

3.5.8 Strictification Theorem

Every weak double category $\mathbb{X}$ has an associated double category $\mathbb{A}$, specially equivalent to it (see 3.5.5), by means of pseudo double functors $F: \mathbb{X} \rightarrow \mathbb{A}$ and $G: \mathbb{A} \rightarrow \mathbb{X}$ with $GF = 1$, $FG \cong 1$ (horizontally isomorphic).

Every pseudo double functor $S: \mathbb{X} \rightarrow \mathbb{Y}$ can be similarly replaced by a strict double functor $S': \mathbb{A} \rightarrow \mathbb{B}$.

Note. Here a weak double category is not assumed to be unitary.

Proof Let us recall that the weak double category $\mathbb{X}$ has a vertical bicategory $\mathbf{Ver}\mathbb{X}$ (see 3.3.4) whose 2-cells $\alpha: u \rightarrow v$ are given by the special cells $\alpha: (u \,{}^1_1\, v)$ of $\mathbb{X}$; its unit and associativity comparisons λ, ρ, κ are those of $\mathbb{X}$. We shall repeatedly use the Coherence Theorem for bicategories (see 2.6.7).

First we replace $\mathbf{Ver}\mathbb{X}$ with the free category $\mathbf{V}$ on the graph of the old vertical arrows. A new vertical arrow $\underline{u} = (u_1, ..., u_n): X \rightarrowtail Y$ is thus a string of consecutive old vertical arrows

$$X = X_0 \rightarrowtail X_1 \rightarrowtail \ \cdots \ \rightarrowtail X_n = Y,$$

including an empty string $\underline{e}_X: X \rightarrowtail X$ for each object; their composition is by concatenation.

The free category $\mathbf{V}$ comes with an evaluation morphism of reflexive graphs $(-)\hat{\ }: \mathbf{V} \to \mathbf{Ver}\mathbb{X}$, that takes the string $\underline{u}$ to

$$\hat{u} = (...((u_1 \otimes u_2) \otimes u_3)...): X \rightarrowtail Y,$$

and $\underline{e}_X$ to the original weak unit e_X. It is actually a unitary pseudo functor of bicategories (defined on a category), with comparison

$$\psi(\underline{u}, \underline{v}): \hat{u} \otimes \hat{v} \to (\underline{u} \otimes \underline{v})\hat{\ }: X \rightarrowtail Z, \tag{3.71}$$

obtained by vertical composition of instances of κ (or of λ and ρ if $\underline{u}$ or $\underline{v}$ is a new vertical identity). The coherence theorem for the bicategory $\mathbf{Ver}\mathbb{X}$ says that we get the same result, no matter how this composition is done, and that (3.71) is indeed coherent with the associativity isocells κ of $\mathbf{Ver}\mathbb{X}$ (and the trivial ones for $\mathbf{V}$)

$$\begin{array}{ccc}
\hat{u} \otimes (\hat{v} \otimes \hat{w}) & \xrightarrow{\kappa} & (\hat{u} \otimes \hat{v}) \otimes \hat{w} \\
{\scriptstyle 1\otimes\psi}\downarrow & & \downarrow{\scriptstyle \psi\otimes 1} \\
\hat{u} \otimes (\underline{v} \otimes \underline{w})\hat{\ } & & (\underline{u} \otimes \underline{v})\hat{\ } \otimes \hat{w} \\
{\scriptstyle \psi}\downarrow & & \downarrow{\scriptstyle \psi} \\
(\underline{u} \otimes \underline{v} \otimes \underline{w})\hat{\ } & = & (\underline{u} \otimes \underline{v} \otimes \underline{w})\hat{\ }
\end{array} \tag{3.72}$$

Now we construct the double category $\mathbb{A}$. It coincides with $\mathbb{X}$ at the level Hor_0, i.e. for objects, horizontal arrows and their composition. The vertical arrows of $\mathbb{A}$ are the previous strings in $\mathbf{V}$. A new double cell $\alpha: \underline{u} \to \underline{v}$, as in the left diagram below, is represented by an old double cell $\hat{\alpha}: (\hat{u} \, {}_g^f \, \hat{v})$

$$\begin{array}{ccccccc}
X \xrightarrow{f} X' & \quad & X = X & \quad & X \xrightarrow{f} X' & \\
{\scriptstyle \underline{u}}\bullet\downarrow \quad \alpha \quad \downarrow\bullet{\scriptstyle \underline{v}} & & {\scriptstyle \underline{u}}\bullet\downarrow \quad 1 \quad \downarrow\bullet{\scriptstyle \underline{u}} & & {\scriptstyle \underline{e}}\bullet\downarrow \quad \underline{e}_f \quad \downarrow\bullet{\scriptstyle \underline{e}} & \\
Y \xrightarrow{g} Y' & & Y = Y & & X \xrightarrow{f} X' &
\end{array} \tag{3.73}$$

In particular the horizontal identity $1_{\underline{u}}$ of a new vertical arrow is represented by $1_{\hat{u}}$, and the vertical identity $\underline{e}_f$ of a horizontal arrow is represented by e_f.

Horizontal composition of double cells in $\mathbb{A}$ is like in $\mathbb{X}$, and forms a category. The vertical composition $\alpha \otimes \beta$ of new double cells is represented by the following old cell $(\alpha \otimes \beta)\hat{\ }$

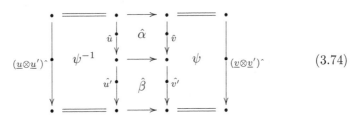

$$(3.74)$$

To prove that $\mathbb{A}$ is a double category, the main point is vertical associativity:

$$\alpha \otimes (\beta \otimes \gamma) = (\alpha \otimes \beta) \otimes \gamma.$$

These new cells are expressed in $\mathbb{X}$ as the composites of the following diagrams, where $u^{\sharp} = (\underline{u} \otimes \underline{u}' \otimes \underline{u}'')\hat{\ }$ and $v^{\sharp} = (\underline{v} \otimes \underline{v}' \otimes \underline{v}'')\hat{\ }$

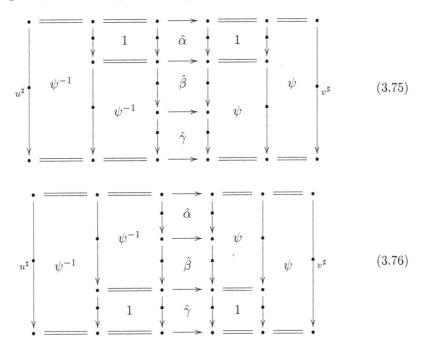

$$(3.75)$$

$$(3.76)$$

(It is understood that in diagram (3.75) we first compose $\hat{\beta} \otimes \hat{\gamma}$, while in diagram (3.76) we first compose $\hat{\alpha} \otimes \hat{\beta}$.)

These cells do coincide. Actually, one can form a three dimensional diagram inserting two associativity isocells of $\mathbb{X}$

$$\kappa \colon \hat{u} \otimes (\hat{u}' \otimes \hat{u}'') \to (\hat{u} \otimes \hat{u}') \otimes \hat{u}'', \qquad \kappa \colon \hat{v} \otimes (\hat{v}' \otimes \hat{v}'') \to (\hat{v} \otimes \hat{v}'') \otimes \hat{v}''.$$

In this way the central prism commutes (by definition of weak double category). The left-hand part of the new diagram, involving the first instance of κ and the left-hand part of the diagrams (3.75) and (3.76), commutes by (3.72); similarly for the right-hand part.

The interchange law in $\mathbb{A}$ is obvious once it is written out (a cell ψ cancels with a cell ψ^{-1}). The identity laws are left to the reader.

The unitary pseudo double functor $G\colon \mathbb{A} \to \mathbb{X}$ has already been constructed:

$$G(\underline{u}) = \hat{u}, \qquad G(\alpha) = \hat{\alpha}, \qquad \underline{G}(\underline{u}, \underline{v}) = \psi(\underline{u}, \underline{v})\colon \hat{u} \otimes \hat{v} \to (\underline{u} \otimes \underline{v})\hat{\ }.$$

The embedding $F\colon \mathbb{X} \to \mathbb{A}$ is obvious: a vertical arrow is sent to the corresponding string of length 1. It preserves vertical identities and composition up to special new isocells $\underline{F}(X)\colon \underline{e}_X \to e_X$ (represented by the double identity $\square_X$) and $\underline{F}(u, v)\colon Fu \otimes Fv \to F(u \otimes v)$ (represented by the horizontal identity of the old composite $u \otimes v$).

It follows that $GF = 1$, while FG is horizontally isomorphic to the identity on $\mathbb{A}$, by special isocells $h\underline{u}\colon (FG(\underline{u}) \, {}^1_1 \, \underline{u})$ represented by the horizontal identity of $\hat{u}$, for each string $\underline{u}$.

Finally, given a pseudo double functor $S\colon \mathbb{X} \to \mathbb{Y}$, with special isocells $\underline{S}X\colon e_{SX} \to S(e_X)$ and $\underline{S}(u, v)\colon Su \otimes Sv \to S(u \otimes v)$, the strictified double functor $S'\colon \mathbb{A} \to \mathbb{B}$ has

$$S'(u_1, ..., u_n) = (Su_1, ..., Su_n),$$

$$S'(\underline{e}_X) = \underline{e}_{SX}, \qquad S'(\alpha)\colon (S'\underline{u} \, {}^{Sf}_{Sg} \, S'\underline{v}),$$

where the cell $S'(\alpha)$ is represented by a modified version of $S(\hat{\alpha})$

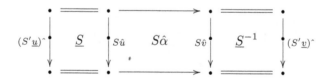

with *generalised* isocells $\underline{S}$ (well defined, by the coherence theorem of bicategories). Let us note that this procedure needs invertible comparisons $\underline{S}$, and cannot be applied to lax or colax functors. $\square$

3.6 Tabulators and span representations

For a vertical arrow u in a weak double category $\mathbb{A}$, the *tabulator* $\top u$, or *cell representer* of u, is a basic limit which will play a relevant role in Chapter 5. The dual notion is called *cotabulator*, or *cell corepresenter* $\bot u$.

Here we begin by introducing their 1-dimensional universal property.

Then we study how to represent weak double categories by spans (or co-spans), when all 1-dimensional tabulators (or cotabulators) exist. (This topic is based on the paper [GP11].)

Tabulators and cotabulators often describe important, well-known con-structions and clarify the structure we are examining – as we have already seen while introducing the weak double category $\mathbb{C}$at of categories, func-tors and profunctors, in 3.4.3. In a 2-category, viewed as a horizontal double category, we get the basic cotensors and tensors $\top e_X = \mathbf{2} \pitchfork X$, $\bot e_X = \mathbf{2} \otimes X$ (see Exercise 3.6.7(b)).

The 1-dimensional tabulator in a (strict) double category $\mathbb{A}$ was intro-duced in [BaE], p. 260, as an 'A-wise limit', and called a 'representation'.

3.6.1 Tabulators

Let $u \colon X \to Y$ be a vertical arrow in the weak double category $\mathbb{A}$. The 1-*dimensional tabulator*, or *cell representer* of u, will be an object $T = \top u$ equipped with a cell $\pi \colon e_T \to u$ (written as π_u when useful)

$$
\begin{array}{ccc}
T & \xrightarrow{\ p\ } & X \\
{\scriptstyle e_T}\downarrow & \pi & \downarrow{\scriptstyle u} \\
T & \xrightarrow{\ q\ } & Y
\end{array}
\qquad (3.77)
$$

such that the pair $(T, \pi \colon e_T \to u)$ is a universal arrow from the functor $e \colon \mathrm{Hor}_0\mathbb{A} \to \mathrm{Hor}_1\mathbb{A}$ to the object u of $\mathrm{Hor}_1\mathbb{A}$. Explicitly, this means that

(tab.1) for every object A and every cell $\varphi \colon e_A \to u$ there is a unique horizontal map $h \colon A \to T$ such that $\varphi = e_h \mid \pi$

$$
\begin{array}{ccc}
A & \xrightarrow{\ f\ } & X \\
{\scriptstyle e}\downarrow & \varphi & \downarrow{\scriptstyle u} \\
A & \xrightarrow{\ g\ } & Y
\end{array}
\quad = \quad
\begin{array}{ccccc}
A & \xrightarrow{\ h\ } & T & \xrightarrow{\ p\ } & X \\
{\scriptstyle e}\downarrow & e_h & {\scriptstyle e}\downarrow & \pi & \downarrow{\scriptstyle u} \\
A & \xrightarrow{\ h\ } & T & \xrightarrow{\ q\ } & Y
\end{array}
\qquad (3.78)
$$

We say that $\mathbb{A}$ *has* 1-*dimensional tabulators* if all these exist, which means that the ordinary functor $e \colon \mathrm{Hor}_0\mathbb{A} \to \mathrm{Hor}_1\mathbb{A}$ has a right adjoint

$$
\top \colon \mathrm{Hor}_1\mathbb{A} \to \mathrm{Hor}_0\mathbb{A}, \qquad e \dashv \top. \qquad (3.79)
$$

If this is the case, the functor e preserves all colimits. Moreover, we shall try to represent $\mathbb{A}$ *as a weak double category of spans*, over the category $\mathrm{Hor}_0(\mathbb{A})$.

Dually we define the 1-*dimensional cotabulator*, or *cell corepresenter* of

u, as an object $C = \perp u$ equipped with a universal cell $\iota\colon u \to e_C$

$$
\begin{array}{ccc}
X & \xrightarrow{\ i\ } & C \\
u \downarrow & \iota & \downarrow e_C \\
Y & \xrightarrow[\ j\ ]{} & C
\end{array}
\qquad (3.80)
$$

$\mathbb{A}$ *has* 1-*dimensional cotabulators* if and only if the functor $e\colon \mathrm{Hor}_0\mathbb{A} \to \mathrm{Hor}_1\mathbb{A}$ has a left adjoint; then e preserves limits and one can try to represent $\mathbb{A}$ as a weak double category of cospans.

The 2-dimensional property of tabulators and cotabulators will be considered in Section 5.3. In this chapter the terms tabulator and cotabulator always refer to the 1-dimensional aspect.

3.6.2 Exercises and complements

Computing tabulators is mostly deferred to the next section, but the following elementary facts can already give some concrete intuition about them.

(a) For a relation $u\colon X \to Y$ in $\mathbb{R}\mathrm{el}\mathbf{Set}$, prove that the tabulator $\top u \subset X{\times}Y$ is the graph or 'tabulation' of the relation u, whereas the cotabulator $\perp u$ is a quotient of $X + Y$, namely the pushout of any span representing the relation.

(b) Extend these facts to $\mathbb{R}\mathrm{el}\mathbf{Ab}$ and $\mathbb{R}\mathrm{el}\mathbf{Gp}$.

3.6.3 Diagonal morphisms

We suppose that the weak double category $\mathbb{A}$ has all tabulators. The following 'diagonal maps' will be used for the comparisons of the span representation, in Theorem 3.6.4, and also in the proof of the Construction Theorem of double limits.

(The present computations are clearer without assuming that $\mathbb{A}$ is unitary.)

(a) The tabulator $\top X = \top e_X$ *of the object* X has a canonical *diagonal morphism* dX, provided by the identity cell of X

$$
dX\colon X \to \top X, \qquad dX \,|\, \pi_X = \square_X. \qquad (3.81)
$$

(b) For a vertical composite $w = u \otimes v\colon X \to Y \to Z$, suppose that there is in $\mathrm{Hor}_0\mathbb{A}$ the pullback $P = \top(u,v) = \top u \times_Y \top v$, represented in the parallelogram below. Then the *diagonal morphism* d_{uv} is defined as follows, by the universal property of $\mathsf{T}w$ (using the cell $\sigma = (\lambda e)^{-1} = (\rho e)^{-1}$)

$$d_{uv} \colon \mathsf{T}(u, v) \to \mathsf{T}w, \qquad d_{uv} \,|\, \pi_w \;=\; \sigma \left| \begin{matrix} p_{uv} \,|\, \pi_u \\ q_{uv} \,|\, \pi_e \end{matrix} \right., \tag{3.82}$$

The theory of double limits, in Chapter 5, will show that $\mathsf{T}(u, v)$ is the limit in $\mathbb{A}$ of the diagram formed by the consecutive vertical arrows u, v. Note also that one cannot apply interchange to $(p_{uv} \,|\, \pi_u) \otimes (q_{uv} \,|\, \pi_v)$.

3.6.4 Theorem and Definition (Span representation)

Let $\mathbb{A}$ be a weak double category. We suppose that:

(a) $\mathbb{A}$ has tabulators,

(b) the ordinary category $\mathbf{C} = \mathrm{Hor}_0(\mathbb{A})$ of objects and horizontal arrows has pullbacks.

Then there is a canonical lax functor, which is trivial in degree zero

$$S \colon \mathbb{A} \to \mathbb{S}\mathrm{pan}(\mathbf{C}), \qquad \mathrm{Hor}_0(S) = \mathrm{id}\mathbf{C}, \tag{3.83}$$

and takes a vertical arrow $u \colon X \nrightarrow Y$ of $\mathbb{A}$ to the span

$$Su = (p, q) \colon X \nrightarrow Y$$

determined by the tabulator $\mathsf{T}u$, with its projections $p \colon \mathsf{T}u \to X$ and $q \colon \mathsf{T}u \to Y$.

The lax functor S will be called the span representation *of $\mathbb{A}$.*

Note. Related results can be found in the paper [Ni].

Proof The action of S on a cell $\alpha \colon (u \,{}^{f}_{g}\, v)$ of $\mathbb{A}$ is described by the following diagram

$$
\begin{array}{ccc}
X & \xrightarrow{f} & X' \\
u \downarrow & \alpha & \downarrow v \\
Y & \xrightarrow{g} & Y'
\end{array}
\qquad
\begin{array}{c}
\mathsf{T}u - \mathsf{T}\alpha \!\!> \mathsf{T}v
\end{array}
\tag{3.84}
$$

where the cell $S\alpha$: $(Su \, {}^{f}_{g} \, Sv)$ is a morphism of spans whose central component is the morphism $\mathsf{T}\alpha$: $\mathsf{T}u \to \mathsf{T}v$ determined by the universal property of the universal cell π_v of the tabulator $\mathsf{T}v$

$$
\mathsf{T}\alpha \,|\, \pi_v = \pi_u \,|\, \alpha \qquad (p'.\mathsf{T}\alpha = fp, \; q'.\mathsf{T}\alpha = gq), \tag{3.85}
$$

$$
\begin{array}{ccc}
\mathsf{T}u \xrightarrow{\mathsf{T}\alpha} \mathsf{T}v \xrightarrow{p'} X' \\
e \downarrow \quad e \downarrow \quad \pi_v \quad \downarrow v \\
\mathsf{T}u \xrightarrow{\mathsf{T}\alpha} \mathsf{T}v \xrightarrow{q'} Y'
\end{array}
=
\begin{array}{ccc}
\mathsf{T}u \xrightarrow{p} X \xrightarrow{f} X' \\
e \downarrow \quad \pi_u \quad e \downarrow \quad \alpha \quad \downarrow v \\
\mathsf{T}u \xrightarrow{q} Y \xrightarrow{g} Y'
\end{array}
$$

To define the laxity comparisons, an object X of $\mathbb{A}$ gives a special cell $\underline{S}(X)$: $e_X \to Se_X$ whose central component is the diagonal morphism defined in (3.81)

$$
d_X : X \to \mathsf{T}X, \qquad d_X \,|\, \pi_X = \square_X. \tag{3.86}
$$

Similarly, for a vertical composite $w = u \otimes v$: $X \twoheadrightarrow Y \twoheadrightarrow Z$, the comparison $\underline{S}(u,v)$: $Su \otimes Sv \to Sw$ is a special cell whose central component is the diagonal morphism d_{uv} defined in (3.82)

$$
d_{uv} : \mathsf{T}u \times_Y \mathsf{T}v \to \mathsf{T}w,
$$
$$
d_{uv} \,|\, \pi w = \sigma \,\left|\, \dfrac{p_{uv} \,|\, \pi_u}{q_{uv} \,|\, \pi_e} \right. . \tag{3.87}
$$

As to the coherence conditions of 3.5.1, we write down the verification of axiom (ldf.3), for the right unitor. For a vertical map u: $X \twoheadrightarrow Y$ and $w = u \otimes e_Y$ we have to check the commutativity of the following diagram in $\mathbf{C}$ (where the pullback $\mathsf{T}u \times_Y Y$ is realised as $\mathsf{T}u$):

$$
\begin{array}{ccc}
\mathsf{T}u \times_Y Y & =\!=\!=\!= & \mathsf{T}u \\
{\scriptstyle (1,d_Y)} \downarrow & & \uparrow {\scriptstyle \mathsf{T}(\rho u)} \\
\mathsf{T}u \times_Y \mathsf{T}e_Y & \xrightarrow{d_{ue}} & \mathsf{T}w
\end{array}
\tag{3.88}
$$

Equivalently, by applying the (cancellable) universal cell π_u and the isocell $\rho = \rho(e_{Tu})$, we show that

$$(\rho \,|\, (1, d_Y) \,|\, d_{ue} \,|\, T(\rho u) \,|\, \pi_u) \;=\; \rho \,|\, \pi_u.$$

In fact, writing $r = p_{ue}$ and $s = q_{ue}$, we have

$$(\rho \,|\, (1, d_Y) \,|\, d_{ue} \,|\, T(\rho u) \,|\, \pi_u) = (\rho \,|\, (1, d_Y) \,|\, d_{ue} \,|\, \pi_w \,|\, \rho u)$$

$$= \left(\rho \,|\, (1, d_Y) \,|\, \sigma \,\Big|\, \frac{r \,|\, \pi_u}{s \,|\, \pi_e} \,\Big|\, \rho u \right) = \left(\frac{(1, d_Y) \,|\, r \,|\, \pi_u}{(1, d_Y) \,|\, s \,|\, \pi_e} \,\Big|\, \rho u \right) \qquad (3.89)$$

$$= \left(\frac{\Box Tu \,|\, \pi_u}{e_q \,|\, \Box_Y} \,\Big|\, \rho u \right) = \left(\frac{\pi_u}{e_q} \,\Big|\, \rho u \right) = (\rho(e_{Tu}) \,|\, \pi_u).$$

The main computations are represented below, from the fourth term to the sixth, with $P = Tu \times_Y Te_Y$ and $d = d_Y$

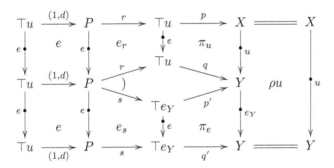

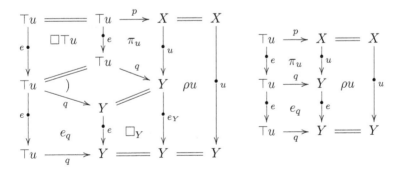

$\Box$

3.6.5 Span and cospan representability

Let $\mathbb{A}$ be a weak double category.

(a) We say that $\mathbb{A}$ is *span representable* if:

- it has tabulators,

- the ordinary category $\mathbf{C} = \mathrm{Hor}_0(\mathbb{A})$ has pullbacks,

- the span-representation lax functor $S\colon \mathbb{A} \to \mathrm{Span}(\mathbf{C})$ of (3.83) is horizontally faithful (see 3.5.3).

The last condition is trivially satisfied when $\mathbb{A}$ is flat.

This property will be studied in Sections 3.7 and 5.4, together with the computation of tabulators. We can remark as of now that, for every category $\mathbf{C}$ with pullbacks, the weak double category $\mathrm{Span}(\mathbf{C})$ is span representable, in a strict sense: the functor $S\colon \mathrm{Span}(\mathbf{C}) \to \mathrm{Span}(\mathbf{C})$ is an isomorphism, and even the identity for the natural choice of the tabulator of a span, namely its central object.

(a*) By horizontal duality, if $\mathbb{A}$ has cotabulators and $\mathbf{C} = \mathrm{Hor}_0(\mathbb{A})$ has pushouts we form a canonical colax functor, called *cospan representation* of $\mathbb{A}$

$$C\colon \mathbb{A} \to \mathbb{C}\mathrm{osp}(\mathbf{C}), \qquad \mathrm{Hor}_0(C) = \mathrm{id}\mathbf{C}, \qquad (3.90)$$

that takes a vertical arrow $u\colon X \to Y$ of $\mathbb{A}$ to the cospan $Cu = (i, j)\colon X \to Y$ determined by the tabulator $\bot u$ and its structural maps $i\colon X \to \bot u$, $j\colon Y \to \bot u$.

In these hypotheses, we say that $\mathbb{A}$ is *cospan representable* if the colax functor C is horizontally faithful. Now, for a category $\mathbf{C}$ with pushouts, $\mathbb{C}\mathrm{osp}(\mathbf{C})$ is 'strictly' cospan representable.

3.6.6 Cell representers in 2-categories

Let $\mathbf{C}$ be a 2-category. 'Weighted limits' in $\mathbf{C}$ will be briefly considered in Section 5.6, where they can be usefully related to double limits. However, we introduce here the basic weighted limit that cannot be obtained as a 'conical limit' (i.e. the limit of a 2-functor with values in $\mathbf{C}$).

In a 2-category $\mathbf{C}$ the *cell representer*, or *path object*, PX of an object X is an object equipped with a 2-cell $\pi\colon p \to q\colon PX \to X$, which is 2-universal for all 2-cells $A \Rightarrow X$. In other words, it satisfies the following 1-dimensional and 2-dimensional universal properties:

(i) for every cell $\alpha\colon A \Rightarrow X$ there exists a unique arrow $h\colon A \to PX$ such that $\alpha = \pi h$,

(ii) for every commutative square $\psi\alpha = \beta\varphi$ of cells $A \Rightarrow X$ (as in the left diagram below) there exists a unique cell $\lambda\colon A \Rightarrow PX$ such that:

$$\alpha = \pi h, \quad \beta = \pi k, \qquad \varphi = p\lambda, \quad \psi = q\lambda, \qquad (3.91)$$

$$f \xrightarrow{\varphi} f'$$

$$\alpha \downarrow \quad = \quad \downarrow \beta \qquad\qquad A \mathrel{\mathop{\rightrightarrows}^{h}_{k}} {\scriptstyle \lambda\downarrow} \; PX \mathrel{\mathop{\rightrightarrows}^{p}_{q}} {\scriptstyle \pi\downarrow} \; X \qquad\qquad (3.92)$$

$$g \xrightarrow[\psi]{} g'$$

In the theory of weighted limits the cell represener PX is usually written as $\mathbf{2} \pitchfork X$ and called the *cotensor* of X by the category $\mathbf{2}$ (according to a more general notion of cotensoring by a small category, deferred to 5.6.2(b)).

The dual notion gives the *cell corepresenter*, or *cylinder object*, more usually called the *tensor* $\mathbf{2} \otimes X$ of the object X by the category $\mathbf{2}$. It comes equipped with a 2-universal cell $X \Rightarrow \mathbf{2} \otimes X$.

3.6.7 Exercises and complements

(a) Prove that, in the 2-category **Cat**, the cell representer of the small category X is the category of morphisms $PX = X^{\mathbf{2}}$, while the cell corepresenter is the cartesian product $IX = \mathbf{2} \times X$. (In an obvious comparison with homotopy theory, the category $\mathbf{2}$ is playing the role of the standard topological interval.)

(b) Let **A** be a 2-category, viewed as a horizontal double category. Prove that the tabulator $\top X = \top e_X$ is the cell representer, or cotensor product $\mathbf{2} \pitchfork X$ defined in 3.6.6. As of now, the reader will only consider the 1-dimensional universal property of the latter, while the 2-dimensional one is deferred to Exercise 5.6.2(c). Dually, the cotabulator $\bot X$ amounts to the tensor product $\mathbf{2} \otimes X$.

*(c) Define the 2-limit of a 2-functor $\mathbf{I} \to \mathbf{C}$, by a 2-universal property. Prove that, if **C** has all cell *corepresenters*, then the ordinary limit of any 2-functor with values in **C** is automatically a 2-limit.

3.7 Computing tabulators and span representations

Tabulators and cotabulators are meant in the 1-dimensional sense, as above. Most of the strict or weak double categories we are considering have them, and are representable by spans or cospans.

On the other hand, the weak double category $\mathbb{R}\mathrm{ng}$ of rings, homomorphisms and bimodules lacks tabulators – a fact related to an anomalous behaviour of colimits, as we shall see in 5.4.8.

After computing some cases, many others are presented as exercises, with solution or hints in Appendix C.

3.7.1 Metric spaces, posets and relations

The double categories

$$\mathbb{R}\mathrm{el}\mathbf{Set} \subset \mathrm{p}\mathbb{O}\mathrm{rd} \subset \mathbb{M}\mathrm{tr},$$

are described in 3.1.2, 3.4.6 and 3.4.7. In these embeddings a set is viewed as a discrete ordered set, while a preordered set X is given a 'distance' with two possible values, 0 or ∞, and $d(x, x') = 0$ means that $x \prec x'$.

We compute now their tabulators and cotabulators, showing that they exist and how they are related. Since these double categories are flat, they are automatically span and cospan representable.

(a) Beginning from $\mathbb{M}\mathrm{tr}$, which contains the other structures, the tabulator of a metric profunctor $u \colon X \nrightarrow Y$ is a sort of 'graph' of u in $X \times Y$ (viewing $u(x, y)$ as a distance, as in 3.4.7)

$$\top u = \{(x, y) \in X \times Y \mid u(x, y) = 0\},$$
$$d((x, y), (x', y')) = d(x, x') \vee d(y, y'),$$
$$(3.93)$$

$$
\begin{array}{ccc}
\top u \xrightarrow{\;p\;} X & \qquad & Z \xrightarrow{\;f\;} X \\
{\scriptstyle e}\downarrow \quad {\geqslant} \quad \downarrow{\scriptstyle u} & & {\scriptstyle e}\downarrow \quad {\geqslant} \quad \downarrow{\scriptstyle u} \\
\top u \xrightarrow[q]{} Y & & Z \xrightarrow[g]{} Y
\end{array}
$$

In fact for every cell $(f, g) \colon e \to u$, as in the right diagram above, we have $u(fz, gz) \leqslant d(z, z) = 0$ and a mapping $Z \to \top u$ of components (f, g).

The cotabulator $\bot u$ is the collage $X +_u Y$ (see 3.4.7), with $d(x, y) = u(x, y)$ and $d(y, x) = \infty$ for $x \in X$, $y \in Y$.

(b) Recalling the description of a preorder-profunctor $u \colon X^{\mathrm{op}} \times Y \to \mathbf{2}$ as a preordered set $X +_u Y$, in 3.4.6, we have

$$\top u = \{(x, y) \mid x \prec y \text{ in } X +_u Y\}, \qquad \bot u = X +_u Y, \qquad (3.94)$$

where $\top u$ has the preorder induced by $X \times Y$. $\mathrm{p}\mathbb{O}\mathrm{rd}$ is closed in $\mathbb{M}\mathrm{tr}$ under both constructions, with respect to the embedding $M \colon \mathrm{p}\mathbb{O}\mathrm{rd} \to \mathbb{M}\mathrm{tr}$ (in (3.60)).

(c) We have already seen that in $\mathbb{R}\mathrm{el}\mathbf{Set}$ the tabulator $\top u \subset X \times Y$ is the graph of the relation u, whereas the cotabulator $\bot u$ is a quotient of the sum $X + Y$, namely the pushout of any span representing the relation (see 3.6.2).

$\mathbb{R}\mathrm{el}\mathbf{Set}$ is closed in $\mathrm{p}\mathbb{O}\mathrm{rd}$ (and therefore in $\mathbb{M}\mathrm{tr}$) under tabulators. On the other hand, the cotabulator of a relation $u \colon X \nrightarrow Y$ is the quotient of its cotabulator in $\mathbb{M}$tr (or in $p\mathbb{O}$rd) that identifies two elements x, y whenever $d(x, y) = 0$ (or $x \prec y$).

3.7.2 Exercises with adjoint pairs

(a) Prove that the double categories $\mathrm{Adj}\mathbf{Cat}$ and $\mathrm{Adj}\mathbf{Ord}$, defined in 3.1.6 and 3.1.7, are span representable.

(b) Prove that they are also cospan representable.

*(c) A reader interested in abelian categories and toposes can also consider the representability of the double categories $\mathrm{Adj}\mathbf{Ac}$ and $\mathrm{Adj}\mathbf{Tp}$ of 3.1.8(e).

3.7.3 Exercises with spans and cospans

(a) For the weak double category $\mathbb{S}\mathrm{pan}(\mathbf{C})$, defined in 3.4.1, consider tabulators, cotabulators and representability, completing what we have seen in 3.6.5.

(b) Prove that $\mathbb{S}\mathrm{pan}\mathbf{Set}$ is *not* cospan representable. How can one extend this fact?

(c) Prove that $\mathbb{C}\mathrm{osp}\mathbf{Set}$ is not span representable.

(d) For a preordered set X, examine the cospan representability of $\mathbb{S}\mathrm{pan}(X)$.

3.7.4 Exercises with profunctors

Study the weak double category $\mathbb{C}\mathrm{at}$ of profunctors, keeping in mind the results of 3.7.1. It may be convenient to prove things in the following order.

(a) $\mathbb{C}\mathrm{at}$ has cotabulators.

(b) $\mathbb{C}\mathrm{at}$ has tabulators.

(c) $\mathbb{C}\mathrm{at}$ is span and cospan representable. (The cospan representation $\mathbb{C}\mathrm{at} \to \mathbb{C}\mathrm{osp}\mathbf{Cat}$ will be further analysed in 4.5.5 and 8.6.7, showing that $\mathbb{C}\mathrm{at}$ is comonadic over $\mathbb{C}\mathrm{osp}\mathbf{Cat}$.)

3.7.5 Exercises on extending functors

The following exercises deal with extending ordinary functors to lax or colax double functors on double categories of spans, or cospans, or relations.

(a) Show that the construction $\mathbb{S}\mathrm{pan}\mathbf{C}$ can be completed to a 2-functor

$$\mathbb{S}\mathrm{pan}\colon \mathbf{Cat}_{\mathrm{pb}} \to \mathrm{CxDbl}, \qquad \mathbf{C} \mapsto \mathbb{S}\mathrm{pan}\mathbf{C}, \qquad (3.95)$$

defined on the full sub-2-category of $\mathbf{Cat}$ containing all categories with (a fixed choice of) pullbacks, with values in the 2-category of weak double categories, *colax* functors and their horizontal transformations (see 3.5.4).

(b) Dually we have a 2-functor

$$\text{Cosp}\colon \mathbf{Cat}_{\text{po}} \to \text{LxDbl}, \qquad \mathbf{C} \mapsto \text{Cosp}\mathbf{C} = \text{Span}(\mathbf{C}^{\text{op}})^{\text{h}}, \qquad (3.96)$$

defined on the full sub-2-category of **Cat** containing all categories with (a choice of) pushouts, with values in the 2-category of weak double categories, lax functors and their horizontal transformations.

(c) Let **A** be a regular category. If **X** has pullbacks, every functor $F\colon \mathbf{X} \to \mathbf{A}$ has a unitary *colax* extension (written as F for simplicity)

$$F\colon \text{Span}\mathbf{X} \to \text{Rel}\mathbf{A}, \qquad F(u', u'') = (Fu')^\sharp \otimes Fu''. \qquad (3.97)$$

When is this extension a strict double functor?

*(d) If **X** has pushouts, every functor $F\colon \mathbf{X} \to \mathbf{Ab}$ has a unitary *lax* extension

$$F\colon \text{Cosp}\mathbf{X} \to \text{Rel}\mathbf{Ab}, \qquad F(u', u'') = Fu' \otimes (Fu'')^\sharp. \qquad (3.98)$$

3.8 Transformations and modifications

We end this chapter with an extension of previous notions, to be used in Chapter 5 for our study of double limits.

We fix two (unitary) weak double categories $\mathbb{I}$ and $\mathbb{A}$, of which the former is small. We define a weak double category $\text{Lx}_{\text{ps}}(\mathbb{I}, \mathbb{A})$ of lax functors $\mathbb{I} \to \mathbb{A}$, pseudo horizontal transformations, pseudo vertical transformations and modifications. *Strict* horizontal transformations and *pseudo* vertical transformations form a weak double subcategory $\text{Lx}(\mathbb{I}, \mathbb{A})$.

3.8.1 Definition (Pseudo horizontal transformations)

Generalising 3.5.4, a *pseudo horizontal transformation* $h\colon F \to G\colon \mathbb{I} \to \mathbb{A}$ of lax functors has the following components:

(a) for every object i in $\mathbb{I}$, a horizontal map $hi\colon Fi \to Gi$ in $\mathbb{A}$,

(b) for every vertical map $u\colon i \rightarrowtail j$ in $\mathbb{I}$, a cell $hu\colon (Fu \, {}^{hi}_{hj} \, Gu)$ in $\mathbb{A}$,

(c) for every horizontal arrow $a\colon i \to i'$ in $\mathbb{I}$, a globular isocell ha in $\mathbb{A}$, the *naturality comparison*

$$
\begin{array}{ccccc}
Fi & \xrightarrow{\;hi\;} & Gi & \xrightarrow{\;Ga\;} & Gi' \\
{\scriptstyle e}\big\downarrow & & ha & & \big\downarrow{\scriptstyle e} \\
Fi & \xrightarrow[\;Fa\;]{} & Fi' & \xrightarrow[\;hi'\;]{} & Gi'
\end{array}
\qquad (3.99)
$$

The following five axioms must be satisfied.

(pht.1) (*Naturality on a cell*) For a cell $\alpha\colon (u\ {}_b^a\ v)\colon \left(\begin{smallmatrix} i & i' \\ j & j' \end{smallmatrix}\right)$ in $\mathbb{I}$

$$
\begin{array}{ccccc}
Fi & \xrightarrow{hi} & Gi & \xrightarrow{Ga} & Gi' \\
{\scriptstyle Fu}\downarrow & {\scriptstyle hu} & \downarrow{\scriptstyle Gu}\ \ {\scriptstyle G\alpha} & & \downarrow{\scriptstyle Gv} \\
Fj & \xrightarrow{-\,hj\,\succ} & Gj & -\,Gb\,\succ & Gj' \\
{\scriptstyle e}\uparrow & {\scriptstyle hb} & & & \downarrow{\scriptstyle e} \\
Fj & \xrightarrow{Fb} & Fj' & \xrightarrow{hj'} & Gj'
\end{array}
\qquad = \qquad
\begin{array}{ccccc}
Fi & \xrightarrow{hi} & Gi & \xrightarrow{Ga} & Gi' \\
{\scriptstyle e}\uparrow & & {\scriptstyle ha} & & \downarrow{\scriptstyle e} \\
Fi & \xrightarrow{-\,Fa\,\succ} & Fi' & -\,hi'\,\succ & Gj' \\
{\scriptstyle Fu}\uparrow & {\scriptstyle F\alpha} & \uparrow{\scriptstyle Fv}\ \ {\scriptstyle hv} & & \downarrow{\scriptstyle Gv} \\
Fj & \xrightarrow{Fb} & Fj' & \xrightarrow{hj'} & Gj'
\end{array}
$$

(pht.2) (*Coherence with vertical identities*) For i in $\mathbb{I}$

$$
\begin{array}{ccccc}
Fi & \xrightarrow{1} & Fi & \xrightarrow{hi} & Gi \\
{\scriptstyle e}\uparrow & {\scriptstyle \underline{F}i} & \uparrow{\scriptstyle Fe}\ \ {\scriptstyle he} & & \uparrow{\scriptstyle Ge} \\
Fi & \xrightarrow{1} & Fi & \xrightarrow{hi} & Gi
\end{array}
\quad = \quad
\begin{array}{ccccc}
Fi & \xrightarrow{hi} & Gi & \xrightarrow{1} & Gi \\
{\scriptstyle e}\uparrow & {\scriptstyle e_{hi}} & \downarrow{\scriptstyle e} & {\scriptstyle \underline{G}i} & \uparrow{\scriptstyle Ge} \\
Fi & \xrightarrow{hi} & Gi & \xrightarrow{1} & Gi
\end{array}
$$

(pht.3) (*Coherence with vertical composition*) For a vertical composite $w = u \otimes v\colon i \twoheadrightarrow j \twoheadrightarrow m$ in $\mathbb{I}$

$$
\begin{array}{ccccc}
Fi & \xrightarrow{\ 1\ } & Fi & \xrightarrow{hi} & Gi \\
{\scriptstyle Fu}\uparrow & & \uparrow{\scriptstyle Fw} & & \downarrow \\
Fj & {\scriptstyle F(u,v)} & \bullet & {\scriptstyle hw} & \downarrow{\scriptstyle Gw} \\
{\scriptstyle Fv}\uparrow & & \downarrow & & \downarrow \\
Fm & \xrightarrow{\ 1\ } & Fm & \xrightarrow{hm} & Gm
\end{array}
\quad = \quad
\begin{array}{ccccc}
Fi & \xrightarrow{hi} & Gi & \xrightarrow{\ 1\ } & Gi \\
{\scriptstyle Fu}\uparrow & {\scriptstyle hu} & \uparrow{\scriptstyle Gu} & & \downarrow \\
Fj & -\,hj\,\succ & Gj & {\scriptstyle \underline{G}(u,v)} & \downarrow{\scriptstyle Gw} \\
{\scriptstyle Fv}\uparrow & {\scriptstyle hv} & \downarrow{\scriptstyle Gv} & & \downarrow \\
Fm & \xrightarrow{hm} & Gm & \xrightarrow{\ 1\ } & Gm
\end{array}
$$

(pht.4) (*Coherence with horizontal identities*) For i in $\mathbb{I}$

$$
\begin{array}{ccccc}
Fi & \xrightarrow{hi} & Gi & \xrightarrow{1} & Gi \\
{\scriptstyle e}\uparrow & & {\scriptstyle h1_i} & & \downarrow{\scriptstyle e} \\
Fi & \xrightarrow{1} & Fi & \xrightarrow{hi} & Gi
\end{array}
\qquad = \qquad e_{hi}.
$$

(pht.5) (*Coherence with horizontal composition*) For a horizontal composite

$ba: i \to i' \to i''$

$$
\begin{array}{ccccccc}
Fi & \xrightarrow{hi} & Gi & \xrightarrow{Ga} & Gi' & \xrightarrow{Gb} & Gi'' \\
{\scriptstyle e}\uparrow & & {\scriptstyle ha} & & {\scriptstyle e}\downarrow & {\scriptstyle e} & {\scriptstyle e}\downarrow \\
Fi & \!-\,Fa\!\!\succ & Fi' & \!-\,hi'\!\!\succ & Gi' & \!-\,Gb\!\succ & Gi'' \\
{\scriptstyle e}\uparrow & {\scriptstyle e} & {\scriptstyle e}\downarrow & & {\scriptstyle hb} & & {\scriptstyle e}\downarrow \\
Fi & \xrightarrow{Fa} & Fi' & \xrightarrow{Fb} & Fi'' & \xrightarrow{hi''} & Gi''
\end{array}
\qquad = \qquad h(ba).
$$

A horizontal transformation h is a pseudo horizontal transformation where all the comparison cells hi are identities. Then these cells become the (redundant) naturality condition (3.66), while (pht.1–3) give the axioms of 3.5.4, and (pht.4, 5) are vacuous.

3.8.2 *Definition* (Pseudo vertical transformations)

Similarly, a *pseudo vertical transformation* $r: F \to G: \mathbb{I} \to \mathbb{A}$ of lax double functors consists of the following components:

(a) for every object i in $\mathbb{I}$, a vertical arrow $ri: Fi \to Gi$ in $\mathbb{A}$,

(b) for every $a: i \to i'$ in $\mathbb{I}$, a cell $ra: (ri \, {}^{Fa}_{Ga} \, ri')$ in $\mathbb{A}$,

(c) for every $u: i \to j$ in $\mathbb{I}$, a special isocell $ru: Fu \otimes rj \to ri \otimes Gu: Fi \to Gj$, the *naturality comparison*.

The following axioms must be satisfied.

(pvt.1) (*Naturality on a cell*) For a cell $\alpha: (u \, {}^a_b \, v)$ in $\mathbb{I}$

$$
\begin{array}{ccccc}
Fi & \xrightarrow{Fa} & Fi' & \xrightarrow{1} & Fi' \\
{\scriptstyle Fu}\downarrow & {\scriptstyle F\alpha} & {\scriptstyle Fv}\downarrow & {\scriptstyle ri'}\downarrow \\
Fj & \!-\,Fb\!\succ & Fj' & {\scriptstyle rv} & Gj' \\
{\scriptstyle rj}\downarrow & {\scriptstyle rb} & {\scriptstyle rj'}\downarrow & {\scriptstyle Gv}\downarrow \\
Gj & \xrightarrow{Gb} & Gj' & \xrightarrow{1} & Gj'
\end{array}
\quad = \quad
\begin{array}{ccccc}
Fi & \xrightarrow{1} & Fi & \xrightarrow{Fa} & Fi' \\
{\scriptstyle Fu}\downarrow & {\scriptstyle ri} & {\scriptstyle ra} & {\scriptstyle ri'}\downarrow \\
Fj & {\scriptstyle ru} & Gi & \!-\,Ga\!\succ & Gj' \\
{\scriptstyle rj}\downarrow & {\scriptstyle Gu}\downarrow & {\scriptstyle G\alpha} & {\scriptstyle Gv}\downarrow \\
Gj & \xrightarrow{1} & Gj & \xrightarrow{Gb} & Gj'
\end{array}
$$

(pvt.2) (*Coherence with horizontal identities*) For i in $\mathbb{I}$, $r(1_i) = 1_{ri}$.

(pvt.3) (*Coherence with horizontal composition*) For $c = ba: i \to i''$ in $\mathbb{I}$, $r(c) = (ra \mid rb)$.

(pvt.4) (*Coherence with vertical identities*) For i in $\mathbb{I}$

$$
\begin{array}{ccc}
Fi \xrightarrow{\ 1\ } Fi & & Fi \xrightarrow{\ 1\ } Fi \xrightarrow{\ 1\ } Fi \\
\scriptstyle ri \downarrow \quad \scriptstyle 1 \quad \downarrow \scriptstyle ri & & \scriptstyle e \downarrow \quad \underline{Fi} \quad \downarrow \scriptstyle Fe \qquad \downarrow \scriptstyle ri \\
Gi \xrightarrow{\ -1\ } Gi & = & Fi \xrightarrow{\ -1\ } Fi \quad {\scriptstyle re_i} \quad Gi \\
\scriptstyle e \downarrow \quad \underline{Gi} \quad \downarrow \scriptstyle Ge & & \scriptstyle ri \downarrow \quad \scriptstyle 1 \quad \downarrow \scriptstyle ri \qquad \downarrow \scriptstyle Ge \\
Gi \xrightarrow{\ 1\ } Gi & & Gi \xrightarrow{\ 1\ } Gi \xrightarrow{\ 1\ } Gi
\end{array}
$$

(pvt.5) (*Coherence with vertical composition*) For $w = u \otimes v \colon i \twoheadrightarrow j \twoheadrightarrow m$ in $\mathbb{I}$, the following pasting, where $\kappa' = \kappa^{-1}$

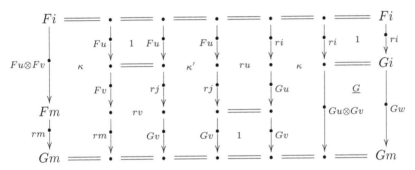

coincides with $(\underline{F}(u,v) \otimes 1_{rm}) \,|\, rw$.

For a consecutive pseudo vertical transformation $s \colon G \twoheadrightarrow H$, the vertical composition $t = r \otimes s$ has obvious components ti and tf, and special isocells tu obtained by pasting the ones of r and s by means of the associativity isocells κ of $\mathbb{A}$, as shown in the diagram below:

$$
ti = ri \otimes si \colon Fi \twoheadrightarrow Hi, \qquad tf = rf \otimes sf \colon (ti \,{}^{Ff}_{Hf}\, ti'),
$$
$$
tu \colon Fu \otimes tj \to ti \otimes Hu \colon Fi \twoheadrightarrow Hj,
$$

$$(3.100)$$

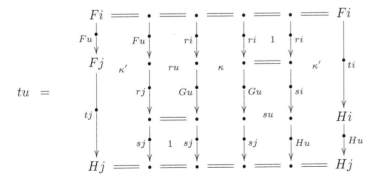

A *vertical transformation* $r: F \to G$ is a pseudo vertical transformation where all the special isocells $ru: Fu \otimes rj \to ri \otimes Gu$ are horizontal identities.

Let us note that, for a pseudo vertical transformation $r: F \to G$ between unitary lax functors, we have $re_i = 1_{ri}$, by (pvt.4), and we can forget about these comparisons (for all i in $\mathbb{I}$).

3.8.3 Definition (Modifications)

A *modification* $\mu: (r \,{}^h_k\, s): ({}^F_G \, {}^{F'}_{G'})$, where h, k are pseudo horizontal transformations and r, s are pseudo vertical transformations (of lax functors of weak double categories), has components μi in $\mathbb{A}$, for every i in $\mathbb{I}$

$$
\begin{array}{ccc}
Fi & \xrightarrow{hi} & F'i \\
{\scriptstyle ri}\downarrow & \mu i & \downarrow{\scriptstyle si} \\
Gi & \xrightarrow[ki]{} & G'i
\end{array}
\qquad \mu i: (ri \,{}^{hi}_{ki}\, si)
\qquad (3.101)
$$

They must satisfy two conditions.

(mod.1) (*Horizontal coherence*) For every $a: i \to i'$ we have

$$
\begin{array}{ccccc}
Fi & \xrightarrow{hi} & F'i & \xrightarrow{F'a} & F'i' \\
{\scriptstyle e}\uparrow & ha & & \downarrow{\scriptstyle e} \\
Fi & \xrightarrow{-Fa} & Fi' & \xrightarrow{-hi'} & F'j' \\
{\scriptstyle ri}\downarrow & ra & {\scriptstyle ri'}\downarrow & \mu i' & \downarrow{\scriptstyle si'} \\
Gi & \xrightarrow[Ga]{} & Gi' & \xrightarrow[ki']{} & G'i'
\end{array}
\quad = \quad
\begin{array}{ccccc}
Fi & \xrightarrow{hi} & F'i & \xrightarrow{F'a} & F'i' \\
{\scriptstyle ri}\downarrow & \mu i & \downarrow{\scriptstyle si} & sa & \downarrow{\scriptstyle si'} \\
Gi & \xrightarrow{-ki} & G'i & \xrightarrow{-G'a} & G'i' \\
{\scriptstyle e}\downarrow & & ka & & \downarrow{\scriptstyle e} \\
Gi & \xrightarrow[Ga]{} & Gi' & \xrightarrow[ki']{} & G'i'
\end{array}
$$

(mod.2) (*Vertical coherence*) For every $u: i \to j$ we have

$$
\begin{array}{ccccc}
Fi & \xrightarrow{1} & Fi & \xrightarrow{ha} & F'i \\
{\scriptstyle Fu}\downarrow & & {\scriptstyle ri}\downarrow & \mu i & \downarrow{\scriptstyle si} \\
Fj & ru & Gi & \xrightarrow{-ki} & G'i \\
{\scriptstyle rj}\downarrow & & {\scriptstyle Gu}\downarrow & ku & \downarrow{\scriptstyle G'u} \\
Gj & \xrightarrow[1]{} & Gj & \xrightarrow[kj]{} & G'j
\end{array}
\quad = \quad
\begin{array}{ccccc}
Fi & \xrightarrow{hi} & F'i & \xrightarrow{1} & F'i \\
{\scriptstyle Fu}\downarrow & hu & {\scriptstyle F'u}\downarrow & & \downarrow{\scriptstyle si} \\
Fj & \xrightarrow{-hj} & F'j & su & G'i \\
{\scriptstyle rj}\downarrow & \mu j & {\scriptstyle sj}\downarrow & & \downarrow{\scriptstyle G'u} \\
Gj & \xrightarrow[kj]{} & G'j & \xrightarrow[1]{} & G'j
\end{array}
$$

Modifications compose horizontally and vertically, by the compositions of cells in $\mathbb{A}$.

3.8.4 Theorem

For weak double categories $\mathbb{I}$ and $\mathbb{A}$ (the former is small) we have a weak double category $\mathbb{L}x_{ps}(\mathbb{I}, \mathbb{A})$: its objects are the lax functors $\mathbb{I} \to \mathbb{A}$, its arrows are pseudo horizontal transformations and pseudo vertical transformations, its double cells are modifications.

We also have a weak double subcategory $\mathbb{L}x(\mathbb{I}, \mathbb{A})$ by restricting the horizontal arrows to the strict horizontal transformations, but keeping all the vertical arrows and all modifications whose horizontal arrows are strict.

The associated 2-categories

$$\mathrm{Lx}_{ps}(\mathbb{I}, \mathbb{A}) = \mathbf{Hor}\, \mathbb{L}x_{ps}(\mathbb{I}, \mathbb{A}),$$
$$\mathrm{Lx}(\mathbb{I}, \mathbb{A}) = \mathbf{Hor}\, \mathbb{L}x(\mathbb{I}, \mathbb{A}),$$

(3.102)

have trivial vertical arrows and globular modifications $\varphi \colon (e_F \overset{h}{\underset{k}{\,}} e_G)$.

Proof By straightforward verification. $\qquad\qquad\qquad\qquad\qquad\qquad\square$

3.8.5 Diagonalisation

An object A of the *unitary* weak double category $\mathbb{A}$ determines a strict double functor constant at A, and denoted by the same letter (or by ΔA when useful)

$$A \colon \mathbb{I} \to \mathbb{A}.$$

(3.103)

Let us note that this strictness depends on the *preunitarity* of $\mathbb{A}$, otherwise we would have a unitary pseudo double functor with an invertible comparison $\underline{\Delta A}(r, s) = \lambda(e_A) = \rho(e_A) \colon e_A \otimes e_A \to e_A$, for all consecutive vertical arrows r, s in $\mathbb{I}$.

This provides a diagonal double functor

$$\Delta \colon \mathbb{A} \to \mathbb{L}x(\mathbb{I}, \mathbb{A}) \subset \mathbb{L}x_{ps}(\mathbb{I}, \mathbb{A}).$$

(3.104)

The horizontal arrow $f \colon A \to A'$ goes to the constant horizontal transformation $\Delta f \colon \Delta A \to \Delta A'$. The vertical arrow $u \colon A \dashrightarrow B$ goes to the constant vertical transformation $\Delta u \colon \Delta A \dashrightarrow \Delta B$, according to which:

- an object X in $\mathbb{A}$ is sent to $u \colon A \dashrightarrow B$,

- a horizontal arrow $h \colon X \to X'$ in $\mathbb{A}$ is sent to the cell $1_u \colon (u \overset{1}{\underset{1}{\,}} u)$,

- a vertical arrow $r \colon X \dashrightarrow Y$ in $\mathbb{A}$ is sent to the special isocell $1_u \colon e_A \otimes u \to u \otimes e_B \colon A \dashrightarrow B$.

Finally a cell $\alpha \colon (u \overset{f}{\underset{g}{\,}} v)$ in $\mathbb{A}$ is sent to the constant modification $\Delta\alpha \colon (\Delta u \overset{\Delta f}{\underset{\Delta g}{\,}} \Delta v)$, whose component on each object X of $\mathbb{A}$ is α itself.

It is also useful to note that a *unitary* lax functor $S\colon \mathbb{A} \to \mathbb{B}$ *preserves diagonalisation*, in the sense that $S.\Delta A = \Delta(SA)$; for a general lax functor S one should proceed in a more complex way: see 5.5.1 and 5.5.7.

3.8.6 Exercises and complements

(a) The double categories **0** and **1** are defined in 3.2.1. Prove that

$$\mathbb{L}\mathrm{x}_{\mathrm{ps}}(\mathbf{0}, \mathbb{A}) = \mathbb{L}\mathrm{x}(\mathbf{0}, \mathbb{A}) = \mathbf{1}.$$

(b) A unitary lax functor $\mathbf{1} \to \mathbb{A}$ is automatically strict and amounts to an object $A = F(0)$. Prove that a lax functor $F\colon \mathbf{1} \to \mathbb{A}$ amounts to a monad in the vertical bicategory **Ver**$\mathbb{A}$, that will be called a *vertical monad* in $\mathbb{A}$.

(c) Describe the weak double category $\mathbb{L}\mathrm{x}(\mathbf{1}, \mathbb{A})$ of vertical monads, horizontal transformations, pseudo vertical transformations and modifications.

3.8.7 The structure of 2-categories

A 2-category is generally viewed here as a horizontal double category, i.e. a double category with trivial vertical arrows. Their functors, transformations and modifications are particular cases of the corresponding items for double categories.

More precisely, a 2-functor $F\colon \mathbf{C} \to \mathbf{D}$ is a double functor between horizontal double categories, and strictly preserves all units and compositions.

A 2-natural (resp. pseudo natural) transformation $h\colon F \to G\colon \mathbf{C} \to \mathbf{D}$ is the same as a horizontal (resp. pseudo horizontal) transformation of such double functors (see 3.8.1). In the strict case h has components $hX\colon FX \to GX$ in $\mathbf{D}$, for every X in $\mathbf{C}$, that satisfy

$$G\alpha.hX = hY.F\alpha \qquad \text{(for } \alpha\colon a \to b\colon X \to Y \text{ in } \mathbf{C}\text{).} \qquad (3.105)$$

In the pseudo case there are also invertible comparisons

$$ha\colon Ga.hX \to hX'.Fa\colon FX \to GX'$$

for every $a\colon X \to X'$ in $\mathbf{C}$, under the axioms (pht.1, 4, 5) of 3.8.1.

A modification $\mu\colon h \to k\colon F \to G\colon \mathbf{C} \to \mathbf{D}$ of pseudo (or strict) natural transformations is the same as a modification $\mu\colon (e_F \, {}^h_k \, e_G)$ of horizontal double categories (see 3.8.3), where e_F and e_G are the vertical identity of the 2-functors. It has components $\mu X\colon hX \to kX\colon FX \to GX$ in $\mathbf{D}$ (for X in $\mathbf{C}$), that satisfy condition (mod.1) of 3.8.3.

The reader may know that 2-categories, 2-functors, 2-natural transformations and their modifications form a 3-category, or want to analyse this fact.

4

Double adjunctions

We deal with (horizontal double) adjunctions $F \dashv G$ between weak double categories, introduced and studied in [GP2]. Their universal property is based on the main composition, the horizontal one.

As already discussed in the Introduction, the general form is a 'colax-lax adjunction', where F is a *colax* functor and G a *lax* one. It will be defined as an orthogonal adjunction (see Section 4.1) *in* the double category $\mathbb{Dbl}$ of weak double categories, lax and colax double functors (introduced in Section 4.2). (One should use a version of $\mathbb{Dbl}$ based on a universe $\mathcal{V}$ to which the ground universe $\mathcal{U}$ belongs; but also here we do not insist on problems of size.)

It is important to note that *these adjunctions cannot be viewed in a 2-category*: a globular structure, of any dimension, can only have one sort of arrows. The same holds for general adjunctions between bicategories, which still are of the colax-lax type.

The last section of this chapter briefly studies pseudo algebras for a 2-monad, and shows that weak double categories can be seen as normal pseudo algebras for an obvious 2-monad on the 2-category of graphs of categories.

Kan extensions *in* or *for* weak double categories form a complex topic, that will not be covered here. It was introduced in [GP3, GP4] and studied also in [Ko1, Ko2].

The components of degree $n = 0, 1$ of a weak double category $\mathbb{A}$, or a (co)lax functor F, or a horizontal transformation h are written as $\mathbb{A}_n = \mathrm{Hor}_n(\mathbb{A})$, $F_n = \mathrm{Hor}_n(F)$, $h_n = \mathrm{Hor}_n(h)$. The infinite dimensional extension of Part II, freely working on degree $n \geqslant 0$, will be simpler and clearer. Also because of this, the study of monads in weak double categories is deferred to the infinite dimensional case, in Chapter 8.

4.1 Companions and adjoints in a weak double category

This section studies the connections between horizontal and vertical morphisms in a weak double category: horizontal morphisms can have vertical *companions* and vertical *adjoints*. Such phenomena, introduced in [GP1, GP2], are interesting in themselves and typical of double categories. Later, orthogonal adjoints will be used to define adjunctions *between* weak double categories.

$\mathbb{A}$ is always a unitary weak double category. $\mathbf{Ver}\mathbb{A}$ denotes its vertical bicategory of objects, vertical arrows and special cells ($u \overset{1}{\underset{1}{\mid}} v$), introduced in 3.3.4.

4.1.1 Orthogonal companions

In the *unitary* weak double category $\mathbb{A}$ the horizontal morphism $f \colon A \to B$ and the vertical morphism $u \colon A \nrightarrow B$ are made *companions* by assigning a pair (η, ε) of cells as below, called the *unit* and *counit*, that satisfy the equations $\eta \mid \varepsilon = e_f$ and $\eta \otimes \varepsilon = 1_u$

$$
\begin{array}{ccc}
A =\!=\!= A & \qquad & A \xrightarrow{\ f\ } B \\
\end{array}
\tag{4.1}
$$

(Without assuming unitarity, the equation $\eta \otimes \varepsilon = 1_u$ should be corrected with the unit comparisons of u.)

Given f, this is equivalent to saying that the pair (u, ε) satisfies the following universal property:

(a) for every cell $\varepsilon' \colon (u' \overset{f}{\underset{g}{\,}} B)$ there is a unique cell $\varphi \colon (u' \overset{A}{\underset{g}{\,}} u)$ such that $\varepsilon' = \varphi \mid \varepsilon$

$$
\begin{array}{ccccc}
A \xrightarrow{\ f\ } B & & A =\!=\!= A \xrightarrow{\ f\ } B \\
\end{array}
\tag{4.2}
$$

In fact, given (η, ε), we can (and must) take $\varphi = \eta \otimes \varepsilon'$; on the other hand, given (a), we define $\eta \colon (A \overset{A}{\underset{f}{\,}} u)$ by the equation $\eta \mid \varepsilon = e_f$ and deduce that $\eta \otimes \varepsilon = 1_u$ because $(\eta \otimes \varepsilon) \mid \varepsilon = (\eta \mid \varepsilon) \otimes \varepsilon = \varepsilon = (1_u \mid \varepsilon)$.

Similarly the pair (u, η) is characterised by a universal property:

(b) for every cell $\eta' \colon (A \overset{g}{\underset{f}{\,}} u')$ there is a unique cell $\psi \colon (u \overset{g}{\underset{B}{\,}} u')$ such that $\eta' = \eta \mid \psi$.

Therefore, if f has a vertical companion, this is determined up to a unique special isocell, *and will often be written as f_**. Companions compose in the obvious (covariant) way: if $g: B \to C$ also has a companion g_* (via η', ε') then $g_* f_*: A \to C$ is companion to $gf: A \to C$, with unit

$$\left(\frac{\eta \,|\, 1}{e \,|\, \eta'} \right) : (A \,\,{}^A_{gf}\,\, g_* f_*). \tag{4.3}$$

Companionship is preserved by *unitary* lax or colax double functors.

We say that $\mathbb{A}$ *has vertical companions* if every horizontal arrow has a vertical companion. For instance, in $\mathbb{C}$at the vertical companion to a functor $f: X \to Y$ is the profunctor

$$f_*: X \to Y, \qquad f_*(x,y) = Y(f(x), y). \tag{4.4}$$

4.1.2 Orthogonal adjoints

Transforming companionship by vertical (or horizontal) duality, the arrows $f: A \to B$ and $v: B \to A$ are made *orthogonal adjoints* by a pair (σ, τ) of cells as below

$$\tag{4.5}$$

with $\sigma \,|\, \tau = e_f$ and $\tau \otimes \sigma = 1_v$. (Let us recall that $\mathbb{A}$ is assumed to be unitary.)

Here f is called the *horizontal adjoint* and v the *vertical* one.

Again, given f these relations can be described by universal properties for (v, τ) or (v, σ)

(a) for every cell $\tau': (v' \,{}^g_f\, B)$ there is a unique cell $\varphi: (v' \,{}^g_A\, v)$ such that $\tau' = \varphi | \tau$,

(b) for every cell $\sigma': (A \,{}^f_g\, v')$ there is a unique cell $\psi: (v \,{}^B_g\, v')$ such that $\sigma' = \sigma | \psi$.

The vertical adjoint of f is determined up to a special isocell and will often be written as f^*; vertical adjoints compose, contravariantly: $(gf)^*$ can be constructed as $f^* g^*$.

We say that $\mathbb{A}$ *has vertical adjoints* if every horizontal arrow has a vertical adjoint. For instance, in $\mathbb{C}$at the vertical adjoint to a functor $f: X \to Y$ is the profunctor

$$f^*: Y \to X, \qquad f^*(y,x) = Y(y, f(x)). \tag{4.6}$$

4.1.3 Exercises

Study vertical companions and vertical adjoints in the strict or weak double categories of Sections 3.1 and 3.4.

4.1.4 Proposition

Let $f: A \to B$ be a horizontal arrow in the weak double category $\mathbb{A}$. The following conditions are equivalent:

(a) f has a vertical companion $u: A \nrightarrow B$ and a vertical adjoint $v: B \nrightarrow A$,

(b) f has a vertical companion $u: A \nrightarrow B$ and $u \dashv v$ in the bicategory $\mathbf{Ver}\mathbb{A}$,

(c) f has a vertical adjoint $v: B \nrightarrow A$ and $u \dashv v$ in the bicategory $\mathbf{Ver}\mathbb{A}$.

Proof To prove the equivalence of (a) and (b), let us have two cells η, ε as in 4.1.1. An orthogonal adjunction (σ, τ) as in 4.1.2 is equivalent to an internal adjunction $(\eta', \varepsilon'): u \dashv v$ in the 2-category $\mathbf{Ver}\mathbb{A}$, by the following formulas

$$\eta' = \eta \otimes \sigma: e_A \to u \otimes v, \quad \varepsilon' = \tau \otimes \varepsilon: v \otimes u \to e_B \quad (\text{in } \mathbf{Ver}\mathbb{A}), \quad (4.7)$$

$$\sigma = \eta' \mid (\varepsilon \otimes 1_v), \qquad \tau = (1_v \otimes \varepsilon) \mid \eta' \quad (\text{in } \mathbb{A}). \quad (4.8)$$

The equivalence of (a) and (c) follows by horizontal duality. $\qquad\square$

4.1.5 Proposition

Let $f: A \to B$ be a horizontal isomorphism in the weak double category $\mathbb{A}$, with $g = f^{-1}$.

If f has a vertical companion $u: A \nrightarrow B$, then its unit and counit are also horizontally invertible and determine each other:

$$(\varepsilon \mid e_g \mid \eta) = \eta \otimes \varepsilon = 1_u,$$
$$\eta^{-1} = \varepsilon \mid e_g, \qquad \varepsilon^{-1} = e_g \mid \eta. \quad (4.9)$$

Conversely, the existence of a horizontally invertible cell $\eta: (A \begin{smallmatrix} A \\ f \end{smallmatrix} u)$ implies that f is horizontally invertible, with companion u and counit $\varepsilon = \eta^{-1} \mid e_f$.

Proof For the first part one can rewrite $(\varepsilon \,|\, e_g \,|\, \eta)$ as follows and apply middle-four interchange

$$
\begin{array}{ccccccc}
A & \xrightarrow{\ f\ } & B & \xrightarrow{\ g\ } & A & =\!=\!= & A \\
e\downarrow & e_f & \downarrow e & e_g & \downarrow e & \eta & \downarrow u \\
A & \xrightarrow{-f\to} & B & \xrightarrow{-g\to} & A & \xrightarrow{-f\to} & B \\
u\downarrow & \varepsilon & \downarrow e & e_g & \downarrow e & e_f & \downarrow e \\
B & =\!=\!= & B & \xrightarrow{\ g\ } & A & \xrightarrow{\ f\ } & B
\end{array}
$$

This implies that η and ε have the inverse cells specified above. The converse is obvious. $\qquad\square$

4.1.6 Theorem and Definition (Sesqui-isomorphism)

Let $f\colon A \to B$ be a horizontal isomorphism in the weak double category $\mathbb{A}$, with $g = f^{-1}\colon B \to A$.

The following conditions are equivalent:

(a) there exist four horizontally invertible cells $\eta, \sigma, \varepsilon, \tau$

$$
\begin{array}{ccc}
A & =\!=\!= & A \\
e\downarrow & \eta & \downarrow u \\
A & \xrightarrow{-f\to} & B \\
e\downarrow & \sigma & \downarrow v \\
A & =\!=\!= & A
\end{array}
\qquad\qquad
\begin{array}{ccc}
B & =\!=\!= & B \\
v\downarrow & \tau & \downarrow e \\
A & \xrightarrow{-f\to} & B \\
u\downarrow & \varepsilon & \downarrow e \\
B & =\!=\!= & B
\end{array}
$$

satisfying the following relations:

$$
\begin{array}{ll}
\varepsilon = \eta^{-1} \,|\, e_f, & \eta = e_f \,|\, \varepsilon^{-1}, \\[4pt]
\tau = \sigma^{-1} \,|\, e_f, & \sigma = e_f \,|\, \tau^{-1},
\end{array}
\tag{4.10}
$$

(b) there exist two horizontally invertible cells η, σ as in the left diagram above,

(c) there exist two horizontally invertible cells ε, τ as in the right diagram above,

(d) f has a vertical companion $u\colon A \nrightarrow B$ with unit $\eta\colon (A\,{}^{A}_{f}\,u)$ and a vertical adjoint $v\colon B \nrightarrow A$ with $\sigma\colon (A\,{}^{f}_{A}\,v)$,

(e) f has a vertical companion $u\colon A \nrightarrow B$ with unit $\eta\colon (A\,{}^{A}_{f}\,u)$ while g has a vertical companion $v\colon B \nrightarrow A$ with unit $\tau^{-1}\colon (B\,{}^{B}_{g}\,v).$

We say that f (or the triple (f, u, v)) is a sesqui-isomorphism *when these conditions hold.*

Note. Here $f \colon A \to B$ is a horizontal isomorphism, while the pair (u, v) is a *vertical adjoint equivalence* between A and B, i.e. an adjoint equivalence in the vertical bicategory **Ver**$\mathbb{A}$.

Proof By 4.1.5 and horizontal duality.   □

4.1.7 Theorem and Definition (Horizontal invariance)

We say that the weak double category $\mathbb{A}$ *is* horizontally invariant *if it satisfies the following equivalent conditions:*

(a) vertical arrows are transferable along horizontal isomorphisms: given two horizontal isomorphisms h, k and a vertical morphism x arranged as below, there always is a horizontally invertible cell φ (a 'filler', as in the well-known Kan extension property)

$$
\begin{array}{ccc}
A' & \xrightarrow{\ h\ } & A \\
{\scriptstyle y}\big\downarrow & \varphi & \big\downarrow{\scriptstyle x} \\
X & \xrightarrow[\ k\ ]{} & B
\end{array}
\qquad\qquad (4.11)
$$

(b) every horizontal isomorphism in $\mathbb{A}$ has a vertical companion,

(c) every horizontal isomorphism in $\mathbb{A}$ has a vertical adjoint,

(d) every horizontal isomorphism in $\mathbb{A}$ is a sesqui-isomorphism *(as defined in 4.1.6).*

Proof First (d) implies (b), by 4.1.6. Conversely, if two inverse horizontal isomorphisms f, g have companions $u = f_*$ and $v = g_*$, we know that all their units and counits are horizontally invertible (by 4.1.5); f and g are easily seen to be sesqui-isomorphisms.

By horizontal duality, (c) also is equivalent to (d).

Finally, the equivalent properties (b) and (c) imply (a), because the filler φ can be obtained as a vertical composite of three horizontally invertible cells, namely $\varepsilon \colon (u \, {}^h_A \, A)$, 1_x and $\tau \colon (v \, {}^B_k \, B)$, where $u \colon A' \nrightarrow A$ is companion to h and $v \colon B \nrightarrow B'$ is vertical adjoint to k.

Conversely, assuming (a), each horizontal iso $f \colon A \to B$ has two horizontally invertible cells $\varepsilon \colon (u \, {}^f_B \, B)$, $\tau \colon (v \, {}^B_f \, B)$, whence it is a sesqui-isomorphism.   □

4.1.8 Exercises, comments and complements

(a) The condition of horizontal invariance is horizontally and vertically selfdual. If it holds, *two objects A, B horizontally isomorphic are always sesqui-isomorphic*, hence vertically equivalent (i.e. equivalent in the vertical bicategory of our weak double category).

(b) Verify that all the examples of Sections 3.1 and 3.4 are horizontally invariant.

(c) This property should be expected of every 'well formed' weak double category. If it fails, the relation of being horizontally isomorphic objects can be poorly related to vertical 'similarity'.

     The reader is invited to consider the 'anomalous behaviour' of the flat double category $\mathbb{T}g$ of topological groups, with algebraic homomorphisms as horizontal maps, continuous mappings as vertical maps and commutative squares (of the underlying mappings) as double cells.

     (The theorem below shows that a horizontally invariant weak double category cannot present similar 'anomalies'.)

4.1.9 Theorem

If the weak double category $\mathbb{A}$ is horizontally invariant, two lax double functors $F, G \colon \mathbb{I} \to \mathbb{A}$ which are horizontally isomorphic are also 'vertically equivalent'.

     More precisely, a horizontal isomorphism $h \colon F \to G$ produces a pseudo vertical transformation $r \colon F \nrightarrow G$ (see 3.8.2) whose general component $ri \colon Fi \nrightarrow Gi$ is a vertical equivalence, vertical companion to $hi \colon Fi \cong Gi$, and determined as such up to special isocells.

Note. In fact, we prove more: F and G are sesqui-isomorphic objects in the weak double category of lax double functors $\mathbb{L}x(\mathbb{I}, \mathbb{A})$ introduced in Section 3.8.

Proof We choose, for every object i in $\mathbb{I}$, a sesqui-isomorphism extending hi

$$
\begin{array}{ccc}
Fi & =\!=\!= & Fi \\
{\scriptstyle e}\big\downarrow & \lambda i & \big\uparrow{\scriptstyle ri} \\
Fi & \xrightarrow{\ hi\ } & Gi \\
{\scriptstyle e}\big\downarrow & \mu i & \big\downarrow{\scriptstyle si} \\
Gi & =\!=\!= & Gi
\end{array}
$$

with horizontally invertible cells λi, μi. Now, the family of vertical equivalences ri can be canonically extended to a pseudo vertical transformation of lax double functors $r\colon F \to G$, as we verify below. Similarly we form a pseudo vertical transformation $s\colon G \to F$; all this makes λ and μ into horizontally invertible modifications.

To complete r, let $rf\colon (ri\ {}^{Ff}_{Gf}\ ri')$ (for $f\colon i \to i'$ in $\mathbb{I}$) be the composed cell

$$
\begin{array}{ccccccc}
Fi & = = = & Fi & \xrightarrow{Ff} & Fi' & = = = & Fi' \\
{\scriptstyle ri}\downarrow & {\scriptstyle (\lambda i)^{-1}} & {\scriptstyle e}\downarrow & {\scriptstyle e} & {\scriptstyle e}\downarrow & {\scriptstyle \lambda i'} & \downarrow{\scriptstyle ri'} \\
Gi & \xrightarrow[(hi)^{-1}]{} & Fi & \xrightarrow[Ff]{} & Fi' & \xrightarrow[hi']{} & Gi'
\end{array}
$$

For a vertical $u\colon i \to j$, the comparison special isocell

$$ru\colon Fu \otimes rj \to ri \otimes Gu\colon Fi \to Gj$$

is defined as the following *normal* composite, with $\lambda^* j = (\lambda j)^{-1}\,|\,e_{hj}$ (see 3.3.3)

$$
\begin{array}{ccc}
Fi & = = = & Fi \\
{\scriptstyle e}\downarrow & {\scriptstyle \lambda i} & \downarrow{\scriptstyle ri} \\
Fi & \xrightarrow{hi} & Gi \\
{\scriptstyle Fu}\downarrow & {\scriptstyle hu} & \downarrow{\scriptstyle Gu} \\
Fj & \xrightarrow{hj} & Gj \\
{\scriptstyle rj}\downarrow & {\scriptstyle \lambda^* j} & \downarrow{\scriptstyle e} \\
Gj & = = = & Gj
\end{array}
$$

$\square$

4.2 The double category $\mathbb{D}$bl and double commas

The strict double category $\mathbb{D}$bl is a crucial structure, first introduced in [GP2], where double adjunctions will live. It consists of weak double categories, with lax and colax double functors and suitable cells. It contains a double category $\mathbb{M}$nc of monoidal categories, with monoidal and comonoidal functors.

Comma weak double categories $U \Downarrow F$ are defined in 4.2.4, when U is colax and F is lax.

Limits and span representation in $\mathbb{D}$bl will be studied in 5.4.6, 5.4.7.

4.2.1 The double category $\mathbb{D}bl$

We do not want to compose lax and colax double functors, as this would destroy their comparisons. But they can be organised in a strict double category $\mathbb{D}bl$, where orthogonal adjunctions will provide our general notion of double adjunction (in the next section), while companion pairs will amount to pseudo double functors (Theorem 4.4.1).

The objects of $\mathbb{D}bl$ are the (small) *weak* double categories $\mathbb{A}, \mathbb{B}, ...$ (not assumed to be unitary). Its horizontal arrows are the *lax* (double) functors $F, G...$; its vertical arrows are the *colax* functors $U, V...$ A cell π

$$
\begin{array}{ccc}
\mathbb{A} & \xrightarrow{F} & \mathbb{B} \\
U \downarrow & \pi & \downarrow V \\
\mathbb{C} & \xrightarrow{G} & \mathbb{D}
\end{array}
\qquad
\begin{array}{ccc}
\mathbb{A} & \xrightarrow{F} & \mathbb{B} \\
U \downarrow & {}_{\nearrow}\pi & \downarrow V \\
\mathbb{C} & \xrightarrow{G} & \mathbb{D}
\end{array}
\qquad (4.12)
$$

is – loosely speaking – a 'horizontal transformation' $\pi \colon VF \dashrightarrow GU$, and *can* be marked with an arrow as at the right, above. Note, however, that – generally – the composites VF and GU are neither lax nor colax (just morphisms of double graphs that respect the horizontal structure): the coherence conditions of π are based on the four 'functors' F, G, U, V and all their comparison cells.

Precisely, the cell π consists of the following data:

(a) a lax functor F with comparison special cells $\underline{F}$ (indexed by the objects A and pairs (u, v) of consecutive vertical arrows of $\mathbb{A}$) and a lax functor G with comparison special cells $\underline{G}$ (similarly indexed by $\mathbb{C}$)

$$F \colon \mathbb{A} \to \mathbb{B}, \quad \underline{F}(A) \colon e_{FA} \to F(e_A), \quad \underline{F}(u, v) \colon Fu \otimes Fv \to F(u \otimes v),$$

$$G \colon \mathbb{C} \to \mathbb{D}, \quad \underline{G}(C) \colon e_{GC} \to G(e_C), \quad \underline{G}(u, v) \colon Gu \otimes Gv \to G(u \otimes v),$$

(b) two colax functors U, V with comparison special cells $\underline{U}, \underline{V}$ (similarly indexed by $\mathbb{A}$ and $\mathbb{B}$)

$$U \colon \mathbb{A} \to \mathbb{C}, \quad \underline{U}(A) \colon U(e_A) \to e_{UA}, \quad \underline{U}(u, v) \colon U(u \otimes v) \to Uu \otimes Uv,$$

$$V \colon \mathbb{B} \to \mathbb{D}, \quad \underline{V}(B) \colon V(e_B) \to e_{VB}, \quad \underline{V}(u, v) \colon V(u \otimes v) \to Vu \otimes Vv,$$

(c) horizontal maps $\pi A \colon VFA \to GUA$ and cells $\pi u \colon VFu \to GUu$ in $\mathbb{D}$ (for A and $u \colon A \dashrightarrow A'$ in $\mathbb{A}$)

$$
\begin{array}{ccc}
VFA & \xrightarrow{\pi A} & GUA \\
VFu \downarrow & \pi u & \downarrow GUu \\
VFA' & \xrightarrow{\pi A'} & GUA'
\end{array}
\qquad (4.13)
$$

These data must satisfy the naturality conditions (c.0), (c.1) (the former is redundant, being implied by the latter) and the coherence conditions (c.2), (c.3)

(c.0) $GUf.\pi A = \pi A'.VFf$ (for $f\colon A \to A'$ in $\mathbb{A}$),

(c.1) $\pi u \,|\, GU\alpha = VF\alpha \,|\, \pi v$ (for $\alpha\colon (u\,{\overset{f}{\underset{g}{}}}\,v)$ in $\mathbb{A}$),

(c.2) $(V\underline{F}A \,|\, \pi e_A \,|\, G\underline{U}A) = (\underline{V}FA \,|\, e_{\pi A} \,|\, \underline{G}UA)$ (for A in $\mathbb{A}$),

(c.3) $(V\underline{F}(u,v) \,|\, \pi w \,|\, G\underline{U}(u,v))$

$\qquad = (\underline{V}(Fu, Fv) \,|\, (\pi u \otimes \pi v) \,|\, \underline{G}(Uu, Uv))$ (for $w = u \otimes v$ in $\mathbb{A}$),

$$
\begin{array}{ccccccc}
VFA & = & VFA & \longrightarrow & GUA & = & GUA \\
\scriptstyle{V(Fu\otimes Fv)}\downarrow & \underline{VF} & \downarrow\scriptstyle{VFw} & \pi w \quad GUw\uparrow & & GU & \downarrow\scriptstyle{G(Uu\otimes Uv)} \\
VFA'' & = & VFA'' & \longrightarrow & GUA'' & = & GUA''
\end{array}
$$

$$
\begin{array}{ccccccc}
VFA & = & VFA & \longrightarrow & GUA & = & GUA \\
 & & \downarrow\scriptstyle{VFu} & \pi u \quad GUu\uparrow & & & \\
\scriptstyle{V(Fu\otimes Fv)}\bullet & \underline{VF} & VFA' & \longrightarrow & GUA' & \underline{GU} & \bullet\scriptstyle{G(Uu\otimes Uv)} \\
 & & \downarrow\scriptstyle{VFv} & \pi v \quad GUv\uparrow & & & \\
VFA'' & = & VFA'' & \longrightarrow & GUA'' & = & GUA''
\end{array}
$$

The horizontal and vertical composition of double cells are both defined *using the horizontal composition* of the weak double category $\mathbb{D}$. Namely, for a consistent matrix of double cells

$$
\begin{array}{ccccc}
\bullet & \overset{F}{\longrightarrow} & \bullet & \overset{F'}{\longrightarrow} & \bullet \\
U\downarrow & \pi & \downarrow V \;\; \rho & & \downarrow W \\
\bullet & \overset{G}{-\!\!\!\rightarrow} & \bullet & \overset{G'}{-\!\!\!\rightarrow} & \bullet \\
U'\downarrow & \sigma & \downarrow V' \;\; \tau & & \downarrow W' \\
\bullet & \underset{H}{\longrightarrow} & \bullet & \underset{H'}{\longrightarrow} & \bullet
\end{array}
\qquad (4.14)
$$

we let:

$$
(\pi \,|\, \rho)(u) = \rho Fu \,|\, G'\pi u, \qquad (\pi \otimes \sigma)(u) = V'\pi u \,|\, \sigma Uu, \qquad (4.15)
$$

$$
\begin{array}{ccc}
WF'FA \longrightarrow G'VFA \longrightarrow G'GUA & \qquad & V'VFA \longrightarrow V'GUA \longrightarrow HU'UA \\
\downarrow \quad \rho Fu \quad \downarrow \quad G'\pi u \quad \downarrow & & \downarrow \quad V'\pi u \quad \downarrow \quad \sigma Uu \quad \downarrow \\
WF'FA' \longrightarrow G'VFA' \longrightarrow G'GUA' & & V'VFA' \longrightarrow V'GUA' \longrightarrow HU'UA'
\end{array}
$$

4.2.2 Theorem

With these definitions $\mathbb{D}$bl *is a strict double category.*

Proof (a) To show that the formulas (4.15) do define two cells of $\mathbb{D}$bl, we write down the verification of condition (c.3) for $\pi \mid \rho$, with respect to a vertical composite $w = u \otimes v$ in $\mathbb{A}$.

Writing a cell as an arrow between its vertical arrows (the horizontal domain and codomain), our property amounts to the commutativity of the outer diagram below, in $\mathrm{Hor}_1 \mathbb{D}$

$$
\begin{array}{ccccc}
WF'Fw & \xrightarrow{\rho Fw} & G'VFw & \xrightarrow{G'\pi w} & G'GUw \\[2pt]
\big\uparrow{\scriptstyle WF'\underline{F}} & & \big\uparrow{\scriptstyle G'V\underline{F}} & & \big\downarrow{\scriptstyle G'G\underline{U}} \\[2pt]
WF'(Fu \otimes Fv) & \xrightarrow{\rho(Fu\otimes Fv)} & G'V(Fu \otimes Fv) & & G'G(Uu \otimes Uv) \\[2pt]
\big\uparrow{\scriptstyle W\underline{F}'F} & & \big\downarrow{\scriptstyle G'\underline{V}F} & & \big\uparrow{\scriptstyle G'\underline{G}U} \\[2pt]
W(F'Fu \otimes F'Fv) & & G'(VFu \otimes VFv) & \xrightarrow{G'(\pi u\otimes\pi v)} & G'(GUu \otimes GUv) \\[2pt]
\big\downarrow{\scriptstyle \underline{W}F'F} & & \big\uparrow{\scriptstyle \underline{G}'VF} & & \big\uparrow{\scriptstyle \underline{G}'GU} \\[2pt]
WF'Fu \otimes WF'Fv & \xrightarrow[\rho Fu\otimes\rho Fv]{} & G'VFu \otimes G'VFv & \xrightarrow[G'\pi u\otimes G'\pi v]{} & G'GUu \otimes G'GUv
\end{array}
$$

In fact the two hexagons commute by condition (c.3) on the double cells π and ρ. The upper rectangle commutes by naturality of ρ on the cell $\underline{F}(u,v)$, and the lower one by naturality of $\underline{G}'$ on the cells $\pi u, \pi v$ (see (ldf.2) in 3.5.1).

(b) The composition laws of cells are strictly associative and unitary, because they are both computed with the horizontal composition of cells in $\mathbb{D}$.

(c) Finally, to verify the interchange law on the four double cells of diagram (4.14), we compute $(\pi \mid \rho) \otimes (\sigma \mid \tau)$ and $(\pi \otimes \sigma \mid \rho \otimes \tau)$ on the vertical arrow $u \colon A \to A'$ of $\mathbb{A}$, and we obtain two cells $W'WF'Fu \to H'V'GUu$, corresponding to the upper or lower path in the following diagram

$$
\begin{array}{ccccc}
W'WF'Fu & \xrightarrow{W'\rho Fu} & W'G'VFu & \xrightarrow{W'G'\pi u} & W'G'GUu \\[2pt]
 & & \big\downarrow{\scriptstyle \tau VFu} & & \big\downarrow{\scriptstyle \tau GUu} \\[2pt]
 & & H'V'VFu & \xrightarrow[H'V'\pi u]{} & H'V'GUu & \xrightarrow[H'\sigma Uu]{} & H'HU'Uu
\end{array}
$$

These two composites coincide because the square commutes: a consequence of axiom (c.1) on the $\mathbb{D}$bl-cell τ, namely its naturality on the cell $\pi u \colon VFu \to GUu$. $\qquad\square$

4.2.3 Exercises and complements

(a) Verify that $\mathbb{D}bl$ is a transpositive double category (see 3.2.5), by the t-contravariant involution $(-)^h$ that turns a weak double category $\mathbb{A}$ into its horizontal dual $\mathbb{A}^h$, exchanging lax and colax functors and transposing double cells

$$(-)^h \colon \mathbb{D}bl \dashrightarrow \mathbb{D}bl, \qquad \mathbb{A} \mapsto \mathbb{A}^h,$$

$$\pi \colon (U \underset{G}{\overset{F}{}} V) \mapsto \pi^h \colon (F^h \underset{V^h}{\overset{U^h}{}} G^h). \tag{4.16}$$

(b) If the colax functors U, V are pseudo functors, prove that a $\mathbb{D}bl$-cell $\pi \colon (U \underset{G}{\overset{F}{}} V)$ is inhabited by a *horizontal transformation of lax functors* $\pi \colon VF \to GU$ (see 3.5.4). Then the arrow mark of a cell, in (4.12), is fully justified.

In particular, restricting $\mathbb{D}bl$ to trivial vertical arrows and globular cells $(1 \underset{G}{\overset{F}{}} 1)$, as in 3.2.4, we get the strict 2-category $\mathbf{Hor}(\mathbb{D}bl) = \mathrm{Lx}\mathbf{Dbl}$ of weak double categories, lax double functors and horizontal transformations introduced in 3.5.4.

(c) Similarly if F, G are pseudo functors, a cell $\pi \colon (U \underset{G}{\overset{F}{}} V)$ contains a *horizontal transformation of colax functors* $\pi \colon VF \to GU$, and again the arrow mark of a cell has a formal meaning.

In particular we have the 2-category $\mathbf{Ver}^*(\mathbb{D}bl) = \mathrm{Cx}\mathbf{Dbl}$ of weak double categories, colax double functors and horizontal transformations.

(d) Viewing bicategories in $\mathrm{Lx}\mathbf{Dbl}$, as vertical weak double categories, we have a 2-category of bicategories, lax functors and *special* transformations $\pi \colon F \to G$, whose components are identity morphisms and special cells

$$\pi A = 1 \colon FA \to GA, \qquad \pi u \colon (Fu \underset{FA'}{\overset{FA}{}} Gu), \tag{4.17}$$

which is only possible if F and G *coincide on the objects*. This agrees with the 2-category considered by Carboni and Rosebrugh to define lax monads of bicategories ([CaR], Prop. 2.1). Note, on the other hand, that lax functors and lax transformations of bicategories (or 2-categories) do *not* form a bicategory.

(e) It is interesting, even unexpected, to note that a double cell $\pi \colon (U \underset{1}{\overset{F}{}} 1)$ gives a notion of *horizontal transformation* $\pi \colon F \dashrightarrow U \colon \mathbb{A} \to \mathbb{B}$ *from a lax to a colax functor,* while a double cell $\pi \colon (1 \underset{G}{\overset{1}{}} V)$ gives a notion of *horizontal transformation* $\pi \colon V \dashrightarrow G \colon \mathbb{A} \to \mathbb{B}$ *from a colax to a lax functor.*

For a fixed pair $\mathbb{A}, \mathbb{B}$ of weak double categories, all the horizontal transformations between lax and colax functors (of the four possible kinds) compose, forming a category $HV(\mathbb{A}, \mathbb{B})$ whose objects are the lax *and* the colax functors $\mathbb{A} \to \mathbb{B}$. (This construction can be extended from $\mathbb{D}bl$ to any double category, see Exercise 4.2.8(a).)

(f) The totally full double subcategory of $\mathbb{D}$bl determined by the discrete double categories can be identified with the double category $\mathbb{Q}$**Set**, whose cells are commutative squares of mappings.

4.2.4 Double commas

Given a *colax* double functor U and a *lax* double functor F with the same codomain, we can construct the *comma* weak double category $U \Downarrow F$, where the projections P and Q are strict double functors, and π is a cell of $\mathbb{D}$bl

$$
\begin{array}{ccc}
U \Downarrow F & \xrightarrow{\ P\ } & \mathbb{A} \\
{\scriptstyle Q}\Big\downarrow & \pi & \Big\downarrow{\scriptstyle U} \\
\mathbb{X} & \xrightarrow[\ F\ ]{} & \mathbb{C}
\end{array}
\tag{4.18}
$$

An object is a triple $(A, X; c \colon UA \to FX)$. A horizontal map $(a, x) \colon (A, X; c) \to (A', X'; c')$ comes from a commutative square of $\mathbb{C}$

$$
\begin{array}{ccc}
UA & \xrightarrow{\ c\ } & FX \\
{\scriptstyle Ua}\Big\downarrow & = & \Big\downarrow{\scriptstyle Fx} \\
UA' & \xrightarrow[\ c'\ ]{} & FX'
\end{array}
\tag{4.19}
$$

Their composition is obvious. A vertical arrow

$$(u, v; \gamma) \colon (A, X; c) \dashrightarrow (B, Y; d)$$

comes from a cell γ of $\mathbb{C}$

$$
\begin{array}{ccc}
UA & \xrightarrow{\ c\ } & FX \\
{\scriptstyle Uu}\Big\updownarrow & \gamma & \Big\updownarrow{\scriptstyle Fv} \\
UB & \xrightarrow[\ d\ ]{} & FY
\end{array}
$$

Their vertical composition $(u, v; \gamma) \otimes (u', v'; \delta)$ is defined using a colaxity comparison of U and a laxity comparison of F

$$
\tag{4.20}
$$

A cell (π, ξ) with the boundary represented below comes from a pair of cells $\pi \colon (u \, {}_b^a \, u')$, $\xi \colon (v \, {}_y^x \, v')$ (in $\mathbb{A}$ and $\mathbb{X}$, respectively) such that the cells $U\pi$ and $F\xi$ are coherent with γ, γ' in $\mathbb{C}$

$$
\begin{array}{ccc}
(A, X; c) & \xrightarrow{\ (a,x)\ } & (A', X'; c') \\
{\scriptstyle (u,v;\gamma)} \downarrow & (\pi, \xi) & \downarrow {\scriptstyle (u',v';\gamma')} \qquad U\pi \,|\, \gamma' = \gamma \,|\, F\xi. \qquad (4.21) \\
(B, Y; d) & \xrightarrow[\ (b,y)\ ]{} & (B', Y'; d')
\end{array}
$$

Their horizontal and vertical compositions are obvious.

The associativity isocell for three consecutive vertical arrows $(u, v; \gamma)$, $(u', v'; \delta)$, $(u'', v''; \varepsilon)$ *is* the pair $(\kappa(\underline{u}), \kappa(\underline{v}))$ of associativity isocells of $\mathbb{A}$ and $\mathbb{X}$ for the triples $\underline{u} = (u, u', u'')$, $\underline{v} = (v, v', v'')$

$$
(\kappa(\underline{u}), \kappa(\underline{v})) \colon (u, v; \gamma) \otimes ((u', v'; \delta) \otimes (u'', v''; \varepsilon)) \to
$$
$$
((u, v; \gamma) \otimes (u', v'; \delta)) \otimes (u'', v''; \varepsilon), \qquad (4.22)
$$
$$
\kappa(\underline{u}) \colon u_1 \to u_2, \qquad\qquad \kappa(\underline{v}) \colon v_1 \to v_2,
$$

where $u_1 = u \otimes (u' \otimes u'')$, $u_2 = (u \otimes u') \otimes u''$ and similarly for v_1, v_2.

To prove its coherence, let us denote by Φ and Φ' the pasted cells of the two diagrams below

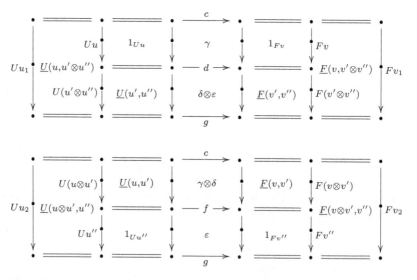

Now the coherence of the previous cell (4.22) is expressed by the equality

$$
(U\kappa(\underline{u}) \,|\, \Phi') = (\Phi \,|\, F\kappa(\underline{v})) \colon Uu_1 \to Fv_2,
$$

which follows from the coherence axioms on U, F and $\mathbb{C}$.

Finally, the strict functors P and Q are projections. The components of π on the objects and vertical arrows are:

$$\pi(A, X; c) = c\colon UA \to FX, \qquad \pi(u, v; \gamma) = \gamma\colon (Uu \, {}^c_d \, Fv). \qquad (4.23)$$

4.2.5 Theorem (Universal properties of commas)

(a) (The horizontal universal property) *For a pair of lax double functors G, H and a cell ξ as below (in $\mathbb{D}$bl) there is a unique lax double functor $L\colon \mathbb{D} \to U \Downarrow F$ such that $G = PL$, $H = QL$ and $\xi = \rho \mid \pi$*

$$
\begin{array}{ccc}
\mathbb{D} \xrightarrow{\;G\;} A & & \mathbb{D} \xrightarrow{\;L\;} U \Downarrow F \xrightarrow{\;P\;} A \\
\left.\begin{array}{c}e\downarrow\\[6pt]\mathbb{D}\end{array}\quad \xi \quad \begin{array}{c}\downarrow U\\[6pt]\end{array}\right. & = & \\
\end{array}
\qquad (4.24)
$$

where the cell ρ is defined by the identity $1\colon QL \to H$ (a horizontal transformation of lax double functors).

Moreover, L is pseudo if and only if both of G and H are.

(b) (The vertical universal property) *A similar property holds for a pair of colax double functors V, W and a cell $\xi'\colon (V \, {}^1_F \, UW)$.*

Proof It is sufficient to prove (a), since $\mathbb{D}$bl is transpositive (by Exercise 4.2.3(a)). L is defined as follows on the items of $\mathbb{D}$: an object D, a horizontal arrow f, a vertical arrow $u\colon D \to\!\!\!\to D'$, a cell φ

$$L(D) = (GD, HD; \xi D\colon UGD \to FHD), \qquad L(f) = (Gf, Hf),$$
$$L(u) = (Gu, Hu; \xi u)\colon LD \to\!\!\!\to LD', \qquad L(\varphi) = (G\varphi, H\varphi).$$

The laxity special cells $\underline{L}$, for D and $w = u \otimes v\colon D \to\!\!\!\to D''$ in $\mathbb{D}$, are constructed with the laxity cells $\underline{G}$ and $\underline{H}$ (and are invertible if and only if the latter are)

$$\underline{L}(D) = (\underline{G}(D), \underline{H}(D))\colon e_{LA} \to L(e_A),$$
$$\underline{L}(u, v) = (\underline{G}(u, v), \underline{H}(u, v))\colon Lu \otimes Lv \to L(u \otimes v)\colon LD \to\!\!\!\to LD''.$$

Here the vertical arrow $Lu \otimes Lv$ of $U \Downarrow F$ is produced by the cell below

$$
\begin{array}{ccccccc}
UGD & =\!=\!= & UGD & \xrightarrow{\;\xi D\;} & FHD & =\!=\!= & FHD \\
\Big\downarrow & & \Big\downarrow{\scriptstyle UGu} & {\scriptstyle \xi u} & \Big\downarrow{\scriptstyle GUu} & & \Big\downarrow \\
{\scriptstyle U(Gu\otimes Gv)}\bullet & & \underline{UG}\;\;UGD' & \longrightarrow & FHD' \quad \underline{FH} & & \bullet{\scriptstyle F(Hu\otimes Hv)} \\
\Big\downarrow & & \Big\downarrow{\scriptstyle UGv} & {\scriptstyle \xi v} & \Big\downarrow{\scriptstyle GUv} & & \Big\downarrow \\
UGD'' & =\!=\!= & UGD'' & \xrightarrow[\;\xi D''\;]{} & FHD'' & =\!=\!= & FHD''
\end{array}
$$

while $L(u \otimes v)$ is produced by ξw.

The coherence condition on $\underline{L}(u,v) = (\underline{G}(u,v), \underline{H}(u,v))$

$$
Lu \otimes Lv \,|\, F\underline{H}(u,v) \;=\; U\underline{G}(u,v)\,|\,\xi w,
$$

follows from the coherence condition of $\xi \colon (\mathbb{D} \, {}^{G}_{FH} \, U)$ as a cell in $\mathbb{D}bl$

$$
(\underline{U}(Gu, Gv) \,|\, \xi u \otimes \xi v \,|\, \underline{FH}(u,v)) \;=\; (U\underline{G}(u,v) \,|\, \xi w).
$$

Uniqueness is obvious. $\qquad\qquad\qquad\qquad\qquad\qquad\qquad\qquad\quad\square$

4.2.6 A double category of monoidal categories

The strict double category $\mathbb{D}bl$ has a totally full double subcategory $\mathbb{M}nc$ of monoidal categories.

As we have seen in 3.3.4, a monoidal category $\mathbf{A}$ is viewed as a vertical weak double category on one formal object $*$, with vertical arrows $A \colon * \rightarrowtail *$ and cells $a \colon A \to A'$; the horizontal composition is the ordinary one, the vertical composition is the tensor product, with identity $E = e_*$.

A horizontal arrow of $\mathbb{M}nc$ is a *monoidal functor* F (lax with respect to tensor product), with comparison arrows $\underline{F} = \underline{F}(*) \colon E \to FE$ and $\underline{F}(A, A') \colon FA \otimes FA' \to F(A \otimes A')$. A vertical arrow is a *comonoidal functor*, which is colax.

A cell $\pi \colon (U \, {}^{F}_{G} \, V)$ as in (4.12) associates to every object A in $\mathbf{A}$ an arrow $\pi A \colon VFA \to GUA$ in $\mathbf{D}$, satisfying the conditions (c.1–3) of 4.2.1. These conditions amount to the commutativity of the diagrams below

$$
\begin{array}{ccc}
VFA & \xrightarrow{\;\pi A\;} & GUA \\
{\scriptstyle VFa}\Big\downarrow & & \Big\downarrow{\scriptstyle GUa} \\
VFA' & \xrightarrow[\;\pi A'\;]{} & GUA'
\end{array}
\qquad\qquad
\begin{array}{ccc}
VFE & \xrightarrow{\;\pi E\;} & GUE \\
{\scriptstyle V\underline{E}}\Big\uparrow & & \Big\downarrow{\scriptstyle G\underline{U}} \\
VE & \xrightarrow[\;\underline{V}\;]{} E \xrightarrow[\;\underline{G}\;]{} & GE
\end{array}
\qquad (4.25)
$$

$$VF(A \otimes A') \xrightarrow{\pi(A \otimes A')} GU(A \otimes A')$$

$$V\underline{F} \Big\uparrow \qquad\qquad\qquad \Big\downarrow GU$$

$$V(FA \otimes FA') \qquad\qquad G(UA \otimes UA') \qquad\qquad (4.26)$$

$$\underline{V}F \Big\downarrow \qquad\qquad\qquad \Big\uparrow \underline{G}U$$

$$VFA \otimes VFA' \xrightarrow[\pi A \otimes \pi A']{} GUA \otimes GUA'$$

The horizontal and vertical composition are defined in (4.15)

$$(\pi \mid \rho)(A) = (G'\pi A).(\rho FA) \colon WF'FA \to G'VFA \to G'GUA,$$
$$(\pi \otimes \sigma)(A) = (\sigma UA).(V'\pi A) \colon V'VFA \to V'GUA \to HU'UA. \qquad (4.27)$$

4.2.7 *Exercise* (Internal monoids)

For this double category $\mathbb{M}\mathrm{nc}$, prove that there is a lax double functor Mon: $\mathbb{M}\mathrm{nc} \to \mathbb{C}\mathrm{at}$ that sends a (small) monoidal category $\mathbf{A}$ to the category Mon($\mathbf{A}$) of its internal monoids.

4.2.8 *Exercises and complements*

(a) *(Unexpected categories within a double category)* The construction of the category $\mathrm{HV}(\mathbb{A}, \mathbb{B})$ in Exercise 4.2.3(e) can be extended, replacing $\mathbb{D}\mathrm{bl}$ with any double category $\mathbb{D}$. We fix two objects A, B in $\mathbb{D}$ and want to construct a category $\mathrm{HV}_{\mathbb{D}}(A, B)$, whose set of objects is the disjoint union of the set of horizontal and vertical arrows of $\mathbb{D}$.

One should now define the new morphisms and their composition, and prove that they do form a category.

(b) We have already seen the particular case where $\mathbb{D} = \mathbb{D}\mathrm{bl}$. The interested reader can work out the case $\mathbb{D} = \mathbb{A}\mathrm{dj}\mathbf{Cat}$, and prove the following general result: the horizontal arrow $f \colon A \to B$ and the vertical arrow $u \colon A \nrightarrow B$ are companions in $\mathbb{D}$ (see 41.1) if and only if they are isomorphic objects of $\mathrm{HV}_{\mathbb{D}}(A, B)$.

4.3 Main definitions

We define the general colax-lax adjunctions between weak double categories, as internal adjunctions in **Dbl**. Corollary 4.3.7 essentially says that a colax double functor can only have a lax right adjoint.

Most of the proofs of this section are deferred to their infinite-dimensional extension, in Section 7.3, where induction on degree works freely.

4.3.1 Double adjunctions

A *(colax-lax) adjunction* $(\eta, \varepsilon)\colon F \dashv G$ between weak double categories is defined as an orthogonal adjunction *in* the double category $\mathbb{D}bl$ (of 4.2.1), between the horizontal arrow G and the vertical arrow F; a situation which occurs naturally in various situations.

Thus the left adjoint $F\colon \mathbb{X} \to \mathbb{A}$ is a colax functor between weak double categories, the right adjoint $G\colon \mathbb{A} \to \mathbb{X}$ is lax, and we have two $\mathbb{D}bl$-cells η, ε, called unit and counit, that satisfy the triangle equations

$$
\begin{array}{ccc}
\mathbb{X} = \mathbb{X} & \qquad \mathbb{A} \xrightarrow{G} \mathbb{X} & \qquad \eta \otimes \varepsilon = 1_F, \\[2pt]
\left\downarrow F \quad \eta \quad \right\| & \qquad \| \quad \varepsilon \quad \downarrow F & \\[2pt]
\mathbb{A} \xrightarrow[G]{} \mathbb{X} & \qquad \mathbb{A} = \mathbb{A} & \qquad \varepsilon \,|\, \eta = e_G.
\end{array}
\tag{4.28}
$$

All this will be analysed below. (Also here the arrow of a colax double functor is marked with a dot *when displayed vertically*, in a diagram of $\mathbb{D}bl$.)

We speak of a *pseudo-lax* (resp. a *colax-pseudo*) adjunction when the left (resp. right) adjoint is pseudo, and of a *pseudo* adjunction when both adjoints are pseudo; the term *pseudo* is replaced with *strict* if it is the case.

From general properties of adjoint arrows (in 4.1.2), we already know that the left adjoint of a lax double functor G is determined up to horizontal isomorphism (a special isocell between vertical arrows in $\mathbb{D}bl$) and that left adjoints compose, contravariantly with respect to their right adjoints.

By an abuse of notation, we may write $\eta\colon 1 \dashrightarrow GF$ and $\varepsilon\colon FG \dashrightarrow 1$, as in 4.2.1. But let us recall once more that there are no composites GF and FG in $\mathbb{D}bl$: the coherence conditions of η and ε are based on the comparison cells of F and G. A general colax-lax adjunction cannot be presented as an internal adjunction in some 2-category, but we shall see in the next section that this is possible in the pseudo-lax or the colax-pseudo case.

4.3.2 A description

A colax-lax adjunction $(\eta, \varepsilon)\colon F \dashv G$ between the weak double categories $\mathbb{X}$ and $\mathbb{A}$ consists thus of:

(a) a colax double functor $F\colon \mathbb{X} \to \mathbb{A}$, with comparison cells

$$\underline{F}X\colon F(e_X) \to e_{FX}, \qquad \underline{F}(u, u')\colon F(u \otimes u') \to Fu \otimes Fu'),$$

(b) a lax double functor $G\colon \mathbb{A} \to \mathbb{X}$, with comparison cells

$$\underline{G}A\colon e_{GA} \to G(e_A), \qquad \underline{G}(v, v')\colon Gv \otimes Gv' \to G(v \otimes v'),$$

(c) two ordinary adjunctions in degree $n = 0, 1$, consistent under domain and codomain and coherent with the comparisons of F and G

$$\eta_n: 1 \to G_n F_n: \mathbf{X}_n \to \mathbf{X}_n, \qquad \varepsilon_n: F_n G_n \to 1: \mathbf{A}_n \to \mathbf{A}_n,$$

$$\varepsilon_n F_n . F_n \eta_n = 1_{F_n}, \qquad G_n \varepsilon_n . \eta_n G_n = 1_{G_n}.$$

Explicitly, point (c) means that we are assigning the following components

- horizontal maps $\eta X: X \to GFX$ and cells $\eta u: (u \; \overset{\eta X}{\underset{\eta X'}{}} \; GFu)$ in $\mathbb{X}$,

- horizontal maps $\varepsilon A: FGA \to A$ and cells $\varepsilon v: (FGv \; \overset{\varepsilon A}{\underset{\varepsilon A'}{}} \; v)$ in $\mathbb{A}$,

under the axioms below (naturality and triangular equations are stated in degree n = 1, which is sufficient).

(ad.1) (*Naturality*) For $\xi: (u \; \overset{f}{\underset{f'}{}} \; u')$ in $\mathbb{X}$ and $\alpha: (v \; \overset{g}{\underset{g'}{}} \; v')$ in $\mathbb{A}$

$$\xi \,|\, \eta u' = \eta u \,|\, GF\xi, \qquad\qquad FG\alpha \,|\, \varepsilon v' = \varepsilon v \,|\, \alpha.$$

(ad.2) (*Triangle equations*) For u in $\mathbb{X}$ and v in $\mathbb{A}$

$$F\eta u \,|\, \varepsilon Fu = 1_{Fu}, \qquad\qquad \eta Gv \,|\, G\varepsilon v = 1_{Gv}.$$

(ad.3) (*Coherence with vertical identities*) For X in $\mathbb{X}$ and A in $\mathbb{A}$

$$\eta e_X \,|\, G\underline{F}X = e_{\eta X} \,|\, \underline{G}FX, \tag{4.29}$$

$$F\underline{G}A \,|\, \varepsilon e_A = \underline{F}GA \,|\, e_{\varepsilon A}. \tag{4.30}$$

(ad.4) (*Coherence with vertical composition*) For $u'' = u \otimes u'$ in $\mathbb{X}$ and $v'' = v \otimes v'$ in $\mathbb{A}$

$$\eta u'' \,|\, G\underline{F}(u, u') \;=\; (\eta u \otimes \eta u') \,|\, \underline{G}(Fu, Fu'), \tag{4.31}$$

$$F\underline{G}(v, v') \,|\, \varepsilon v'' \;=\; \underline{F}(Gv, Gv') \,|\, (\varepsilon v \otimes \varepsilon v'), \tag{4.32}$$

$$FGA \;=\!=\; FGA \longrightarrow A$$

$$\begin{array}{c} F(Gv\otimes Gv') \downarrow \quad FG \quad \downarrow \quad \varepsilon v'' \quad \downarrow v'' \\ FGA'' \;=\!=\; FGA'' \longrightarrow A'' \end{array} = \begin{array}{c} FGA \;=\!=\; FGA \longrightarrow A \\ \downarrow \quad FGv \downarrow \quad \varepsilon v \quad \downarrow v \\ \quad \underline{FG}\; FGA' \longrightarrow A' \\ \downarrow \quad FGv' \downarrow \quad \varepsilon v' \quad \downarrow v' \\ FGA'' \;=\!=\; FGA'' \longrightarrow A'' \end{array}$$

4.3.3 Exercises and complements

(a) (*Composition*) Prove that two consecutive adjunctions can be composed, by a suitable pasting of cells in $\mathbb{Dbl}$.

(b) In the colax-lax adjunction $F \dashv G$ the comparison cells of G and the unit η determine the comparison cells of F.

(c) (*Unitarity*) We say that the adjunction is *unitary* if the 'functors' F and G are. Prove that:

(i) if G is unitary, then F is unitary if and only if $\eta e_X = e_{\eta X}$ (for all X in $\mathbb{X}$),

(i*) if F is unitary, then G is unitary if and only if $\varepsilon e_A = e_{\varepsilon A}$ (for all A in $\mathbb{A}$).

4.3.4 Theorem (Characterisation by hom-sets)

An adjunction $(\eta, \varepsilon)\colon F \dashv G$ can equivalently be given by a colax double functor $F\colon \mathbb{X} \to \mathbb{A}$, a lax double functor $G\colon \mathbb{A} \to \mathbb{X}$ and two functorial isomorphisms L_0 and L_1

$$L_n\colon \mathbf{A}_n(F_n-,.) \to \mathbf{X}_n(-,G_n.)\colon \mathbf{X}_n^{\mathrm{op}} \times \mathbf{A}_n \to \mathbf{Set} \qquad (n=0,1), \quad (4.33)$$

whose components are consistent with domain, codomain and the vertical structure (through the comparison cells of F and G), i.e. satisfy the following conditions:

(adh.0) L_0 has components $L(X, A)\colon \mathbf{A}_0(FX, A) \to \mathbf{X}_0(X, GA)$,

(adh.1) L_1 has components $L(u, v)\colon \mathbf{A}_1(Fu, v) \to \mathbf{X}_1(u, Gv)$, that take a cell $\alpha\colon (Fu \,{}^{g}_{g'}\, v)$ to a cell $L\alpha\colon (u \,{}^{Lg}_{Lg'}\, Gv)$,

(adh.2) $L(\underline{F}X \,|\, e_g) = e_{Lg} \,|\, \underline{G}A$,

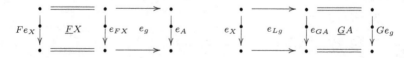

(adh.3) $L(\underline{F}(u, u') \mid \alpha \otimes \alpha') = (L\alpha \otimes L\alpha') \mid \underline{G}(v, v')$,

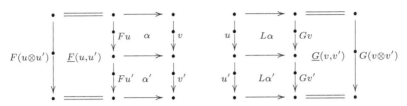

Proof See the infinite-dimensional extension of this result, in 7.3.4. □

4.3.5 Corollary (Characterisation by commas)

An adjunction amounts to an isomorphism L of weak double categories, over the product $\mathbb{A} \times \mathbb{B}$

$$F \Downarrow \mathbb{B} \xrightarrow{\quad L \quad} \mathbb{A} \Downarrow G$$
$$\searrow \qquad \swarrow$$
$$\mathbb{A} \times \mathbb{B}$$

(4.34)

Proof It is a straightforward consequence of the previous theorem. □

4.3.6 Theorem (Right adjoint by universal properties)

Given a colax double functor $F: \mathbb{X} \to \mathbb{A}$, the existence (and choice) of a right adjoint lax double functor G amounts to two conditions (rad.0, 1) *(including two choices):*

(rad.0) *for every object A in $\mathbb{A}$ there is a universal arrow*

$$(GA, \varepsilon A: FGA \to A)$$

from the functor F_0 to the object A (and we choose one),

(rad.1) *for every vertical map $v: A \nrightarrow A'$ in $\mathbb{A}$ there is a universal arrow*

$$(Gv, \varepsilon v), \qquad Gv: GA \nrightarrow GA', \qquad \varepsilon v: (FGv \; {}^{\varepsilon A}_{\varepsilon A'} \; v),$$

from the functor F_1 to the object v of $\mathbf{A}_1$ (and we choose one, consistently with the previous choice).

Explicitly, these universal properties mean that:

(i) for each X in $\mathbb{X}$ and $g: FX \to A$ there is a unique $f: X \to GA$ such that $g = \varepsilon A.Ff: FX \to A$,

(ii) for each $u\colon X \rightarrowtail X'$ in $\mathbb{X}$ and each cell $\alpha\colon (Fu \begin{smallmatrix} g \\ g' \end{smallmatrix} v)$ in $\mathbb{A}$ there is a unique cell $\xi\colon (u \begin{smallmatrix} f \\ f' \end{smallmatrix} Gv)$ in $\mathbb{X}$ such that $\alpha = F\xi \,|\, \varepsilon v$.

The comparison special cells of G

$$\underline{G}(A)\colon e_{GA} \to G(e_A), \qquad \underline{G}(v,v')\colon Gv \otimes Gv' \to G(v \otimes v'),$$

are thus determined by those of F, as the unique solution of the equations (4.30) and (4.32), respectively

$$F\underline{G}A \,|\, \varepsilon e_A \;=\; \underline{F}GA \,|\, e_{\varepsilon A},$$
$$F\underline{G}(v,v') \,|\, \varepsilon v'' \;=\; \underline{F}(Gv,Gv') \,|\, (\varepsilon v \otimes \varepsilon v'). \tag{4.35}$$

Proof See 7.3.6.　　　　　　　　　　　　　　　　　　　　　　　　□

4.3.7 Corollary (Right adjoint by components)

A colax double functor $F\colon \mathbb{X} \to \mathbb{A}$ has a lax right adjoint if and only if each component $F_n\colon \mathrm{Hor}_n(\mathbb{X}) \to \mathrm{Hor}_n(\mathbb{A})$ $(n = 0,1)$ has an (ordinary) right adjoint $G_n\colon \mathrm{Hor}_n(\mathbb{A}) \to \mathrm{Hor}_n(\mathbb{X})$, consistently with vertical domain and codomain:

$$G_1(v)\colon GA \rightarrowtail GA', \qquad \text{for } v\colon A \rightarrowtail A' \text{ in } \mathbb{A},$$
$$G_1(\alpha)\colon (G_1 v \begin{smallmatrix} Gg \\ Gg' \end{smallmatrix} G_1 v'), \qquad \text{for } \alpha\colon (v \begin{smallmatrix} g \\ g' \end{smallmatrix} v') \text{ in } \mathbb{A}. \tag{4.36}$$

Note. Loosely speaking, we are saying that a colax double functor can only have a lax right adjoint.

Proof A straightforward consequence of the previous theorem.　　　□

4.3.8 Theorem (Graph factorisation of adjunctions)

Let $F \dashv G$ be a colax-lax adjunction between $\mathbb{X}$ and $\mathbb{A}$. Using the isomorphism of double categories $L\colon F \Downarrow \mathbb{A} \to \mathbb{X} \Downarrow G$ (of Corollary 4.3.5) we can factorise the adjunction as a composite of three adjunctions:

$$\mathbb{X} \; \underset{P}{\overset{F'}{\rightleftarrows}} \; F \Downarrow \mathbb{A} \; \underset{L^{-1}}{\overset{L}{\rightleftarrows}} \; \mathbb{X} \Downarrow G \; \underset{G'}{\overset{Q}{\rightleftarrows}} \; \mathbb{A} \tag{4.37}$$

$$F = QLF', \qquad\qquad G = PL^{-1}G',$$

- *a coreflective colax-strict adjunction $F' \dashv P$ (with unit $PF' = 1$),*
- *an isomorphism $L \dashv L^{-1}$,*

- *a reflective strict-lax adjunction $Q \dashv G'$ (with counit $QG' = 1$),*
where the comma-projections P and Q are strict double functors.

Proof See 7.3.8. □

4.4 Double adjunctions and pseudo double functors

We now consider adjunctions where the left or right adjoint is pseudo, and
we introduce adjoint equivalences of weak double categories.

Again the proofs are deferred to their infinite-dimensional extension, in
Section 7.4.

4.4.1 Theorem (Companions in $\mathbb{D}$bl)

*A lax double functor G has an orthogonal companion F in $\mathbb{D}$bl if and only
if it is pseudo. Then one can define $F = G_*$ as the colax double functor
which coincides with G except for comparison cells $\underline{F} = \underline{G}^{-1}$, horizontally
inverse to those of G.*

Proof See 7.4.2. □

4.4.2 Particular double adjunctions

(a) As defined in 4.3.1, a *pseudo-lax* adjunction $F \dashv G$ is a colax-lax adjunction between weak double categories where the left adjoint F is pseudo.
The comparison cells of F are then horizontally invertible and the composites GF and FG are lax double functors, while the unit and counit are
horizontal transformations of such functors (as remarked in 4.2.3(b)).

Therefore, *a pseudo-lax adjunction gives an adjunction in the 2-category*
Lx**Dbl** of weak double categories, lax double functors and horizontal transformations; we shall prove that *these two facts are actually equivalent* (Theorem 4.4.3).

(b) Dually, a *colax-pseudo* adjunction, where the right adjoint G is pseudo,
will amount to an adjunction in the 2-category Cx**Dbl** of pseudo double
categories, colax double functors and horizontal transformations.

(c) Finally, a *pseudo* adjunction, where both F and G are pseudo, will be
the same as an adjunction within the 2-category Ps**Dbl** whose arrows are
the pseudo double functors.

4.4.3 Theorem

(a) (Pseudo-lax adjunctions) *For every adjunction $F \dashv G$ in the 2-category* LxDbl *the functor F is pseudo and we have a pseudo-lax double adjunction, in the sense of 4.3.1 and 4.4.2.*

(b) (Colax-pseudo adjunctions) *For every adjunction $F \dashv G$ in the 2-category* CxDbl *the functor G is pseudo and we have a colax-pseudo double adjunction, in the sense of 4.3.1 and 4.4.2.*

Note. More formally, (a) can be rewritten saying that, in $\mathbb{D}$bl, if the horizontal arrow $G \colon \mathbb{A} \to \mathbb{X}$ has a 'horizontal left adjoint' $H \colon \mathbb{X} \to \mathbb{A}$ (within the horizontal 2-category $\mathbf{Hor}\mathbb{D}$bl $=$ LxDbl), then it also has an orthogonal adjoint $F \colon \mathbb{X} \rightarrowtail \mathbb{A}$ (colax). (Applying Proposition 4.1.4, it would follow that H and F are companions, whence H is pseudo, by 4.4.1, and isomorphic to F.)

Proof See 7.4.3.   □

4.4.4 Equivalences of weak double categories

An *adjoint equivalence* between two weak double categories $\mathbb{X}$ and $\mathbb{A}$ is defined as a pseudo adjunction $(\eta, \varepsilon) \colon F \dashv G$ where the horizontal transformations $\eta \colon 1_X \to GF$ and $\varepsilon \colon FG \to 1_A$ are invertible.

Adjoint equivalences are characterised below. We recall, from 3.5.3, that a *pseudo* functor $F \colon \mathbb{X} \to \mathbb{A}$ is horizontally full (resp. faithful) if and only if $\mathrm{Hor}_1 F$ is full (resp. faithful): this implies that $\mathrm{Hor}_0 F$ is also.

Moreover we say that F is *essentially surjective on vertical arrows* if $\mathrm{Hor}_1 F$ is essentially surjective on objects: for every vertical arrow $v \colon A \rightarrowtail A'$ in $\mathbb{A}$ there is some $u \colon X \rightarrowtail X'$ in $\mathbb{X}$ and some cell $F(u) \to v$, horizontally invertible in $\mathbb{A}$. It follows that $\mathrm{Hor}_0 F$ is also essentially surjective on objects (because F is pseudo).

4.4.5 Theorem (Characterisations of equivalences)

Let $F \colon \mathbb{X} \to \mathbb{A}$ be a pseudo double functor between two horizontally invariant weak double categories. The following conditions are equivalent:

(i) $F \colon \mathbb{X} \to \mathbb{A}$ belongs to an adjoint equivalence of weak double categories,

(ii) $F \colon \mathbb{X} \to \mathbb{A}$ is an equivalence of weak double categories (see 3.5.5),

(iii) F is horizontally full and faithful, and essentially surjective on vertical arrows (see 4.4.4),

(iv) the ordinary functors $\mathrm{Hor}_0 F$ and $\mathrm{Hor}_1 F$ are equivalences of categories,

(v) the ordinary functor $\mathrm{Hor}_1 F$ *is an equivalence of categories.*

Proof Conditions (i) to (iv) are proved to be equivalent in Theorem 7.4.5, the infinite-dimensional extension of this result. Conditions (iv) and (v) are equivalent by 1.2.5 and 4.4.4. □

4.5 Computing double adjunctions

The (strict or weak) double categories of Sections 3.1 and 3.4 are connected by various adjunctions. Many of them are presented here as exercises. The non-obvious solutions can be found in Appendix C, or here when they are to be used in the next chapters.

The 'pushout-pullback adjunction' between the weak double categories of spans and cospans is an interesting example, studied in detail in 4.5.6–4.5.7; it will be extended to infinite dimension in Section 7.1, and reconsidered in Section 8.5.

4.5.1 Exercises on preordered sets and L-metric spaces

We have seen in Section 3.4 the double categories

- pOrd of preordered sets, increasing mappings and preorder profunctors,

- $\mathbb{M}\mathrm{tr} = \mathbb{R}_+\mathrm{Cat}$ of L-metric spaces, weak contractions and profunctors.

The canonical embedding $M \colon \mathrm{pOrd} \to \mathbb{M}\mathrm{tr}$ identifies a preordered set with an L-metric space having distance in $\{0, \infty\}$; similarly a preorder profunctor $u \colon X \nrightarrow Y$ becomes a profunctor of L-metric spaces with values in $\{0, \infty\}$.

Prove that this embedding is *reflective and lax coreflective*, i.e. has a strict left adjoint P (the reflector) and a lax right adjoint Q (the lax coreflector)

$$M \colon \mathrm{pOrd} \leftrightarrows \mathbb{M}\mathrm{tr} \colon P, Q, \qquad P \dashv M \dashv Q. \qquad (4.38)$$

4.5.2 Exercises on profunctors and spans

We have seen in 1.5.3(e) that in dimension one there is a chain of ordinary adjunctions between **Cat** and **Set**

$$F = \mathrm{Ob} \colon \mathbf{Cat} \to \mathbf{Set}, \qquad \pi_0 \dashv D \dashv F \dashv C. \qquad (4.39)$$

(a) The reader will extend part of this chain to double adjunctions between the weak double categories $\mathbb{C}$at (of categories, functors and profunctors) and $\mathbb{S}$pan$\mathbf{Set}$ (of sets, mappings and spans)

$$F\colon \mathbb{C}\text{at} \to \mathbb{S}\text{pan}\mathbf{Set}, \qquad \pi_0 \dashv D \dashv F. \qquad (4.40)$$

(b) Show that F cannot have a right adjoint.

(c) Show that the strict double functor D (a left adjoint!) *does not preserve cotabulators* (which are colimits). Show that the lax functor F is not pseudo unitary.

We shall see, in Theorem 5.5.6, that these two facts are linked; on the other hand, unitary adjunctions are well related to double limits (by Theorem 5.5.5).

4.5.3 Exercises on relations and spans

We have already seen in (3.70) the split *lax embedding* S

$$S\colon \mathbb{R}\text{el}\mathbf{Set} \to \mathbb{S}\text{pan}\mathbf{Set}, \quad R\colon \mathbb{S}\text{pan}\mathbf{Set} \to \mathbb{R}\text{el}\mathbf{Set} \quad (RS = 1). \qquad (4.41)$$

S is the *lax* double functor that takes a relation $u \subset X \times Y$ to the jointly monic span $Su = (X \leftarrow u \to Y)$. In fact, Su is the span associated to the tabulator of u, so that the lax functor S is the span representation of $\mathbb{R}\text{el}\mathbf{Set}$ (see 3.7.1). R is the obvious (strict) functor taking a span to the associated relation.

The reader will verify that this pair forms a strict-lax adjunction $R \dashv S$, trivial on the objects, with counit $\varepsilon = 1\colon RS \to 1$ and the obvious unit linking a span to the associated jointly monic span

$$\eta\colon \text{id}(\mathbb{S}\text{pan}\mathbf{Set}) \to SR, \qquad \eta X = 1_X, \quad \eta u\colon u \to SRu. \qquad (4.42)$$

The (idempotent) lax monad $T = SR$ will be extended in Section 8.5 to the weak double category $\mathbb{S}\text{pan}\mathbf{C}$ of spans over a regular category $\mathbf{C}$, yielding $\mathbb{R}\text{el}(\mathbf{C})$ as its double category of 'strictly algebraic' spans.

4.5.4 Exercises on order profunctors and profunctors

Prove that the adjunction $(\text{po}, U)\colon \mathbf{Cat} \nrightarrow \mathbf{pOrd}$ of 1.5.3(f) can be extended to a double adjunction between the weak double categories $\mathbb{C}$at and $\mathbb{p}\mathbb{O}$rd.

4.5.5 Exercises on profunctors and cospans

We have seen in Exercise 3.7.4(a) that $\mathbb{C}$at is cospan representable, by a colax functor C which is the identity in degree 0 and essentially defined by the cotabulator $\perp p$ of a profunctor $p \colon X \nrightarrow Y$, in degree 1

$$C \colon \mathbb{C}\mathrm{at} \to \mathbb{C}\mathrm{osp}\mathbf{Cat}, \quad C(p) = (X \to X +_p Y \leftarrow Y). \tag{4.43}$$

The reader can describe the colaxity comparisons of C, and prove that C has a right adjoint. The solution will be used in the next chapters, and is given here.

Solution. We recall that $\perp p = X +_p Y$ is the collage of X and Y along p, with *additional maps* $\lambda \colon x \dashrightarrow y$ in the set $p(x, y)$. The profunctor p is represented as a cospan of categories and functors

$$X \xrightarrow{p'} U \xleftarrow{p''} Y \qquad (U = X +_p Y), \tag{4.44}$$

of a particular kind, characterised by the following conditions:

(i) p', p'' are disjoint embeddings of full subcategories that cover all the objects of U,

(ii) U has no arrows from an object of $p''(Y)$ to an object of $p'(X)$.

For a composed profunctor $m = p \otimes q \colon X \nrightarrow Y \nrightarrow Z$, the central object of the *composed cospan* $C(p) \otimes C(q)$ is the pushout

$$W = (X +_p Y) +_Y (Y +_q Z),$$

while the central object $X +_m Z$ of the *composed profunctor* m is the full subcategory of W determined by $\mathrm{Ob}(X) \cup \mathrm{Ob}(Z)$. The colaxity comparison 'is' the embedding of this subcategory

$$\begin{aligned} &\underline{C}(p, q) \colon C(p \otimes q) \to C(p) \otimes C(q), \\ &X +_m Z \to (X +_p Y) +_Y (Y +_q Z), \end{aligned} \tag{4.45}$$

and the associator for ternary compositions is a restriction of the associator of cospans.

C has a (lax) right adjoint P, which is trivial again in degree 0 and defined as follows, on a cospan of categories $u = (u', u'') = (X \to U \leftarrow Y)$ and a cell $(f, h, g) \colon (u', u'') \to (v', v'')$

$$P \colon \mathbb{C}\mathrm{osp}\mathbf{Cat} \to \mathbb{C}\mathrm{at},$$

$$P(u) = U(u'(-), u''(=)) \colon X^{\mathrm{op}} \times Y \to \mathbf{Set}, \tag{4.46}$$

$$P(f, h, g) \colon P(u) \to P(v), \quad P(f, h, g)(x, y)(\lambda) = h(\lambda).$$

Take a composed cospan $w = u \otimes v \colon X \to Y \to Z$, computed by the pushout category W

$$
\begin{array}{c}
W \\
m' \nearrow \quad \nwarrow m'' \\
U \qquad V \\
u' \nearrow \quad \nwarrow u'' \quad v' \nearrow \quad \nwarrow v'' \\
X \qquad Y \qquad Z
\end{array}
\qquad
\begin{array}{l}
w' = m'u', \\[1.5em]
w'' = m''v''.
\end{array}
$$

Here the comparison $\underline{P}$ works as follows, on formal arrows $\lambda \colon x \dashrightarrow y$ and $\mu \colon y \dashrightarrow z$ (i.e. $\lambda \colon u'(x) \to u''(y)$ in U and $\mu \colon v'(y) \to v''(z)$ in V):

$$
\underline{P}(u, v) \colon P(u) \otimes P(v) \to P(w),
$$
$$
(\lambda \otimes \mu \colon x \dashrightarrow z) \mapsto (m''(\mu).m'(\lambda) \colon w'(x) \dashrightarrow w''(z)).
\tag{4.47}
$$

Now the composite PC is the identity of $\mathbb{C}\mathrm{at}$, since

$$
PC(p) = (X +_p Y)(i'(-), i''(=)) = p.
$$

The unit of the adjunction is $\mathrm{id} \colon 1 \to PC \colon \mathbb{C}\mathrm{at} \to \mathbb{C}\mathrm{at}$. The counit ε is trivial in degree zero and defined as follows on a cospan $u = (u', u'')$

$$
\varepsilon \colon CP \to 1 \colon \mathbb{C}\mathrm{osp}\mathbf{Cat} \to \mathbb{C}\mathrm{osp}\mathbf{Cat},
$$
$$
\varepsilon u = (1_X, \varepsilon_u, 1_Y) \colon (X \to X +_u Y \leftarrow Y) \to (X \to U \leftarrow Y),
\tag{4.48}
$$
$$
\varepsilon_u|_X = u', \quad \varepsilon_u|_Y = u'', \quad \varepsilon_u(\lambda \colon x \dashrightarrow y) = \lambda.
$$

The (strictly idempotent) colax comonad $S = CP$ will be dealt with in Section 8.6; it determines profunctors as 'strictly coalgebraic' cospans.

4.5.6 Exercises and complements (Pushout-pullback adjunction)

Let $\mathbf{C}$ be a category with pullbacks and pushouts. Extending an example sketched in Section 0.4 of the general Introduction, the reader can prove that the weak double categories $\mathbb{S}\mathrm{pan}\mathbf{C}$ and $\mathbb{C}\mathrm{osp}\mathbf{C}$ of spans and cospans over $\mathbf{C}$ are connected by a colax-lax adjunction, computed by pushouts and pullbacks

$$
F \colon \mathbb{S}\mathrm{pan}\mathbf{C} \rightleftarrows \mathbb{C}\mathrm{osp}\mathbf{C} \colon G, \qquad \eta \colon 1 \dashrightarrow GF, \quad \varepsilon \colon FG \dashrightarrow 1,
\tag{4.49}
$$

It will be called the *pushout-pullback adjunction* of $\mathbf{C}$. The adjunction is unitary and trivial in degree zero. The solution is below, for future use.

Solution. By Theorem 3.6.4 we have a lax functor G, namely the span representation of cospans, and a colax functor F, the cospan representation of spans. (Generally such functors do not give representability, as we have seen in 3.7.3.)

In degree 0 (for sets and mappings), everything is an identity. In degree 1 (for vertical arrows and cells), F operates by pushout over spans and cells, and G by pullbacks. The special cell $\eta u \colon u \to GFu$ comes from the universal property of the pullback $G(Fu)$, as in the left diagram below

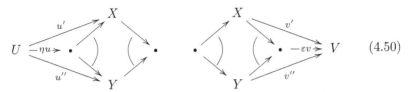

$$(4.50)$$

The counit $\varepsilon v \colon FGv \to v$ turns up in a dual way, as in the right diagram above.

The comparison cell $\underline{F}(u,v) \colon F(u \otimes v) \to Fu \otimes Fv$ is given by the natural morphism from the pushout W'' of $u \otimes v = (u'z', v''z'')$ to the vertex W' of the cospan $Fu \otimes Fv$

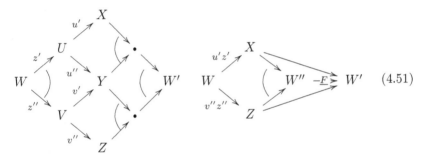

$$(4.51)$$

Since we agreed to follow the unitarity constraint for the choice of pullbacks and pushouts in $\mathbf{C}$, the adjunction is *unitary*, in the sense that this property holds for the weak double categories $\mathbb{Span}\mathbf{C}$, $\mathbb{Cosp}\mathbf{C}$ and the colax-lax functors F, G.

Now, it is easy to see, in the diagram below, that the spans u and GFu have the same cocones (i.e. the same commuting cospans)

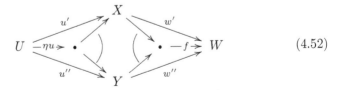

$$(4.52)$$

In fact, if the cospan (w', w'') commutes with the span GFu, then it obviously commutes with u; conversely, if it commutes with u, then it factorises through its pushout Fu, and commutes with the pullback GFu of the latter.

It follows that the spans u and GFu have the same pushout, and $F(\eta u)$ is invertible. It will be useful to describe this fact saying that two spans $u, \underline{u}$ between the same objects are *equivalent* when they have the same cocones, i.e. the same pushout; then any special cell $u \to \underline{u}$ is made invertible by F. Similarly we speak of *equivalent* cospans $v, \underline{v}$ when they have the same pullback.

Thus the ordinary adjunction of degree 1 is idempotent (see Section 1.7). More precisely, the *double adjunction* $F \dashv G \colon \mathbb{S}\mathrm{pan}\mathbf{C} \rightharpoondown \mathbb{C}\mathrm{osp}\mathbf{C}$ is idempotent, following the definition of Section 8.4. (This double adjunction will be extended to an infinite-dimensional *multiple adjunction* between cubical spans and cospans, in 7.1.2.)

4.5.7 Spans, cospans and bispans

Still working on the previous adjunction (in (4.49)), it is interesting to note that its graph factorisation (see (4.37)) can be realised replacing the two double commas with an (isomorphic) weak double category $\mathbb{B}\mathrm{isp}\mathbf{C}$

$$\mathbb{S}\mathrm{pan}\mathbf{C} \; \underset{P}{\overset{F'}{\rightleftarrows}} \; \mathbb{B}\mathrm{isp}\mathbf{C} \; \underset{G'}{\overset{Q}{\rightleftarrows}} \; \mathbb{C}\mathrm{osp}\mathbf{C} \tag{4.53}$$

$$F = QF', \qquad G = PG'.$$

Its category of objects and horizontal morphisms is $\mathbf{C}$, while a vertical arrow $X \rightarrowtail Y$ is a *bispan*, or *diamond* $u \colon \lozenge \to \mathbf{C}$, defined on the 'formal diamond' below

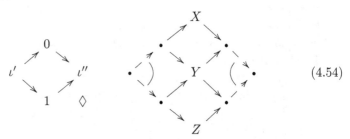

$$\tag{4.54}$$

In fact, the category $\lozenge$ is isomorphic to the formal square $\mathbf{2} \times \mathbf{2}$, *but* we are using its structure as a formal interval, with faces $0, 1 \colon \{*\} \to \lozenge$ that determine the domain and codomain of a bispan. A vertical composition $X \rightarrowtail Y \rightarrowtail Z$ is computed by a pullback and a pushout, as in the right diagram above. Finally, F' acts by pushout, G' by pullback, while P and Q are projections.

Of course, $\mathbb{B}\mathrm{isp}\mathbf{C}$ has little to do with the double category $\mathbb{Q}\mathbf{C}$ of commutative squares of $\mathbf{C}$.

4.5.8 Extending adjunctions

(a) We start from an adjunction $F \dashv G$ between ordinary categories

$$F: \mathbf{X} \rightleftarrows \mathbf{A}: G, \qquad \eta: 1 \to GF, \quad \varepsilon: FG \to 1. \qquad (4.55)$$

If $\mathbf{X}$ and $\mathbf{A}$ have (a fixed choice of) pullbacks, this adjunction can be extended in a natural way to a unitary *colax-pseudo* adjunction between the weak double categories of spans

$$\mathbb{S}\mathrm{pan}(F): \mathbb{S}\mathrm{pan}(\mathbf{X}) \rightleftarrows \mathbb{S}\mathrm{pan}(\mathbf{A}): \mathbb{S}\mathrm{pan}(G), \qquad (4.56)$$

by applying to (4.55) the 2-functor $\mathbb{S}\mathrm{pan}: \mathbf{Cat}_{\mathrm{pb}} \to \mathrm{CxDbl}$ of 3.7.5.

(b) Similarly, starting from the adjunction (4.55) and supposing that $\mathbf{X}$ and $\mathbf{A}$ have pushouts, one can apply the 2-functor $\mathbb{C}\mathrm{osp}: \mathbf{Cat}_{\mathrm{po}} \to \mathrm{LxDbl}$ of 3.7.5, to obtain a *pseudo-lax* adjunction of weak double categories

$$\mathbb{C}\mathrm{osp}(F): \mathbb{C}\mathrm{osp}(\mathbf{X}) \rightleftarrows \mathbb{C}\mathrm{osp}(\mathbf{A}): \mathbb{C}\mathrm{osp}(G). \qquad (4.57)$$

(c) Generalising another example sketched at the beginning of the general Introduction, we shall prove in A4.6 that an adjunction $F \dashv G$ between abelian categories, or more generally Puppe-exact categories, can be extended to a colax-lax adjunction $\mathbb{R}\mathrm{el}F \dashv \mathbb{R}\mathrm{el}G$ between double categories of relations. Moreover $\mathbb{R}\mathrm{el}F$ (resp. $\mathbb{R}\mathrm{el}G$) is a double functor if and only if the functor F (resp. G) is exact.

This can be applied to the exponential-law adjunction in the category $\mathbf{A} = R\,\mathbf{Mod}$ of modules on a commutative ring, for a fixed R-module A (see 2.2.1(c))

$$F: \mathbf{A} \rightleftarrows \mathbf{A}: G, \qquad F(X) = X \otimes_R A, \quad G(Y) = \mathrm{Hom}_R(A, Y). \qquad (4.58)$$

*Here the functor F is exact when A is a *flat* R-module, while G is exact when A is *projective*. In $\mathbf{Ab}$, flat means torsion-free, while projective is equivalent to free (as an abelian group). All this can be found in any text on Homological Algebra, e.g. in [M2].*

4.5.9 *The cocomma-comma adjunction

Let $\mathbf{C}$ be a 2-category with pullbacks, pushouts, comma and cocomma squares (defined by the universal property mentioned in 1.5.6 and its dual). The weak double categories $\mathbb{S}\mathrm{pan}\mathbf{C}$ and $\mathbb{C}\mathrm{osp}\mathbf{C}$ are defined as usual (and only depend on the 1-dimensional structure of $\mathbf{C}$).

There is now a second colax-lax adjunction, computed by cocommas and commas

$$H: \mathbb{S}\mathrm{pan}\mathbf{C} \rightleftarrows \mathbb{C}\mathrm{osp}\mathbf{C}: K, \qquad \eta: 1 \dashrightarrow KH, \quad \varepsilon: HK \dashrightarrow 1, \qquad (4.59)$$

where *neither H nor K* is unitary.

In degree 0 (for objects and morphisms of **C**), everything is an identity. In degree 1 (for vertical arrows and cells), H operates by cocommas (ϑ) over spans and cells, and K by commas (κ) over cospans; the special cells ηu and εv are determined by the equations $\eta u \,|\, \kappa = \vartheta$ and $\vartheta' | \varepsilon v = \kappa'$

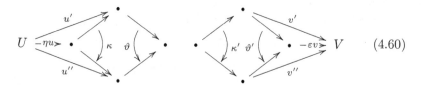

$$(4.60)$$

The reader can check that H is a colax double functor (while K is lax), with comparison cells $\underline{H}(u, v)$ for vertical composition given by the natural morphism from the cocomma $H(u'x', v''x'')$ to the cospan $Hu \otimes Hv$

$$\underline{H}(u, v) \colon H(u \otimes v) \to Hu \otimes Hv,$$
$$\underline{H}(u, v).\vartheta(u'x', v''x'') = (y'.\vartheta u.x') \otimes (y''.\vartheta v.x''),$$

$$(4.61)$$

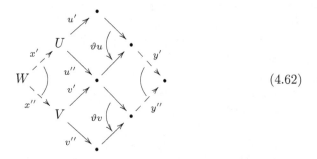

$$(4.62)$$

The factorisation (4.37) can be realised much as above, in 4.5.7, replacing commutative squares with quintets, in the vertical arrows of the central weak double category. Again this structure has little to do with the double category $\mathbb{Q}\mathbf{C}$ of quintets of **C** (defined in 3.1.4).

4.6 *Weak double categories as pseudo algebras

For a 2-monad T on a 2-category **C** we build a double category of pseudo T-algebras, with lax and colax morphisms as horizontal and vertical arrows. Then we show how strict and weak double categories can be seen, respectively, as algebras and normal pseudo algebras for an obvious 2-monad on graphs of categories.

Algebras and pseudo algebras for a 2-monad are studied, for instance, in [Bur, BlKP, Fi].

4.6.1 Pseudo algebras for a 2-monad

We have a 2-monad $T = (T, h, m)$ on the 2-category $\mathbf{C}$. This means a 2-functor $T\, \mathbf{C} \to \mathbf{C}$ with 2-natural transformations $h\colon 1 \to T$ and $m\colon T^2 \to T$ satisfying the usual axioms.

A *pseudo algebra* in $\mathbf{C}$ is a quadruple (A, a, ω, κ) consisting of an object A of $\mathbf{C}$, a map $a\colon TA \to A$ (the *structure*) and two vertically invertible cells (the *comparisons*)

$$\omega\colon 1_A \to a.hA \qquad\qquad \text{(the \textit{normaliser})},$$
$$\kappa\colon a.Ta \to a.mA \qquad \text{(the \textit{extended associator})}, \tag{4.63}$$

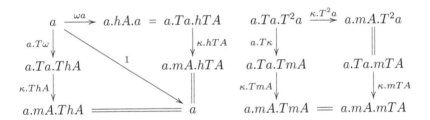

These data have to satisfy three conditions of coherence:

$$a.T\omega \otimes \kappa.ThA = 1_a = \omega a \otimes \kappa.hTA,$$
$$\kappa.T^2a \otimes \kappa.mTA = a.T\kappa \otimes \kappa.TmA, \tag{4.64}$$

A (strict) *morphism of pseudo algebras* $f\colon (A, a, \omega, \kappa) \to (B, b, \omega, \kappa)$ is a morphism $f\colon A \to B$ of $\mathbf{C}$ which preserves the structure (note that the comparisons of pseudo algebras are always denoted by the same letters, ω and κ):

$$b.Tf = f.a, \qquad \omega f = f\omega, \qquad \kappa.T^2f = f\kappa. \tag{4.65}$$

We write as $\mathbf{Psa}(T)$ the category of pseudo algebras and their strict morphisms.

(Obviously, an algebra for T is a pseudo algebra whose comparisons are identities.)

4.6.2 Normal pseudo algebras

We say that a pseudo algebra (A, a, ω, κ) is *normal* if the normaliser ω is the identity, and therefore $a.hA = 1_A$.

Normal pseudo algebras are important, and should not be called 'unitary', as the following example shows. We consider the 2-monad $T \colon \mathbf{Cat} \to \mathbf{Cat}$, where $T(A)$ is the free strict monoidal category over the category A; its objects are the finite families $(x_1, ..., x_n)$ of objects of A.

A strict monoidal structure $a \colon TA \to A$ over a (small) category A

$$a(x_1, ..., x_n) = \otimes x_i, \tag{4.66}$$

gives all finite tensor products in A; the identity object $E = a(\underline{e})$ comes from the empty family $\underline{e}$.

A normal pseudo algebra (A, a, κ) amounts here to an unbiased monoidal category, *with a trivial unary tensor* $\otimes x = x$ (of single objects of A); the two unitors and the binary associator all come from the extended (or *unbiased*) associator κ, that operates on finite tensor products

$$\otimes (\otimes \underline{e}, \otimes x) \to \otimes x, \qquad \otimes (\otimes x, \otimes \underline{e}) \to \otimes x,$$
$$\otimes (\otimes x, \otimes(y, z)) \to \otimes(x, y, z) \leftarrow \otimes (\otimes(x, y), \otimes z) . \tag{4.67}$$

In a general pseudo algebra each object x has an associated object $\otimes x$, isomorphic to x; when the procedure is idempotent, $\otimes x$ can be viewed as a 'normal form' of x.

4.6.3 Lax and colax morphism

Let us come back to a 2-monad $T = (T, h, m)$ on the 2-category $\mathbf{C}$ and its pseudo algebras.

(a) A *lax morphism of pseudo algebras*

$$\mathbf{f} = (f, \varphi) \colon (A, a, \omega, \kappa) \to (B, b, \omega, \kappa)$$

is a morphism $f \colon A \to B$ of $\mathbf{C}$ with a comparison cell φ such that:

$$\varphi \colon b.Tf \to f.a, \qquad \omega f \otimes \varphi.hA = f\omega,$$
$$\kappa.T^2 f \otimes \varphi.mA = b.T\varphi \otimes \varphi.Ta \otimes f\kappa. \tag{4.68}$$

$$
\begin{array}{ccc}
f & \xrightarrow{\ f\omega\ } & fa.hA \\
{\scriptstyle \omega f}\downarrow & & \uparrow{\scriptstyle \varphi.hA} \\
b.hB.f & = & b.Tf.hA
\end{array}
\qquad
\begin{array}{ccccc}
b.Tb.T^2 f & \xrightarrow{\ b.T\varphi\ } & b.Tf.Ta & \xrightarrow{\ \varphi.Ta\ } & fa.Ta \\
{\scriptstyle \kappa.T^2 f}\downarrow & & & & \downarrow{\scriptstyle f\kappa} \\
b.mB.T^2 f & = & b.Tf.mA & \xrightarrow[\ \varphi.mA\ ]{} & fa.mA
\end{array}
$$

For a consecutive lax morphism $(g, \gamma) \colon (B, b, \omega, \kappa) \to (C, c, \omega, \kappa)$ we let

$$(g, \gamma).(f, \varphi) = (gf, \gamma.Tf \otimes g\varphi), \quad \gamma.Tf \otimes g\varphi \colon c.T(gf) \to gfa. \quad (4.69)$$

We have thus a category $\mathrm{LxPsa}(T)$, with identities

$$\mathrm{id}(A, a, \omega, \kappa) = (\mathrm{id}A, 1_a).$$

A *pseudo morphism* is a lax morphism (f, φ) where the cell φ is vertically invertible.

(b) A *colax morphism of pseudo algebras*

$$\mathbf{r} = (r, \rho) \colon (A, a, \omega, \kappa) \to (B, b, \omega, \kappa)$$

is a morphism $r \colon A \to B$ of $\mathbf{C}$ with a comparison cell ρ such that:

$$\rho \colon r.a \to b.Tr, \qquad r\omega \otimes \rho.hA = \omega r,$$
$$r\kappa \otimes \rho.mA = \rho.Ta \otimes b.T\rho \otimes \kappa.T^2 r, \qquad (4.70)$$

$$
\begin{array}{ccc}
r & \xrightarrow{\ r\omega\ } & ra.hA \\
{\scriptstyle \omega r} \downarrow & & \downarrow {\scriptstyle \rho.hA} \\
b.hB.r & =\!=\!= & b.Tr.hA
\end{array}
\qquad
\begin{array}{ccccc}
ra.Ta & \xrightarrow{\ \rho.Ta\ } & b.Tr.Ta & \xrightarrow{\ b.T\rho\ } & b.Tb.T^2 r \\
{\scriptstyle r\kappa} \downarrow & & & & \downarrow {\scriptstyle \kappa.T^2 r} \\
ra.mA & \xrightarrow[\ \rho.mA\ ]{} & b.Tr.mA & =\!=\!= & b.mB.T^2 r
\end{array}
$$

Given a second colax morphism $(s, \sigma) \colon (B, b, \omega, \kappa) \to (C, c, \omega, \kappa)$, we let:

$$(s, \sigma).(r, \rho) = (sr, \sigma.Tr \otimes s\rho), \quad \sigma.Tr \otimes s\rho \colon sra \to c.T(sr). \quad (4.71)$$

This gives a category $\mathrm{CxPsa}(T)$, with identities as above.

4.6.4 A double category of pseudo algebras

We form now a double category $\mathbb{P}\mathrm{sa}(T)$ *of pseudo algebras* of T, with lax morphisms (horizontally) and colax morphisms (vertically). The construction is similar to that of $\mathbb{D}\mathrm{bl}$ in Section 4.2, and we will see that it extends it.

Its objects are the pseudo T-algebras $\mathbb{A} = (A, a, \omega, \kappa)$, $\mathbb{B} = (B, b, \omega, \kappa)$, ...; its horizontal arrows are the *lax* morphisms $\mathbf{f} = (f, \varphi)$, $\mathbf{g} = (g, \gamma)$, ...; its vertical arrows are the *colax* morphisms $\mathbf{r} = (r, \rho)$, $\mathbf{s} = (s, \sigma)$, ...

A double cell $\boldsymbol{\pi} \colon (\mathbf{r}\ {}^{\mathbf{f}}_{\mathbf{g}}\ \mathbf{s})$ consists of four morphisms as above together with a 2-cell $\pi \colon sf \to gr \colon A \to D$ in $\mathbf{C}$ (under a coherence condition)

$$
\begin{array}{ccc}
A & \xrightarrow{\ f\ } & \mathbb{B} \\
{\scriptstyle r} \downarrow & \boldsymbol{\pi} & \downarrow {\scriptstyle s} \\
\mathbb{C} & \xrightarrow[\ g\ ]{} & \mathbb{D}
\end{array}
\qquad
\begin{array}{ccc}
A & \xrightarrow{\ f\ } & B \\
{\scriptstyle r} \downarrow & {\scriptstyle \nearrow\ \pi} & \downarrow {\scriptstyle s} \\
C & \xrightarrow[\ g\ ]{} & D
\end{array}
\qquad (4.72)
$$

$$f: A \to B, \quad \varphi: b.Tf \to f.a, \qquad g: C \to D, \quad \gamma: d.Tg \to g.c,$$

$$r: A \to C, \quad \rho: r.a \to c.Tr, \qquad s: B \to D, \quad \sigma: s.b \to d.Ts.$$

Let us note that sf and gr are morphisms of **C** (not of algebras), and that $\pi: (r\ {}^{f}_{g}\ s)$ lives in the double category $\mathbb{Q}\mathbf{C}$ of quintets over the 2-category **C** (see 3.1.4).

These data must satisfy this coherence condition, in the 2-category **C**

(coh) $s\varphi \otimes \pi a \otimes g\rho = \sigma.Tf \otimes d.T\pi \otimes \gamma.Tr,$

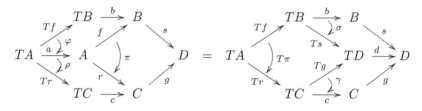

The horizontal and vertical composition of double cells are defined using the corresponding operations in $\mathbb{Q}\mathbf{C}$. Namely, for a consistent matrix of double cells

$$\begin{array}{ccccc}
\bullet & \xrightarrow{\ f\ } & \bullet & \xrightarrow{\ f'\ } & \bullet \\
{\scriptstyle r}\downarrow & \pi & {\scriptstyle s}\downarrow & \vartheta & \downarrow{\scriptstyle t} \\
\bullet & -\!\mathrm{g}\!\to & \bullet & -\mathrm{g}'\!> & \bullet \\
{\scriptstyle r'}\downarrow & \zeta & {\scriptstyle s'}\downarrow & \omega & \downarrow{\scriptstyle t'} \\
\bullet & \xrightarrow{\ h\ } & \bullet & \xrightarrow{\ h'\ } & \bullet
\end{array}$$ (4.73)

$\pi \mid \vartheta$ and $\pi \otimes \zeta$ are represented by the corresponding composites of double cells in $\mathbb{Q}\mathbf{C}$. The latter are computed by vertical composition of 2-cells of **C**

$$\pi \mid \vartheta = \vartheta f \otimes g'\pi, \qquad \frac{\pi}{\zeta} = s'\pi \otimes \zeta r. \qquad (4.74)$$

The coherence of these cells is the subject of an exercise below. It follows that these composition laws are strictly associative and unitary. Moreover, they satisfy the middle-four interchange law because this holds in the double category $\mathbb{Q}\mathbf{C}$.

We have thus a forgetful double functor, which is cellwise faithful

$$U: \mathbb{P}\mathrm{sa}(T) \to \mathbb{Q}\mathbf{C}, \qquad U(\mathbb{A}) = A,$$
$$U(\mathbf{f}) = f, \quad U(\mathbf{r}) = r, \quad U(\boldsymbol{\pi}) = \pi. \qquad (4.75)$$

We write as $\mathbb{P}\mathrm{sa}_N(T)$ the totally full double subcategory of normal pseudo algebras of T.

4.6.5 Exercises

(a) Prove that the double cells defined in (4.74) satisfy the coherence axiom (coh).

(b) Prove that a lax morphism of pseudo algebras which is invertible in the category $\mathrm{Lx}\mathbb{P}\mathrm{sa}(T)$ is necessarily a pseudo morphism.

4.6.6 Graphs of categories

To examine double categories in the present framework, we let $\mathbf{C} = \mathrm{Gph}\mathbf{Cat}$ be the 2-category of graphs $A = (A_i, \partial^\alpha)$ in $\mathbf{Cat}$

$$\partial^\alpha \colon A_1 \rightrightarrows A_0, \tag{4.76}$$

where a 2-cell $\varphi \colon F \to G \colon A \to B$ is a pair of natural transformations of ordinary functors, consistent with the faces

$$\varphi_i \colon F_i \to G_i \colon A_i \to B_i, \quad \partial^\alpha \varphi_1 = \varphi_0 \partial^\alpha \quad (i = 0, 1, \ \alpha = \pm). \tag{4.77}$$

A double category $\mathbb{D}$ has an underlying graph $U(\mathbb{D}) = (\mathrm{Hor}_i(\mathbb{D}), \partial^\alpha)$, formed by the category $\mathrm{Hor}_0(\mathbb{D})$ of objects and horizontal arrows and the category $\mathrm{Hor}_1(\mathbb{D})$ of vertical arrows and double cells (both with horizontal composition), linked by two ordinary functors, the vertical faces ∂^α.

We have a forgetful 2-functor

$$U \colon \mathbf{Dbl} \to \mathrm{Gph}\mathbf{Cat} = \mathbf{C}, \tag{4.78}$$

on the 2-category of (small) double categories, double functors and horizontal transformations.

4.6.7 Theorem (Strict double categories as algebras)

The 2-functor U defined above is 2-monadic: it gives a comparison 2-isomorphism $K \colon \mathbf{Dbl} \to \mathbf{Alg}(T)$ with the 2-category of T-algebras for the associated 2-monad.

Proof A graph of categories $A = (A_i, \partial^\alpha)$ generates a free double category DA, described as follows.

(a) $\mathrm{Hor}_0(DA)$ is the category A_0.

(b) $\mathrm{Ver}_0(DA)$ is the free category generated by the graph of sets $\mathrm{Ob}A = (\mathrm{Ob}A_i, \partial^\alpha)$; its arrows give the vertical arrows $(u_1, ..., u_n)$ of DA, including the vertical unit $e(x)$ on an object x of A_0 (the *empty path* at x).

(c) $\mathrm{Ver}_1(DA)$ is the free category generated by the graph of sets $\mathrm{Mor}A =$

$(\mathrm{Mor}A_i, \partial^{\alpha})$; its arrows give the double cells $(a_1, ..., a_n)$ of DA, including the vertical unit $e(f)$ on a morphism f of A_0

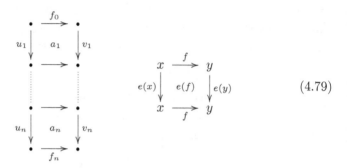

(4.79)

(d) The horizontal composition of these double cells is a concatenation of compositions in A_1, which we write as $a_i \mid b_i$

$$(a_1, ..., a_n) \mid (b_1, ..., b_n) = ((a_1 \mid b_1), ..., (a_n \mid b_n)).$$

The obvious embedding $hA: A \to UDA$ gives the 2-universal arrow from A to U. This gives the left 2-adjoint $D: \mathbf{C} \to \mathbf{Dbl}$ and the associated 2-monad (T, h, m) on $\mathbf{C}$, with $T = UD$.

The comparison K is plainly an isomorphism of 2-categories. □

4.6.8 Theorem (Weak double categories as normal pseudo algebras)

The double categories $\mathbb{P}\mathrm{sa}_N(T)$ and $\mathbb{D}\mathrm{bl}$ are correlated by an adjoint equivalence of double categories (see 4.4.4)

$$V: \mathbb{P}\mathrm{sa}_N(T) \rightleftarrows \mathbb{D}\mathrm{bl} : J,$$

$$VJ = 1, \quad \varepsilon: JV \cong 1, \qquad V\varepsilon = 1, \quad \varepsilon J = 1. \tag{4.80}$$

Proof (a) A normal pseudo algebra $\mathbb{A} = (A, c, \kappa)$ for T is a graph of categories $A = (A_i, \partial^{\alpha})$ with an assigned vertical composition of finite paths of vertical arrows and cells

$$u_1 \otimes ... \otimes u_n = c(u_1, ..., u_n), \qquad a_1 \otimes ... \otimes a_n = c(a_1, ..., a_n),$$
$$e_x = c(e(x)), \qquad e_f = c(e(f)), \tag{4.81}$$

and an (invertible) extended associator $\kappa: c.Dc \to c.mA$.

As a first consequence of normality, the unary vertical composition is trivial: $c(u) = u$ and $c(a) = a$, for all items of the category A_1. Second, $\mathbb{A}$

is trivial in degree 0, in the sense that $(TA)_0 = \mathrm{Hor}_0(DA) = A_0$, while the functor c_0 and the natural transformation κ_0

$$c_0 \colon (TA)_0 \to A_0, \quad \kappa_0 \colon c_0.Dc_0 \to c_0.(mA)_0 \colon (T^2A)_0 \to A_0, \qquad (4.82)$$

are identities: this follows easily from the coherence conditions (4.64), where wA, $(hA)_0$ and $(mA)_0$ are identities.

(b) $\mathbb{A}$ can be viewed as an 'unbiased' weak double category, where all finite vertical compositions are assigned. The fact that κ_0 is trivial says that the comparison cells of the unbiased associator κ_1 are special, i.e. their horizontal arrows are identities – as required in the axioms of weak double categories for the binary associator and the unitors.

The normal pseudo algebra $\mathbb{A}$ has un underlying weak double category $V(\mathbb{A})$, obtained by extracting the binary and zeroary vertical operations and their comparisons. We get a canonical double functor V, that sends:

- a lax or colax morphism of normal pseudo algebras to the corresponding lax or colax functor of weak double categories, reducing the unbiased comparisons of finite vertical composition to the 'biased ones', of binary and zeroary composition,

- a double cell π to the corresponding double cell in $\mathbb{D}\mathrm{bl}$, with the same components on objects and vertical arrows.

(c) The other way round, we construct a double functor J such that $VJ = 1$, by choosing a 'bracketing' of n-ary compositions. Namely, a weak double category $\mathbb{D}$ can be extended to a normal pseudo algebra $J(\mathbb{D})$ by defining the n-ary vertical composition (of vertical arrows or cells) as

$$x_1 \otimes \ldots \otimes x_n = (\ldots((x_1 \otimes x_2) \otimes x_3)\ldots \otimes x_n), \qquad (4.83)$$

and extending the comparisons. A lax or colax double functor becomes a lax or colax morphism, by extending its comparisons. For a double cell we just note that its coherence with the unbiased comparisons implies coherence with the biased ones.

(d) Finally, a normal pseudo algebra $\mathbb{A} = (A, c, \kappa)$ produces an object $JV(\mathbb{A}) = (A, c', \kappa')$ with modified unbiased vertical operations and modified unbiased comparisons. The identity $\mathrm{id}A$ of the underlying graph extends to an invertible *pseudo morphism* $\varepsilon\mathbb{A} \colon JV(\mathbb{A}) \to \mathbb{A}$ satisfying the triangular conditions, in (4.80). $\qquad \square$

5

Double limits

This chapter is about limits in weak double categories, and is based on [GP1]. Unless differently specified, *limit* means 'horizontal double limit': the structural 'arrows' are horizontal morphisms and double cells, universal by horizontal composition.

Limits in weak double categories are defined in Sections 5.1 and 5.2. The Construction Theorem, in 5.3.6, proves that all double limits can be constructed from *products*, *equalisers* and *tabulators*.

If $\mathbf{C}$ is a 2-category, viewed as a horizontal double category, we will see in Sections 5.6 and 5.7 that weighted limits in $\mathbf{C}$ are the same as double limits.

Marginally, *vertical* double limits are also considered in strict double categories (see 5.2.4), but we have few instances of them. *Transpositive* double limits make sense in a transpositive double category, and are even more exceptional among our examples (see 5.4.6, 5.4.7).

For a strict double category $\mathbb{A}$, Bastiani–Ehresmann [BaE] considered '$\mathbb{A}$-wise limits', which would be called here 1-dimensional limits for double functors $F\colon \mathbb{V} \to \mathbb{A}$ defined on a vertical double category. Our construction of double limits, restricted to this case, coincides with the construction of $\mathbb{A}$-wise limits given in [BaE] (p. 265, Proposition 3), by means of one-dimensional tabulators (called 'representations') and limits of ordinary functors in the category $\mathrm{Hor}_0\mathbb{A}$ of objects and horizontal arrows of $\mathbb{A}$.

Notation. $\mathbb{I}$ and $\mathbb{A}$ are always unitary weak double categories, and $\mathbb{I}$ is small. $\mathbb{L}\mathrm{x}(\mathbb{I}, \mathbb{A})$ is the weak double category of lax functors $\mathbb{I} \to \mathbb{A}$, with strict horizontal transformations, *pseudo* vertical transformations and modifications (see 3.8.4). The diagonal double functor $\Delta\colon \mathbb{A} \to \mathbb{L}\mathrm{x}(\mathbb{I}, \mathbb{A})$ was introduced in 3.8.5. The structural functors of vertical faces and degeneracy of $\mathbb{A}$ are always written as $\partial^{\pm}\colon \mathrm{Hor}_1\mathbb{A} \rightrightarrows \mathrm{Hor}_0\mathbb{A} : e$.

5.1 Functorial limits

We begin by a global approach to limits: the lax limit functor of $\mathbb{A}$ over $\mathbb{I}$ is defined as a *unitary* right adjoint $L\colon \mathbb{L}x(\mathbb{I}, \mathbb{A}) \to \mathbb{A}$ to the diagonal double functor $\Delta\colon \mathbb{A} \to \mathbb{L}x(\mathbb{I}, \mathbb{A})$.

5.1.1 The importance of unitarity

Limits and colimits are well related to *unitary* lax (or colax) functors, and to unitary adjunctions.

A basic reason was already suggested in 3.5.2. An object of the weak double category $\mathbb{A}$ can be viewed as a (strict) functor $A\colon \mathbf{1} \to \mathbb{A}$, and a horizontal arrow as a functor $\mathbf{2} \to \mathbb{A}$. All this is preserved by composing with a *unitary* lax (or colax) functor $S\colon \mathbb{A} \to \mathbb{B}$.

The interest of requiring the lax limit functor $L\colon \mathbb{L}x(\mathbb{I}, \mathbb{A}) \to \mathbb{A}$ to be unitary already appears in 5.1.5: the terminal object of a double category is defined by means of a *unitary* right adjoint of the terminal double functor $\mathbb{A} \to \mathbf{1}$. A non-unitary right adjoint gives a 'terminal pair' with the structure of a vertical monad (see 5.1.6(d)); *it is better treated as an exception,* both from the theoretical and the concrete point of view.

Unitarity is even more crucial in the relationship of limits with adjunctions, dealt with in Section 5.5. We have already encountered the strict double functor $D\colon \mathbb{S}\text{pan}\mathbf{Set} \to \mathbb{C}\text{at}$ which does not preserve cotabulators, even though it has a lax right adjoint F (see 4.5.2(c)). We shall see that this can only happen because F is not unitary (in Theorem 5.5.6).

5.1.2 Definition

We say that $\mathbb{A}$ has *lax functorial* $\mathbb{I}$-limits if the diagonal double functor $\Delta\colon \mathbb{A} \to \mathbb{L}x(\mathbb{I}, \mathbb{A})$ of 3.8.5 has a unitary right adjoint, called the *lax limit functor* of $\mathbb{A}$ over $\mathbb{I}$

$$L\colon \mathbb{L}x(\mathbb{I}, \mathbb{A}) \to \mathbb{A}, \qquad \Delta \dashv L,$$
$$p\colon \Delta L \to 1\colon \mathbb{L}x(\mathbb{I}, \mathbb{A}) \to \mathbb{L}x(\mathbb{I}, \mathbb{A}), \tag{5.1}$$

where we write as p the counit of the adjunction.

We say that $\mathbb{A}$ has *pseudo functorial* $\mathbb{I}$-limits when L is pseudo, and we speak of *strict functorial* $\mathbb{I}$-limits when there exists a strict limit functor L.

The lax limit functor is determined up to horizontal isomorphism. If $\mathbb{A}$ is horizontally invariant, its values LF are also determined up to sesqui-isomorphism in $\mathbb{A}$ (see Section 4.1).

5.1.3 Functorial limits by components

Applying Corollary 4.3.7 on double adjunctions, the existence of lax functorial limits over $\mathbb{I}$ amounts to saying that:

(a) each component $\Delta_n \colon \mathrm{Hor}_n(\mathbb{A}) \to \mathrm{Hor}_n(\mathbb{Lx}(\mathbb{I}, \mathbb{A}))$, for $n = 0, 1$, has a right adjoint

$$L_n \colon \mathrm{Hor}_n(\mathbb{Lx}(\mathbb{I}, \mathbb{A})) \to \mathrm{Hor}_n(\mathbb{A}),$$

(b) one can fix two functors L_0, L_1 strictly preserved by the faces $\partial^\pm \colon \mathrm{Hor}_1 \to \mathrm{Hor}_0$ and the degeneracy $e \colon \mathrm{Hor}_0 \to \mathrm{Hor}_1$.

The last condition means that:

(i) $Lr \colon LF \to LG$, $L\xi \colon (Lr \, {}^{Lh}_{Lk} \, Lr')$ (*coherence with faces*),

(ii) $L(e_F) = e_{LF}$, $L(e_h) = e_{Lh}$ (*unitarity*),

for every pseudo vertical transformation $r \colon F \to G$ of lax functors $\mathbb{I} \to \mathbb{A}$, every modification $\xi \colon (r \, {}^h_k \, r')$ and every horizontal transformation $h \colon F \to F'$.

We shall see in 5.2.7 that, when $\mathbb{A}$ is horizontally invariant, these strict preservation properties are equivalent to the corresponding weak ones.

5.1.4 Functorial limits by universal properties

Applying Theorem 4.3.6, the existence in $\mathbb{A}$ of lax functorial limits over $\mathbb{I}$ amounts to the following two conditions:

(lfl.0) for every lax functor $F \colon \mathbb{I} \to \mathbb{A}$ there is a universal arrow $(LF, pF \colon \Delta LF \to F)$ from the functor Δ_0 to the object F of $\mathrm{Hor}_0(\mathbb{Lx}(\mathbb{I}, \mathbb{A}))$,

(lfl.1) for every pseudo vertical transformation $r \colon F \to G \colon \mathbb{I} \to \mathbb{A}$ of lax functors there is a universal arrow $(Lr, pr \colon \Delta Lr \to r)$ from the functor Δ_1 to the object r of $\mathrm{Hor}_1(\mathbb{Lx}(\mathbb{I}, \mathbb{A}))$, *consistently with the previous choice* (as specified in 5.1.3(i), (ii)).

More explicitly, the universal properties of the components pF and pr mean that:

(i) for each object A in $\mathbb{A}$ and each horizontal transformation $h \colon \Delta A \to F \colon \mathbb{I} \to \mathbb{A}$, there is a unique horizontal arrow $f \colon A \to LF$ in $\mathbb{A}$ such that $h = pF.\Delta f \colon \Delta A \to F$,

(ii) for each vertical arrow $u \colon A \to B$ in $\mathbb{A}$ and each modification $\xi \colon (\Delta u \, {}^h_k \, r)$, there is a unique cell $\alpha \colon (u \, {}^f_g \, Lr)$ in $\mathbb{A}$ such that $\xi = (\Delta \alpha \,|\, pr) \colon \Delta r \to r$.

5.1.5 Terminal pairs and terminal objects

The significance of unitary limits is evident while considering the terminal object.

We are taking $\mathbb{I} = \mathbf{0}$, the empty double category, so that $\mathbb{L}x(\mathbb{I}, \mathbb{A}) = \mathbf{1}$ is the singleton double category and $\Delta \colon \mathbb{A} \to \mathbf{1}$ is the (unique) double functor with values in $\mathbf{1}$.

Now, a lax right adjoint $\mathbf{1} \to \mathbb{A}$ to Δ is a *terminal pair* (T, t), formed of an object T and a vertical arrow $t \colon T \nrightarrow T$ such that

(ter.0) for every object A there is precisely one map $t \colon A \to T$ (also written as t_A),

(tp.1) for every vertical map $u \colon A \nrightarrow B$ there is precisely one cell $\tau \colon u \to t$ (also written as τ_u)

$$
\begin{array}{ccc}
A & \xrightarrow{\;t\;} & T \\
u \downarrow & \quad \tau \quad & \downarrow t \\
B & \xrightarrow[\;t\;]{} & T
\end{array}
\qquad (5.2)
$$

The object T is said to be a *terminal object* if (T, e_T) is a terminal pair, which means that the double functor $\mathbb{A} \to \mathbf{1}$ has a *unitary lax right adjoint*. Now T satisfies (ter.0) and

(ter.1) for every vertical map $u \colon A \nrightarrow B$ there is precisely one cell $\tau \colon u \to e_T$.

Property (ter.0) means that T is a terminal object of the ordinary category $\mathrm{Hor}_0\mathbb{A}$, and will be expressed saying that T is a *1-dimensional terminal object* of $\mathbb{A}$.

Property (ter.1) means that the vertical identity e_T is a terminal object of the ordinary category $\mathrm{Hor}_1\mathbb{A}$, and implies (ter.0), by applying it to the vertical identity e_A. As a partial converse, *if $\mathbb{A}$ has 1-dimensional cotabulators* (see Section 3.6), then the ordinary functor $e \colon \mathrm{Hor}_0\mathbb{A} \to \mathrm{Hor}_1\mathbb{A}$ preserves limits, and (ter.0) implies (ter.1).

Since cotabulators exist in all our structures of real interest, a terminal pair which is not a terminal object is a rare phenomenon: see the examples of 5.4.5 and 5.4.8.

5.1.6 Exercises and complements

(a) Find the terminal object of the double categories $\mathrm{Adj}\mathbf{Cat}$ and $\mathbb{R}\mathrm{el}\mathbf{C}$, for a regular category $\mathbf{C}$.

(b) Find the terminal object of the following weak (or strict) double categories, defined in Section 3.4: $\mathbb{S}\mathrm{pan}\mathbf{Set}$, $\mathbb{C}\mathrm{osp}\mathbf{Set}$, $\mathbb{C}\mathrm{at}$, $\mathrm{p}\mathbb{O}\mathrm{rd}$, $\mathbb{M}\mathrm{tr}$. Consider also $\mathbb{S}\mathrm{pan}\mathbf{C}$ and $\mathbb{C}\mathrm{osp}\mathbf{C}$, under suitable conditions on $\mathbf{C}$.

(c) We know that the terminal object of a horizontally invariant weak double category is determined up to sesqui-isomorphism, and therefore up to horizontal isomorphism and vertical equivalence.

In fact, the terminal object of the *non-invariant* double category $\mathbb{T}g$ of 4.1.8(c) is also vertically determined, but the reader can easily see that this fails for general products (i.e. double limits over discrete categories).

(d) For a terminal pair (T, t), use the properties (ter.0) and (tp.1) of 5.1.5 to construct directly a lax functor $F \colon \mathbf{1} \to \mathbb{A}$, i.e. a vertical monad in $\mathbb{A}$ (see 3.8.6).

5.2 Double cones and double limits

After the global approach of the previous section we examine now the local aspects: the limit of a lax functor $F \colon \mathbb{I} \to \mathbb{A}$ and the limit of a pseudo vertical transformation $r \colon F \to G \colon \mathbb{I} \to \mathbb{A}$. We have already seen that *we need both to construct lax functorial $\mathbb{I}$-limits.*

We also define pseudo limits, that will be used in Sections 5.6 and 5.7.

$\mathbb{I}$ and $\mathbb{A}$ are always unitary weak double categories, and $\mathbb{I}$ is small.

5.2.1 Cones and pseudo cones

The weak double categories $\mathbb{C}\mathrm{one}(F)$ and $\mathrm{Ps}\mathbb{C}\mathrm{one}(F)$, *of cones* and *pseudo cones of F*, are defined as the following two double commas $\Delta \Downarrow F$ (see 4.2.4)

$$
\begin{array}{ccc}
\mathbb{C}\mathrm{one}(F) \longrightarrow & \mathbb{A} & \qquad \mathrm{Ps}\mathbb{C}\mathrm{one}(F) \longrightarrow \mathbb{A} \\
\downarrow \qquad \qquad \downarrow {\scriptstyle \Delta} & & \downarrow \qquad \qquad \qquad \downarrow {\scriptstyle \Delta} \\
\mathbf{1} \xrightarrow{\; F \;} \mathrm{Lx}(\mathbb{I}, \mathbb{A}) & & \mathbf{1} \xrightarrow{\; F \;} \mathrm{Lx}_{\mathrm{ps}}(\mathbb{I}, \mathbb{A})
\end{array} \qquad (5.3)
$$

Dually, the weak double categories $\mathbb{C}\mathrm{ocone}(F)$ and $\mathrm{Ps}\mathbb{C}\mathrm{ocone}(F)$, *of cocones* and *pseudo cocones of F*, are defined as double commas $F \Downarrow \Delta$

$$
\begin{array}{ccc}
\mathbb{C}\mathrm{ocone}(F) \longrightarrow & \mathbf{1} & \qquad \mathrm{Ps}\mathbb{C}\mathrm{ocone}(F) \longrightarrow \mathbf{1} \\
\downarrow \qquad \qquad \downarrow {\scriptstyle F} & & \downarrow \qquad \qquad \qquad \downarrow {\scriptstyle F} \\
\mathbb{A} \xrightarrow{\; \Delta \;} \mathrm{Lx}(\mathbb{I}, \mathbb{A}) & & \mathbb{A} \xrightarrow{\; \Delta \;} \mathrm{Lx}_{\mathrm{ps}}(\mathbb{I}, \mathbb{A})
\end{array} \qquad (5.4)
$$

5.2.2 Cones, arrows and cells

We give now an explicit description of $\mathrm{Ps}\mathbb{C}\mathrm{one}(F)$, and then of $\mathbb{C}\mathrm{one}(F)$ as a substructure of the former.

(a) A *(horizontal) pseudo cone* of the lax functor $F\colon \mathbb{I} \to \mathbb{A}$ is a pair $(A, h\colon A \to F)$ comprising an object A of $\mathbb{A}$ (the *vertex* of the cone) and a pseudo horizontal transformation of lax functors $h\colon A \to F\colon \mathbb{I} \to \mathbb{A}$.

As defined in 3.8.1, h is defined by the following components in $\mathbb{A}$:

- a horizontal map $hi\colon A \to Fi$ for every object i in $\mathbb{I}$,

- a cell $hu\colon (A\,{}^{hi}_{hj}\,Fu)$ for every vertical arrow $u\colon i \rightarrowtail j$ in $\mathbb{I}$,

- a globular isocell ha for every horizontal arrow $a\colon i \to i'$ in $\mathbb{I}$,

$$
\begin{array}{ccc}
A \xrightarrow{\ hi\ } Fi \\
{\scriptstyle e}\Big\downarrow \quad hu \quad \Big\downarrow{\scriptstyle Fu} \\
A \xrightarrow[\ hj\ ]{} Fj
\end{array}
\qquad
\begin{array}{ccc}
A \xrightarrow{\ hi\ } Fi \xrightarrow{\ Fa\ } Fi' \\
{\scriptstyle e}\Big\downarrow \quad\quad ha \quad\quad \Big\downarrow{\scriptstyle e} \\
A \xrightarrow[\ hi'\ ]{} Fi'
\end{array}
\qquad (5.5)
$$

under the axioms (pht.1–5) of naturality and coherence.

It is a *cone* when all the *comparison cells* ha are vertical identities.

(b) A *horizontal morphism* $f\colon (A, h) \to (A', k)$ of pseudo cones of $F\colon \mathbb{I} \to \mathbb{A}$ is a horizontal arrow $f\colon A \to A'$ in $\mathbb{A}$ that commutes with the cone elements, as follows:

(i)   for every i in $\mathbb{I}$: $\qquad hi = ki.f\colon A \to Fi$,

(ii)   for every $u\colon i \rightarrowtail j$: $\qquad hu = (e_f \,|\, ku)\colon e_A \to Fu$,

(iii)   for every $a\colon i \to i'$: $\qquad ha = (e_f \,|\, ka)\colon e_A \to e_{Fi'}$,

$$
\begin{array}{ccc}
A \xrightarrow{\ hi\ } Fi \\
{\scriptstyle e}\Big\downarrow \quad hu \quad \Big\downarrow{\scriptstyle Fu} \\
A \xrightarrow[\ hj\ ]{} Fj
\end{array}
=
\begin{array}{ccccc}
A \xrightarrow{\ f\ } A' \xrightarrow{\ ki\ } Fi \\
{\scriptstyle e}\Big\downarrow \quad e_f \quad \Big\downarrow{\scriptstyle Fu} \quad ku \quad \Big\downarrow{\scriptstyle Fu} \\
A \xrightarrow[\ f\ ]{} A' \xrightarrow[\ kj\ ]{} Fj
\end{array}
$$

$$
\begin{array}{ccc}
A \xrightarrow{\ hi\ } Fi \xrightarrow{\ Fa\ } Fi' \\
{\scriptstyle e}\Big\downarrow \quad ha \quad \Big\downarrow{\scriptstyle e} \\
A \xrightarrow[\ hi'\ ]{} Fi'
\end{array}
=
\begin{array}{ccccc}
A \xrightarrow{\ f\ } A' \xrightarrow{\ ki\ } Fi \xrightarrow{\ Fa\ } Fi' \\
{\scriptstyle e}\Big\downarrow \quad e_f \quad \Big\downarrow{\scriptstyle e} \quad ka \quad \Big\downarrow{\scriptstyle e} \\
A \xrightarrow[\ f\ ]{} A' \xrightarrow[\ ki'\ ]{} Fi'
\end{array}
$$

Horizontal morphisms compose, forming a category.

(c) A *vertical morphism* $(u, \xi)\colon (A, h) \rightarrowtail (B, k)$ of pseudo cones of $F\colon \mathbb{I} \to \mathbb{A}$ comprises a vertical arrow $u\colon A \rightarrowtail B$ (that gives a vertical transformation $\Delta u\colon \Delta A \rightarrowtail \Delta B$, constant at u) and a modification $\xi\colon (\Delta u\,{}^{h}_{k}\,e_F)$, also written as $\xi\colon (u\,{}^{h}_{k}\,e_F)$.

We have thus, for every i in $\mathbb{I}$, a cell $\xi i\colon (u\,{}^{hi}_{ki}\,Fi)$ in $\mathbb{A}$ that satisfies the conditions (mod.1, 2) of Definition 3.8.3.

(d) A double cell of pseudo cones

$$
\begin{array}{ccc}
(A,h) & \xrightarrow{\ f\ } & (A',h') \\
{\scriptstyle(u,\xi)}\big\downarrow & \alpha & \big\downarrow{\scriptstyle(v,\zeta)} \\
(B,k) & \xrightarrow[\ g\ ]{} & (B',k')
\end{array}
\tag{5.6}
$$

'is' a cell $\alpha\colon (u \underset{g}{\overset{f}{\to}} v)$ in $\mathbb{A}$ such that, for every i in $\mathbb{I}$, $\xi i = \alpha\,|\,\zeta i$

$$
\begin{array}{ccc}
A & \xrightarrow{\ hi\ } & Fi \\
{\scriptstyle u}\big\downarrow & \xi i & \big\downarrow{\scriptstyle e} \\
B & \xrightarrow[ki]{} & Fi
\end{array}
\;=\;
\begin{array}{ccccc}
A & \xrightarrow{\ f\ } & A' & \xrightarrow{\ h'i\ } & Fi \\
{\scriptstyle u}\big\downarrow & \alpha & \big\downarrow{\scriptstyle v} & \zeta i & \big\downarrow{\scriptstyle e} \\
B & \xrightarrow[g]{} & B' & \xrightarrow[k'i]{} & Fi
\end{array}
\tag{5.7}
$$

They compose 'as' in $\mathbb{A}$. This concludes the definition of $\mathrm{Ps}\mathbb{C}\mathrm{one}(F)$.

The weak double category $\mathbb{C}\mathrm{one}(F)$ *of cones of* F is the totally full substructure of $\mathrm{Ps}\mathbb{C}\mathrm{one}(F)$ determined by the strict cones, i.e. the horizontal transformations $A \to F$. Note that a vertical morphism (u,ξ) still contains a *pseudo* vertical transformation u.

5.2.3 Limits and pseudo limits

A (horizontal) *double limit* of a lax functor $F\colon \mathbb{I} \to \mathbb{A}$ is defined as a terminal object (see 5.1.5) of the weak double category $\mathbb{C}\mathrm{one}(F)$, while a *pseudo limit* of F is a terminal object of $\mathrm{Ps}\mathbb{C}\mathrm{one}(F)$.

Explicitly, leaving Δ understood, we have a strict (resp. pseudo) cone $(L, p\colon L \to F)$ such that:

(dl.1) for every strict (resp. pseudo) cone $(A, h\colon A \to F)$ there is a unique horizontal morphism $f\colon A \to L$ in $\mathbb{A}$ such that $h = pf\colon A \to L \to F$, in the weak double category $\mathbb{L}\mathrm{x}(\mathbb{I}, \mathbb{A})$,

(dl.2) for every vertical arrow $(u,\xi)\colon (A, h\colon A \to F) \to\!\!\!\to (B, k\colon B \to F)$ of strict (resp. pseudo) cones of F there is a unique cell $\alpha\colon (u \underset{g}{\overset{f}{\to}} L)$ in $\mathbb{A}$ such that $\xi = (\alpha\,|\,e_p)\colon u \to e_F$

$$
\begin{array}{ccc}
A & \xrightarrow{\ hi\ } & Fi \\
{\scriptstyle u}\big\downarrow & \xi i & \big\downarrow{\scriptstyle e} \\
B & \xrightarrow[ki]{} & Fi
\end{array}
\;=\;
\begin{array}{ccccc}
A & \xrightarrow{\ f\ } & L & \xrightarrow{\ pi\ } & Fi \\
{\scriptstyle u}\big\downarrow & \alpha & \big\downarrow{\scriptstyle e} & e & \big\downarrow{\scriptstyle e} \\
B & \xrightarrow[g]{} & L & \xrightarrow[pi]{} & Fi
\end{array}
\qquad (i \text{ in } \mathbb{I}).
\tag{5.8}
$$

In the strict case, axiom (dl.1) means that (L, p) is a universal arrow from $\Delta_0\colon \mathrm{Hor}_0\mathbb{A} \to \mathrm{Hor}_0\mathbb{L}\mathrm{x}(\mathbb{I}, \mathbb{A})$ to the object F, or equivalently that L represents the contravariant functor $(\Delta(-), F)\colon \mathrm{Hor}_0\mathbb{A} \dashrightarrow \mathbf{Set}$, via $p\colon L \to F$. Axiom (dl.2) means that (e_L, e_p) is a universal arrow from $\Delta_1\colon \mathrm{Hor}_1\mathbb{A} \to \mathrm{Hor}_1\mathbb{L}\mathrm{x}(\mathbb{I}, \mathbb{A})$ to the object (u, ξ), or equivalently that e_L represents the contravariant functor $(\Delta(-), F)\colon \mathrm{Hor}_1\mathbb{A} \dashrightarrow \mathbf{Set}$ via $e_p\colon e_L \to F$.

When u is a vertical identity, one can easily verify that the globular cell ξ is vertically invertible if and only if α is.

Dually the *colimit* and the *pseudo colimit* of a lax functor $F\colon \mathbb{I} \to \mathbb{A}$ are defined as an initial object of $\mathbb{C}\mathrm{ocone}(F)$ or $\mathrm{Ps}\mathbb{C}\mathrm{ocone}(F)$.

We mostly deal with strict limits and colimits. We shall prove in Section 5.3 that all limits in a weak double category can be constructed from products, equalisers (of pairs of horizontal arrows) and tabulators; the 1-dimensional version of the latter has already been introduced in Section 3.6.

5.2.4 Comments and complements

(a) The conditions (dl.1, 2) will be called, respectively, the 1-*dimensional* and the 2-*dimensional* universal property of the limit. We speak of a 1-*dimensional limit* when only the first is assumed. Also here (dl.2) implies (dl.1).

(b) More generally, a *limit pair* $(L, l\colon L \nrightarrow L)$ of F will be a terminal pair of $\mathbb{C}\mathrm{one}(F)$, as defined in 5.1.5. It is a limit when $l = e_L$. Limit pairs which are not limits rarely occur.

(c) If F and G have a limit, a horizontal transformation $h\colon F \to G\colon \mathbb{I} \to \mathbb{A}$ determines a horizontal arrow $\mathrm{Lim}\, h\colon \mathrm{Lim}\, F \to \mathrm{Lim}\, G$. Vertical transformations are considered below.

(d) In a *strict* double category $\mathbb{A}$ we also consider *vertical limits*, defined as horizontal limits in the transpose double category $\mathbb{A}^{\mathrm{t}}$. Examples will be given in 5.4.6, 5.4.7, A2.7.

5.2.5 Vertical functoriality

The *limit of a pseudo vertical transformation* $r\colon F \nrightarrow G\colon \mathbb{I} \to \mathbb{A}$ is defined as a pair $(u, \pi\colon \Delta u \to r)$, universal with respect to the horizontal composition of modifications, in the weak double category $\mathbb{L}\mathrm{x}(\mathbb{I}, \mathbb{A})$.

In other words, we have a vertical arrow $u\colon L \nrightarrow M$ and a modification $\pi\colon (\Delta u \,{}^p_q\, r)$ such that every modification $\xi\colon (\Delta v \,{}^h_k\, r)$ factorises uniquely as

$\Delta\varphi \mid \pi$

$$
\begin{array}{ccc}
A \xrightarrow{hi} Fi \\
v\downarrow \quad \xi i \quad \downarrow ri \\
B \xrightarrow[ki]{} Gi
\end{array}
=
\begin{array}{ccccc}
A \xrightarrow{f} L \xrightarrow{pi} Fi \\
v\downarrow \quad \varphi \quad \downarrow u \quad \pi i \quad \downarrow ri \\
B \xrightarrow[g]{} M \xrightarrow[qi]{} Gi
\end{array}
\qquad (i \text{ in } \mathbb{I}). \qquad (5.9)
$$

The solution π is unique up to horizontal composition with a horizontally invertible cell φ. The latter is necessarily special, if we restrict to the solutions π whose horizontal arrows $p\colon \Delta L \to F$, $q\colon \Delta M \to G$ are two *fixed* (even one-dimensional) double limits of F and G.

The universal property (dl.2) amounts to saying that the vertical identity of the 1-dimensional limit of F is the limit of the vertical identity e_F.

5.2.6 Proposition

The weak double category $\mathbb{A}$ *has functorial* $\mathbb{I}$-*limits if and only if:*

(a) every lax double functor $F\colon \mathbb{I} \to \mathbb{A}$ *has a 1-dimensional limit* (LF, pF),

(b) every pseudo vertical transformation $r\colon F \nrightarrow G\colon \mathbb{I} \to \mathbb{A}$ *has a limit*

$$
Lr\colon LF \nrightarrow LG, \qquad\qquad pr\colon (\Delta Lr \ {}^{pF}_{pG}\ r), \qquad\qquad (5.10)
$$

consistently with the previous choice: the conditions (i) and (ii) of 5.1.3 are satisfied.

Note. Because of the unitarity condition 5.1.3(ii), each limit cone (LF, pF) is a limit, in the full 2-dimensional sense.

Proof This statement is a rewriting of 5.1.3 and 5.1.4, based on the definitions of local limits in this section. $\qquad\qquad\qquad\qquad\qquad\qquad\square$

5.2.7 Theorem (Limits and horizontal invariance)

Let the weak double category A be horizontally invariant.

(a) The existing $\mathbb{I}$-*limits of vertical transformations are also horizontally invariant, in the sense that we can modify their domain and codomain up to horizontal isomorphism.*

More precisely, let a pseudo vertical transformation $r\colon F \nrightarrow G\colon \mathbb{I} \to \mathbb{A}$ *be given, with double limits* (L, p) *of* F *and* (M, q) *of* G, *and a consistent limit* (u, π) *of* r, *with* $u\colon L \nrightarrow M$ *and* $\pi\colon (\Delta u \ {}^{p}_{q}\ r)$. *If also* (L', h) *and* (M', k) *are double limits of* F *and* G, *there is a limit* (v, ρ) *of* r *consistent with them.*

(b) If $\mathbb{A}$ has all the $\mathbb{I}$-limits specified in Proposition 5.2.6 (in (a) and (b)) and these are preserved as limits by the structural functors

$$\partial^{\pm} : \mathrm{Hor}_1 \mathbb{A} \underset{\longrightarrow}{\overset{\longrightarrow}{\rightleftarrows}} \mathrm{Hor}_0 \mathbb{A} : e,$$

then $\mathbb{A}$ has functorial $\mathbb{I}$-limits.

Proof (a) It is sufficient to prove that, in the limit of r, the given limit (M, q) of G can be replaced by any other limit (M', k); similarly, one can modify (L, p).

First, there is a unique horizontal iso $m \colon M \to M'$ such that $q = \Delta m \,|\, k$. By horizontal invariance, this m can be embedded in a cell λ, horizontally invertible, with inverse λ'

$$
\begin{array}{ccccc}
L & \xrightarrow{\ pi\ } & Fi & & \\
u \uparrow & \pi i & \downarrow ri & & \\
M & \xrightarrow{-\,m\,\rightarrow} & M' & \xrightarrow{-\,ki\,\rightarrow} & Gi \\
y \downarrow & \lambda & \downarrow e \quad e & \downarrow e & \\
M' & \xrightarrow[\ 1\ ]{} & M' & \xrightarrow[\ ki\ ]{} & Gi
\end{array}
\tag{5.11}
$$

As shown above, we have cells $\lambda i = \lambda \,|\, e_{ki}$ and $pi = \pi i \otimes \lambda i \colon u \otimes y \to ri$. The latter form a modification $\rho \colon (\Delta(u \otimes y) \overset{p}{\underset{k}{\to}} r)$.

We want to prove that $u \otimes y \colon L \twoheadrightarrow M'$ is a limit of r, with projections (pi). Take a cone $\alpha i \colon x \to ri$, with factorisation $\alpha i = \alpha \,|\, \pi i$

$$
\begin{array}{ccc}
X' \xrightarrow{\ fi\ } Fi & & X' \xrightarrow{\ f\ } L \xrightarrow{\ pi\ } Fi \\
x \downarrow \quad \alpha i \quad \downarrow ri & = & x \downarrow \quad \alpha \quad \downarrow u \quad \pi i \quad \downarrow ri \\
X'' \xrightarrow[\ gi\ ]{} Gi & & X'' \xrightarrow[\ g\ ]{} M \xrightarrow[\ qi\ ]{} Gi
\end{array}
\tag{5.12}
$$

There is a unique cell μ such that $e_{gi} = \mu \,|\, \lambda i$ (for all i), that is $\mu = e_{mg} \,|\, \lambda'$

$$
\begin{array}{ccc}
X'' \xrightarrow{\ gi\ } Gi & & X'' \xrightarrow{\ g\ } M \xrightarrow{\ qi\ } Gi \\
e \downarrow \quad e \quad \downarrow e & = & e \downarrow \quad \mu \quad \downarrow y \quad \lambda i \quad \downarrow e \\
X'' \xrightarrow[\ gi\ ]{} Gi & & X'' \xrightarrow[\ mg\ ]{} M' \xrightarrow[\ ki\ ]{} Gi
\end{array}
\tag{5.13}
$$

Now the vertical pasting of the diagrams (5.12) and (5.13) provides the solution of our problem, namely $(\alpha \otimes \mu) \colon x \to u \otimes y$. Its uniqueness is similarly proved.

Finally, after (a), point (b) is a rewriting of Proposition 5.2.6. $\qquad\square$

5.3 The Construction Theorem

We prove now that all double limits can be constructed with products, equalisers and tabulators.

5.3.1 Level limits

We begin from the elementary case of a horizontal double category $\mathbb{I}$, associated to an ordinary small category $\mathbf{I}$.

Here, a (strict) double functor $F: \mathbb{I} \to \mathbb{A}$ is the same as an ordinary functor $F_0 = \mathrm{Hor}_0 F: \mathbf{I} \to \mathbf{A} = \mathrm{Hor}_0 \mathbb{A}$, and a pseudo vertical transformation $r: F \to G: \mathbb{I} \to \mathbb{A}$ is the same as an ordinary functor $\mathbf{I} \to \mathrm{Hor}_1 \mathbb{A}$ (because $r(e_i) = 1_{ri}$ for all objects i in $\mathbf{I}$, as remarked at the end of 3.8.2).

(a) A double cone $(A, h: A \to F)$ of F is the same as an ordinary cone of F_0, and the first universal property (dl.1) for $\mathrm{Lim}\, F = (A, p: A \to F)$ amounts to saying that $(A, p) = \mathrm{Lim}\, F_0$. When all these exist we say that $\mathbb{A}$ has *1-dimensional level* $\mathbf{I}$*-limits*.

(b) A double limit of F is an ordinary limit of F_0 which also satisfies (dl.2), i.e. is preserved by the functor $e: \mathrm{Hor}_0 \mathbb{A} \to \mathrm{Hor}_1 \mathbb{A}$. (This is automatically true when $\mathbb{A}$ has 1-dimensional cotabulators.) When all these exist we say that $\mathbb{A}$ has *level* $\mathbf{I}$*-limits*.

(c) Finally $\mathbb{A}$ has *lax functorial level* $\mathbf{I}$*-limits* if:

(lfll) the categories $\mathrm{Hor}_0 \mathbb{A}$ and $\mathrm{Hor}_1 \mathbb{A}$ have (ordinary) $\mathbf{I}$-limits, preserved by the three structural functors $\partial^{\pm}: \mathrm{Hor}_0 \mathbb{A} \rightrightarrows \mathrm{Hor}_1 \mathbb{A} : e$.

5.3.2 Products and equalisers

(a) In particular, consider the ordinary product in $\mathrm{Hor}_0 \mathbb{A}$, of a (small) family $(X_i)_{i \in I}$ of objects of $\mathbb{A}$ (our base double category $\mathbb{I}$ is discrete).

The cone $(X, (pi: X \to X_i)_i)$ is the (double) *product* of the family in $\mathbb{A}$ if and only if:

(dp.2) given two cones $(A, (hi))$, $(B, (ki))$, a vertical arrow $u: A \rightarrowtail B$ and a family of cells $\xi i: (u \,{}^{hi}_{ki}\, eX_i)$, there is precisely one cell φ such that $\xi = \varphi \,|\, e_p$

$$
\begin{array}{ccc}
A \xrightarrow{hi} X_i & & A \xrightarrow{f} X \xrightarrow{pi} X_i \\
u\downarrow \quad \xi i \quad \downarrow e & = & u\downarrow \quad \varphi \quad \downarrow e \quad e \quad \downarrow e \\
B \xrightarrow[ki]{} X_i & & B \xrightarrow[g]{} X \xrightarrow[pi]{} X_i
\end{array}
\qquad (i \in I). \qquad (5.14)
$$

Spelling out the condition for the existence of lax functorial products in $\mathbb{A}$ is also easy.

(b) Consider now the ordinary equaliser $p\colon L \to X$ in $\mathrm{Hor}_0\mathbb{A}$ of a pair of horizontal maps $f, g\colon X \rightrightarrows Y$. Then (L, p) is the (double) *equaliser* of these maps in $\mathbb{A}$ if and only if

(de.2) for every cell $\xi\colon (u \,{}^x_y\, e_X)$ of $\mathbb{A}$ which equalises f and g (i.e. $\xi \mid f = \xi \mid g$) there is precisely one cell $\varphi\colon u \to e_L$ such that $\xi = \varphi \mid e_p$.

The existence and functoriality of (co)products and (co)equalisers in our main structures will be studied in the exercises of the next two sections.

5.3.3 *Proposition* (Level limits)

The weak double category $\mathbb{A}$ has all small (resp. finite) level limits if and only if it has all small (resp. finite) products and equalisers; the construction is the standard one.

If $\mathbb{A}$ has lax (resp. pseudo, strict) functorial products and lax (resp. pseudo, strict) functorial equalisers, this construction provides a similar functor of level $\mathbf{I}$-limits, for every small category $\mathbf{I}$.

Proof It is a straightforward consequence of the construction and preservation theorem for ordinary limits, taking into account 5.3.1. $\qquad\square$

5.3.4 *Two-dimensional tabulators*

We have already studied 1-dimensional tabulators, in Sections 3.6 and 3.7. The (double) *tabulator* of a vertical arrow $u\colon X_0 \to X_1$ is defined as the limit of the associated double functor $u\colon \mathbf{2}^{\mathbf{t}} \to \mathbb{A}$ (which is strict because $\mathbb{A}$ is unitary).

The double limit is thus an object $T = \mathsf{T}u$ equipped with two horizontal maps $pi\colon T \to X_i$ ($i = 0, 1$) and a cell $\pi\colon (e_T \,{}^{p0}_{p1}\, u)$ satisfying the following conditions:

(dt.1) for every object A and every cell $\varphi\colon e_A \to u$ there is a unique horizontal map $f\colon A \to T$ such that $\varphi = e_f \mid \pi$

$$
\begin{array}{ccc}
A \xrightarrow{h0} X_0 & & A \xrightarrow{\ f\ } T \xrightarrow{p0} X_0 \\
e \downarrow \ \ \varphi \ \ \downarrow u & = & e \downarrow \ \ e_f \ \ \downarrow e \ \ \pi \ \ \downarrow u \\
A \xrightarrow[h1]{} X_1 & & A \xrightarrow[\ f\ ]{} T \xrightarrow[p1]{} X_1
\end{array}
\qquad (5.15)
$$

(dt.2) (the *tetrahedron property*) for every vertical arrow

$$(x, \xi)\colon (A, \varphi) \to (B, \psi)$$

of cones, i.e. for every vertical arrow $x\colon A \to B$ and every modification

$\xi\colon (x \overset{h}{\underset{k}{\to}} u)$ (with $\xi_0\,|\,\psi = \varphi\,|\,\xi_1$) there is a unique cell $\alpha\colon (x \overset{f}{\underset{g}{\to}} T)$ in $\mathbb{A}$ such that

$$\varphi = e_f\,|\,\pi, \qquad \psi = e_g\,|\,\pi, \qquad \xi_i = \alpha\,|\,e_{pi} \qquad (i = 0,1), \qquad (5.16)$$

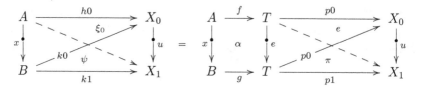

$\mathbb{A}$ has lax functorial tabulators if and only if:

(i) it has 1-dimensional tabulators of vertical arrows,

(ii) it has the limit of every pseudo vertical transformation of vertical arrows, consistently with the previous 1-dimensional limits, with respect to vertical faces and degeneracy.

5.3.5 Remarks

Taking into account a remark at the end of 3.8.2, a pseudo vertical transformation $u \to v\colon \mathbf{2}^t \to \mathbb{A}$ amounts to two vertical arrows $r_i\colon X_i \to Y_i$ and a special isocell r (the naturality comparison on the vertical arrow of $\mathbf{2}^t$)

$$
\begin{array}{l}
\begin{array}{ccc}
X_0 & & \\
 & \searrow^{r_0} & \\
u\downarrow & & Y_0 \\
 & r & \downarrow v \\
X_1 & & \\
 & \searrow_{r_1} & \\
 & & Y_1
\end{array}
\qquad
\begin{array}{l}
(r_0, r_1, r)\colon u \to v\colon \mathbf{2}^t \to \mathbb{A}, \\[2mm]
r\colon u \otimes r_1 \to r_0 \otimes v\colon X_0 \to Y_1.
\end{array}
\end{array}
\qquad (5.17)
$$

Thus (dt.2) says that:

- for every vertical morphism $(x, \xi)\colon (A, \varphi) \to (B, \psi)$ of cones of $u\colon \mathbf{2}^t \to \mathbb{A}$ there is a unique cell $\alpha\colon (x \overset{f}{\underset{g}{\to}} e_T)$ in $\mathbb{A}$ such that $\xi = \alpha\,|\,e_\pi$

In other words, the limit of the vertical identity of $u\colon \mathbf{2}^t \to \mathbb{A}$ is the vertical identity of the 1-dimensional tabulator $\pi\colon e_T \to u$ (viewed as a horizontal transformation $e_T \to u\colon \mathbf{2}^t \to \mathbb{A}$).

5.3.6 Theorem (The Construction Theorem of double limits)

(i) The weak double category $\mathbb{A}$ has all small (resp. finite) limits if and only if it has small (resp. finite) products, equalisers and tabulators. The construction is explicitly described in the proof.

(ii) If $\mathbb{A}$ has lax (or pseudo, strict) functorial limits of these basic kinds (i.e. products, equalisers and tabulators), our construction provides a lax (or pseudo, strict) functor of $\mathbb{I}$-limits, for any small $\mathbb{I}$.

Proof See 5.3.7 and 5.3.8. □

5.3.7 Proof of the Construction Theorem, Part I

(*The double limit of a double diagram.*) Of course we have only to prove the 'sufficiency part' of the statement, for a lax double functor $F\colon \mathbb{I} \to \mathbb{A}$. We consider first the 'free case', which is considerably simpler, contains various interesting non-level cases and allows a more economical algorithm.

We assume thus that $\mathbb{I}$ is just a double graph (or, equivalently, that $\mathbb{I}$ is the free double category generated by a double graph). In this case, the cones $h\colon \Delta A \to F$ are just subject to *one* naturality condition (ht.1), namely the naturality on a cell, which simplifies the problem.

The solution is based on the following steps: we construct an ordinary graph **S** and the associated horizontal double graph $\mathbb{S}$, then we turn F into a morphism of double graphs $G\colon \mathbb{S} \to \mathbb{A}$, and we take its limit. The procedure is similar to computing the end of a functor $H\colon \mathbf{C}^{\mathrm{op}} \times \mathbf{C} \to D$ as the limit of an associated functor based on Kan's *subdivision category* of $\mathbf{C}$ (see [K3], Section 1.10, or [M4], Section IX.5).

(A) We form the graph **S**, called the *horizontal subdivision* of $\mathbb{I}$, by replacing every vertical arrow of $\mathbb{I}$ with a new object, that *simulates its tabulator*. Precisely, **S** is formed by the following objects and arrows (and is finite whenever $\mathbb{I}$ is):

(a) all the objects and horizontal arrows of $\mathbb{I}$,

(b) for every vertical map $u\colon i_u \rightarrowtail j_u$ of $\mathbb{I}$, a new formal object $u\hat{}$, also written as u, together with two new arrows, $p_u\colon u\hat{} \to i_u$ and $q_u\colon u\hat{} \to j_u$,

(c) for every cell $\alpha\colon (u\,{}^f_g\,v)$ of $\mathbb{I}$, a new arrow $\alpha\hat{}\colon u\hat{} \to v\hat{}$, also written as α.

(B) Viewing **S** as a horizontal double graph $\mathbb{S}$ with no vertical arrows, we extend $F\colon \mathbb{I} \to \mathbb{A}$ to a morphism $G\colon \mathbb{S} \to \mathbb{A}$.

(a) G coincides with F on the objects and horizontal arrows of $\mathbb{I}$.

(b) For every vertical map $u\colon i_u \twoheadrightarrow j_u$ of $\mathbb{I}$, we set

$$Gu = \top(Fu), \qquad (5.18)$$

while $Gp_u\colon Gu \to Gi_u$ and $Gq_u\colon Gu \to Gj_u$ are the structural arrows of $\top(Fu)$ in $\mathbb{A}$; these morphisms will again be written as p_u and q_u, while we write as $\pi_u\colon (e\, {}^{p_u}_{q_u}\, Fu)$ the structural cell of the tabulator.

(c) For every cell $\alpha\colon (u\, {}^f_g\, v)$ of $\mathbb{I}$, $G\alpha$ is the horizontal map of $\mathbb{A}$ such that

$$G\alpha\colon \top(Fu) \to \top(Fv), \qquad G\alpha \,|\, \pi_v = \pi_u \,|\, F\alpha, \qquad (5.19)$$

according to the universal property of π_v.

The limit of this horizontal diagram $G\colon \mathbb{S} \to \mathbb{A}$ exists, by hypotheses and the previous Proposition 5.3.3. We want to prove that it gives the limit of F; in fact, we construct an isomorphism between the double categories of cones $\Delta \Downarrow F$ and $\Delta' \Downarrow G$, whose terminal objects yield our two limits. Here, $\Delta'\colon \mathbb{A} \to \mathbb{L}\mathrm{x}(\mathbb{S}, \mathbb{A})$ is the new diagonal double functor.

(C) The canonical double functor $\Delta \Downarrow F \to \Delta' \Downarrow G$.

(a) Let $(A, h\colon \Delta A \to F)$ be a double cone of F. Its 'ordinary part' $(A, (hi)_i)$ can be extended to an ordinary cone $(A, h'\colon \Delta'A \to G)$ of G, using the non-ordinary part $(hu)_u$: we define $h'u\colon A \to \top(Fu)$ as the horizontal map of $\mathbb{A}$ determined by the cell hu, via the tabulator-property

$$e_{h'u} \,|\, \pi_u = hu. \qquad (5.20)$$

Now h' is indeed a cone, as it is coherent with the new arrows p_u, q_u, $G\alpha$:

$$p_u h'u = h(i_u), \qquad q_u h'u = h(j_u), \qquad G\alpha.h'u = h'v, \qquad (5.21)$$

where the first two properties follow from (5.20), and the third from the cancellation property of π_v

$$(h'u \,|\, G\alpha \,|\, \pi_v) = (h'u \,|\, \pi_u \,|\, F\alpha) = hu \,|\, F\alpha = hv = h'v \,|\, \pi_v. \qquad (5.22)$$

(b) A horizontal map $f\colon (A, h\colon \Delta A \to F) \to (B, k\colon \Delta B \to F)$ of $\Delta \Downarrow F$ determines a horizontal map $f\colon (A, h') \to (B, k')$ of $\Delta' \Downarrow G$, since

$$(f \,|\, k'u \,|\, \pi_u) = f \,|\, ku = hu = h'u \,|\, \pi_u. \qquad (5.23)$$

(c) A vertical map $(r, \xi)\colon (A, h\colon \Delta A \to F) \twoheadrightarrow (B, k\colon \Delta B \to F)$ of $\Delta \Downarrow F$, where $r\colon A \twoheadrightarrow B$ is vertical in $\mathbb{A}$ and $\xi\colon (\Delta r\, {}^h_k\, e_F)$ is an $\mathbb{X}$-cell, determines a vertical map $(r, \xi')\colon (A, h') \to (B, k')$ of $\Delta' \Downarrow G$; here the cell ξ is extended to $\xi'\colon (\Delta' r \, {}^{h'}_{k'} \, e_G)$, where $\xi' u$ satisfies the following relation (letting $\eta u = \xi i_u \otimes ku = hu \otimes \xi j_u$)

$$
\begin{array}{ccc}
\begin{array}{ccc}
A & \xrightarrow{hi_u} & Fi_u \\
{\scriptstyle r}\big\downarrow & {\scriptstyle \eta u} & \big\downarrow{\scriptstyle Fu} \\
B & \xrightarrow[kj_u]{} & Fj_u
\end{array}
& = &
\begin{array}{ccccccc}
A & \xrightarrow{h'u} & \mathsf{T}(Fu) & \xrightarrow{p_u} & Fi_u \\
{\scriptstyle r}\big\downarrow & {\scriptstyle \xi'u} & \big\downarrow{\scriptstyle e} & {\scriptstyle \pi_u} & \big\downarrow{\scriptstyle Fu} \\
B & \xrightarrow[k'u]{} & \mathsf{T}(Fu) & \xrightarrow[q_u]{} & Fj_u
\end{array}
\end{array}
\qquad (5.24)
$$

(d) Finally, it follows that a cell of $\Delta \Downarrow F$ determines one of $\Delta' \Downarrow G$.

(D) In the opposite direction, one constructs a canonical double functor $\Delta' \Downarrow G \to \Delta \Downarrow F$ inverse to the former, by similar arguments.

We just specify its action on the objects. Given a cone of G

$$
(A, (hi\colon A \to Fi)_i, (h'u\colon A \to Gu)_u),
$$

one forms a double cone $(A, h\colon \Delta A \to F)$ of F by letting

$$
hu = h'u \,|\, \pi_u, \qquad (5.25)
$$

which satisfies (ht.1) since, for $\alpha\colon (u \, {}^{f}_{g} \, v)$ in $\mathbb{I}$

$$
hu \,|\, F\alpha = (h'u \,|\, \pi_u \,|\, F\alpha) = (h'u \,|\, G\alpha \,|\, \pi_v) = h'v \,|\, \pi_v = hv. \qquad (5.26)
$$

(E) To verify the statement 5.3.6(ii) we assume that $\mathbb{A}$ has lax (or pseudo, strict) functorial limits for the basic cases, and we construct a similar functor for $\mathbb{I}$-limits; by Proposition 5.2.6, this can be reduced (both on hypotheses and conclusion) to two more elementary choices.

The first choice is given by the 'one-dimensional part' of what we have already proven. As to the second, a pseudo vertical transformation

$$
r = ((ri), (rf), (ru))\colon F \nrightarrow F'\colon \mathbb{I} \to \mathbb{A}
$$

has a natural extension to a pseudo vertical transformation $r\colon G \nrightarrow G'\colon$ $\mathbf{S} \to \mathbb{A}$, which is defined on the new objects $u\,\hat{}$ through the fact that tabulators have been assigned a choice of limits of pseudo vertical transformations

$$
r(u\,\hat{})\colon \mathsf{T}(Fu) \nrightarrow \mathsf{T}(F'u) \qquad \text{(for } u \text{ vertical in } \mathbb{I}\text{)}, \qquad (5.27)
$$

so that also the value on the new arrows $\alpha\,\hat{}\colon u\,\hat{} \to v\,\hat{}$ is uniquely determined. But this pseudo vertical transformation of horizontal functors $r\colon G \nrightarrow G'$ has an assigned limit, because of 5.3.3. Finally, since both steps respect the structural functors $\partial^{\pm}$ and e, so does their result.

5.3.8 Proof of the Construction Theorem, Part II

(*The general case.*) Now let $\mathbb{I}$ be an arbitrary weak double category. The previous construction of the graph $\mathbf{S}$ (step (A) of 5.3.7), of the morphism G (step (B)) and of the isomorphism $\Delta \Downarrow F \cong \Delta' \Downarrow G$ (steps (C) and (D)) is supplemented as follows.

(A) $\mathbf{S}$ has some supplementary objects and arrows:

- for every i in $\mathbb{I}$, a new arrow $d_i \colon i \to e_i\hat{\ }$ (simulating the diagonal map of 3.6.3(a)),

- for every vertical composition $w = u \otimes v$ in $\mathbb{I}$, a new object (u, v) and three arrows

$$p_{uv} \colon (u, v) \to u\hat{\ }, \qquad q_{uv} \colon (u, v) \to v\hat{\ }, \qquad d_{uv} \colon (u, v) \to w\hat{\ },$$

(simulating the object $\mathsf{T}(u, v)$ of 3.6.3(b) and its arrows).

(B) G is extended to these objects and arrows, by the objects and maps of $\mathbb{A}$ that they simulate:

- $G(d_i)$ is the diagonal map $dFi \colon Fi \to \mathsf{T}Fi = G(e_i\hat{\ })$ (in 3.6.3(a)); again, for the sake of simplicity, we write $G(d_i)$ as d_i, and π_i for the structural cell of the tabulator $\mathsf{T}(Fi)$ (i.e. π_u for $u = e_i$),

- $G_{uv} = \mathsf{T}(Fu, Fv) = \mathsf{T}Fu \times_{Fj} \mathsf{T}Fv$ is the limit of the composable pair Fu, Fv (in 3.6.3(b)); the arrows p_{uv}, q_{uv}, d_{uv} of $\mathbf{S}$ are taken by G to the projections and the diagonal of G_{uv}, which we simply write as

$$p_{uv} \colon G_{uv} \to Gu, \qquad q_{uv} \colon G_{uv} \to Gv, \qquad d_{uv} \colon G_{uv} \to G(u \otimes v).$$

Note that (G_{uv}, p_{uv}, q_{uv}) is the pullback of (q_u, p_v) in $\mathrm{Hor}_0 \mathbb{A}$.

(C) Given a double cone $(A, h \colon \Delta A \to F)$, we extend our previous $h' \colon \Delta' A \to G$ (of 5.3.7(C)) to the new objects (u, v) by letting $h'_{uv} \colon A \to G_{uv}$ be defined by the pullback-property of G_{uv}

$$p_{uv} . h'_{uv} = h'u \colon A \to Gu, \qquad q_{uv} . h'_{uv} = h'v \colon A \to Gv. \tag{5.28}$$

In order to prove that the new h' is a cone, it suffices to prove its coherence with the new arrows $d_i, p_{uv}, q_{uv}, d_{uv}$; two conditions hold by definition (5.28) above, the remaining two follow from the definition of the diagonal maps d_i, d_{uv} and the structural cells π_i, π_u (together with their usual cancellation property). In the first case we have

$$(hi \,|\, d_i \,|\, \pi_i) = hi \,|\, \square_{Fi} = e_{hi} = h(e_i) = h'(e_i) \,|\, \pi_i. \tag{5.29}$$

(D) Given an ordinary cone $(A, h' \colon \Delta' A \to G)$, we have to prove that the old associated double cone $(A, h \colon \Delta A \to F)$, defined by letting $hu = h'u \,|\, \pi_u$, is indeed a double cone for the new situation, i.e. satisfies also the conditions (ht.2, 3) concerning the vertical composition in $\mathbb{I}$. This proceeds much as above (letting $w = u \otimes v$, in the second case)

$$e_{hi} = (hi \,|\, d_i \,|\, \pi_i) = h'(e_i) \,|\, \pi_i = h(e_i), \tag{5.30}$$

$$\frac{hu}{hv} \,\Big|\, \underline{F}(u,v) \;=\; \frac{h'u \,|\, \pi_u}{h'v \,|\, \pi_v} \,\Big|\, \underline{F}(u,v) \;=\; h'_{uv} \,\Big|\, \frac{p_{uv} \,|\, \pi_u}{q_{uv} \,|\, \pi_v} \,\Big|\, \underline{F}(u,v)$$

$$= (h'_{uv} \,|\, d_{uv} \,|\, \pi_w) \;=\; h'w \,|\, \pi_w \;=\; hw. \tag{5.31}$$

(E) For the lax functoriality part, take a pseudo vertical transformation $r = ((ri), (rf), (ru)) \colon F \to F' \colon \mathbb{I} \to \mathbb{A}$. The extended $r \colon G \to G' \colon \mathbb{S} \to \mathbb{A}$ is defined on the new objects (u, v)

$$r(u,v) \colon \top(Fu, Fv) \to \top(F'u, F'v), \tag{5.32}$$

by the limit of pseudo vertical transformations of tabulators and pullbacks. Again, the extension to the new arrows

$$d_i \colon i \to e_i^{\,\hat{}}, \quad p_{uv} \colon (u,v) \to u^{\,\hat{}}, \quad q_{uv} \colon (u,v) \to v^{\,\hat{}}, \quad d_{uv} \colon (u,v) \to w^{\,\hat{}},$$

is uniquely determined.

5.4 Computing limits and colimits

The following exercises and examples deal with the limits and colimits of the strict or weak double categories of Sections 3.1 and 3.4. Solutions or hints are given in Appendix C.

Lax functoriality of limits and colax functoriality of colimits are obviously important, also because they imply the 2-dimensional property.

In some cases we also examine their *pseudo functoriality*, but this property is not of much interest here; more complete results on this point can be found in [GP1], Section 6.

5.4.1 Exercises on enriched profunctors

Study the flat double categories $\mathbb{M}tr \supset p\mathbb{O}rd \supset \mathbb{R}el\mathbf{Set}$ of enriched profunctors (see 3.1.2, 3.4.6, 3.4.7), proving that all of them have lax functorial limits and colax functorial colimits.

More precisely:

- the double category $\mathbb{M}tr$ of L-metric spaces has functorial sums and cotabulators,

- the double subcategory $p\mathbb{O}rd$ of preordered sets (identified with L-metric spaces with distance in $\{0, \infty\}$) is closed in $\mathbb{M}$tr under limits and colimits; it has functorial products, sums and cotabulators,

- the double subcategory $\mathbb{R}$el$\mathbf{Set}$ (where a set has distance $d(x, x') = 0$ if $x = x'$, and ∞ otherwise) is closed in both under limits, sums and coequalisers, while cotabulators are quotients of the corresponding ones in $\mathbb{M}$tr and $\mathrm{p}\mathbb{O}$rd; products and sums are functorial.

5.4.2 Exercises on adjoint functors

Study limits and colimits in the double category $\mathbb{A}$dj$\mathbf{Cat}$ of categories, functors and adjunctions and in the double subcategories $\mathbb{A}$dj$\mathbf{Ord}$ and $\mathbb{A}$dj$_0\mathbf{Ord}$ (see 3.1.6 and 3.1.7).

5.4.3 Exercises on spans and cospans

(a) Study limits and colimits in the weak double category $\mathbb{S}$pan$(\mathbf{C})$, where $\mathbf{C}$ is a category with pullbacks.

(a*) Dual results hold in $\mathbb{C}$osp$(\mathbf{C})$, for a category $\mathbf{C}$ with pushouts.

5.4.4 Exercises on profunctors

Study the weak double category $\mathbb{C}$at of categories, functors and profunctors (introduced in 3.4.3), proving that it has lax functorial limits and colax functorial colimits.

5.4.5 Examples of non-unitary limits

We have seen in 5.1.5 that the existence of non-unitary limits requires the failure of cotabulators.

Let us take, for instance, a monoidal category $\mathbf{C} = (\mathbf{C}, \otimes, E)$, consider it as a bicategory on one object, and then as a (vertical) weak double category $\mathbb{D}$. The latter has one object $*$ and one horizontal arrow 1_*; its vertical arrows $X \colon * \nrightarrow *$ are the objects of $\mathbf{C}$, with vertical identity $e_* = E$; its double cells $f \colon X \to Y$ are the morphisms of $\mathbf{C}$; its comparisons are those of $\mathbf{C}$.

It is easy to see – and can be proved as an exercise – that the unique object $*$ is a double terminal object in $\mathbb{D}$ if and only if the tensor unit E is a terminal object in $\mathbf{C}$ (as happens, for instance, in the cartesian case), if and only if $\mathbb{D}$ has cotabulators.

(In fact $*$ is trivially a 1-dimensional terminal object; but a terminal vertical arrow of $\mathbb{D}$ is the same as a terminal object of $\mathbf{C}$. The object $*$ is thus a terminal object of $\mathbb{D}$ if and only if the tensor unit E is terminal in $\mathbf{C}$. This is equivalent to the existence of cotabulators in $\mathbb{D}$: indeed a vertical arrow $X \colon * \nrightarrow *$ admits the only object $*$ as a cotabulator if and only if there is a unique map $X \to E$ in $\mathbf{C}$.)

Dually $*$ is a double initial object if and only if E is initial in $\mathbf{C}$, if and only if $\mathbb{D}$ has tabulators. For instance it is the case in the monoidal category $\mathbf{Rng}$ of rings, with the usual tensor product and tensor unit $\mathbb{Z}$, examined in Exercises 2.2.4 (i), (j). As remarked there, this structure is not cocartesian, unless we restrict to commutative rings.

5.4.6 Exercises on level limits in $\mathbb{D}$bl

The *transpositive* double category $\mathbb{D}$bl (see 4.2.3) has some horizontal level limits: all products and some equalisers.

It is important to note that these limits *can be transposed to vertical limits*. These 'transpositive limits' will be further investigated in a more adequate framework: the *triple* category $\mathrm{Stc}\mathbb{D}$bl (see Section 6.1) where we add the strict double functors in an additional 'transversal direction'. This embedding can be generalised (see 6.1.7) and can be thought to give *the right way of studying transpositive limits*.

5.4.7 Exercises on tabulators in $\mathbb{D}$bl

(a) Compute horizontal tabulators in $\mathbb{D}$bl.

(b) Prove that $\mathbb{D}$bl is horizontally span-representable.

(c) Compute vertical tabulators and show that $\mathbb{D}$bl is vertically span-representable.

5.4.8 Exercises on bimodules

The weak double category $\mathbb{R}$ng of (unitary) rings, homomorphisms and bimodules (see 3.4.5) presents a defective situation, with all level limits and cotabulators, but few level colimits and tabulators.

(a) Prove the existence of level limits and cotabulators in $\mathbb{R}$ng.

(b) Prove that there is an initial pair and no initial object, in the full 2-dimensional sense. Colimits of rings give 1-dimensional level colimits in $\mathbb{R}$ng, which are *not* 2-dimensional colimits, generally.

(c) Prove that $\mathbb{R}$ng cannot have all tabulators, even in the 1-dimensional sense.

(d) Give a similar study of the weak double category of monoids, homo-morphisms and bimodules, introduced in 3.4.8.

5.4.9 *Complements on bimodules*

For a bimodule $X\colon R \nrightarrow S$, we constructed in [GP1], Section 5.3, a ring T of *triangular matrices* $\left(\begin{smallmatrix} r & x \\ 0 & s \end{smallmatrix}\right)$, with obvious sum and 'matrix product' (for $r, r' \in R$, $s, s' \in S$, $x, x' \in X$)

$$\begin{pmatrix} r & x \\ 0 & s \end{pmatrix} \cdot \begin{pmatrix} r' & x' \\ 0 & s' \end{pmatrix} = \begin{pmatrix} rr' & rx' + xs' \\ 0 & ss' \end{pmatrix}. \tag{5.33}$$

This ring was wrongly presented as the tabulator of X, which cannot be (as we have seen that all of them cannot exist, in 5.4.8(c)). In fact, we do have two obvious ring homomorphisms

$$p\colon T \to R, \qquad q\colon T \to S, \tag{5.34}$$

but the projection $T \to X$ does not form with them a homomorphism of bimodules.

In order 'to make sense' of this construction, we can replace the vertical identity $T\colon T \nrightarrow T$ with the vertical endomorphism $X\colon T \nrightarrow T$, where the abelian group X has the obvious structure of (T, T)-bimodule

$$\begin{pmatrix} r' & x' \\ 0 & s' \end{pmatrix} \cdot x \cdot \begin{pmatrix} r'' & x'' \\ 0 & s'' \end{pmatrix} = r'xs''. \tag{5.35}$$

Now the identity of X gives a homomorphism of bimodules $(T, X, T) \to (R, X, S)$, i.e. a cell π

$$\begin{array}{ccc} T & \xrightarrow{\ p\ } & R \\ X \downarrow & \pi & \downarrow X \\ T & \xrightarrow[\ q\ ]{} & S \end{array} \tag{5.36}$$

$$\begin{pmatrix} r' & x' \\ 0 & s' \end{pmatrix} . x . \begin{pmatrix} r'' & x'' \\ 0 & s'' \end{pmatrix} \;\mapsto\; r'xs'' = p\begin{pmatrix} r' & x' \\ 0 & s' \end{pmatrix} . x . q\begin{pmatrix} r'' & x'' \\ 0 & s'' \end{pmatrix}.$$

We can characterise the pair $((T, X, T), \pi)$ as the universal cell $\pi\colon \bullet \to (R, X, S)$ whose domain is a vertical *endomorphism*. In fact, for every cell $\varphi\colon (A, Y, A) \to (R, X, S)$ with $f\colon A \to R$, $g\colon A \to S$ there is a unique cell

ψ such that $\psi \,|\, \pi = \varphi$

$$
\begin{array}{ccccc}
A & \overset{h}{\longrightarrow} & T & \overset{p}{\longrightarrow} & R \\
Y\downarrow & \psi & X\downarrow & \pi & \downarrow X \\
A & \underset{h'}{\longrightarrow} & T & \underset{q}{\longrightarrow} & S
\end{array}
\qquad (5.37)
$$

$$
h(a) = \begin{pmatrix} fa & \varphi a \\ 0 & ga \end{pmatrix}, \qquad\qquad \psi(y) = \varphi(y),
$$

$$
\psi(ayb) = \varphi(ayb) = fa.\varphi y.gb = ha.\varphi y.ha = ha.\psi y.ha.
$$

More formally, we are replacing the degeneracy embedding $e\colon \mathrm{Hor}_0\mathbb{A} \to \mathrm{Hor}_1\mathbb{A}$ with the embedding

$$
U\colon \mathrm{Ve}\mathbb{A} \to \mathrm{Hor}_1\mathbb{A}, \qquad (5.38)
$$

of the full subcategory of vertical endomorphisms.

For $\mathbb{A} = \mathbb{R}\mathrm{ng}$, the pair $((T, X, T), \pi)$ is a universal arrow from this embedding to the 'object' (R, X, S) of $\mathrm{Hor}_1\mathbb{R}\mathrm{ng}$, and we have constructed the right adjoint

$$
\mathrm{Hor}_1\mathbb{R}\mathrm{ng} \to \mathrm{Ve}\mathbb{R}\mathrm{ng}, \qquad (R, X, S) \mapsto (T, X, T), \qquad (5.39)
$$

with counit π.

5.5 Adjunctions, limits and unitarity

We prove that the right adjoint $G\colon \mathbb{A} \to \mathbb{X}$ of a *unitary* colax-lax adjunction $F \dashv G$ preserves limits (Theorem 5.5.5). When F is not unitary, this can fail even for a strict G. (But it is still true that G must preserve level limits, see 5.5.7.)

We are given a lax functor $T\colon \mathbb{I} \to \mathbb{A}$ between unitary weak double categories. We begin by reformulating the definition of cones and limit of T in terms of cells in $\mathbb{D}\mathrm{bl}$, whose vertical arrows are strict functors; such a cell is inhabited by a horizontal transformation of lax functors (as remarked in 4.2.3(b)), that will be marked by an arrow – here fully justified.

5.5.1 Transforming cones

Let $G\colon \mathbb{A} \to \mathbb{X}$ be a lax functor, and consider the diagonal functors

$$
\Delta\colon \mathbb{A} \to \mathbb{L}\mathrm{x}(\mathbb{I}, \mathbb{A}), \qquad \Delta'\colon \mathbb{X} \to \mathbb{L}\mathrm{x}(\mathbb{I}, \mathbb{X}). \qquad (5.40)
$$

We have a horizontal transformation

$$\gamma \colon \Delta'(GA) \to G.\Delta A \colon \mathbb{I} \to \mathbb{X},$$

$$\gamma i = 1_{GA} \colon GA \to GA, \qquad \gamma u = \underline{G}(A) \colon e_{GA} \to G(e_A), \tag{5.41}$$

which produces a lax functor of cones

$$\hat{G} \colon \mathbb{C}\text{one}(T) \to \mathbb{C}\text{one}(GT),$$

$$(A, h \colon \Delta A \to T) \mapsto (GA, Gh.\gamma \colon \Delta'(GA) \to GT). \tag{5.42}$$

When G is a unitary lax functor, γ is the identity and G preserves diagonalisation: $G.\Delta A = \Delta'(GA)$, as we have already seen in 3.8.5. Then the associated lax functor $\hat{G}$ on cones acts simply as:

$$(A, h \colon \Delta A \to T) \mapsto (GA, Gh \colon \Delta'(GA) \to GT). \tag{5.43}$$

We say that the lax functor G *preserves* (the existing) $\mathbb{I}$-*limits* if

(i) whenever (A, h) is the limit cone of a lax functor $T \colon \mathbb{I} \to \mathbb{A}$, then $\hat{G}(A, h)$ is the limit cone of GT,

(ii) a similar property holds for the limit of a pseudo vertical transformation $r \colon T \to S \colon \mathbb{I} \to \mathbb{A}$.

We begin by treating the unitary case, because of two main reasons.

First, we have already seen in 4.5.2(c) that the discrete embedding $D \colon \mathbb{S}\text{pan}\mathbf{Set} \to \mathbb{C}\text{at}$, viewing a span as a set-profunctor, has a non-unitary right adjoint *and* does not preserve cotabulators.

Second, a lax functor G takes an object to a vertical monad, and a family of objects A_i ($i \in I$) to a family of vertical monads $G.A_i = (G(A_i), t_i, \eta_i, \mu_i)$. If we say that G 'preserves products', it is not clear whether we mean that it takes a product-object $A = \prod A_i$

- to a product-object $X = \prod G(A_i)$,

- or to the limit of the family of vertical monads $G.A_i$.

The general case will only be dealt with at the end of this section, in 5.5.6 and 5.5.7, as a marginal topic. In particular, 5.5.7 is precisely concerned with the preservation of products.

5.5.2 Theorem (Cones as cells)

(a) A cone $(A, h \colon \Delta A \to T)$ of T amounts to a cell in $\mathbb{D}\text{bl}$

$$
\begin{array}{ccc}
\mathbb{I} & \xrightarrow{\;Q\;} & 1 \\
{\scriptstyle 1}\downarrow & {\llap{\raisebox{-0.5ex}{$\scriptstyle h$}}\nearrow} & \downarrow{\scriptstyle A} \\
\mathbb{I} & \xrightarrow[\;T\;]{} & \mathbb{A}
\end{array}
\tag{5.44}
$$

i.e. a horizontal transformation $h\colon AQ \to T$ *of lax functors (because* $A\colon$
$1 \to \mathbb{A}$ *is strict).*

(b) A vertical morphism $(u, \xi)\colon (A_0, h_0) \to (A_1, h_1)$ *of cones of* T *amounts
to a cell* ξ *in* $\mathbb{D}$bl, *as in the left diagram below, that satisfies the following
condition, for* $n = 0, 1$

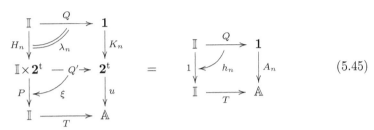

$$(5.45)$$

where H_n *and* K_n *denote the obvious embeddings at the level* $(-, n)$ *of*
$- \times \mathbf{2}^{\mathrm{t}}$, *and* $\lambda_n \colon K_n Q = Q' H_n$ *is a trivial cell.*

Proof (a) Obvious, because $\Delta A = AQ \colon \mathbb{I} \to \mathbb{A}$.

(b) We rewrite (A_0, h_0) and (A_1, h_1) as (A', h') and (A'', h''), respectively.

As defined in 5.2.2, a vertical morphism $(u, \xi) \colon (A', h') \rightarrowtail (A'', h'')$ of
cones of T comprises a vertical arrow $u \colon A' \rightarrowtail A''$ (that gives a vertical transformation $\Delta u \colon \Delta A' \rightarrowtail \Delta A''$, constant at u) and a modification
$\xi \colon (\Delta u \, {}^{h'}_{h''} \, e_T)$.

The latter means two horizontal transformations $h' \colon A'Q \to T$ and
$h'' \colon A''Q \to T$ of lax functors and, for every i in $\mathbb{I}$, a cell $\xi i \colon (u \, {}^{h'i}_{h''i} \, e_{Ti})$ in
$\mathbb{A}$ that satisfies the conditions (mod.1, 2) of Definition 3.8.3.

Here (mod.1) says that $\xi i \,|\, Ta = \xi i'$, for every $a \colon i \to i'$ in $\mathbb{I}$; (mod.2)
says that

$$\xi i \otimes h'' r = h' r \otimes \xi j \qquad \text{(for } r \colon i \rightarrowtail j \text{ in } \mathbb{I}). \qquad (5.46)$$

On the other hand the cell ξ in (5.45) 'is' a horizontal transformation
$\xi \colon uQ' \to TP$. This gives the horizontal transformations h', h'' (obtained
as above), together with a family of cells in $\mathbb{A}$

$$\xi i \colon (u \, {}^{h'i}_{h''i} \, e_{Ti}) \qquad \text{(for } i \text{ in } \mathbb{I}), \qquad (5.47)$$

that corresponds to the vertical arrow $(i, z) \colon (i, 0) \rightarrowtail (i, 1)$, given by the
vertical arrow $z \colon 0 \rightarrowtail 1$ of $\mathbf{2}^{\mathrm{t}}$.

Computing the horizontal transformation $\xi \colon uQ' \to TP$ on the vertical
arrow $(r, z) \colon (i, 0) \rightarrowtail (j, 1)$ we have, taking into account that $(i, z) \otimes (r, 1) =
(r, z) = (r, 0) \otimes (j, z)$

$$\xi i \otimes h'' r = \xi(r, z) = h' r \otimes \xi j,$$

which is condition (5.46). The remaining points are straightforward. $\qquad\square$

5.5.3 A remark

When we pre-compose a cone $(A, h \colon \Delta A \to T)$ with a horizontal arrow $f \colon A' \to A$ we get a cone $(A', h.\Delta f \colon \Delta A' \to T)$ which is expressed as the pasting $h \,|\, f$ of cells in $\mathbb{D}bl$

$$
\begin{array}{ccccc}
\mathbb{I} & \xrightarrow{\;Q\;} & \mathbf{1} & =\!=\!= & \mathbf{1} \\
{\scriptstyle 1}\downarrow & {\scriptstyle h} & {\scriptstyle A}\downarrow\;\;{\scriptstyle f} & & \downarrow{\scriptstyle A'} \\
\mathbb{I} & \xrightarrow[\;T\;]{} & \mathbb{A} & =\!=\!= & \mathbb{A}
\end{array}
\tag{5.48}
$$

The horizontal composition $h \,|\, f$ appears thus in reverse order with respect to the composition $\Delta A' \to \Delta A \to T$ of horizontal transformations, which may be confusing. Marking the cells of $\mathbb{D}bl$ with an arrow will help us to manage this drawback.

5.5.4 Limits as universal cells

As we have seen, a cone $(A, h \colon A \to T)$ is a cell of $\mathbb{D}bl$, as in the left diagram below

$$
\begin{array}{ccc}
\mathbb{I} \xrightarrow{\;Q\;} \mathbf{1} & \qquad & \mathbb{I}\times\mathbf{2}^{\mathrm{t}} \xrightarrow{\;Q'\;} \mathbf{2}^{\mathrm{t}} \\
{\scriptstyle 1}\downarrow\;{\scriptstyle h}\;\downarrow{\scriptstyle A} & & {\scriptstyle P}\downarrow\;{\scriptstyle \xi}\;\downarrow{\scriptstyle u} \\
\mathbb{I} \xrightarrow[\;T\;]{} \mathbb{A} & & \mathbb{I} \xrightarrow[\;T\;]{} \mathbb{A}
\end{array}
\tag{5.49}
$$

and a vertical morphism of cones is a cell as in the right diagram above (whose faces are computed as in (5.45)).

Now, a cone (L, p) is the 1-dimensional limit of T if and only if (i) holds:

(i) for every object A in $\mathbb{A}$, the mapping $f \mapsto h = p \,|\, f$ (a horizontal composition of cells, in $\mathbb{D}bl$) gives a bijection between $\mathbb{A}$-horizontal arrows $f \colon A \to L$ and cones (A, h) of T

$$
\begin{array}{ccccc}
\mathbb{I} & \xrightarrow{\;Q\;} & \mathbf{1} & =\!=\!= & \mathbf{1} \\
{\scriptstyle 1}\downarrow & {\scriptstyle p} & {\scriptstyle L}\downarrow\;\;{\scriptstyle f} & & \downarrow{\scriptstyle A} \\
\mathbb{I} & \xrightarrow[\;T\;]{} & \mathbb{A} & =\!=\!= & \mathbb{A}
\end{array}
\tag{5.50}
$$

(L, p) is the double limit if and only if (ii) also holds:

(ii) for every vertical arrow $u \colon A' \nrightarrow A''$ in $\mathbb{A}$, the mapping

$$
\tau \mapsto \xi = (\lambda \otimes p) \,|\, \tau
$$

gives a bijection between $\mathbb{A}$-cells $\tau\colon u \to e_L$ and $\mathbb{D}$bl-cells ξ whose boundary is the outer perimeter of the following diagram (where λ is the obvious 'commutative cell')

$$\tag{5.51}$$

5.5.5 *Theorem* (Limits and unitary adjunctions)

Let $(\eta, \varepsilon)\colon F \dashv G$ be a unitary colax-lax adjunction. Then the lax functor $G\colon \mathbb{A} \to \mathbb{X}$ preserves all the (existing) limits of lax double functors with values in $\mathbb{A}$.

Proof Take a lax double functor $T\colon \mathbb{I} \to \mathbb{A}$, and suppose that $(A, h\colon AQ \to T)$ is a limit of T. To prove that $G(A, h) = (GA, Gh)\colon (GA)Q \to GT$ is a limit of GT, we follow the standard procedure, for the 1-dimensional property, as rewritten above, in 5.5.4(i).

Given a cone $(X, k\colon XQ \to GT)$ of GT, the pasting h' of $\mathbb{D}$bl-cells displayed at the left

$$\tag{5.52}$$

gives a cone $(FX, h'\colon FX.Q \to T)$ of T (*by the unitarity of F*). Therefore, there is a unique $g\colon FX \to A$ such that $h' = (h \,|\, g)\colon FX.Q \to T$. Now, the adjoint morphism $f = Gg.\eta X\colon X \to GFX \to GA$ is the unique horizontal arrow of $\mathbb{X}$ such that $k = (h \,|\, f)\colon XQ \to GT$ (as one sees pasting the $\mathbb{D}$bl-cell $\eta\colon (F \,{}^1_G\, 1)$ at the right of both diagrams above).

For the 2-dimensional property we are given a vertical arrow $(v, \xi)\colon (X', k') \to (X'', k'')$ of cones of GT. The pasting ξ' of $\mathbb{D}$bl-cells displayed at the left

$$
\begin{array}{ccc}
\mathbb{I}\times\mathbf{2}^t \xrightarrow{\ Q\ } \mathbf{2}^t & & \\
\end{array}
\qquad = \qquad
\begin{array}{c}
\end{array}
\qquad (5.53)
$$

gives a vertical arrow $(Fv, \xi') \colon (FX', h') \to (FX'', h'')$ of T (by the unitarity of F). Therefore, there is a unique $\alpha \colon Fv \to e_A$ such that $\xi' = (e_h \,|\, \alpha) \colon Fv.Q \to TP$.

Now, the adjoint cell $\varphi = (\eta v \,|\, G\alpha) \colon v \to GFv \to Ge_A$ is the unique cell of $\mathbb{X}$ such that $\xi = (\varphi \,|\, Ge_h) \colon vQ \to GT$. $\qquad\square$

5.5.6 Theorem (Tabulators, adjunctions and unitarity)

All tabulators are meant in the 1-dimensional sense. Let us have a general colax-lax adjunction between weak double categories

$$
F \colon \mathbb{X} \rightleftarrows \mathbb{A} \colon G, \qquad (\eta, \varepsilon) \colon F \dashv G. \qquad (5.54)
$$

(a) If F is pseudo unitary, then G preserves all the existing tabulators of $\mathbb{A}$, in the following sense: given a vertical arrow $v \colon A \rightarrowtail A'$ having tabulator $(V, \tau \colon (V \,{}^p_q\, v))$ in $\mathbb{A}$, the object GV is the tabulator-object of Gv in $\mathbb{X}$, via the obvious cell π from e_{GV} to Gv

$$
\pi = (\underline{G}(V) \,|\, G\tau) \colon (e_{GV} \,{}^p_q\, Gv). \qquad (5.55)
$$

(b) Take X in $\mathbb{X}$ and assume that $v = F(e_X) \colon FX \rightarrowtail FX$ has a tabulator in $\mathbb{A}$, preserved by G; then the colaxity cell $\underline{F}X \colon Fe_X \to e_{FX}$ has a horizontal retraction $\zeta \colon e_{FX} \to F(e_X)$, with $\underline{F}X \,|\, \zeta = 1$.

Notes. In (a), taking F strictly unitary gives a less clear argument in the proof. In (b), one cannot prove $\underline{F}X$ to be invertible; to wit, the terminal double functor $\mathbb{R}\mathrm{ng} \to \mathbf{1}$ trivially preserves all the existing tabulators, but we have seen in 5.4.8(b) that it has a colax left adjoint which is not pseudo unitary.

Proof (a) Assume that the comparison cell $\underline{F}X$ has a horizontal inverse ζ; we want to prove that GV is the tabulator of Gv via π. Take an $\mathbb{X}$-cell $\xi \colon (e_X \,{}^f_g\, Gv)$, and its adjoint cell $\xi' = (F\xi \,|\, \varepsilon v) \colon (Fe_X \,{}^{f'}_{g'}\, v)$ in $\mathbb{A}$. There is precisely one $k \colon FX \to V$ in $\mathbb{A}$ such that

$$k \mid \tau = (\zeta \mid \xi') \colon (e_{FX} \overset{f'}{\underset{g'}{\to}} v). \qquad (5.56)$$

Its adjoint $h = Gk.\eta X \colon X \to GV$ satisfies the condition $h \mid \pi = \xi$ in $\mathbb{X}$

$$h \mid \pi = (h \mid \underline{GV} \mid G\tau) = (\eta X \mid Gk \mid \underline{GV} \mid G\tau) = (\eta X \mid \underline{GFX} \mid Ge_k \mid G\tau)$$
$$= (\eta(e_X) \mid G(\underline{F}X) \mid G\zeta \mid G\xi') = \eta(e_X) \mid G\xi' = \xi,$$

$$
\begin{array}{ccccccc}
X & \overset{h}{\longrightarrow} & GV & =\!=\!= & GV & \overset{Gp}{\longrightarrow} & GA \\
{\scriptstyle e}\downarrow & {\scriptstyle e_h} & {\scriptstyle e}\downarrow & \underline{GV} & {\scriptstyle Ge}\downarrow \ \ G\tau & & \downarrow{\scriptstyle Gv} \\
X & \underset{h}{\longrightarrow} & GV & =\!=\!= & GV & \underset{Gq}{\longrightarrow} & GA'
\end{array}
\qquad (5.57)
$$

Conversely, if $h \mid \pi = \xi$, then the adjoint map $k = \varepsilon V.Fh \colon FX \to V$ gives, in $\mathbb{A}$

$$(\underline{F}X \mid e_k \mid \tau) = (\underline{F}X \mid e_{Fh} \mid e_{\varepsilon V} \mid \tau) = (Fe_h \mid \underline{F}G(V) \mid e_{\varepsilon V} \mid \tau)$$
$$= (Fe_h \mid F\underline{GV} \mid \varepsilon e_V \mid \tau) = (Fe_h \mid F\underline{GV} \mid FG\tau \mid \varepsilon v)$$
$$= (Fe_h \mid F\pi \mid \varepsilon v) = F\xi \mid \varepsilon v = \xi'.$$

This means that the cell $\underline{F}X \mid e_k$ is uniquely determined by ξ. Since $\underline{F}X$ is invertible, k is uniquely determined as well.

(b) Take an object X in $\mathbb{X}$. Recall that the colaxity cell $\underline{F} = \underline{F}X \colon F(e_X) \to e_{FX}$ corresponds to $\underline{F}' = e_{\eta X} \mid \underline{GFX}$ (by 4.3.3(b)). In diagram (5.57), let $A = A' = FX$ and $v = F(e_X) \colon FX \nrightarrow FX$, with tabulator $(V, \tau \colon (V \overset{p}{\underset{q}{\to}} v))$ preserved by G: GV is the tabulator of Gv via $\pi = \underline{GV} \mid G\tau$.

The unit of the adjunction yields a cell

$$\xi = (1_v)' = \eta(e_X) \colon (e_X \overset{\eta X}{\underset{\eta X}{\to}} GF(e_X)),$$

whence one map $h \colon X \to GV$ in $\mathbb{X}$ such that $e_h \mid \pi = \eta(e_X)$. The map h corresponds to $k \colon FX \to V$ and it suffices to verify that the cell $\zeta = (e_k \mid \tau) \colon (FX \overset{pk}{\underset{qk}{\to}} v) \colon e_{FX} \to v$ is a retraction of $\underline{F}$:

$$(\underline{F} \mid e_k \mid \tau)' = (\underline{F}' \mid Ge_k \mid G\tau) = (e_{\eta X} \mid \underline{GFX} \mid Ge_k \mid G\tau)$$
$$= (e_{\eta X} \mid e_{Gk} \mid \underline{GV} \mid G\tau) = e_h \mid \pi = \eta(e_X) = (1_v)'.$$

$\square$

5.5.7 Exercises and complements

(a) (*Level limits and adjunctions*) Given a colax-lax double adjunction $F \dashv G$, the lax functor G preserves level limits, while the colax functor F preserves level colimits (independently of the unitarity of F or G).

(b) (*Extending unary products*) Study the limit of a strict or lax functor $F \colon \mathbf{1} \to \mathbb{A}$, as described in Exercise 3.8.6(b).

(c) Prove that a lax functor $S \colon \mathbb{A} \to \mathbb{B}$ always preserves the limit of a strict functor $A \colon \mathbf{1} \to \mathbb{A}$, in a suitable sense.

5.6 *Weighted limits in 2-categories as double limits

This section is addressed to a reader with some knowledge of the theory of 2-categories.

After reviewing the definition of weighted limits in a 2-category $\mathbf{C}$, we prove that a W-weighted limit in $\mathbf{C}$, for a given 2-functor $W \colon \mathbf{I} \to \mathbf{Cat}$, is the same as a double limit based over a suitable double category $\mathbb{El}(W)$ (introduced in [Pa1, GP14]). Using the construction theorem of double limits, it follows that *weighted limits in* $\mathbf{C}$ *amount to double limits in* $\mathbf{C}$, viewed as a horizontal double category.

In this section $\mathbf{I}$ is always a small 2-category equipped with a 2-functor $W \colon \mathbf{I} \to \mathbf{Cat}$, its *weight*. We write as $[\mathbf{I}, \mathbf{Cat}]$ (resp. $[\mathbf{I}, \mathbf{Cat}]_{\mathrm{ps}}$) the 2-category of 2-functors $\mathbf{I} \to \mathbf{Cat}$, their 2-natural (resp. pseudo natural) transformations, and modifications (see 3.8.7).

A reader interested in the theory of weighted limits (also called indexed limits) is referred to [St2, Gra2, Ke2, Ke3, BKPS].

5.6.1 Weighted limits

The limit of a 2-functor $F \colon \mathbf{I} \to \mathbf{C}$ can be defined by a 2-universal property (see 2.3.5), but something more general is needed, to include important universal constructions as the cell representer $\mathbf{2} \pitchfork X$ (see 3.6.6), and others recalled below.

The W-*weighted pseudo limit* (L, p), or *pseudo* W-*limit*, of a 2-functor $F \colon \mathbf{I} \to \mathbf{C}$ with respect to the 2-functor $W \colon \mathbf{I} \to \mathbf{Cat}$, is an object $L = \mathrm{psLim}_W F$ of $\mathbf{C}$ equipped with a pseudo natural transformation

$$p \colon W \to \mathbf{C}(L, F(-)) \colon \mathbf{I} \to \mathbf{Cat}, \qquad (5.58)$$

that gives, for every A in $\mathbf{C}$, an isomorphism of categories

$$\mathbf{C}(A, L) \cong [\mathbf{I}, \mathbf{Cat}]_{\mathrm{ps}}(W, \mathbf{C}(A, F)). \qquad (5.59)$$

This means that the following 1-*dimensional* and 2-*dimensional* universal properties are satisfied:

(i) for every similar pair $(A, h\colon W \to \mathbf{C}(A, F))$ there is a unique morphism $f\colon A \to L$ in $\mathbf{C}$ such that:

$$h = \mathbf{C}(L, f).p\colon W \to \mathbf{C}(L, F) \to \mathbf{C}(A, F), \tag{5.60}$$

(ii) for every modification $\xi\colon h \to k\colon W \to \mathbf{C}(A, F)$ there is a unique 2-cell $\alpha\colon f \to g\colon A \to L$ in $\mathbf{C}$ such that:

$$\xi = \mathbf{C}(L, \alpha).p\colon W \to \mathbf{C}(L, F) \to \mathbf{C}(A, F). \tag{5.61}$$

The (strict) *W-limit* of F, written as $\mathrm{Lim}_W F$, is similarly defined, replacing 'pseudo natural' by 2-natural and $[\mathbf{I}, \mathbf{Cat}]_{\mathrm{ps}}$ by $[\mathbf{I}, \mathbf{Cat}]$.

The 2-category $\mathbf{C}$ is said to be *2-complete* if it has all weighted limits. (Then it also has all weighted pseudo limits, as proved in [BKPS].)

5.6.2 Exercises and complements

We examine now the basic cases of weighted limit in the 2-category $\mathbf{C}$.

(a) The trivial weight $W\colon \mathbf{I} \to \mathbf{Cat}$, constant at the terminal category $\mathbf{1}$, gives the ordinary 2-limit of the 2-functor $F\colon \mathbf{I} \to \mathbf{C}$, also called a *conical limit* in the general theory of weighted limits. This is a 2-universal cone $(A, h\colon A \to F)$ (see 2.3.5).

The reader can write its 1-dimensional and 2-dimensional universal properties and prove that all conical limits in $\mathbf{C}$ can be constructed from 2-products and 2-equalisers.

(b) Another important weighted limit is the *cotensor* $\mathbf{S} \pitchfork X$, where X is an object of $\mathbf{C}$: this is defined as the limit L of the functor $X\colon \mathbf{1} \to \mathbf{C}$ weighted by a small category $\mathbf{S}\colon \mathbf{1} \to \mathbf{Cat}$, and comes equipped with a functor $p\colon \mathbf{S} \to \mathbf{C}(L, X)$. The reader can write the universal properties of $(\mathbf{S} \pitchfork X, p)$, for a small category $\mathbf{S}$, and verify that $\mathbf{2} \pitchfork X$ coincides with the cell representer defined in 3.6.6.

(c) Prove that the cotensor $\mathbf{2} \pitchfork X$ is the same as the tabulator $\mathsf{T}e_X$ in $\mathbf{C}$, viewed as a horizontal double category. (The 1-dimensional aspect has already been considered, in Exercise 3.6.7(b).)

(d) In $\mathbf{Cat}$ the cotensor $\mathbf{S} \pitchfork X$ can be realised as a category of functors.

5.6.3 From weighted 2-categories to double categories

For a fixed 2-functor $W\colon \mathbf{I} \to \mathbf{Cat}$, the double category $\mathbb{El}(W)$ *of elements of W* is defined as the following double comma $\mathbf{1} \Downarrow W$ (see 4.2.4), of strict functors of horizontal double categories

$$
\begin{array}{ccc}
\mathbb{E}l(W) & \xrightarrow{\ P\ } & \mathbf{1} \\
Q \downarrow \ \ \swarrow{\scriptstyle \pi} & & \downarrow U \\
\mathbf{I} & \xrightarrow[\ W\ ]{} & \mathbf{Cat}
\end{array}
\qquad (5.62)
$$

Concretely, an object of $\mathbb{E}l(W)$ is a pair (i, X) where i is in $\mathbf{I}$ and X is an object of the category $W(i)$ (viewed as a functor $X: \mathbf{1} \to W(i)$).

A horizontal arrow $a = (a, X): (i, X) \to (i', X')$ comes from an $\mathbf{I}$-morphism $a: i \to i'$ such that $(Wa)(X) = X'$; these arrows compose as in $\mathbf{I}$. A vertical arrow $x = (i, x): (i, X) \to (i, Y)$ comes from a $W(i)$-morphism $x: X \to Y$; they compose as in the category $W(i)$.

A double cell $\xi: (x\,{}^{a}_{b}\,y)$

$$
\begin{array}{ccc}
(i, X) & \xrightarrow{\ a\ } & (i', X') \\
x \downarrow & \xi & \downarrow y \\
(i, Y) & \xrightarrow[\ b\ ]{} & (i', Y')
\end{array}
\qquad (5.63)
$$

comes from a 2-cell $\xi: a \to b: i \to i'$ of $\mathbf{I}$ such that $(W\xi)(x) = y$, where $(W\xi)(x): (Wa)X \to (Wb)Y$ is the diagonal of the commutative square

$$
\begin{array}{ccc}
(Wa)X & \xrightarrow{\ (W\xi)X\ } & (Wb)X \\
{\scriptstyle (Wa)x}\downarrow & & \downarrow{\scriptstyle (Wb)y} \\
(Wa)Y & \xrightarrow[\ (W\xi)Y\ ]{} & (Wb)Y
\end{array}
\qquad (5.64)
$$

A 2-functor $F: \mathbf{I} \to \mathbf{C}$ between 2-categories has an associated double functor $F(W)$ with values in the horizontal double category $\mathbf{C}$

$$
F(W): \mathbb{E}l(W) \to \mathbf{C},
$$

$$
(i, X) \mapsto Fi, \qquad (a: (i, X) \to (i', X')) \mapsto Fa: Fi \to Fi', \qquad (5.65)
$$

$$
(x: (i, X) \to (i, Y)) \mapsto e_{Fi}, \qquad (\xi: (x\,{}^{a}_{b}\,y)) \mapsto (F\xi: (Fi\,{}^{Fa}_{Fb}\,Fi')).
$$

5.6.4 *Cones and limits*

The double category $\mathrm{Ps}\mathbb{C}\mathrm{one}(F(W))$ of the pseudo cones of the double functor $F(W): \mathbb{E}l(W) \to \mathbf{C}$ has been defined in 5.2.1 as the double comma

$\Delta \Downarrow F(W)$ of the diagonal functor Δ

$$
\begin{array}{ccc}
\mathbb{P}s\mathbb{C}one(F(W)) & \xrightarrow{\ \ P\ \ } & \mathbf{C} \\
\Big\downarrow\ \ {}_{\pi} & & \Big\downarrow{}^{\Delta} \\
1 & \xrightarrow[F(W)]{} & \mathbb{L}x_{ps}(\mathbb{E}l(W), \mathbb{A})
\end{array}
\qquad (5.66)
$$

It can be analysed as follows.

(a) A pseudo cone $(A, h\colon A \to F(W))$ is an object A of $\mathbf{C}$ equipped with:

- a map $h(i, X)\colon A \to Fi$, for every i in $\mathbf{I}$ and every $X \in W(i)$,

- a 2-cell $h(i, x)\colon h(i, X) \to h(i, Y)\colon A \to Fi$, for every i in $\mathbf{I}$ and every $x\colon X \to Y$ in $W(i)$,

- an invertible 2-cell $h(a, X)$, for every $a\colon i \to j$ in $\mathbf{I}$ and every $X \in W(i)$

$$
A \quad \overset{Fa.h(i,X)}{\underset{h(j, Wa(X))}{\rightrightarrows}} {\scriptstyle \Downarrow h(a,X)} \quad Fj
\qquad (5.67)
$$

under the axioms (pht.1–5) of naturality and coherence (in 3.8.1).

It is a *cone* when all the comparison cells $h(a, X)$ are vertical identities.

Speaking of a W-*pair* (i, X), or (i, x), or (a, X), below, we mean a pair as above.

(b) A horizontal morphism $f\colon (A, h) \to (A', h')$ of pseudo cones is a horizontal arrow $f\colon A \to A'$ in $\mathbf{C}$ that commutes with the cone elements, as follows

(i) $\quad h(i, X) = h'(i, X).f\colon A \to Fi,$ $\qquad$ for every W-pair (i, X),

(ii) $\quad h(i, x) = h'(i, x).f\colon A \Rightarrow Fi,$ $\qquad$ for every W-pair (i, x),

(iii) $\quad h(a, X) = h'(a, X).f\colon A \Rightarrow Fj,$ $\qquad$ for every W-pair (a, X).

Horizontal morphisms compose, forming a category.

(c) A vertical morphism $\xi\colon (A, h) \rightarrowtail (A, k)$ of pseudo cones is a modification $\xi\colon (A \overset{h}{\underset{k}{\to}} F(W))$.

We have thus, for every W-pair (i, X), a 2-cell in $\mathbf{C}$

$$
\xi(i, X)\colon h(i, X) \to k(i, X)\colon A \to Fi,
$$

that satisfies the conditions (mod.1, 2) of 3.2.7.

(d) A double cell of cones

$$
\begin{array}{ccc}
(A,h) & \xrightarrow{\ f\ } & (A',h') \\
\xi\downarrow & \alpha & \downarrow\varsigma \\
(A,k) & \xrightarrow[\ g\ ]{} & (A',k')
\end{array}
\qquad (5.68)
$$

is a 2-cell $\alpha\colon f \to g\colon A \to A'$ in $\mathbf{C}$ such that, for every W-pair (i,X)

$$
A \underset{g}{\overset{f}{\Rightarrow}} {\downarrow\alpha}\ A' \underset{k'(i,X)}{\overset{h'(i,X)}{\Rightarrow}} {\downarrow\varsigma(i,X)}\ Fi \quad = \quad \xi(i,X). \qquad (5.69)
$$

Spelling out the conditions of 5.2.3, for a pseudo cone (L,p) of $F(W)$ to be the pseudo limit of the latter, we have:

(dl.1) for every pseudo cone $(A, h\colon A \to F(W))$ there is a unique morphism $f\colon A \to L$ in $\mathbf{C}$ such that

$$
\begin{aligned}
h(i,X) &= p(i,X).f\colon A \to Fi & \text{(for } i \text{ in } \mathbb{I},\ X \text{ in } Wi), \\
h(i,x) &= p(i,x).f\colon A \Rightarrow Fi & \text{(for } i \text{ in } \mathbb{I},\ x\colon X \to Y \text{ in } Wi),
\end{aligned}
\qquad (5.70)
$$

(dl.2) for every vertical morphism $\xi\colon (A,h) \to (A,k)$ of pseudo cones there is a unique 2-cell $\alpha\colon f \to g\colon A \to L$ in $\mathbf{C}$ such that, for i in $\mathbb{I}$ and X in Wi

$$
A \underset{g}{\overset{f}{\Rightarrow}} {\downarrow\alpha}\ L \xrightarrow{p(i,X)} Fi \quad = \quad \xi(i,X). \qquad (5.71)
$$

5.6.5 Theorem (From weighted 2-limits to double limits)

For every 2-functor $F\colon \mathbb{I} \to \mathbf{C}$, the weighted limit $(\mathrm{Lim}\,_W F, p)$ is the same as the double limit of the associated double functor $F(W)\colon \mathbb{E}\mathrm{l}(W) \to \mathbf{C}$ (i.e. they solve the same universal problem).

Similarly the weighted pseudo limit $(\mathrm{psLim}\,_W F, p)$ is the same as the pseudo limit of $F(W)$.

Proof The analytic descriptions of these 'limits', in 5.6.1 and 5.6.4, amount to the same thing. □

5.6.6 Theorem

(a) The existence of all weighted limits in a 2-category $\mathbf{C}$ amounts to that of all double limits in the associated horizontal double category.

(b) All weighted limits in $\mathbf{C}$ can be constructed from conical limits and cotensors $\mathbf{2} \pitchfork X$.

Note. Point (b) is well known in the theory of weighted limits: see [St2], Theorem 10.

Proof (a) If $\mathbf{C}$ has all weighted limits, then it has all conical limits and cotensors $\mathbf{2} \pitchfork X$, which means all level double limits and all tabulators $\mathsf{T}e_X$, whence all double limits by Theorem 5.3.6. The converse follows from Theorem 5.6.5. Point (b) is proved in the same way. $\qquad\square$

5.6.7 Definition (Weighted cones)

Theorem 5.6.5 allows us to define the *double* categories of W-weighted cones and pseudo cones of the 2-functor $F \colon \mathbf{I} \to \mathbf{C}$ as

$$\mathbb{C}\mathrm{one}_W(F) = \mathbb{C}\mathrm{one}(F(W)),$$
$$\mathrm{Ps}\mathbb{C}\mathrm{one}_W(F) = \mathrm{Ps}\mathbb{C}\mathrm{one}(F(W)). \tag{5.72}$$

The terminal objects of these *double* categories give the W-weighted limit of F, strict or pseudo, respectively.

Note that one cannot express the 2-dimensional universal property of weighted (strict or pseudo) limits by terminality *in a 2-category*. This is probably why weighted cones are rarely considered in the theory of 2-categories.

5.6.8 A direct construction of weighted cones

Let V be the 2-functor

$$V \colon \mathbf{C} \to [\mathbf{I}, \mathbf{Cat}], \qquad V(A) = \mathbf{C}(A, F(-)). \tag{5.73}$$

Without going through $\mathbb{E}\mathrm{l}(W)$ and $F(W)$, the double categories of weighted cones and pseudo cones can be constructed, up to isomorphism, as the following double commas

$$
\begin{array}{ccc}
\mathbb{C}\mathrm{one}_W(F) \longrightarrow \mathbf{1} & \qquad & \mathrm{Ps}\mathbb{C}\mathrm{one}_W(F) \longrightarrow \mathbf{1} \\
\Big\downarrow {\scriptstyle \swarrow}{\scriptstyle \pi} \quad \Big\downarrow {\scriptstyle W} & & \Big\downarrow {\scriptstyle \swarrow}{\scriptstyle \pi'} \quad \Big\downarrow {\scriptstyle W} \\
\mathbf{C} \xrightarrow{\ V\ } [\mathbf{I}, \mathbf{Cat}] & & \mathbf{C} \xrightarrow{\ V\ } [\mathbf{I}, \mathbf{Cat}]_{\mathrm{ps}}
\end{array}
\tag{5.74}
$$

In fact all the items (including compositions) of these double categories amount to the corresponding ones in the double categories analysed in 5.6.4.

It would be interesting to consider whether any double limit in $\mathbf{C}$, based on a double category $\mathbb{I}$, can be obtained as a *single* weighted limit for an associated weight $W: \mathbf{I} \to \mathbf{Cat}$ (defined on an associated 2-category).

5.6.9 Exercises and complements

We consider here some elementary weighted limits in 2-categories, and the corresponding double limits in weak double categories.

(a) Given two arrows $f: X' \to Y$ and $g: X'' \to Y$ in the 2-category $\mathbf{C}$, the *comma-object* $f \downarrow g$ is the W-limit of the 2-functor $(f, g): \mathbf{I} \to \mathbf{C}$, where $\mathbf{I}$ is the cospan-category represented at the left, below (with trivial cells) and the functor $W: \mathbf{I} \to \mathbf{Cat}$ takes the values shown at the right:

$$a: 0' \to 1 \leftarrow 0'' : b \qquad\qquad 1 \to 2 \leftarrow 1$$
$$\tag{5.75}$$
$$W(a): 0 \mapsto 0, \qquad\qquad W(b): 0 \mapsto 1.$$

Describe the 1-dimensional universal property of the comma object, construct the double category $\mathbb{El}(W)$ and interpret the limits that it produces in a weak double category.

(b) Same exercise for the *inserter* of two parallel arrows $f, g: X \to Y$, defined by the following weight

$$a, b: 0 \rightrightarrows 1 \qquad\qquad 1 \rightrightarrows 2$$
$$\tag{5.76}$$
$$W(a): 0 \mapsto 0, \qquad W(b): 0 \mapsto 1.$$

(c) Same exercise for the *equifier* of two 2-cells $\varphi, \psi: f \to g: X \to Y$, defined by the weight W represented below

$$0 \quad \overset{a}{\underset{b}{\overrightarrow{\alpha \downarrow\downarrow \beta}}} \quad 1 \qquad\qquad 1 \quad \overset{a'}{\underset{b'}{\overrightarrow{\downarrow}}} \quad 2 \tag{5.77}$$

$$W(a) = a': 0 \mapsto 0, \qquad W(b) = b': 0 \mapsto 1, \qquad W(\alpha) = W(\beta).$$

(d) The 2-category $\mathbf{C}$ is 2-complete if and only if the associated double category $\mathbb{Q}\mathbf{C}$ of quintets (see 3.1.4) has all limits. In this case the double category $\mathbb{Q}\mathbf{C}$ is span representable.

5.7 *Hints at persistent double limits

Persistent double limits where first presented in a talk by Paré [Pa1]; they are related to flexible weighted limits in 2-categories, introduced in [BKPS].

The following is a brief review, without proofs, of a part of our study in [GP13, GP14].

$\mathbb{I}$ is small weak double category.

5.7.1 Persistent limits

Persistent double limits are defined in [GP13], Definition 6.1, after a complex analysis of invariance properties. Here we follow a shortcut, using a characterisation proved in [GP13], Theorem 6.6.

We say that the weak double category $\mathbb{I}$ *parametrises persistent limits*, or that $\mathbb{I}$-*limits are persistent* if, for every weak double category $\mathbb{A}$ and every double functor $F \colon \mathbb{I} \to \mathbb{A}$ having a limit $(A, h \colon A \to F)$ and a pseudo limit $(A', h' \colon A' \to F)$, the canonical morphism $f \colon A \to A'$ (determined by $h = h'f$) is a horizontal equivalence.

If this happens for a given weak double category $\mathbb{A}$, we say that $\mathbb{I}$-*limits are persistent in* $\mathbb{A}$.

5.7.2 Grounded double categories

The general property of persistence can be characterised by a sort of initiality property. Namely, we say that the weak double category $\mathbb{I}$ is *grounded* if every connected component of the ordinary category $\mathrm{Hor}_0\mathbb{I}$ has a *natural weak initial object*.

By this we mean a functor $\Phi \colon \mathrm{Hor}_0\mathbb{I} \to \mathrm{Hor}_0\mathbb{I}$ which is trivial on every horizontal arrow $a \colon i \to i'$ of $\mathbb{I}$, and is equipped with a natural transformation $\varphi \colon \Phi \to 1$.

In elementary terms, for every object i there is an object Φi and a morphism $\varphi i \colon \Phi i \to i$ such that, for every $a \colon i \to i'$ in $\mathbb{I}$ we have:

$$\Phi i = \Phi i', \qquad \varphi i' = a.\varphi i, \qquad (5.78)$$

Note that the first condition in (5.78) is redundant.

Note also that each object Φi comes equipped with an idempotent endomorphism $p_i = \varphi(\Phi i) \colon \Phi i \to \Phi i$. (This is the identity if and only if Φi is the initial object of its connected component in $\mathrm{Hor}_0\mathbb{I}$, if and only if such an object exists.)

5.7.3 Persistence Theorem

The following conditions on the weak double category $\mathbb{I}$ *are equivalent:*

(i) $\mathbb{I}$-*limits are persistent,*

(ii) $\mathbb{I}$-*limits are persistent in the weak double category* $\mathbb{C}$at *of categories, functors and profunctors,*

(iii) $\mathbb{I}$ *is grounded,*

(iv) for every double functor $F \colon \mathbb{I} \to \mathbb{A}$, *every pseudo cone* $(A, h \colon A \to F)$ *is vertically isomorphic to a strict cone, in the weak double category* PsCone(F).

Proof See [GP13], Theorems 6.4 and 6.6. □

5.7.4 Theorem (Persistent double limits and flexible 2-limits)

A 2-functor $W \colon \mathbf{I} \to \mathbf{Cat}$ *is a flexible weight [BKPS] if and only if the double category* $\mathbb{E}l(W)$ *is grounded, if and only if it parametrises persistent limits in* $\mathbf{Cat}$ *(or equivalently in* $\mathbb{C}$at, *or in every weak double category).*

Proof See [GP14], Theorem 3.2. □

5.7.5 A partial converse

Verity's thesis [Ve] gives a partial converse to the dual of this result.

As proved in Theorem 2.7.1 therein, the class of persistent weighted colimits in the 2-category $\mathbf{Cat}$ is closed (in the sense of [AK]) and generated by sums, coinserters, coequifiers and idempotent-splittings. It coincides thus with the so-called closed class of (PIES)*-colimits, which precisely amounts to the class of flexible colimits in $\mathbf{Cat}$, as proved in [BKPS].

Part II
Multiple categories

6

Weak and lax multiple categories

The second part of this book is devoted to the study of multiple categories, that extend double categories to higher dimension, with one strict composition in direction 0 and various – possibly weak – compositions in positive directions.

In this chapter we start from a simple example, a triple category built on the double category $\mathbb{D}$bl of weak double categories (Section 6.1). In Sections 6.2 to 6.5 we define strict, weak and lax multiple categories of infinite dimension, their functors and transformations. Our main structure, a *chiral*, or *χ-lax*, multiple category (introduced in 6.4.8) is partially lax, in the sense that the weak composition laws in directions i, j have a directed interchanger χ_{ij} (for $i < j$) which need not be invertible. The even laxer notion of *intercategory* is examined in Sections 6.8 and 6.9.

These lax structures come in two forms, transversally dual to each other, according to the direction of interchangers; these forms are named 'left' and 'right', respectively, as explained in 6.4.8. We mainly work in the right-hand case.

Various 3-dimensional or infinite-dimensional examples are presented in Sections 6.1, 6.3, 6.4 and 6.6. Some of them are 'of cubical type', which – loosely speaking – means that the positive directions are interchangeable (see 6.3.1, 6.4.5). In most of these structures we compute tabulators, defined in Section 6.7 after a preliminary analysis in Section 6.1.

Chiral multiple categories can be organised in a multiple category, as we shall see in Section 7.5.

As always, $\mathbb{N}$ is the ordered set of natural numbers, and $\mathbb{N}^*$ the subset of the positive ones. Multiple categories, of any kind of laxity, are denoted as A, B.... The boolean variables α, β take values in the cardinal set $2 = \{0, 1\}$, also written as $\{-, +\}$.

The formal span and cospan categories are always written as in (1.4)

$$\vee: \quad 0 \leftarrow \iota \rightarrow 1 \qquad\qquad \wedge: \quad 0 \rightarrow \iota \leftarrow 1.$$

Again, solutions and hints to the exercises can be found in Apendix C.

6.1 Some examples of triple categories

Formally, a double category is a category object in **Cat**, and a *triple category* is a category object in the category of double categories and double functors. An explicit definition of multiple categories of any dimension will be given in Section 6.2.

This introductory section constructs some simple triple categories, and gives a first motivation for studying them and their limits.

6.1.1 An augmented triple category

We start from the (strict) double category $\mathbb{D}bl$ of weak double categories, lax and colax double functors (see Section 4.2) and we embed it in a triple category $S = Stc\mathbb{D}bl$, by adding new arrows, namely the strict double functors, in an additional *transversal direction*.

(a) The set S_*, of *objects* of S, consists of the (small) weak double categories $\mathbb{A}, \mathbb{B}, \dots$.

(b) The set S_i of *i-arrows*, or *i-directed arrows* $\mathbb{A} \rightarrow_i \mathbb{B}$ (for $i = 0, 1, 2$) has the following elements:

- a *0-arrow*, or *transversal arrow*, is a (strict double) functor,

- a *1-arrow* is a lax functor,

- a *2-arrow* is a colax functor.

Each set S_i has a degeneracy and two faces

$$\begin{aligned}
e_i &: S_* \rightarrow S_i, & e_i(\mathbb{A}) &= \mathrm{id}\mathbb{A}, \\
\partial_i^\alpha &: S_i \rightarrow S_*, & \partial_i^- &= \mathrm{Dom}, & \partial_i^+ &= \mathrm{Cod}.
\end{aligned} \tag{6.1}$$

(c) The sets S_{12}, S_{01}, S_{02} of *double cell* of S consist of the following items:

- a *12-cell* $\pi: (U \,{}^F_G\, V)$ is an arbitrary double cell of $\mathbb{D}bl$ (see 4.2.1), with lax (resp. colax) functors in direction 1 (resp. 2) and components

$$\pi A: VF(A) \rightarrow GU(A), \qquad \pi u: VF(u) \rightarrow GU(u),$$

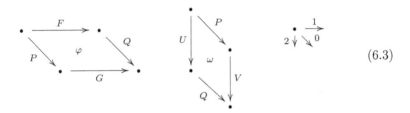

$$(6.2)$$

- a 01-cell, as shown in the left diagram below, is a double cell of $\mathbb{D}$bl with lax functors in direction 1, strict functors in direction 0 and a horizontal transformation $\varphi\colon QF \to GP$ (of lax functors)

$$(6.3)$$

- a 02-cell, as shown in the right diagram above, is a double cell of $\mathbb{D}$bl with strict functors in direction 0, colax functors in direction 2 and a horizontal transformation $\omega\colon VP \to QU$ (of colax functors).

Each set S_{ij} (for $0 \leqslant i < j \leqslant 2$) has two degeneracies and four faces, that are obvious

$$e_i\colon S_j \to S_{ij}, \qquad e_j\colon S_i \to S_{ij},$$
$$\partial_i^\alpha\colon S_{ij} \to S_j, \qquad \partial_j^\alpha\colon S_{ij} \to S_i. \tag{6.4}$$

Thus $e_1\colon S_2 \to S_{12}$ assigns to a 2-arrow U the identity cell $e_1(U)$ of the original double category for the 1-directed (i.e. horizontal) composition, while the i-faces $\partial_i^\alpha \pi$ of the 12-cell π are its domain and codomain for the i-directed composition (and are j-directed arrows, with $j \neq i$)

$$\partial_1^-(\pi) = U, \quad \partial_1^+(\pi) = V, \quad \partial_2^-(\pi) = F, \quad \partial_2^+(\pi) = G. \tag{6.5}$$

(d) Finally S_{012} is the set of *triple cells* of Stc$\mathbb{D}$bl: such an item Π is a 'commutative cube' determined by its six faces; the latter are double cells of the previous three types

$$(6.6)$$

The commutativity condition means the following equality of pasted double cells in $\mathbb{Dbl}$ (the non-labelled ones are inhabited by horizontal transformations that are identities):

$$
(6.7)
$$

More explicitly, the commutativity condition amounts to the following equality of components (horizontal composites of double cells in the weak double category $\mathbb{B}$):

$$
(YQFu \xrightarrow{Y\varphi u} YGPu \xrightarrow{\rho Pu} KVPu \xrightarrow{K\omega u} KRUu)
$$
$$
= (YQFu \xrightarrow{\zeta Fu} SXFu \xrightarrow{S\pi u} SHUu \xrightarrow{\psi Uu} KRUu),
$$

$$
(6.8)
$$

where u is any vertical arrow in the weak double category $\mathbb{A}$.

(e) All composition laws are derived from those of $\mathbb{Dbl}$, using the fact that the additional 0-directed structure is a particular case of the 1- and 2-directed ones. Therefore all these laws are strictly associative and unitary, and each pair of them satisfies the interchange law.

The fact that any triple cell of StcDbl is determined by its boundary (i.e. its six faces) can be expressed saying that the triple category StcDbl is *box-like*.

6.1.2 Comments

Embedding the double category $\mathbb{Dbl}$ into the triple category StcDbl can be motivated by the fact that:

(a) the horizontal and vertical limits existing in $\mathbb{Dbl}$ remain as *transversal limits* in StcDbl, where their projections are duly recognised as *strict* double functors,

(b) (more interestingly) new transversal limits appear in StcDbl, for which there is 'no sufficient room' in the original double category.

We anticipate now a sketch of tabulators, studied in Section 6.7, showing point (a) in 6.1.3, 6.1.4 and point (b) in 6.1.5, 6.1.6.

6.1.3 Horizontal tabulators in $\mathbb{D}$bl

As we have seen in 5.4.7, in the double category $\mathbb{D}$bl every vertical arrow $U \colon \mathbb{A} \nrightarrow \mathbb{B}$ has a *horizontal tabulator* $(\mathbb{T}, P, Q, \tau)$, with a horizontally universal cell τ as in the left diagram below

$$
\begin{array}{ccc}
\mathbb{T} \xrightarrow{\ P\ } \mathbb{A} & \quad & \mathbb{S} \xrightarrow{\ F\ } \mathbb{T} \xrightarrow{\ P\ } \mathbb{A} \\
e \downarrow \quad \tau \quad \downarrow U & \quad & e \downarrow \quad e_F \quad e \downarrow \quad \tau \quad \downarrow U \\
\mathbb{T} \xrightarrow[\ Q\ ]{} \mathbb{B} & \quad & \mathbb{S} \xrightarrow[\ F\ ]{} \mathbb{T} \xrightarrow[\ Q\ ]{} \mathbb{B}
\end{array}
\qquad (6.9)
$$

The weak double category $\mathbb{T} = U \downdownarrows \mathbb{B}$ has objects

$$
(A, B, b \colon U A \to B),
$$

with A in $\mathbb{A}$ and b horizontal in $\mathbb{B}$. The universal property says that every similar double cell $\tau' \colon (e_{\mathbb{S}} \ {}_{Q'}^{P'} \ U)$ factorises as $\tau' = e_F \,|\, \tau$, by a unique horizontal arrow $F \colon \mathbb{S} \to \mathbb{T}$, as in the right diagram above: the lax functor F is defined on the objects as

$$
F(S) = (P'(S), Q'(S), \tau'S \colon UP'(S) \to Q'(S)),
$$

and is strict whenever P' and Q' are.

Since P and Q are strict double functors, *this construction also gives the tabulator, or e_2-tabulator, of the 2-arrow U of Stc$\mathbb{D}$bl:* it will be defined in 6.7.1 as an object $\mathsf{T}_2 U$ with a universal 02-cell $\tau \colon e_2(\mathsf{T}_2 U) \nrightarrow_0 U$.

Now the universal property says that every 02-cell $\tau' \colon e_2(\mathbb{S}) \nrightarrow_0 U$ factorises as $\tau' = \tau.e_2(F)$, by a unique 0-arrow $F \colon \mathbb{S} \nrightarrow_0 \mathbb{T}$. (Note that now $\tau' \colon (e_2\mathbb{S} \ {}_{Q'}^{P'} \ U)$ is a double cell whose horizontal arrows P', Q' are strict functors, so that F is strict as well.)

6.1.4 Vertical tabulators in $\mathbb{D}$bl

Similarly we have seen in 5.4.7 that every horizontal arrow $F \colon \mathbb{A} \to \mathbb{B}$ has a *vertical tabulator* $(\mathbb{T}, P, Q, \tau)$, providing a vertically universal cell τ as below

$$
\begin{array}{ccc}
\mathbb{T} \xrightarrow{\ 1\ } \mathbb{T} \\
P \downarrow \quad \tau \quad \downarrow Q \\
\mathbb{A} \xrightarrow[\ F\ ]{} \mathbb{B}
\end{array}
\qquad (6.10)
$$

Now, the weak double category $\mathbb{T} = \mathbb{B} \Downarrow F$ has objects $(A, B, b \colon B \to FA)$, with A in $\mathbb{A}$ and b a horizontal arrow of $\mathbb{B}$.

Again, P and Q are strict double functors, and *this construction also gives the tabulator*, or e_1-*tabulator*, of the 1-arrow F of StcDbl: it will be defined in 6.7.1 as an object $\mathsf{T}_1 F$ with a universal 01-cell $\tau \colon e_1(\mathsf{T}_1 F) \to_0 F$.

6.1.5 Exercises and complements (Higher tabulators, I)

We deal now with a double cell π of $\mathbb{D}$bl

$$\begin{array}{ccc} \mathbb{A} & \xrightarrow{F} & \mathbb{B} \\ U\downarrow & \pi & \downarrow V \\ \mathbb{C} & \xrightarrow{G} & \mathbb{D} \end{array} \tag{6.11}$$

viewed as a 12-cell of the triple category StcDbl.

In the latter the reader can construct the *total tabulator*, or e_{12}-*tabulator*, of π. This is defined as an object $\mathbb{T} = \mathsf{T}\pi = \mathsf{T}_{12}\pi$ with a universal 012-cell $\Pi \colon e_{12}(\mathbb{T}) \to_0 \pi$, where $e_{12} = e_1 e_2 = e_2 e_1$

$$\tag{6.12}$$

6.1.6 Exercises and complements (Higher tabulators, II)

We conclude this topic considering two other higher tabulators $\mathsf{T}_i \pi$ of a 12-cell in StcDbl: the solutions are now *1-dimensional* cells – a lax and a colax double functor – instead of an object as above.

(a) Prove that π has an e_1-*tabulator*, defined as a 2-arrow $\mathsf{T}_1 \pi$ with a universal 012-cell $e_1(\mathsf{T}_1 \pi) \to_0 \pi$. (Note that the e_1-tabulator of π is 2-directed, like the 1-faces of π.)

(b) Prove that π has an e_2-*tabulator*, defined as a 1-arrow $\mathsf{T}_2 \pi$ with a universal 012-cell $e_2(\mathsf{T}_2 \pi) \to_0 \pi$.

(c) Tabulators are *preserved by faces and degeneracies*

$$\partial_i^\alpha(\mathsf{T}_j \pi) = \mathsf{T}_j(\partial_i^\alpha \pi), \quad \mathsf{T}_j(e_i X) = e_i(\mathsf{T}_j X) \qquad (i \neq j). \tag{6.13}$$

(d) By composing universal arrows, the total e_{12}-tabulator of π can be obtained as

$$\mathsf{T}_{12}\pi = \mathsf{T}_2\mathsf{T}_1\pi = \mathsf{T}_1\mathsf{T}_2\pi. \tag{6.14}$$

6.1.7 An extension

(a) Generalising the construction of StcDbl, let $\mathbb{D}$ be a (strict) double category whose arrows are displayed in directions 1 (horizontal) and 2 (vertical).

Furthermore we have an assigned category **S** of so-called *strong arrows* embedded as a subcategory in $\mathrm{Hor}_0\mathbb{D}$ and $\mathrm{Ver}_0\mathbb{D}$, and every strong arrow p comes with a pair (η, ε) of cells as below, its *unit* and *counit*

$$
\begin{array}{ccc}
\begin{array}{ccc}
A & \!\!=\!\!=\!\! & A \\
e\downarrow & \eta & \downarrow p \\
A & \xrightarrow{\ p\ } & B
\end{array}
&
\begin{array}{ccc}
A & \xrightarrow{\ p\ } & B \\
p\downarrow & \varepsilon & \downarrow e \\
B & \!\!=\!\!=\!\! & B
\end{array}
\end{array}
\tag{6.15}
$$

that satisfy the equations $\eta \,|\, \varepsilon = e_p$ and $\eta \otimes \varepsilon = 1_p$. In other words p, as a horizontal arrow, is companion to itself as a vertical one (see 4.1.1).

(b) We now embed $\mathbb{D}$ in a triple category $\mathsf{S} = \mathrm{Stg}\mathbb{D}$ (or $\mathrm{Stg_s}\mathbb{D}$) with the same objects, adding the strong arrows in the transversal direction. So far, degeneracies of objects and faces of arrows derive from the original ones

$$\partial_i^\alpha \colon S_i \to S_*, \qquad\qquad e_i \colon S_* \to S_i. \tag{6.16}$$

(c) The sets S_{12}, S_{01}, S_{02} of *double cells* of S consist of the following items:
- a 12-cell is an arbitrary double cell of $\mathbb{D}$,
- a 01-cell, as shown in the *left* diagram below, is a double cell $\varphi \colon (p \,{}^f_g\, q)$ of $\mathbb{D}$ whose arrows p, q (vertical in $\mathbb{D}$) are strong

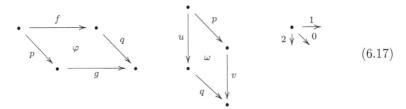

$$\tag{6.17}$$

- a 02-cell, as shown in the *right* diagram above, is a double cell $\omega \colon (u \,{}^p_q\, v)$ of $\mathbb{D}$ whose arrows p, q (horizontal in $\mathbb{D}$) are strong.

Each S_{ij} (for $0 \leqslant i < j \leqslant 2$) has four faces, evident from the diagrams above, and two obvious degeneracies

$$
\begin{array}{ll}
\partial_i^\alpha \colon S_{ij} \to S_j, & \partial_j^\alpha \colon S_{ij} \to S_i, \\
e_i \colon S_j \to S_{ij}, & e_j \colon S_i \to S_{ij}.
\end{array}
\tag{6.18}
$$

For instance $e_1 \colon S_0 \to S_{01}$ assigns to a transversal arrow p the horizontal identity cell $1_p \colon (p \, {}_1^1 \, p)$ of the original double category.

(d) Finally S_{012} is the set of *triple cells* of $\mathsf{Stg}\mathbb{D}$: such an item Π is a 'commutative cube' determined by its six faces; the latter are double cells of the previous three types

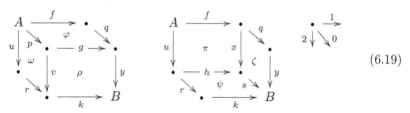

$$(6.19)$$

The commutativity condition means the following equality of pasted double cells in $\mathbb{D}$

$$(6.20)$$

(e) All composition laws are derived from those of $\mathbb{D}$, using the fact that the additional 0-directed structure is a particular case of the 1- and 2-directed ones. Therefore all these laws are strictly associative, unitary and satisfy interchange.

Again the triple category $\mathsf{Stg}\mathbb{D}$ is box-like.

(f) Applying this procedure to $\mathbb{D}$bl and choosing as strong arrows the strict functors, we recover the previous construction $\mathsf{Stc}\mathbb{D}$bl. Another possible choice is the pseudo functors: this would give a larger triple category.

Applying our procedure to the double category $\mathbb{LC}\mathbf{Cat}$ (resp. $\mathbb{LR}\mathbf{Ac}$) defined in 3.1.4, we would choose as strong arrows the functors that preserve limits and colimits (resp. the exact functors).

6.1.8 Some triple categories of cubical type

We end this section with some examples where the positive indices, 1 and 2, 'can be interchanged', in a sense that will be made precise in 6.3.1.

(a) Starting from a double category $\mathbb{A}$ one can always form a triple category $\mathsf{Cub_3}(\mathbb{A})$, in a way different from the previous one.

The objects are those of $\mathbb{A}$. The transversal arrows $f, g, \ldots$, displayed horizontally in the diagram below, are the horizontal maps of $\mathbb{A}$, while the 1-directed and 2-directed arrows are the vertical maps of $\mathbb{A}$.

A general triple cell is a diagram in $\mathbb{A}$ as below, with $\varphi \otimes \rho = \pi \otimes \psi$

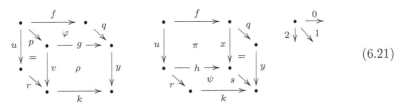

$$(6.21)$$

As evident from this diagram, 01-cells and 02-cells are double cells of $\mathbb{A}$, while 12-cells are commutative squares of vertical arrows. Composition in direction 0 (in dimension 1, 2, or 3) works with the horizontal composition of $\mathbb{A}$ (of arrows or double cells); composition in direction 1 or 2 works with the vertical composition of $\mathbb{A}$. All these laws are strictly associative, unitary and satisfy interchange.

The triple category $\mathsf{Cub_3}(\mathbb{A})$ is of interest when the horizontal arrows of $\mathbb{A}$ 'preserve the structure of the objects'; or – more formally – when $\mathbb{A}$ has all horizontal limit and colimits. This is not the case of $\mathbb{Dbl}$, for which we used a different procedure.

(b) Starting from a 2-category $\mathbf{C}$ one can form a triple category $\mathsf{Q_3}(\mathbf{C})$ *of quintets*, based on the double category $\mathbb{Q}\mathbf{C}$ of 3.1.4.

The objects and arrows are those of $\mathbf{C}$ (and $\mathbb{Q}\mathbf{C}$). The 2-dimensional cells are quintets, i.e. double cells of $\mathbb{Q}\mathbf{C}$. A triple cell is a 'commutative cube' of quintets, with $\omega \,|\, (\varphi \otimes \rho) = (\pi \otimes \psi) \,|\, \zeta$

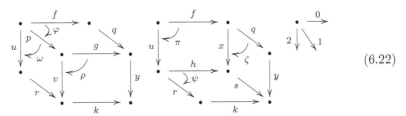

$$(6.22)$$

All compositions are derived from those of $\mathbb{Q}\mathbf{C}$. The triple category $\mathsf{Q_3}(\mathbf{C})$ is also box-like.

Let us note that the triple category $\mathsf{Cub_3}(\mathbb{Q}\mathbf{C})$ is a substructure of $\mathsf{Q_3}(\mathbf{C})$, where all 12-cells are commutative quintets.

(c) If $\mathbf{C}$ is an ordinary category, $\mathbb{Q}\mathbf{C}$ is the double category of commutative squares of $\mathbf{C}$. Both previous constructions give the triple category

$\mathsf{Cub}_3(\mathbb{Q}\mathbf{C}) = \mathsf{Q}_3(\mathbf{C})$ *of commutative 3-cubes* of $\mathbf{C}$, where a triple cell is precisely such a diagram in $\mathbf{C}$.

6.2 Strict multiple categories

We give now an explicit definition of a (strict) multiple category.

A reader having some knowledge of cubical categories can think of an extension of this notion where all the directions are of different *sorts*, represented by indices $i \in \mathbb{N}$.

6.2.1 Hinting at the geometry

An index $i \in \mathbb{N}$ will represent a *sort* or *direction* of our structure, including the *transversal* direction $i = 0$ (that will be treated differently, from 6.2.5 on).

Basically, a multiple category A (to be defined below) will have:

- a set A_* of objects,

- a set A_i of i-arrows, or i-*directed* arrows, for every index $i \geqslant 0$ (with faces in A_*),

- a set A_{ij} of 2-dimensional ij-cells, for indices $i < j$ (with faces in A_i and A_j),

- and generally a set $A_\mathbf{i} = A_{i_1 i_2 \dots i_n}$ of n-dimensional $\mathbf{i}$-cells, for every *multi-index* $\mathbf{i}$ of n natural indices

$$0 \leqslant i_1 < i_2 < \dots < i_n \qquad (n \geqslant 0),$$

with faces in each set $A_{i_1 \dots \hat{i}_j \dots i_n}$,

- together with categorical composition laws, with respect to these faces, that satisfy interchange 'by pairs'.

6.2.2 Multiple sets

A *multi-index* $\mathbf{i}$ is a finite subset of $\mathbb{N}$, possibly empty. Writing $\mathbf{i} \subset \mathbb{N}$ it will be understood that $\mathbf{i}$ is finite; writing $\mathbf{i} = \{i_1, \dots, i_n\}$ we always mean that $\mathbf{i}$ has n *distinct* elements, written in the natural order $i_1 < i_2 < \dots < i_n$; the integer $n \geqslant 0$ is called the *dimension* of $\mathbf{i}$.

We shall use the following notation

$$
\begin{aligned}
\mathbf{i}j = j\mathbf{i} = \mathbf{i} \cup \{j\} && (\text{for } j \in \mathbb{N} \setminus \mathbf{i}), \\
\mathbf{i}|j = \mathbf{i} \setminus \{j\} && (\text{for } j \in \mathbf{i}), && (6.23) \\
\mathbf{n} = \{0, \dots, n-1\}, \quad [n] = \{1, \dots, n\}, && (\text{for } n \geqslant 0).
\end{aligned}
$$

A *multiple set* is a system of sets and mappings $X = ((X_{\mathbf{i}}), (\partial_i^\alpha), (e_i))$, under the following two assumptions.

(mls.1) For every multi-index $\mathbf{i} = \{i_1, ..., i_n\}$, $X_{\mathbf{i}}$ is a set whose elements are called $\mathbf{i}$-*cells* of X and said to be of *dimension n*. For the sake of simplicity, we write X_*, X_i, X_{ij}... for $X_\emptyset$, $X_{\{i\}}$, $X_{\{i,j\}}$, ... Thus X_* is of dimension 0 while X_0, X_1, ... are of dimension 1.

(mls.2) For $i \in \mathbf{i}$ and $\alpha = \pm$ we have mappings, called *faces* and *degeneracies* of $X_{\mathbf{i}}$

$$\partial_i^\alpha \colon X_{\mathbf{i}} \to X_{\mathbf{i}|i}, \qquad e_i \colon X_{\mathbf{i}|i} \to X_{\mathbf{i}}, \qquad (6.24)$$

that satisfy the *multiple relations*

$$\partial_i^\alpha . \partial_j^\beta = \partial_j^\beta . \partial_i^\alpha \quad (i \neq j), \qquad e_i . e_j = e_j . e_i \quad (i \neq j),$$
$$\partial_i^\alpha . e_j = e_j . \partial_i^\alpha \quad (i \neq j), \qquad \partial_i^\alpha . e_i = \text{id}. \qquad (6.25)$$

Faces commute and degeneracies commute, but ∂_i^α and e_j only commute for $i \neq j$. These relations look much simpler than the cubical ones because here an index i stands for a particular sort, instead of a mere position, and is never 'shifted'.

Note also that ∂_i^α acts on $X_{\mathbf{i}}$ if i belongs to the multi-index $\mathbf{i}$ (and cancels i), while e_i acts on $X_{\mathbf{i}}$ if i does not belong to $\mathbf{i}$ (and inserts i); therefore the composites $\partial_i^\alpha . \partial_i^\beta$, and $e_i . e_i$ make no sense, here: one cannot cancel or insert twice the same index.

If $\mathbf{i} = \mathbf{j} \cup \mathbf{k}$ is a disjoint union and $\boldsymbol{\alpha} = (\alpha_1, ..., \alpha_r)$ is a mapping $\boldsymbol{\alpha} \colon \mathbf{k} = \{k_1, ..., k_r\} \to \{-, +\}$, we have an *iterated face* and an *iterated degeneracy* (both independent of the order of composition)

$$\partial_{\mathbf{k}}^{\boldsymbol{\alpha}} = \partial_{k_1}^{\alpha_1} ... \partial_{k_r}^{\alpha_r} \colon X_{\mathbf{i}} \to X_{\mathbf{j}}, \qquad e_{\mathbf{k}} = e_{k_1} ... e_{k_r} \colon X_{\mathbf{j}} \to X_{\mathbf{i}}. \qquad (6.26)$$

There are thus, in particular, 2^n *total* $\mathbf{i}$-*faces*, or *vertices*, and one *total* $\mathbf{i}$-*degeneracy*

$$\partial_{\mathbf{i}}^{\boldsymbol{\alpha}} = \partial_{i_1}^{\alpha_1} ... \partial_{i_n}^{\alpha_n} \colon X_{\mathbf{i}} \to X_*, \qquad e_{\mathbf{i}} = e_{i_1} ... e_{i_n} \colon X_* \to X_{\mathbf{i}}. \qquad (6.27)$$

A *morphism* $f \colon X \to Y$ of multiple sets is a family of mappings $f_{\mathbf{i}} \colon X_{\mathbf{i}} \to Y_{\mathbf{i}}$ (its components) that commute with faces and degeneracies.

(To be pedantic, let us note that a face $\partial_i^\alpha \colon X_{\mathbf{i}} \to X_{\mathbf{i}|i}$ also depends on the multi-index $\mathbf{i}$. It could be written in a more complete form, e.g. as $\partial_i^{\mathbf{i}\alpha}$, but generally there will be no need of this. The same can be said of degeneracies.)

6.2.3 Cubical sets and multiple sets

Cubical sets form a presheaf category $\mathbf{Set}^{\underline{I}^{op}}$ (that will be analysed more completely in Appendix B).

The *cubical site* $\underline{I} \subset \mathbf{Set}$ [GM, G2] has for objects the powers 2^n of the cardinal $2 = \{0, 1\}$ (for $n \geqslant 0$), with elements $t = (t_1, ..., t_n) \colon [n] \to 2$.

A morphism $2^m \to 2^n$ takes out some coordinates and inserts some 0's or 1's, without modifying the order of the remaining coordinates. Such morphisms are generated by the following cofaces and codegeneracies (for $1 \leqslant i \leqslant n$ and $\alpha \in 2$):

$$\partial_i^\alpha \colon 2^{n-1} \to 2^n, \qquad e_i \colon 2^n \to 2^{n-1},$$

$$\partial_i^\alpha(t_1, ..., t_{n-1}) = (t_1, ..., t_{i-1}, \alpha, t_i, ..., t_{n-1}), \qquad (6.28)$$

$$e_i(t_1, ..., t_n) = (t_1, ..., \hat{t}_i, ..., t_n),$$

under the *cocubical relations*

$$\partial_j^\beta . \partial_i^\alpha = \partial_{i+1}^\alpha . \partial_j^\beta \quad (j \leqslant i), \qquad e_i . e_j = e_j . e_{i+1} \quad (j \leqslant i),$$

$$e_j . \partial_i^\alpha = \partial_{i-1}^\alpha . e_j \ (j < i), \quad e_i . \partial_i^\alpha = 1, \quad e_j . \partial_i^\alpha = \partial_i^\alpha . e_{j-1} \ (j > i). \qquad (6.29)$$

Modifying all this, the *multiple site* $\underline{M}$ has an object $2^{\mathbf{i}} = \mathbf{Set}(\mathbf{i}, 2)$ for every multi-index $\mathbf{i} \subset \mathbb{N}$, with elements $t \colon \mathbf{i} \to 2$. The category $\underline{M} \subset \mathbf{Set}$ is generated by the following mappings (for $i \in \mathbf{i}$ and $\alpha \in 2$)

$$\partial_i^\alpha \colon 2^{\mathbf{i}|i} \to 2^{\mathbf{i}}, \qquad (\partial_i^\alpha t)(j) = t(j), \quad \partial_i^\alpha(t)(i) = \alpha \quad (j \neq i),$$

$$e_i \colon 2^{\mathbf{i}} \to 2^{\mathbf{i}|i}, \qquad (e_i t)(j) = t(j) \qquad\qquad\qquad (j \neq i), \qquad (6.30)$$

under the *comultiple relations*, dual to the multiple relations of (6.25). (Since all commutativity conditions stay the same, the only comultiple relation different from the previous ones is: $e_i \partial_i^\alpha = 1$.)

The category of multiple sets is thus the presheaf category $\mathbf{Set}^{\underline{M}^{op}}$.

There is a canonical (covariant) functor

$$F \colon \underline{M} \to \underline{I}, \qquad\qquad F(2^{\mathbf{i}}) = 2^n,$$

$$F(\partial_{i_j}^\alpha \colon 2^{\mathbf{i}|i_j} \to 2^{\mathbf{i}}) = \partial_j^\alpha \colon 2^{n-1} \to 2^n, \qquad (6.31)$$

$$F(e_{i_j} \colon 2^{\mathbf{i}} \to 2^{\mathbf{i}|i_j}) = e_j \colon 2^n \to 2^{n-1},$$

where $\mathbf{i} = \{i_1, ..., i_j, ..., i_n\}$.

F associates to a cubical set $K \colon \underline{I}^{op} \to \mathbf{Set}$ a multiple set $KF^{op} \colon \underline{M}^{op} \to \mathbf{Set}$.

We say that the multiple set X is *of cubical type* if it can be obtained in this way, which happens if and only if it is 'invariant under renaming indices, in the same order'.

Precisely, X has to satisfy the following relations, where the multi-index

$\mathbf{i} = \{i_1, ..., i_j, ..., i_n\} \subset \mathbb{N}$ is replaced with the 'normalised' multi-index $[n] = \{1, ..., j, ..., n\}$ (for $n \geqslant 0$)

$$X_{\mathbf{i}} = X_{[n]},$$

$$(\partial_{i_j}^\alpha : X_{\mathbf{i}} \to X_{i_1...\hat{i}_j...i_n}) = (\partial_j^\alpha : X_{[n]} \to X_{[n]|j}), \qquad (6.32)$$

$$(e_{i_j} : X_{i_1...\hat{i}_j...i_n} \to X_{\mathbf{i}}) = (e_j : X_{[n]|j} \to X_{[n]}).$$

This notion is equivalent to the classical notion of a cubical set, by a *rewriting of multi-indices*. In fact, normalising the multi-index $[n]|j = \{1, ..., \hat{j}, ..., n\}$ to $[n-1]$, the multiple relations become cubical relations and our multiple set of cubical type becomes a cubical set, of components $X_n = X_{[n]}$.

Here we prefer to avoid such rewritings and stay within multiple sets.

More generally, we have a *multiple site* $\underline{M}(N)$ *based on any set* N, and N-*indexed multiple objects* $X : \underline{M}(N) \to \mathbf{C}$ in any category $\mathbf{C}$; a multi-index is now a finite subset $\mathbf{i} \subset N$. We shall generally use a *totally ordered pointed set* $N = (N, 0)$, together with the associated ordered set $N^* = N \setminus \{0\}$. Besides $\mathbb{N}$, we shall frequently use the ordinal $\mathbf{n}$ and the ordered set $\mathbb{Z}$ of all integers.

6.2.4 Multiple categories

We are now ready for a formal definition of our main strict structure.

(mlt.1) A *multiple category* A is, first of all, a multiple set of components $A_{\mathbf{i}}$ whose elements will be called $\mathbf{i}$-*cells*. As above $\mathbf{i}$ is any multi-index, i.e. any finite subset of $\mathbb{N}$, and we write $A_*, A_i, A_{ij}, ...$ for $A_\emptyset, A_{\{i\}}, A_{\{i,j\}}, ...$

(mlt.2) Given two $\mathbf{i}$-cells x, y which are i-*consecutive* (i.e. $\partial_i^+ x = \partial_i^- y$, with $i \in \mathbf{i}$), the i-*composite* $x +_i y \in A_{\mathbf{i}}$ is defined, satisfying the following relations with faces and degeneracies

$$\partial_i^-(x +_i y) = \partial_i^- x, \qquad \partial_i^+(x +_i y) = \partial_i^+ y,$$
$$\partial_j^\alpha(x +_i y) = \partial_j^\alpha x +_i \partial_j^\alpha y, \qquad e_j(x +_i y) = e_j x +_i e_j y \quad (j \neq i). \qquad (6.33)$$

(mlt.3) For $j \notin \mathbf{i}$ these data form a category $\mathrm{cat}_{\mathbf{i},j}(\mathsf{A})$ with objects in $A_{\mathbf{i}}$, arrows in $A_{\mathbf{i}j}$, faces ∂_j^α, identities e_j and composition $+_j$.

(mlt.4) For $i < j$ we have

$$(x +_i y) +_j (z +_i u) = (x +_j z) +_i (y +_j u) \qquad (\textit{binary } ij\textit{-interchange}), \quad (6.34)$$

whenever these composites make sense.

Let us note that the lower interchanges are already expressed above:

the last condition of (6.33) is interchange between i-composition and j-identities, while the zeroary interchange is already written in the axioms of multiple sets: $e_i e_j = e_j e_i$, for $j \neq i$.

Again, we can more generally consider *N-indexed multiple categories*, where N is a totally ordered set, pointed at 0. If N is the ordinal $\mathbf{n} = \{0, ..., n-1\}$ we obtain the n-dimensional version of a multiple category, called an *n-tuple category*. The 0-, 1- and 2-dimensional versions amount – respectively – to a set, a category or a double category.

We have already studied some triple categories in Section 6.1. Infinite dimensional examples of cubical type will be seen in Section 6.3. In Section 7.7 we shall construct the exponential of multiple categories.

6.2.5 Transversal categories

The *transversal* direction, corresponding to the index $i = 0$, plays a special role. It will be used for the transformations of multiple functors and for the structural arrows of limits and colimits; its composition will stay strict, in all the weak or lax extensions we shall consider. We think of it as the 'dynamic' or 'structural' direction, along which 'transformation occurs', while the positive directions are viewed as the 'static' or 'geometric' ones (like the vertical direction of polarised double categories, in 3.2.6).

A *positive* multi-index $\mathbf{i} = \{i_1, ..., i_n\} \subset \mathbb{N}^*$ (with $n \geqslant 0$ positive indices) has an 'augmented' multi-index $0\mathbf{i} = \{0, i_1, ..., i_n\}$. The *transversal category of* $\mathbf{i}$*-cubes of* A

$$\mathsf{A}_{\mathbf{i}} = \mathrm{tv}_{\mathbf{i}}(\mathsf{A}) = \mathrm{cat}_{\mathbf{i},0}(\mathsf{A}), \qquad (6.35)$$

- has objects in $A_{\mathbf{i}}$, called $\mathbf{i}$-*cubes* and viewed as n-dimensional objects of type $\mathbf{i}$,

- has arrows $f \colon x \to_0 x'$ (or $f \colon x \to x'$) in $A_{0\mathbf{i}}$, called $\mathbf{i}$-*maps*, with faces $\partial_0^-(f) = x$, $\partial_0^+(f) = x'$,

- has identities $1_x = \mathrm{id}x = e_0(x) \colon x \to_0 x$ and composition $gf = f +_0 g$.

All these items, including $\mathbf{i}$ and $0\mathbf{i}$, are said to be *of degree n* (while their *dimension* is either n or $n+1$): the degree always refers to the number of positive indices. A *transversal isomorphism* is an isomorphism in a transversal category.

In all of our 'concrete' examples, the 0-composition $gf = f +_0 g$ is realised by the usual composition of mappings, while the geometric compositions (also called *concatenations*) of cubes $x +_i y$ and of transversal maps $f +_i g$ are often (but not always!) computed by means of operations (sums, products, tensor products, pushouts, pullbacks...) where reversing the order of the operands would have no motivation.

The unique positive multi-index of degree 0, namely $\emptyset$, gives the category $\mathrm{tv}_*(A)$ of *objects* of A (i.e. *-cells) and their transversal maps (i.e. *-maps, or 0-cells). A 1-dimensional $\{i\}$-cube is also called an *i-arrow*.

Faces and degeneracies give (ordinary) functors, for $i \in \mathbf{i}$ and $\alpha = \pm$

$$\partial_i^\alpha \colon \mathrm{tv}_{\mathbf{i}}(A) \to \mathrm{tv}_{\mathbf{i}|i}(A), \qquad e_i \colon \mathrm{tv}_{\mathbf{i}|i}(A) \to \mathrm{tv}_{\mathbf{i}}(A). \tag{6.36}$$

Forgetting the positive compositions, a multiple category A has an underlying *premultiple category* $|A|$, i.e. a multiple object in **Cat**, indexed by the (non-pointed) set $\mathbb{N}^*$.

More generally, for an N-indexed multiple (or premultiple) category a 'positive' multi-index $\mathbf{i}$ should be understood as a finite subset $\mathbf{i} \subset N^* = N \setminus \{0\}$.

6.2.6 Multiple functors and transversal transformations

A *multiple functor* $F \colon A \to B$ between multiple categories is a morphism of multiple sets $F = (\hat{F}_{\mathbf{i}} \colon A_{\mathbf{i}} \to B_{\mathbf{i}})$ that preserves all the composition laws. Its *transversal* $\mathbf{i}$-*component*, from $\mathbf{A_i} = \mathrm{tv}_{\mathbf{i}}(A)$ to $\mathbf{B_i} = \mathrm{tv}_{\mathbf{i}}(B)$

$$F_{\mathbf{i}} = \mathrm{tv}_{\mathbf{i}}(F) \colon \mathbf{A_i} \to \mathbf{B_i} \qquad (\mathbf{i} \subset \mathbb{N}^*), \tag{6.37}$$

is an ordinary functor. Its action on an $\mathbf{i}$-map $f \colon x \to_0 y$ is written as $F(f) \colon F(x) \to_0 F(y)$, or $F_{\mathbf{i}}(f) \colon F_{\mathbf{i}}(x) \to_0 F_{\mathbf{i}}(y)$.

(The functor $F_{\mathbf{i}}$ is determined by the mappings $\hat{F}_{\mathbf{i}}$ and $\hat{F}_{0\mathbf{i}}$; we generally work with the transversal components $F_{\mathbf{i}}$ and do not use the components $\hat{F}_{\mathbf{i}}$ and $\hat{F}_{0\mathbf{i}}$, pertaining to multiple sets.)

A *transversal transformation* $\varphi \colon F \to G \colon A \to B$ between multiple functors consists of a family of $\mathbf{i}$-maps in B, for every $\mathbf{i} \subset \mathbb{N}^*$ and every $\mathbf{i}$-cube x in A, that agrees with positive faces

$$\begin{aligned} \varphi x \colon Fx \to_0 Gx && (\varphi_{\mathbf{i}} x \colon F_{\mathbf{i}}(x) \to_0 G_{\mathbf{i}}(x)), \\ \varphi(\partial_i^\alpha x) = \partial_i^\alpha(\varphi x) && (i \in \mathbf{i} \subset \mathbb{N}^*). \end{aligned} \tag{6.38}$$

The following axioms of naturality and coherence are assumed:

(trt.1) for all $f \colon x \to_0 y$ in A, we have a commutative diagram of transversal maps in B

$$\begin{array}{ccc} Fx & \xrightarrow{\varphi x} & Gx \\ {\scriptstyle Ff}\downarrow & & \downarrow{\scriptstyle Gf} \\ Fy & \xrightarrow[\varphi y]{} & Gy \end{array}$$

(trt.2) φ commutes with positive degeneracies and compositions:

$$\varphi(e_i t) = e_i(\varphi t), \qquad\qquad \varphi(x +_i y) = \varphi x +_i \varphi y,$$

where $i \in \mathbf{i} \subset \mathbb{N}^*$, t is an $\mathbf{i}|i$-cube, x and y are i-consecutive $\mathbf{i}$-cubes.

Again, φ has *transversal components*

$$\varphi_{\mathbf{i}} = \mathrm{tv}_{\mathbf{i}}(\varphi) \colon F_{\mathbf{i}} \to G_{\mathbf{i}} \colon \mathbf{A}_{\mathbf{i}} \to \mathbf{B}_{\mathbf{i}} \qquad\qquad (\mathbf{i} \subset \mathbb{N}^*), \qquad (6.39)$$

which are natural transformations of ordinary functors (this fact comprises axiom (trt.1)).

6.2.7 Multiple categories and duality

Given two multiple categories A, B, we have thus the category $\mathbf{Mlt}(\mathsf{A}, \mathsf{B})$ of their multiple functors and transversal transformations. All these form the 2-category $\mathbf{Mlt}$, in an obvious way.

Multiple categories have dualities, generated by reversing each direction i and permuting directions; they form an infinite-dimensional hyperoctahedral group.

We are mainly interested in the *transversal dual* A^{tr} that reverses all transversal faces ∂_0^α and all transversal compositions, so that $\mathrm{tv}_{\mathbf{i}}(\mathsf{A}^{\mathrm{tr}}) = (\mathrm{tv}_{\mathbf{i}}(\mathsf{A}))^{\mathrm{op}}$; for two $\mathbf{i}$-maps $f \colon x \to_0 x'$ and $g \colon x' \to_0 x''$ in A, we have corresponding maps f^*, g^* in A^{tr} with

$$f^* \colon x' \to_0 x, \qquad\qquad f^* . g^* = (g.f)^*. \qquad (6.40)$$

The prefix *co-* always refers to transversal duality. This corresponds to ordinary duality in dimension 1 (for categories) and horizontal duality in dimension 2 (for double categories).

Similarly one defines the 2-category $\mathrm{p}\mathbf{Mlt}$ of *premultiple categories, premultiple functors and transversal transformations*, with a forgetful 2-functor

$$| - | \colon \mathbf{Mlt} \to \mathrm{p}\mathbf{Mlt}, \qquad (6.41)$$

whose action on the objects has been introduced at the end of 6.2.5.

More generally for a totally ordered pointed set $N = (N, 0)$ we have the 2-categories $\mathbf{Mlt}_N$ and $\mathrm{p}\mathbf{Mlt}_N$ of N-indexed multiple, or premultiple, categories.

6.2.8 Truncation and triple categories

As we already remarked, restricting all indices to the subsets of the ordinal set $\mathbf{n} = \{0, ..., n-1\}$ we obtain an n-dimensional version of a multiple category, called an *n-tuple category*, or n-multiple category, where the highest cells have dimension n. The 0-, 1- and 2-dimensional versions amount – respectively – to a set, or a category, or a double category.

There is thus a truncation 2-functor with values in the 2-category $\mathbf{Mlt_n}$ of n-tuples categories, which has both adjoints

$$\mathrm{trc}_n\colon \mathbf{Mlt} \to \mathbf{Mlt_n}, \qquad \mathrm{sk}_n \dashv \mathrm{trc}_n \dashv \mathrm{cosk}_n. \qquad (6.42)$$

The left adjoint (called *skeleton*) adds degenerate items of all missing types $\mathbf{i} \not\subset \mathbf{n}$. The right adjoint (called *coskeleton*) is more complex: for instance, if $\mathbf{C}$ is a category and $\mathbf{i}$ a positive multi-index, an $\mathbf{i}$-cube of $\mathrm{cosk}_1(\mathbf{C})$ is a functor $x\colon 2^{\mathbf{i}} \to \mathbf{C}$ where $2 = \{0,1\}$ is discrete (so that x is a family of objects of $\mathbf{C}$ indexed by the set $2^{\mathbf{i}}$); an $\mathbf{i}$-map is a natural transformation of these functors.

We are particularly interested in the 3-dimensional notion, called a *triple category*, already explored in Section 6.1. Its cells, corresponding to multi-indices $\mathbf{i} \subset \{0,1,2\}$, are:

- objects, of one sort (for $\mathbf{i} = \emptyset$),

- arrows of three sorts, in directions 0 (*transversal*), 1 and 2,

- 2-dimensional cells of three sorts, in directions 01, 02 and 12,

- 3-dimensional cells of one sort, in direction 012.

Other kinds of truncation, by degree or dimension, will be mentioned in 6.3.4 and 6.3.5.

6.3 Multiple categories of cubical type

Cubical categories are here viewed as *multiple categories of cubical type*. An independent definition will be given in Appendix B.

The notion of cubical category which we use here was defined in [G2, G4]: it includes transversal maps, which can be of a different sort from the other arrows, and are crucial for the weak and lax extensions. It differs on this point from the notion of [AlBS], that was called a 'reduced cubical category' in [G2, G4].

6.3.1 The cubical type

We say that the multiple category A is *of cubical type* if its components, faces, degeneracies and compositions are *invariant under renaming* positive *indices, in the same order.*

In other words, for all $\mathbf{i} = \{i_1, ..., i_n\} \subset \mathbb{N}^*$

$$\mathrm{tv}_{\mathbf{i}}(\mathsf{A}) = \mathrm{tv}_{[n]}(\mathsf{A}),$$

$$(\partial_{i_j}^\alpha \colon A_{\mathbf{i}} \to A_{\mathbf{i}|i_j}) = (\partial_j^\alpha \colon A_{[n]} \to A_{[n]|j}),$$

$$(e_{i_j} \colon A_{\mathbf{i}|i_j} \to A_{\mathbf{i}}) = (e_j \colon A_{[n]|j} \to A_{[n]}),$$

$$(+_{i_j} \colon A_{\mathbf{i}} \underset{\bullet}{\times} A_{\mathbf{i}} \to A_{\mathbf{i}}) = (+_j \colon A_{[n]} \underset{\bullet}{\times} A_{[n]} \to A_{[n]}),$$

(6.43)

where $- \underset{\bullet}{\times} -$ denotes the appropriate composition pullback.

(With respect to multiple sets (in 6.2.3), we are using here a different form of normalised multi-index, that only intervenes on positive indices and preserves both dimension and degree. In other words, a component $A_{0\mathbf{i}}$ of $\mathbf{i}$-transversal maps is normalised as $A_{0[n]}$, not as $A_{[n+1]}$.)

A multiple category of cubical type is equivalent to a cubical category, as defined in [G2, G4] and Appendix B. Generally, we shall make no difference between these two notions.

In a multiple category of cubical type an $\mathbf{i}$-cube $x \in A_{\mathbf{i}} = A_{[n]}$ is also called an *n-cube*, and an $\mathbf{i}$-map $f \colon x \to y$ (belonging to $A_{0\mathbf{i}} = A_{0[n]}$) is also called an *n-map*.

In the truncated case the previous invariance condition is trivially satisfied up to dimension 2 (corresponding to sets, categories and double categories), since a subset $\mathbf{i} \subset \{0, 1\}$ is automatically normalised in the present sense; on the other hand, a triple category can be of cubical type or not: for instance the examples of 6.1.8 are of this type, while the structure $\mathrm{Stc}\mathbb{D}\mathrm{bl}$ of 6.1.1 is not, since its 1- and 2-arrows are of different kinds.

Infinite-dimensional globular categories, usually called ω-categories, can be analysed as cubical categories of a globular type: see [AlBS] and [GP5], Section 2.

6.3.2 Symmetries

As we shall soon see, many of our examples of cubical categories are of an even more particular, 'symmetric' kind – with *positive faces, degeneracies and compositions related by symmetries* – so that the structure in direction 1, for instance, determines all the positive ones. Again, symmetric cubical categories, in their own right, will be dealt in Appendix B; here they can be viewed as follows.

A multiple category *of symmetric cubical type* is a multiple category of cubical type A (as defined above) with an assigned action of the symmetric group S_n on each transversal category $\mathrm{tv}_{[n]}(\mathsf{A})$, generated by *transposition* functors

$$s_i \colon \mathrm{tv}_{[n]}(\mathsf{A}) \to \mathrm{tv}_{[n]}(\mathsf{A}), \qquad i = 1, ..., n-1 \qquad (n \geqslant 2). \qquad (6.44)$$

These transpositions satisfy the well-known Moore relations of the symmetric group (see (B.26)). Moreover s_i exchanges the i-indexed structure with the $(i+1)$-indexed one, leaving the rest unchanged. This will be made precise in Appendix B.

In the truncated case the symmetric structure, that only works on *positive* indices, *is trivial up to dimension* 2: for sets, categories and *double categories as well.* (One should not confuse this notion with the *transpositive* property of strict double categories, that works on the indices 0, 1; see 3.2.5.) On the other hand, a triple category of cubical type A is made symmetric (*if this is possible*) by assigning two involutions $s_1 \colon A_{12} \to A_{12}$ and $s_1 \colon A_{012} \to A_{012}$ that satisfy the axioms above.

6.3.3 Commutative cubes on a category

The simplest example is the symmetric cubical category $\mathsf{Cub}(\mathbf{C})$ *of commutative cubes* of a category $\mathbf{C}$, that extends the 3-dimensional structure of 6.1.8(a).

An n-cube is a functor $x \colon \mathbf{2}^n \to \mathbf{C}$ ($n \geqslant 0$), where $\mathbf{2}$ is the arrow category. An n-map $f \colon x \to y$ is a natural transformation of these functors, and can be seen as a functor $f \colon \mathbf{2}^n {\times} \mathbf{2} \to \mathbf{C}$ (this is the same as a cube of dimension $n+1$: transversal maps are here redundant).

Without using the cubical framework, an **i**-cube is a functor $x \colon \mathbf{2}^\mathbf{i} \to \mathbf{C}$, where **i** is a subset $i_1 < ... < i_n$ of $\mathbb{N}^*$; it can be viewed as an n-cube 'displayed' in n directions $i_1, ..., i_n$, which need not be the standard directions $1, ..., n$ of $\mathbf{i} = [n]$. The difference is formal, but the machinery of faces and degeneracies is different.

Applications of this multiple category (and its truncations) to algebraic K-theory can be found in Shimakawa [Sh].

6.3.4 Cubes on a double category

More generally, we start now from a double category $\mathbb{A}$ and construct a multiple category $\mathsf{Cub}(\mathbb{A})$ that extends the double category itself and the triple category $\mathsf{Cub}_3(\mathbb{A})$ of 6.1.8(a). It is of symmetric cubical type, and gives the previous $\mathsf{Cub}(\mathbf{C})$ when applied to the double category $\mathbb{Q}\mathbf{C}$ of commutative squares of the category $\mathbf{C}$.

For a positive multi-index $\mathbf{i} = \{i_1, ..., i_n\} \subset \mathbb{N}^*$, an **i**-cube of $\mathsf{Cub}(\mathbb{A})$ is a functor $x\colon \mathbf{2^i} \to \mathrm{Ver}_0(\mathbb{A})$, i.e. a commutative n-cube of vertical arrows of $\mathbb{A}$, displayed in the directions $i_1, ..., i_n$.

An **i**-map $f\colon x \to y$ is a functor $f\colon \mathbf{2^i} \to \mathrm{Ver}_1(\mathbb{A})$, i.e. an n-cube of double cells of $\mathbb{A}$, commutative under vertical composition.

Thus, in degree 2 (and dimension 3), a $0ij$-cell (for $0 < i < j$) is a diagram in $\mathbb{A}$ as in (6.21), with $\varphi \otimes \rho = \pi \otimes \psi$

$$(6.45)$$

The transversal composition of these cells works with the horizontal composition in $\mathbb{A}$; all the geometric compositions work with the vertical composition in $\mathbb{A}$.

Every transversal map of degree $n - 1 \geqslant 3$ is an n-dimensional cubical diagram whose iterated faces in direction $0ij$ (of degree 2) are as above.

*Note. The last step is a coskeletal extension 'by degree', right adjoint to a kind of truncation that, in a multiple category, keeps all the items of degree < 3. The 'stronger' truncation $\mathrm{trc}_3\colon \mathbf{Mlt} \to \mathbf{Mlt_3}$ that we have considered in (6.42) only keeps the **i**-cells with $\mathbf{i} \subset \{0, 1, 2\}$, and gives a different coskeletal extension.*

6.3.5 Higher quintets on a 2-category

We start now from a 2-category $\mathbf{C}$ and construct a multiple category $\mathsf{M} = \mathsf{Q}(\mathbf{C})$ *of (higher) quintets* over $\mathbf{C}$, extending the double category $\mathbb{Q}\mathbf{C}$ of quintets, see 3.1.4, and the triple category $\mathsf{Q}_3(\mathbf{C})$ of 6.1.8(b). It is of symmetric cubical type, in a strong sense because here everything is invariant under renaming *all* indices, in the same order; the transversal direction plays no special role.

(a) The objects of M are those of $\mathbf{C}$; in every direction $i \geqslant 0$, an i-cell $f\colon A \to_i B$ is a $\mathbf{C}$-morphism. They form the category $\mathrm{cat}_i(\mathsf{M})$, underlying the 2-category $\mathbf{C}$ (and independent of i).

(b) In dimension 2, an ij-cell (for $0 \leqslant i < j$) is a *quintet* of $\mathbf{C}$, consisting of four morphisms and a 2-cell φ

$$
\begin{array}{ccc}
A & \xrightarrow{f} & X' \\
p \downarrow & \searrow \varphi & \downarrow q \\
C & \xrightarrow{g} & D
\end{array}
\qquad \varphi\colon qf \to gp\colon A \to D. \qquad\qquad (6.46)
$$

These cells have two obvious composition laws, in directions i and j, and form a double category $\mathrm{dbl}_{ij}(\mathsf{M})$; it is the same as the double category $\mathbb{Q}\mathbf{C}$, 'displayed' in directions i and j.

(c) In dimension 3 an ijk-cell (for $0 \leqslant i < j < k$) is a 'commutative cube' Π of quintets

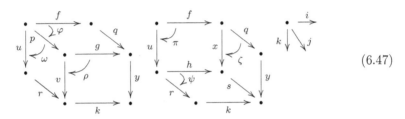

$$(6.47)$$

More precisely, we have six quintets

$$
\begin{array}{lll}
\varphi\colon qf \to gp, & \psi\colon sh \to kr & \text{(two } ij\text{-cells, the faces } \partial_k^\alpha \Pi), \\
\pi\colon xf \to hu, & \rho\colon yg \to kv & \text{(two } ik\text{-cells, the faces } \partial_j^\alpha \Pi), \\
\omega\colon vp \to ru, & \zeta\colon yq \to sx & \text{(two } jk\text{-cells, the faces } \partial_i^\alpha \Pi),
\end{array}
$$

which must satisfy the following commutativity relation, in $\mathbf{C}$

$$
y\varphi \otimes \rho p \otimes k\omega = \zeta f \otimes s\pi \otimes \psi u\colon yqf \to kru. \qquad\qquad (6.48)
$$

The compositions in directions i, j, k amount to compositions of faces in the double categories $\mathrm{dbl}_{ij}(\mathsf{M})$, $\mathrm{dbl}_{ik}(\mathsf{M})$ and $\mathrm{dbl}_{jk}(\mathsf{M})$. We have thus a triple category $\mathrm{trp}_{ijk}(\mathsf{M})$.

(d) Every cell of dimension $n > 3$ is an n-dimensional cube whose 2-dimensional faces are quintets, under the condition that each 3-dimensional face in direction ijk be an ijk-cell, as defined above.

Note. The last step is a coskeletal extension 'by dimension', right adjoint to a kind of truncation that, in a multiple category, keeps all the items of dimension $\leqslant 3$; then a 3-dimensional cube has no transversal maps, which is not a problem in the strict case.

6.3.6 Other examples

Extending the double categories of Section 3.1.9, we shall construct in 6.6.5:

- a symmetric cubical category $\mathsf{Pmap}(\mathbf{C})$ *of cubical partial maps of* $\mathbf{C}$,
- a symmetric cubical category $\mathsf{Piso}(\mathbf{C})$ *of cubical partial isomorphisms.*

6.4 Weak and chiral multiple categories

We now extend multiple categories to the *weak* case. As in the strict case, the basic structure of a weak multiple category A can be defined as a *multiple set of components* $A_{\mathbf{i}}$, for $\mathbf{i} \subset \mathbb{N}$, or as *a multiple object in* $\mathbf{Cat}$ *of components* $\mathbf{A}_{\mathbf{i}} = \mathrm{tv}_{\mathbf{i}}(\mathsf{A})$, for $\mathbf{i} \subset \mathbb{N}^*$. The former approach was followed in [GP8]; here we follow the latter, that fits better with the theory of adjunctions and monads developed below.

The prime examples are based on (cubical) spans and cospans, in 6.4.6; they will allow to study many other structures by span or cospan representability, in Section 6.7.

We end with a more general notion, partially lax: a *chiral*, or *χ-lax*, *multiple category* (see 6.4.8): it has the same structure of a weak multiple category, except for the fact that the interchangers χ_{ij} of the i- and j-compositions (for $0 < i < j$) are not supposed to be invertible.

6.4.1 The basic structure

We begin by the basic framework; the comparisons for unitarity, associativity and interchange will be introduced in 6.4.2 and their coherence conditions in 6.4.3, 6.4.4.

(wmc.1) A *weak multiple category* A is, first of all, a family of categories $\mathbf{A}_{\mathbf{i}} = \mathrm{tv}_{\mathbf{i}}(\mathsf{A})$, for $\mathbf{i} \subset \mathbb{N}^*$, called the *transversal categories*, or *transversal components* of A.

The objects of $\mathbf{A}_{\mathbf{i}}$ are called $\mathbf{i}$-cubes (or $\mathbf{i}$-cells) and form a set $A_{\mathbf{i}}$; the morphisms $f \colon x^- \to x^+$ are called (transversal) $\mathbf{i}$-maps (or $0\mathbf{i}$-cells) and form a set $M_{\mathbf{i}} = A_{0\mathbf{i}}$. The categorical structure has faces $\partial_0^\alpha(f) = x^\alpha$ ($\alpha = \pm$), identities $e_0(x) = \mathrm{id}(x) = 1_x$ and a composition law $c_0(f, g) = f +_0 g = gf$ defined on the set of consecutive pairs (f, g) of $\mathbf{i}$-maps ($\partial_0^+ f = \partial_0^- g$)

$$\partial_0^\alpha \colon M_{\mathbf{i}} \; \rightrightarrows \; A_{\mathbf{i}} \colon e_0, \qquad c_0 \colon M_{\mathbf{i}} \underset{\bullet}{\times} M_{\mathbf{i}} \to M_{\mathbf{i}}. \tag{6.49}$$

(wmc.2) These categories form a premultiple category, i.e. a multiple object in $\mathbf{Cat}$, indexed by the finite subsets of $\mathbb{N}^*$. In other words, we have functors of faces and degeneracies

$$\partial_i^\alpha \colon \mathrm{tv}_{\mathbf{i}}(\mathsf{A}) \rightleftarrows \mathrm{tv}_{\mathbf{i}|i}(\mathsf{A}) \colon e_i \qquad (i \in \mathbf{i} \subset \mathbb{N}^*, \, \alpha = \pm), \tag{6.50}$$

that satisfy the multiple relations of 6.2.2. (We also have a multiple set (A_i) indexed by the finite subsets of $\mathbb{N}$.)

(wmc.3) Furthermore i-cubes and i-maps have, in every (positive) direction $i \in \mathbf{i}$, a *geometric* i-composition written in additive notation and diagrammatic order

$$x +_i y, \qquad f +_i g \qquad (\partial_i^+ x = \partial_i^- y, \ \partial_i^+ f = \partial_i^- g), \qquad (6.51)$$

which commutes with the transversal faces, degeneracies and composition of (6.49). In other words it is a functor $+_i \colon \mathbf{A_i} \times_i \mathbf{A_i} \to \mathbf{A_i}$, defined on the pullback-category of pairs of i-consecutive i-cubes (and i-maps).

(wmc.4) Including the transversal composition, we have now a composition law $x +_i y$ of consecutive i-*cells* $x, y \in A_i$, for $i \in \mathbf{i} \subset \mathbb{N}$. The following interactions with faces and degeneracies are assumed, for $i \neq j$ (even though for $i = 0$ or $j = 0$ part of these relations follow from the previous axioms)

$$\partial_i^- (x +_i y) = \partial_i^- (x), \qquad\qquad \partial_i^+ (x +_i y) = \partial_i^+ (y),$$

$$\partial_j^\alpha (x +_i y) = \partial_j^\alpha (x) +_i \partial_j^\alpha (y), \quad e_j (x +_i y) = e_j (x) +_i e_j (y). \qquad (6.52)$$

For i-*cells* $x, y \in A_i$, when we write $\partial_i^\alpha x$ or $x +_i y$ it is understood that $i \in \mathbf{i}$ and $\partial_i^+ x = \partial_i^- y$; when we write $e_i x$ it is understood that $i \notin \mathbf{i}$. The last condition in (6.52) is a *strict interchange between i-composition and j-identities.*

For a positive multi-index $\mathbf{i}$, a transversal i-map $f \colon x \to_0 y$ is said to be *i-special*, or *special in direction* $i \in \mathbf{i}$, if its two i-faces are transversal identities

$$\partial_i^\alpha f = e_0 \partial_i^\alpha x = e_0 \partial_i^\alpha y \qquad (\alpha = \pm). \qquad (6.53)$$

This, of course, implies that the i-cubes x, y have the same i-faces: $\partial_i^\alpha x = \partial_i^\alpha y$ (in $A_{\mathbf{i}|i}$). We say that f is *ij-special* if it is special in both directions i, j.

For $\mathbf{i} = (i_1, ..., i_n)$ we say that a transversal i-map $f \colon x \to y$ is *special on vertices* (or *special*, for short) if all its total faces, namely the 2^n maps $\partial_{i_1}^{\alpha_1} ... \partial_{i_n}^{\alpha_n} (f)$ of degree 0, are identities. Every i-special map is of this kind.

6.4.2 *Comparisons*

Now we require that the positive compositions in A are unitary, associative and satisfy interchange *up to* invertible transversal comparisons: left unitors, right unitors, associators and interchangers. The letter $\mathbf{i}$ denotes a positive multi-index, with $i \in \mathbf{i}$. (In the diagrams below a line segment represents a cell and a double one represents a cell degenerate in the direction of the line itself.)

(wmc.5) For every **i**-cube x we have an invertible i-special **i**-map $\lambda_i x$, which is natural on **i**-maps and has the following faces (for $j \neq i$ in **i**)

$$\lambda_i x \colon (e_i \partial_i^- x) +_i x \to_0 x \qquad (\textit{left } i\textit{-unitor}),$$

$$\partial_j^\alpha \lambda_i x = \lambda_i \partial_j^\alpha x \qquad (\partial_i^\alpha \lambda_i x = e_0 \partial_i^\alpha x), \tag{6.54}$$

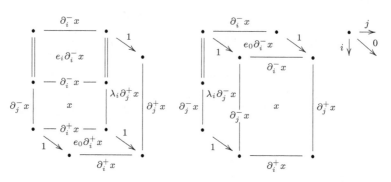

The condition in parentheses says again that $\lambda_i x$ is i-special; this repetition will be omitted below. The naturality condition means that for every **i**-map $f \colon x \to_0 x'$ the following square of **i**-maps commutes (in the transversal category $\mathbf{A_i}$)

$$
\begin{array}{ccc}
(e_i \partial_i^- x) +_i x & \xrightarrow{\;\lambda_i x\;} & x \\
\scriptstyle (e_i \partial_i^- f) +_i f \downarrow & & \downarrow \scriptstyle f \\
(e_i \partial_i^- x') +_i x' & \xrightarrow[\;\lambda_i x'\;]{} & x'
\end{array}
\tag{6.55}
$$

(wmc.6) For every **i**-cube x we have an invertible i-special **i**-map $\rho_i x$, which is natural on **i**-maps and has the following faces (for $j \neq i$ in **i**)

$$\rho_i x \colon x +_i (e_i \partial_i^+ x) \to_0 x \qquad (\textit{right } i\textit{-unitor}),$$

$$\partial_j^\alpha \rho_i x = \rho_i \partial_j^\alpha x. \tag{6.56}$$

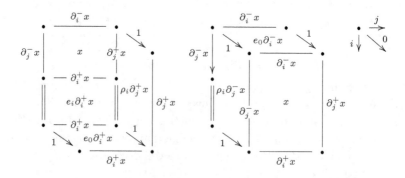

(wmc.7) For three i-consecutive i-cubes x, y, z we have an invertible i-special i-map $\kappa_i(x, y, z)$, which is natural on i-maps and has the following faces (for $j \neq i$ in i)

$$\kappa_i(x, y, z) \colon x +_i (y +_i z) \to_0 (x +_i y) +_i z \qquad (i\text{-associator}),$$

$$\partial_j^\alpha \kappa_i(x, y, z) = \kappa_i(\partial_j^\alpha x, \partial_j^\alpha y, \partial_j^\alpha z)$$

(6.57)

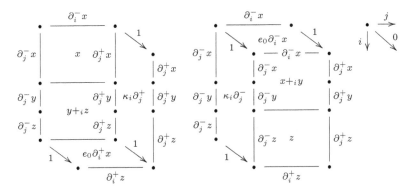

(wmc.8) For $i < j$ in i and a consistent matrix of i-cubes $\left(\begin{smallmatrix} x & y \\ z & u \end{smallmatrix}\right)$ (with i-consecutive rows and j-consecutive columns), we have an invertible ij-special i-map $\chi_{ij}(x, y, z, u)$, the ij-interchanger, which is natural on i-maps and has the following k-faces (for $k \neq i, j$ in i)

$$\chi_{ij}(x, y, z, u) \colon (x +_i y) +_j (z +_i u) \to_0 (x +_j z) +_i (y +_j u),$$

$$\partial_k^\alpha \chi_{ij}(x, y, z, u) = \chi_{ij}(\partial_k^\alpha x, \partial_k^\alpha y, \partial_k^\alpha z, \partial_k^\alpha u),$$

(6.58)

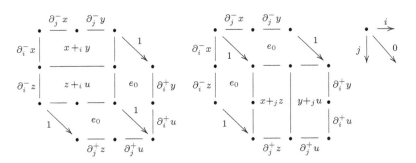

(wmc.9) Finally these comparisons must satisfy conditions of coherence, *written below* in 6.4.3, 6.4.4.

We say that A is *unitary* if all the comparisons λ, ρ are identities, and *pre-unitary* if every unitor of type $\lambda(e_i z) = \rho(e_i z) \colon e_i z +_i e_i z \to_0 e_i z$ is an identity (see (6.59)).

The *transversal dual* A^{tr} of a weak multiple category A reverses the transversal faces and compositions (as in 6.2.7), and has inverted comparisons $\lambda_i^*(x) = ((\lambda_i x)^{-1})^*$, etc.

As in 6.2.4, we also consider *N-indexed weak multiple categories*, where the term *positive* multi-index means *never* 0. The truncated case is considered below, in 6.4.7.

6.4.3 Coherence conditions, I

The coherence axiom (wmc.9) means that the comparisons satisfy various conditions; for future reference we split them in two parts, deferring to the next point 6.4.4 all the conditions involving the interchanger χ_{ij}.

The following diagrams of transversal maps must commute (assuming that all the compositions and degeneracies in direction $i > 0$ make sense).

(wmc.9.i) *Coherence pentagon of the i-associator* $\kappa = \kappa_i$

$$
\begin{array}{ccc}
 & (x +_i y) +_i (z +_i u) & \\
 \nearrow^{\kappa} & & \searrow^{\kappa} \\
x +_i (y +_i (z +_i u)) & & ((x +_i y) +_i z) +_i u \\
\downarrow{\scriptstyle 1+_i\kappa} & & \nearrow{\scriptstyle \kappa+_i 1} \\
x +_i ((y +_i z) +_i u) & \xrightarrow{\ \kappa\ } & (x +_i (y +_i z)) +_i u
\end{array}
$$

(wmc.9.ii) *Coherence condition of the i-associator and i-unitors*:

$$
\begin{array}{ccc}
x +_i (e_i \partial_i^- y +_i y) & \xrightarrow{\ \kappa\ } & (x +_i e_i \partial_i^+ x) +_i y \\
 \searrow{\scriptstyle 1+_i\lambda} & & \swarrow{\scriptstyle \rho+_i 1} \\
 & x +_i y &
\end{array}
$$

These conditions amount to asking that, for every positive multi-index $\mathbf{j}$ and $i \notin \mathbf{j}$, the $\mathbf{j}$-cubes of A form a weak double category $\mathrm{dbl}_{\mathbf{j},i}(\mathsf{A})$ with horizontal arrows in $A_{0\mathbf{j}}$, vertical arrows in $A_{\mathbf{j}i}$ and double cells in $A_{0\mathbf{j}i}$.

In particular, from (3.37), we have:

$$
\lambda_i(e_i z) = \rho_i(e_i z) : e_i z +_i e_i z \to_0 e_i z. \tag{6.59}
$$

6.4.4 Coherence conditions, II

Finally we list the conditions involving the interchangers χ_{ij} (for $0 < i < j$). The following diagrams of transversal maps must commute, for $0 < i < j < k$. All positive compositions are assumed to be legitimate.

(wmc.9.iii) *Coherence hexagon of* $\chi = \chi_{ij}$ *and* κ_i

$$(x +_i (y +_i z)) +_j (x' +_i (y' +_i z')) \xrightarrow{\kappa_i +_j \kappa_i} ((x +_i y) +_i z) +_j ((x' +_i y') +_i z')$$

$$\chi \downarrow \qquad\qquad \downarrow \chi$$

$$(x +_j x') +_i ((y +_i z) +_j (y' +_i z')) \qquad\qquad ((x +_i y) +_j (x' +_i y')) +_i (z +_j z')$$

$$1 +_i \chi \downarrow \qquad\qquad \downarrow \chi +_i 1$$

$$(x +_j x') +_i ((y +_j y') +_i (z +_j z')) \xrightarrow[\kappa_i]{} ((x +_j x') +_i (y +_j y')) +_i (z +_j z')$$

(wmc.9.iv) *Coherence hexagon of* $\chi = \chi_{ij}$ *and* κ_j

$$(x +_i x') +_j ((y +_i y') +_j (z +_i z')) \xrightarrow{\kappa_j} ((x +_i x') +_j (y +_i y')) +_j (z +_i z')$$

$$1 +_j \chi \downarrow \qquad\qquad \downarrow \chi +_j 1$$

$$(x +_j x') +_j ((y +_j z) +_i (y' +_j z')) \qquad\qquad ((x +_j y) +_i (x' +_j y')) +_j (z +_i z')$$

$$\chi \downarrow \qquad\qquad \downarrow \chi$$

$$(x +_j (y +_j z)) +_i (x' +_j (y' +_j z')) \xrightarrow[\kappa_j +_i \kappa_j]{} ((x +_j y) +_j z) +_i ((x' +_j y') +_j z')$$

(wmc.9.v) *Coherence conditions of* $\chi = \chi_{ij}$ *with* $\lambda = \lambda_i$ *and* $\rho = \rho_i$

$$(e_i \partial_i^- x +_i x) +_j (e_i \partial_i^- y +_i y) \xrightarrow{\lambda +_j \lambda} x +_j y \xleftarrow{\rho +_j \rho} (x +_i e_i \partial_i^+ x) +_j (y +_i e_i \partial_i^+ y)$$

$$\chi \downarrow \qquad\qquad \| \qquad\qquad \downarrow \chi$$

$$(e_i \partial_i^- x +_j e_i \partial_i^- y) +_i (x +_j y) \qquad\qquad (x +_j y) +_i (e_i \partial_i^+ x +_j e_i \partial_i^+ y)$$

$$\| \qquad\qquad \| \qquad\qquad \|$$

$$e_i \partial_i^- (x +_j y) +_i (x +_j y) \xrightarrow{\lambda} x +_j y \xleftarrow{\rho} (x +_j y) +_i e_i \partial_i^+ (x +_j y)$$

(wmc.9.vi) *Coherence conditions of* $\chi = \chi_{ij}$ *with* $\lambda = \lambda_j$ *and* $\rho = \rho_j$

$$e_j \partial_j^- (x +_i y) +_j (x +_i y) \xrightarrow{\lambda} x +_j y \xleftarrow{\rho} (x +_i y) +_j e_j \partial_j^+ (x +_i y)$$

$$\| \qquad\qquad \| \qquad\qquad \|$$

$$(e_j \partial_j^- x +_i e_j \partial_j^- y) +_j (x +_i y) \qquad\qquad (x +_i y) +_j (e_j \partial_j^+ x +_i e_j \partial_j^+ y)$$

$$\chi \downarrow \qquad\qquad \| \qquad\qquad \downarrow \chi$$

$$(e_j \partial_j^- x +_j x) +_i (e_j \partial_j^- y +_j y) \xrightarrow[\lambda +_i \lambda]{} x +_j y \xleftarrow[\rho +_i \rho]{} (x +_j e_j \partial_j^+ x) +_i (y +_j e_j \partial_j^+ y)$$

(wmc.9.vii) *Coherence hexagon of the interchangers* χ_{ij}, χ_{jk} *and* χ_{ik}

$$((x +_i y) +_j (z +_i u)) +_k ((x' +_i y') +_j (z' +_i u'))$$

$\chi_{jk} \downarrow$

$$((x +_i y) +_k (x' +_i y')) +_j ((z +_i u) +_k (z' +_i u')) \qquad \Big| \chi_{ij} +_k \chi_{ij}$$

$\chi_{ik} +_j \chi_{ik} \Big| \qquad ((x +_j z) +_i (y +_j u)) +_k ((x' +_j z') +_i (y' +_j u'))$

$$((x +_k x') +_i (y +_k y')) +_j ((z +_k z') +_i (u +_k u')) \qquad \Big| \chi_{ik}$$

$\chi_{ij} \Big| \qquad ((x +_j z) +_k (x' +_j z')) +_i ((y +_j u) +_k (y' +_j u'))$

$\qquad\qquad\qquad\qquad\qquad \downarrow \chi_{jk} +_i \chi_{jk}$

$$((x +_k x') +_j (z +_k z')) +_i ((y +_k y') +_j (u +_k u'))$$

6.4.5 Cubical type

Extending the definition of 6.3.1, we say that the weak multiple category A is *of cubical type* if its components, faces, degeneracies, compositions *and comparisons* are invariant under renaming positive indices, in the same order, so that everything is determined by the *n-cubes* of $A_{1,...,n}$ and the *n-maps* of $A_{0,1,...,n}$, i.e. by the transversal components $\mathrm{tv}_{[n]}A$.

Now the formulas (6.43) must be completed for the invariance of comparisons: for instance the left unitor

$$\lambda_i x \colon (e_i \partial_i^- x) +_i x \to_0 x \qquad (x \in A_{i_1,...,i_n}), \tag{6.60}$$

where $i = i_j$ is the j-th index, must be the same as the left unitor

$$\lambda_j x \colon (e_j \partial_j^- x) +_j x \to_0 x$$

of the 'normalised form' $x \in A_{1,...,n}$.

We speak of *symmetric cubical type* when we further have a left action of the symmetric groups S_n on $A_{1,...,n}$ and $A_{0,1,...,n}$ as described in 6.3.2, that is also coherent with the comparisons $\lambda_i, \rho_i, \kappa_i, \chi_{ij}$. Namely, for every permutation $s \colon [n] \to [n]$ that takes i to i' and j to j' we have:

$$s.\lambda_i = \lambda_{i'}.s, \qquad s.\rho_i = \rho_{i'}.s, \qquad s.\kappa_i = \kappa_{i'}.s,$$
$$s.\chi_{ij} = \chi_{i'j'}.s \qquad (i \neq j), \tag{6.61}$$

where we introduce $\chi_{ji} = (\chi_{ij})^{-1}$ for $i < j$.

All this is better understood from the examples below. Again, we generally make no difference between a 'weak multiple category of symmetric cubical type' and a 'weak symmetric cubical category'.

6.4.6 Examples

(a) The weak symmetric cubical category $\mathsf{Cosp}(\mathbf{C})$ *of cubical cospans* over a category $\mathbf{C}$ with (a fixed choice of) pushouts has been constructed in [G2], to deal with higher-dimensional cobordism.

A detailed construction will be given in Section B4, taking advantage of the cubical machinery developed in Appendix B, but the interested reader can begin to work it out now, from the following inputs.

An n-cube is a functor $x \colon \wedge^n \to \mathbf{C}$, where $\wedge$ is the formal-cospan category

$$
0 \longrightarrow \iota \longleftarrow 1
$$
$$
\wedge
$$

$$
\begin{array}{ccc}
00 \longrightarrow \iota 0 \longleftarrow 10 \\
\downarrow \quad\quad \downarrow \quad\quad \downarrow \\
0\iota \longrightarrow \iota\iota \longleftarrow 1\iota \\
\uparrow \quad\quad \uparrow \quad\quad \uparrow \\
01 \longrightarrow \iota 1 \longleftarrow 11 \quad \wedge^2
\end{array}
\qquad
\begin{array}{c}
\bullet \xrightarrow{1} \\
\downarrow 2 \\
{}
\end{array}
\qquad (6.62)
$$

A transversal map of n-cubes is a natural transformation of these functors. The transversal component of degree n of $\mathsf{Cosp}(\mathbf{C})$ is thus the category

$$
\mathrm{Cosp}_n(\mathbf{C}) = \mathbf{Cat}(\wedge^n, \mathbf{C}). \qquad (6.63)
$$

The concatenation $x +_i y$ of two n-cubes which are i-consecutive (i.e. $\partial_i^+(x) = \partial_i^-(y)$) is computed in the obvious way, by 3^{n-1} pushouts 'along' the common face. The comparisons come from the universal property of the latter.

Concatenation can be given a formal definition. It is based on the *model of binary composition* (for ordinary cospans), namely the category $\wedge_2$ displayed below, with one 'marked square'

$$
\begin{array}{c}
\iota \\
a \quad\quad c \\
0 \quad\quad b \quad\quad 1 \quad \wedge_2
\end{array}
\qquad (6.64)
$$

and equipped with the embedding $k \colon \wedge \to \wedge_2$ determined by the labelling of the objects $0, \iota, 1$ in $\wedge_2$.

Now, given two consecutive cospans x, y in $\mathbf{C}$, their concatenation is defined as

$$x +_1 y = [x, y].k \colon \wedge \to \wedge_2 \to \mathbf{C}, \qquad (6.65)$$

using the obvious functor $[x, y] \colon \wedge_2 \to \mathbf{C}$ that takes the marked square to a distinguished pushout in $\mathbf{C}$.

1-directed concatenation of n-cubes is based on the cartesian product $\wedge_2 \times \wedge^{n-1}$, represented below in dimension two

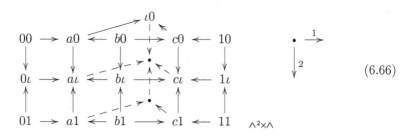

$$(6.66)$$

The 2-dimensional truncation $\mathsf{Cosp}_2(\mathbf{C})$ is the weak double category $\mathbb{C}\mathrm{osp}\mathbf{C}$ constructed in Section 3.4.

(b) Similarly we have the weak symmetric cubical category $\mathsf{Span}(\mathbf{C})$ *of cubical spans* over a category $\mathbf{C}$ with (a fixed choice of) pullbacks. An n-cube is a functor $x \colon \vee^n \to \mathbf{C}$, where $\vee = \wedge^{\mathrm{op}}$ is the formal-span category

$$
\begin{array}{ccc}
& & 00 \leftarrow \iota 0 \to 10 \\
0 \leftarrow \iota \to 1 & & 0\iota \leftarrow \iota\iota \to 1\iota \\
& & 01 \leftarrow \iota 1 \to 11 \\
\vee & & \vee^2
\end{array}
\qquad (6.67)
$$

Again, a transversal map of n-cubes is a natural transformation of these functors. $\mathsf{Span}(\mathbf{C})$ is transversally dual to $\mathsf{Cosp}(\mathbf{C}^{\mathrm{op}})$, with positive compositions computed by pullbacks.

The 2-dimensional truncation $\mathsf{Span}_2(\mathbf{C})$ is the weak double category $\mathbb{S}\mathrm{pan}\mathbf{C}$ constructed in Section 3.4.

(c) The weak symmetric cubical category of *cubical bispans*, or *cubical diamonds* $\mathsf{Bisp}(\mathbf{C})$, over a category $\mathbf{C}$ with pullbacks and pushouts, is similarly constructed over the formal diamond category of (4.54). The 2-dimensional truncation $\mathsf{Bisp}_2(\mathbf{C})$ is the weak double category $\mathbb{B}\mathrm{isp}(\mathbf{C})$ described in 4.5.7.

(d) Multiple structures of cubical relations and partial mappings will be introduced in Section 6.6.

6.4.7 Weak n-tuple categories

As in the strict case (in 6.2.8), the n-dimensional structure of a *weak n-multiple category*, or *weak n-tuple category*, is obtained by restricting all multi-indices to the subsets of the ordinal $\mathbf{n} = \{0, 1, ..., n-1\}$.

The 0- and 1-dimensional versions still amount to a set or a category, but the 2-dimensional notion is now a weak double category. We are particularly interested in the 3-dimensional case, a *weak triple category*.

Starting from an (unbounded) weak multiple category A, its truncation with multi-indices $\mathbf{i} \subset \mathbf{n}$ gives a weak n-tuple category $\mathrm{trc}_n A$, also written as $A_\mathbf{n}$.

Thus $\mathsf{Span}_3(\mathbf{C})$ is the weak triple category of 2-cubical spans (over a category with pullbacks), where the highest-dimensional 'objects' are 2-cubes $x\colon \vee^2 \to \mathbf{C}$, i.e. spans of spans, and their transversal maps are 3-dimensional.

Similarly we have the weak n-tuple categories $\mathsf{Span}_\mathbf{n}(\mathbf{C})$, $\mathsf{Cosp}_\mathbf{n}(\mathbf{C})$ and $\mathsf{Bisp}_\mathbf{n}(\mathbf{C})$ of $(n-1)$-dimensional cubical spans, cospans and bispans (with their n-dimensional transformations).

6.4.8 Chiral multiple categories

A *chiral multiple category*, or *χ-lax multiple category*, or *cm-category*, is a partially lax generalisation of a weak multiple category. (The term 'chiral' refers to something that – generally – cannot be superposed to its mirror image.)

This notion is no longer transversally selfdual and has two instances. A *right chiral multiple category* has the same structure and satisfies the same axioms considered above in the weak multiple case, except for the fact that the ij-interchanger, for $0 < i < j$

$$\chi_{ij}(x, y, z, u)\colon (x +_i y) +_j (z +_i u) \to_0 (x +_j z) +_i (y +_j u), \qquad (6.68)$$

is *no longer supposed to be invertible*.

By transversal duality, a *left chiral multiple category* has an ij-interchange comparison directed the other way round, for $0 < i < j$

$$\chi_{ij}(x, y, z, u)\colon (x +_j z) +_i (y +_j u) \to_0 (x +_i y) +_j (z +_i u), \qquad (6.69)$$

with the obvious modification of the coherence axioms.

Both structures still have the *strict* degenerate interchanges mentioned in 6.2.4 and 6.4.1, which it will be useful to write in the following form, for

$0 < i < j$:

$$e_j e_i(x) = e_i e_j(x),$$

$$e_i x +_j e_i y = e_i(x +_j y), \qquad e_j(x +_i y) = e_j(x) +_i e_j(y). \tag{6.70}$$

We generally work in the *right* chiral case, that will just be called 'chiral'. In the truncated n-dimensional case every left chiral n-tuple category can be turned into a right chiral one just by reversing the positive indices, $i \mapsto n - i$; in this way we avoid resorting to transversal duality, which would turn transversal limits into colimits. In the infinite dimensional case this also works if we replace the natural indices with the integral ones, or with any (infinite) *selfdual* totally ordered pointed set.

A *chiral triple category* is the 3-dimensional truncated notion, with multi-indices $\mathbf{i} \subset \{0, 1, 2\}$. Our main example of this kind is the (right) chiral triple category $\mathsf{SC}(\mathbf{C}) = \mathsf{S}_1 \mathsf{C}_1(\mathbf{C})$ *of spans and cospans* over a category $\mathbf{C}$ (with pushouts and pullbacks), that will be described in Section 6.6, together with other structures of higher dimension, including the *unbounded chiral multiple category* $\mathsf{S}_{-\infty} \mathsf{C}_\infty(\mathbf{C})$, indexed by the ordered set $\mathbb{Z}$ of integers. These examples motivate our terminology for the alternative right/left: in the right-hand case the *lower* composition laws use limits (i.e. right adjoints), and come *before* the *upper* ones that use colimits: for instance, pullbacks *before* pushouts in $\mathsf{SC}(\mathbf{C})$.

In Section 6.8 we shall briefly sketch *intercategories*, a further generalisation of chiral multiple categories introduced in [GP6, GP7], where not only χ but also the three strict interchanges listed above, in (6.70), are laxified.

6.4.9 Theorem and Definition (The flat case)

(a) Extending the 2-dimensional case, a chiral multiple category A *is said to be (transversally)* flat *if it satisfies the following equivalent conditions:*

(i) for every positive multi-index $\mathbf{i}$ *of degree* $n \geqslant 1$, *each* $\mathbf{i}$-*map* f *is determined by the family of its faces* $\partial_i^\alpha f$, *for* $i \in 0\mathbf{i}$,

(ii) for every positive multi-index $\mathbf{i} = \{i_1, ..., i_n\}$ *of any degree* $n \geqslant 1$, *each* $\mathbf{i}$-*map* $f: x \to y$ *is determined by its 0-faces* x, y *together with the family of its total faces, the* 2^n *transversal maps* $\partial_{i_1}^{\alpha_1} ... \partial_{i_n}^{\alpha_n} f$ *of degree 0.*

(b) In the flat case two transversal maps $f, g: x \to y$ *that are special on vertices (see 6.4.1) always coincide.*

(c) In the flat case the coherence axiom (wmc.9) (of 6.4.3, 6.4.4) is redundant and can be omitted.

Note. In (ii) one can equivalently add the value $n = 0$, which gives a void condition: a $*$-map $f: x \to y$ has one total face, namely itself.

Proof The proof is easy, and can be written as an exercise.

(a) Plainly (ii) $\Rightarrow$ (i), because the faces $\partial_i^\alpha(f)$ determine the total faces $\partial_{i_1}^{\alpha_1} \ldots \partial_{i_n}^{\alpha_n}(f)$, which are supposed to be sufficient to determine f.

We prove the converse by induction on the degree $n \geqslant 1$. For $n = 1$ and $\mathbf{i} = \{i\}$ both conditions say the same thing; therefore we suppose that (i) holds in degree $n \geqslant 2$ and (ii) in degree $n - 1$, and prove that (ii) also holds in degree n. Let us take two transversal $\mathbf{i}$-maps $f, g \colon x \to y$ with the same total $\mathbf{i}$-faces. For every $i \in \mathbf{i}$, $\partial_i^\alpha(f)$ and $\partial_i^\alpha(g)$ are $\mathbf{i}|i$-maps with the same 0-faces (namely $\partial_i^\alpha(x)$ and $\partial_i^\alpha(y)$) and the same 2^{n-1} total faces (namely the total faces of f and g whose 'exponent' for the index i is α). Therefore $\partial_i^\alpha(f) = \partial_i^\alpha(g)$, by (ii) in degree $n - 1$, and we conclude that $f = g$ by (i) in degree n.

(b) If $n = 0$, f and g coincide with their total faces and must be the identity of $x = y$. The case $n > 0$ follows immediately from (a): if the $\mathbf{i}$-map $f \colon x \to y$ is special on vertices, all its total faces are transversal identities of $\partial_{i_1}^{\alpha_1} \ldots \partial_{i_n}^{\alpha_n}(x) = \partial_{i_1}^{\alpha_1} \ldots \partial_{i_n}^{\alpha_n}(y)$.

(c) Finally we assume that A satisfies the axioms (wmc.1–8) and is flat; point (b) still holds. Each coherence condition of 6.4.3, 6.4.4 is about the coincidence of two transversal maps f, g with the same 0-faces; moreover such maps are i-special in at least one positive direction i, whence they are special on vertices and must coincide. $\square$

6.5 Lax functors and transversal transformations

We introduce lax, colax and pseudo multiple functors between chiral multiple categories, together with their transversal transformations.

More generally, we are also interested in the intermediate cases, dealt with in 6.5.7–6.5.8, of *mixed-laxity functors*, colax up to a certain degree and lax above. (The opposite arrangement makes no sense.) These 'functors' were introduced in [GP6] for the 3-dimensional case and extended in [GP12] to infinite dimension.

6.5.1 Strict multiple functors

Extending a previous definition (in 6.2.6), a *(multiple) functor* $F \colon \mathsf{A} \to \mathsf{B}$ between chiral multiple categories is a morphism of multiple sets which preserves all the composition laws and all the comparisons (listed in 6.4.2):

$$F(\lambda_i x) = \lambda_i(Fx), \qquad\qquad F(\rho_i x) = \rho_i(Fx),$$
$$F(\kappa_i(x, y, z)) = \kappa_i(Fx, Fy, Fz), \qquad\qquad (6.71)$$
$$F(\chi_{ij}(x, y, z, u)) = \chi_{ij}(Fx, Fy, Fz, Fu).$$

Again it is convenient to view F as a morphism of multiple objects in **Cat**, with transversal components that are ordinary functors

$$F_i = tv_i(F) \colon tv_i(A) \to tv_i(B) \qquad (i \subset \mathbb{N}^*). \qquad (6.72)$$

A *transversal transformation* $\varphi \colon F \to G \colon A \to B$ between multiple functors of chiral multiple categories is a face-consistent family of natural transformations of ordinary functors

$$\varphi_i = tv_i(\varphi) \colon tv_i(F) \to tv_i(G) \colon tv_i(A) \to tv_i(B) \qquad (i \subset \mathbb{N}^*),$$
$$\varphi x \colon Fx \to_0 Gx, \qquad\qquad \varphi(\partial_i^\alpha x) = \partial_i^\alpha(\varphi x), \qquad (6.73)$$

subject to the same axiom of coherence (trt.2) of the strict case (in 6.2.6).

Given two chiral multiple categories A and B we have thus the category $\mathbf{StCmc}(A, B)$ of their strict multiple functors and transversal transformations. All these form the 2-category $\mathbf{StCmc}$. An *isomorphism* of chiral multiple categories is an invertible arrow in the latter: this means a multiple functor $F \colon A \to B$ whose transversal components $F_i \colon tv_i(A) \to tv_i(B)$ are ordinary isomorphisms.

6.5.2 Lax multiple functors

Basically, a *lax (multiple) functor* $F \colon A \to B$ between (right) chiral multiple categories has components $F_i = tv_i(F) \colon tv_i(A) \to tv_i(B)$ (for all positive multi-indices i) that are functors and agree with all faces, as in the strict case.

Moreover F is equipped with i-special *comparison* i-maps $\underline{F}_i$, for $i \in i \subset \mathbb{N}^*$, t in $A_{i|i}$ and i-consecutive cubes x, y in A_i

$$\underline{F}_i(t) \colon e_i F(t) \to_0 F(e_i t),$$
$$\underline{F}_i(x, y) \colon Fx +_i Fy \to_0 F(x +_i y), \qquad (6.74)$$
$$\partial_j^\alpha \underline{F}_i(x) = \underline{F}_i(\partial_j^\alpha x), \qquad \partial_j^\alpha \underline{F}_i(x, y) = \underline{F}_i(\partial_j^\alpha x, \partial_j^\alpha y) \quad (j \neq i).$$

These comparisons have to satisfy the following axioms of naturality and coherence, again for $i \in i \subset \mathbb{N}^*$.

(lmf.1) (*Naturality of unit comparisons*) For every $\mathbf{i}|i$-map $f: x \to_0 y$ in A we have:

$$Fe_i(f).\underline{F}_i(x) = \underline{F}_i(y).e_i(Ff): e_iFx \to_0 F(e_iy).$$

(lmf.2) (*Naturality of composition comparisons*) For two i-consecutive $\mathbf{i}$-maps $f: x \to_0 x'$ and $g: y \to_0 y'$ in A we have:

$$F(f +_i g).\underline{F}_i(x,y) = \underline{F}_i(x',y').(Ff +_i Fg): Fx +_i Fy \to_0 F(x' +_i y').$$

(lmf.3) (*Coherence with unitors*) For an $\mathbf{i}$-cube x with i-faces $\partial_i^- x = z$ and $\partial_i^+ x = u$ (preserved by F) we have two commutative diagrams of $\mathbf{i}$-maps:

$$
\begin{array}{ccc}
e_i(Fz) +_i Fx & \xrightarrow{\lambda_i Fx} & Fx \\
\scriptstyle \underline{F}_i z +_i 1 \downarrow & & \uparrow \scriptstyle F(\lambda_i x) \\
F(e_iz) +_i Fx & \xrightarrow[\underline{F}_i]{} & F(e_iz +_i x)
\end{array}
\qquad
\begin{array}{ccc}
Fx +_i e_i(Fu) & \xrightarrow{\rho_i Fx} & Fx \\
\scriptstyle 1 +_i \underline{F}_i u \downarrow & & \uparrow \scriptstyle F(\rho_i x) \\
Fx +_i F(e_iu) & \xrightarrow[\underline{F}_i]{} & F(x +_i e_iu)
\end{array}
$$

(lmf.4) (*Coherence with associators*) For three i-consecutive $\mathbf{i}$-cubes x, y, z in A we have a commutative diagram of $\mathbf{i}$-maps:

$$
\begin{array}{ccc}
Fx +_i (Fy +_i Fz) & \xrightarrow{\kappa_i F} & (Fx +_i Fy) +_i Fz \\
\scriptstyle 1 +_i \underline{F}_i \downarrow & & \downarrow \scriptstyle \underline{F}_i +_i 1 \\
Fx +_i F(y +_i z) & & F(x +_i y) +_i Fz \\
\scriptstyle \underline{F}_i \downarrow & & \downarrow \scriptstyle \underline{F}_i \\
F(x +_i (y +_i z)) & \xrightarrow[F\kappa_i]{} & F((x +_i y) +_i z)
\end{array}
$$

(lmf.5) (*Coherence with interchangers*) For $i < j$ in $\mathbf{i}$ and a consistent matrix of $\mathbf{i}$-cubes $\left(\begin{smallmatrix} x & y \\ z & u \end{smallmatrix}\right)$ (with i-consecutive rows and j-consecutive columns) we have a commutative diagram of $\mathbf{i}$-maps:

$$
\begin{array}{ccc}
(Fx +_i Fy) +_j (Fz +_i Fu) & \xrightarrow{\chi_{ij} F} & (Fx +_j Fz) +_i (Fy +_j Fu) \\
\scriptstyle \underline{F}_i +_j \underline{F}_i \downarrow & & \downarrow \scriptstyle \underline{F}_j +_i \underline{F}_j \\
F(x +_i y) +_j F(z +_i u) & & F(x +_j y) +_i F(z +_j u) \\
\scriptstyle \underline{F}_j \downarrow & & \downarrow \scriptstyle \underline{F}_i \\
F((x +_i y) +_j (z +_i u)) & \xrightarrow[F\chi_{ij}]{} & F((x +_j z) +_i (y +_j u))
\end{array}
$$

The lax multiple functor F is said to be *unitary* if all its unit comparisons $\underline{F}_i(x)$ are transversal identities. As in the theory of double categories, this unitarity condition is *at a more basic level* than unitarity of weak or chiral multiple categories: indeed *only in this case* F commutes with all degeneracies and is a morphism of multiple sets.

Lax multiple functors compose, in a categorical way, with the following composed comparisons for GF

$$\underline{GF}_i(x) = G\underline{F}_i(x).\underline{G}_i(Fx)\colon e_i(GFx) \to GF(e_ix),$$

$$\underline{GF}_i(x, y) = G\underline{F}_i(x, y).\underline{G}_i(Fx, Fy)\colon \tag{6.75}$$
$$GFx +_i GFy \to GF(x +_i y).$$

A *colax (multiple) functor* F has comparisons in the opposite direction

$$\underline{F}_i(x)\colon F(e_ix) \to e_i(Fx), \qquad \underline{F}_i(x, y)\colon F(x +_i y) \to Fx +_i Fy, \tag{6.76}$$

under axioms (cmf.1–5) transversally dual to the previous ones (see 6.4.2). For instance:

(cmf.1) (*Naturality of unit comparisons*) For every $i|i$-map $f\colon x \to_0 y$ in A we have:

$$e_i(Ff).\underline{F}_i(x) = \underline{F}_i(y).Fe_i(f)\colon F(e_ix) \to_0 e_i(Fy). \tag{6.77}$$

A *pseudo (multiple) functor* is a lax functor whose comparisons are invertible; it is made colax by the inverse comparisons.

Full and faithful lax (or colax) multiple functors are defined in 7.4.4, together with multiple equivalences.

6.5.3 Transversal transformations

A *transversal transformation* $\varphi\colon F \to G\colon \mathsf{A} \to \mathsf{B}$ between lax multiple functors of chiral multiple categories is a face-consistent family of natural transformations of ordinary functors

$$\varphi_i = tv_i(\varphi)\colon tv_i(F) \to tv_i(g)\colon tv_i(\mathsf{A}) \to tv_i(\mathsf{B}) \qquad (i \subset \mathbb{N}^*),$$
$$\varphi x\colon Fx \to_0 Gx, \qquad \varphi(\partial_i^\alpha x) = \partial_i^\alpha(\varphi x), \tag{6.78}$$

under an extended *coherence axiom* (trt.2L) that involves the comparison maps of F and G

(trt.2L) for a positive multi-index $\mathbf{i}$, $i \in \mathbf{i}$, t in $A_{\mathbf{i}|i}$ and $z = x +_i y$ in $A_\mathbf{i}$, we have commutative diagrams in B:

$$
\begin{array}{ccc}
e_iF(t) & \xrightarrow{\ e_i(\varphi t)\ } & e_iG(t) \\
{\scriptstyle \underline{F}_i(t)}\downarrow & & \downarrow{\scriptstyle \underline{G}_i(t)} \\
F(e_it) & \xrightarrow[\ \varphi(e_it)\ ]{} & G(e_it)
\end{array}
\qquad\qquad
\begin{array}{ccc}
Fx +_i Fy & \xrightarrow{\ \varphi x +_i \varphi y\ } & Gx +_i Gy \\
{\scriptstyle \underline{F}_i(x,y)}\downarrow & & \downarrow{\scriptstyle \underline{G}_i(x,y)} \\
Fz & \xrightarrow[\ \varphi z\ ]{} & Gz
\end{array}
$$

We have now the 2-category LxCmc of chiral multiple categories, lax multiple functors and transversal transformations, with the sub-2-category PsCmc of pseudo functors. A lax functor which is (transversally) isomorphic to a pseudo functor is pseudo as well.

Similarly one defines the 2-category CxCmc, for the colax case; the last axiom above has now a transversally dual form (trt.2C), with reversed vertical arrows.

The forgetful 2-functor

$$|-|: \text{StCmc} \to \text{pMlt}, \tag{6.79}$$

takes a chiral multiple category A to the associated premultiple category $|A|$ (see 6.2.7). The underlying premultiple 'functors' and transformations will be written without marks: F rather than $|F|$, and so on.

6.5.4 Proposition and Definition (Pseudo isomorphism)

An invertible arrow in LxCmc *is the same as a* pseudo *functor* $F: A \to B$ *between chiral multiple categories where all the components*

$$F_\mathbf{i} = \text{tv}_\mathbf{i}(F): \mathbf{A_i} \to \mathbf{B_i}$$

are invertible (i.e. isomorphisms of ordinary categories).

It will be called a pseudo isomorphism *of chiral multiple categories.*

F is made colax by inverting its comparisons; this gives a general invertible arrow in CxCmc.

Proof We suppose that the lax functor F has an inverse $G: B \to A$ in LxCmc, and prove that F is a pseudo functor. In fact, for every positive multi-index $\mathbf{i}$, $G_\mathbf{i}$ is inverse to $F_\mathbf{i}$. Moreover, composing their unit comparisons, as in (6.75), we get (for x in $A_{\mathbf{i}|i}$ and y in $B_{\mathbf{i}|i}$)

$$G\underline{F}_i(x).\underline{G}_i(Fx) = \text{id}(e_i x), \qquad F\underline{G}_i(y).\underline{F}_i(Gy) = \text{id}(e_i y). \tag{6.80}$$

Applying the functor $F_\mathbf{i}$ to the first equation and computing the second in $y = Fx$ we get

$$\underline{F}_i(x).F\underline{G}_i(Fx) = \text{id}(Fe_i x), \qquad F\underline{G}_i(Fx).\underline{F}_i(x) = \text{id}(e_i Fx), \tag{6.81}$$

so that $\underline{F}_i(x): e_i Fx \to_0 F(e_i x)$ is an invertible transversal map. The same holds for the composition comparisons. The converse is obvious. $\qquad\square$

6.5.5 *Transversal invariance*

We now extend the notion of horizontal invariance of double categories (see 4.1.7), obtaining a property that – again – should be expected of any 'well formed' multiple category.

We say that the chiral multiple category A is *transversally invariant* if its cubes are 'transferable' along transversally invertible maps. Precisely:

(i) given an **i**-cube x of degree n and a family of $2n$ *invertible* transversal maps $f_i^\alpha : y_i^\alpha \to_0 \partial_i^\alpha x$ ($i \in \mathbf{i}$, $\alpha = \pm$) *with consistent positive faces* (and otherwise arbitrary domains y_i^α)

$$\partial_i^\alpha(f_j^\beta) = \partial_j^\beta(f_i^\alpha) \qquad\qquad (i \neq j \text{ in } \mathbf{i}), \qquad\qquad (6.82)$$

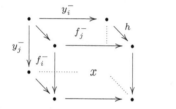

$$(h = \partial_i^+(f_j^-) = \partial_j^-(f_i^+)),$$

there exists an invertible **i**-map $f: y \to_0 x$ (a *filler*) with positive faces $\partial_i^\alpha f = f_i^\alpha$ (and therefore $\partial_i^\alpha y = y_i^\alpha$).

Of course, this property can be equivalently stated for a family of invertible maps $g_i^\alpha : \partial_i^\alpha x \to_0 y_i^\alpha$.

6.5.6 *Proposition* (Transversal modification of a functor)

Let $F: \mathsf{X} \to \mathsf{A}$ *be a lax functor of chiral multiple categories.*

(a) *Given a family of transversal isomorphisms* $\varphi x: Fx \to Gx$ *in* A, *indexed by the cubes of* X, *with consistent positive faces and degeneracies:*

$$\partial_i^\alpha(\varphi_{\mathbf{i}} x) = \varphi_{\mathbf{i}|i}(\partial_i^\alpha x), \qquad e_i(\varphi_{\mathbf{i}|i}\, y) = \varphi_{\mathbf{i}}(e_i y)$$
$$(i \in \mathbf{i},\, x \text{ in } X_{\mathbf{i}},\, y \text{ in } X_{\mathbf{i}|i}), \qquad\qquad (6.83)$$

there is precisely one way of extending these data to a lax functor $G: \mathsf{X} \to \mathsf{A}$ *and a transversal isomorphism* $\varphi: F \to G$.

(b) *If* A *is transversally invariant, every family of ∗-isomorphisms* $\varphi x: Fx \to Gx$ *in* A, *indexed by the objects of* X, *can be (non-uniquely) extended to a family of transversal isomorphisms* $\varphi x: Fx \to Gx$ *in* A, *indexed by all the cubes of* X, *that satisfies* (6.83). *(Then we can form a lax functor* $G: \mathsf{X} \to \mathsf{A}$ *and a transversal isomorphism* $\varphi: F \to G$.)

The same holds for colax functors, pseudo functors, unitary (co)lax functors.

Proof (a) This point is an obvious extension of Exercise 1.2.6(i). There is precisely one way of extending G to all transversal maps $f: x \to y$ and making the family φ natural:

$$Gf = \varphi y.Ff.(\varphi x)^{-1}: Gx \to Fx \to Fy \to Gy. \tag{6.84}$$

The comparisons of G are determined by those of F, as required by axiom (trt.2L) of transversal transformations for the family φ; their coherence follows as well.

(b) We extend the given family (φx) by induction on the degree of **i**-cubes of X: degree 0 is given, and we suppose we have chosen $\varphi_i x$ for all cubes of degree $< n$, fulfilling (6.83).

For an **i**-cube x of degree n we already have a family of transversal isomorphisms in A

$$f_i^\alpha = \varphi_{\mathbf{i}|i}(\partial_i^\alpha x): \partial_i^\alpha(Fx) \to G(\partial_i^\alpha x) \qquad (i \in \mathbf{i}, \, \alpha = \pm), \tag{6.85}$$

with consistent faces, by hypothesis. By transversal invariance we can choose an invertible filler

$$\varphi_i x: Fx \to Gx, \qquad \partial_i^\alpha(\varphi_i x) = \varphi_{\mathbf{i}|i}(\partial_i^\alpha x), \tag{6.86}$$

which is thus coherent with faces.

This choice is easily made consistent with degeneracies: if $x = e_i y$ for some $i \in \mathbf{i}$ and $y \in X_{\mathbf{i}|i}$, we take $\varphi_i x = e_i(\varphi_{\mathbf{i}|i} y)$. If x is also j-degenerate, then $x = e_i e_j z = e_j e_i z$; therefore, by the inductive assumption, both procedures give the same result:

$$e_i(\varphi y) = e_i e_j(\varphi z) = e_j e_i(\varphi z) = e_j(\varphi e_i z).$$

□

6.5.7 Mixed laxity functors

We have already considered two main kinds of functors between chiral multiple categories, namely the lax functors and the colax ones (in 6.5.2).

Yet the theory will require the introduction of a countable family of intermediate kinds, depending on an extended natural index p, for $1 \leqslant p \leqslant \infty$.

A *p-mixed functor* $F: \mathsf{A} \to_p \mathsf{B}$ between chiral multiple categories will be a *mixed-laxity functor* which is colax in all positive directions $i < p$ (if any) and lax in all finite directions $i \geqslant p$ (if any). In particular, this will be a lax functor for $p = 1$ and a colax functor for $p = \infty$.

Again, F has components $F_{\mathbf{i}} = \mathrm{tv_{\mathbf{i}}}(F)\colon \mathbf{A_i} \to \mathbf{B_i}$, for all positive multi-indices $\mathbf{i}$, that are ordinary functors and commute with the faces: $\partial_i^\alpha F_{\mathbf{i}} = F_{\mathbf{i}|i}\,\partial_i^\alpha$ (for $i \in \mathbf{i}$).

Moreover F is equipped with i-special *comparison* $\mathbf{i}$-maps $\underline{F}_i$ (for $t \in A_{\mathbf{i}|i}$ and $z = x +_i y$ in $A_{\mathbf{i}}$), *either* in the lax direction for $p \leqslant i < \infty$

$$\underline{F}_i(t)\colon e_i F(t) \to_0 F(e_i t), \qquad \underline{F}_i(x,y)\colon Fx +_i Fy \to_0 Fz, \qquad (6.87)$$

or in the colax direction for $0 < i < p$

$$\underline{F}_i(t)\colon F(e_i t) \to_0 e_i F(t), \qquad \underline{F}_i(x,y)\colon F(z) \to_0 Fx +_i Fy. \qquad (6.88)$$

All these comparisons must commute with faces (for $j \neq i$ in $\mathbf{i}$)

$$\partial_j^\alpha \underline{F}_i(t) = \underline{F}_i(\partial_j^\alpha t), \qquad \partial_j^\alpha \underline{F}_i(x,y) = \underline{F}_i(\partial_j^\alpha x, \partial_j^\alpha y). \qquad (6.89)$$

Furthermore they have to satisfy the axioms of naturality and coherence (see 6.5.2), either in the lax or in the colax form

- (lmf.1–4) for $i \geqslant p$,   - (cmf.1–4) for $i < p$.

Finally there is an axiom of coherence with the interchanger χ_{ij} ($0 < i < j$), which has three forms: the first corresponds to (lmf.5), the last to (cmf.5) and (b) is an intermediate case

(a) for $p \leqslant i < j$ (so that F is i- and j-lax), we have commutative diagrams of transversal maps:

$$
\begin{array}{ccc}
(Fx +_i Fy) +_j (Fz +_i Fu) & \xrightarrow{\ \chi_{ij} F\ } & (Fx +_j Fz) +_i (Fy +_j Fu) \\
{\scriptstyle \underline{F}_i +_j \underline{F}_i}\Big\downarrow & & \Big\downarrow{\scriptstyle \underline{F}_j +_i \underline{F}_j} \\
F(x +_i y) +_j F(z +_i u) & & F(x +_j y) +_i F(z +_j u) \\
{\scriptstyle \underline{F}_j}\Big\downarrow & & \Big\downarrow{\scriptstyle \underline{F}_i} \\
F((x +_i y) +_j (z +_i u)) & \xrightarrow[\ F\chi_{ij}\ ]{} & F((x +_j z) +_i (y +_j u))
\end{array}
$$

(b) for $0 < i < p \leqslant j$ (so that F is i-colax and j-lax), we have commutative diagrams:

$$
\begin{array}{ccc}
(Fx +_i Fy) +_j (Fz +_i Fu) & \xrightarrow{\ \chi_{ij} F\ } & (Fx +_j Fz) +_i (Fy +_j Fu) \\
{\scriptstyle \underline{F}_i +_j \underline{F}_i}\Big\uparrow & & \Big\downarrow{\scriptstyle \underline{F}_j +_i \underline{F}_j} \\
F(x +_i y) +_j F(z +_i u) & & F(x +_j y) +_i F(z +_j u) \\
{\scriptstyle \underline{F}_j}\Big\downarrow & & \Big\uparrow{\scriptstyle \underline{F}_i} \\
F((x +_i y) +_j (z +_i u)) & \xrightarrow[\ F\chi_{ij}\ ]{} & F((x +_j z) +_i (y +_j u))
\end{array}
$$

(c) for $0 < i < j < p$ (so that F is i- and j-colax), we have commutative diagrams as in (a), with vertical arrows reversed.

The composition of p-mixed functors is easily defined: their comparisons are separately composed, as in (6.75) for all the lax directions and in the dual way for the colax ones.

In the truncated n-dimensional case, for chiral n-tuple categories indexed by the ordinal $\mathbf{n} = \{0, ..., n-1\}$, the p-mixed functors depend on an index $p \in [n] = \{1, ..., n\}$.

In dimension three there are thus three kinds: lax functors (for $p = 1$), colax functors (for $p = 3$), and the intermediate case of *colax-lax functors*, called colax-lax morphisms in [GP6].

Let us note that the 'lax-colax' case makes no sense (in any dimension): modifying the diagram in (a) by reversing all arrows $\underline{F}_j$ would give a diagram where no pairs of arrows compose.

6.5.8 Transversal transformations

A *transversal transformation* $\varphi \colon F \to G \colon \mathsf{A} \to \mathsf{B}$ of p-mixed functors (between chiral multiple categories) consists of a face-consistent family of transversal maps in B

$$\varphi(x) = \varphi_{\mathbf{i}}(x) \colon Fx \to_0 Gx, \qquad \text{(for every } \mathbf{i}\text{-cube } x \text{ of } \mathsf{A}),$$
$$\partial_i^\alpha . \varphi_{\mathbf{i}} = \varphi_{\mathbf{i}|i} . \partial_i^\alpha \qquad \text{(for } i \in \mathbf{i}), \qquad (6.90)$$

so that each component $\varphi_{\mathbf{i}} \colon F_{\mathbf{i}} \to G_{\mathbf{i}} \colon \mathsf{A}_{\mathbf{i}} \to \mathsf{B}_{\mathbf{i}}$ is a natural transformation of ordinary functors:

(nat) for all $f \colon x \to_0 y$ in A, $Gf.\varphi x = \varphi y.Ff$.

Moreover φ has to satisfy the following two coherence conditions with the comparisons of F and G, for a degenerate cube $e_i(t)$ (with t in $A_{\mathbf{i}|i}$) and a composite $z = x +_i y$ in $A_{\mathbf{i}}$.

(coh.a) If $p \leqslant i$ (so that F and G are both lax in direction i), we have commutative diagrams:

$$
\begin{array}{ccc}
e_iF(t) & \xrightarrow{e_i(\varphi t)} & e_iG(t) \\
\underline{F}_i(t) \downarrow & & \downarrow \underline{G}_i(t) \\
F(e_it) & \xrightarrow[\varphi(e_it)]{} & G(e_it)
\end{array}
\qquad
\begin{array}{ccc}
Fx +_i Fy & \xrightarrow{\varphi x +_i \varphi y} & Gx +_i Gy \\
\underline{F}_i(x,y) \downarrow & & \downarrow \underline{G}_i(x,y) \\
Fz & \xrightarrow[\varphi z]{} & Gz
\end{array}
$$

(coh.b) If $i < p$ (so that F and G are colax in direction i), we have commutative diagrams as above, with vertical arrows reversed.

The operations of these transversal transformation, namely the transversal composition and the whisker composition with p-mixed functors, are defined componentwise for every positive multi-index $\mathbf{i} \subset \mathbb{N}$.

We have now a family of 2-categories $\mathrm{Mx}_p\mathbf{Cmc}$ of chiral multiple categories, p-mixed functors and transversal transformations. In particular

$$\mathbf{LxCmc} = \mathrm{Mx}_1\mathbf{Cmc}, \qquad \mathbf{CxCmc} = \mathrm{Mx}_\infty\mathbf{Cmc}. \qquad (6.91)$$

In Chapter 7 we shall construct the *double* category $\mathbb{C}\mathrm{mc}$ of chiral multiple categories, with lax and colax multiple functors, and a larger *multiple* category Cmc (indexed by the ordinal $\omega + 1 = \{0, 1, ..., \infty\}$), where the arrows in a positive direction p are the p-mixed functors.

6.6 Examples of chiral and premultiple categories

In this section, $\mathbf{C}$ is a category equipped with a choice of pullbacks and pushouts.

The weak double category $\mathbb{S}\mathrm{pan}\mathbf{C}$, *of arrows and spans* of $\mathbf{C}$, can be 'amalgamated' with the weak double category $\mathbb{C}\mathrm{osp}\mathbf{C}$, *of arrows and co-spans* of $\mathbf{C}$, to form a 3-dimensional structure: the chiral triple category $\mathsf{SC}(\mathbf{C})$ *of spans and cospans* of $\mathbf{C}$. It has been studied in [GP7], Subsection 6.4, with notation $\mathsf{SpanCosp}(\mathbf{C})$.

Interchanging the positive directions one gets the *left* chiral triple category $\mathsf{CS}(\mathbf{C})$ of *cospans and spans* of $\mathbf{C}$. Higher dimensional examples are considered in 6.6.4. For the sake of simplicity, we assume that, in our choices, *the pullback or pushout of an identity along any map is an identity.* (Omitting this convention would simply give non-trivial invertible unitors λ and ρ for 1- and 2-composition.)

In the second part of this section we introduce some multiple categories of partial maps, and a premultiple category RelSet of cubical relations.

A structure similar to $\mathsf{SC}(\mathbf{C})$, where $\mathbf{C}$ is the category of graphs, has been used in papers on sequential and parallel composition: see [KaSW, ChSW, GiKS] and their references.

6.6.1 A long exercise

We want to form a chiral triple category $\mathsf{SC}(\mathbf{C})$ *of spans and cospans* of $\mathbf{C}$, where the 0, 1- and 2-arrows are the arrows, spans and cospans of $\mathbf{C}$. The 12-cubes will be functors $\pi \colon \vee \times \wedge \to \mathbf{C}$, and the highest dimensional cells will be their natural transformations, so that:

$$\mathrm{tv}_{12}\mathsf{SC}(\mathbf{C}) = \mathbf{Cat}(\vee \times \wedge, \mathbf{C}). \qquad (6.92)$$

An interested reader can autonomously work out the details. The basic part can be found below, in 6.6.2 and 6.6.3.

6.6.2 A triple set with compositions

We begin by constructing a triple set A enriched with composition laws.

(a) The set A_* consists of the objects of $\mathbf{C}$.

(b) The set A_0 is formed of the maps of $\mathbf{C}$, written as $p\colon X \to_0 Y$ or $p\colon X \to Y$; they compose as in $\mathbf{C}$, forming the category $\mathrm{tv}_*(\mathsf{A})$. This composition will be written as qp, or $q.p$ when useful.

(b$'$) The set A_1 consists of the spans of $\mathbf{C}$, written as $f\colon X \to_1 Y$ or $(f', f'')\colon (X \leftarrow \bullet \to Y)$; their composition $f +_1 g$ works by pullback. Formally, f is a functor $\vee \to \mathbf{C}$ defined on the formal-span category $\vee$ (in (6.64)).

(b$''$) The set A_2 consists of the cospans of $\mathbf{C}$, written as $u\colon X \to_2 Y$ or $(u', u'')\colon (X \to \bullet \leftarrow Y)$; their composition $u +_2 v$ works by pushout. Formally, u is a functor $\wedge \to \mathbf{C}$ defined on the formal-cospan category $\wedge = \vee^{\mathrm{op}}$.

Each set A_i (for $i = 0, 1, 2$) has two faces $\partial_i^\alpha\colon A_i \to A_*$ (implicitly used in the previous composition laws) and a degeneracy $e_i\colon A_* \to A_i$.

(c) A 01-cell $\varphi \in A_{01}$, as in the left diagram below, is a natural transformation $\varphi\colon f \to g\colon \vee \to \mathbf{C}$ of spans, or equivalently a commutative diagram of $\mathbf{C}$ as in the right diagram below

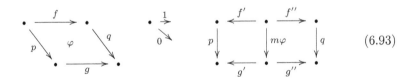

$$(6.93)$$

Their 0-composition, written as $\varphi'\varphi$, is obvious (that of natural transformations) and gives the category $\mathrm{tv}_1(\mathsf{A})$, with obvious units $e_0\colon A_1 \to A_{01}$. Their 1-composition $\varphi +_1 \psi$ is computed by two pullbacks in $\mathbf{C}$.

(c$'$) A 02-cell $\omega \in A_{02}$, as in the left diagram below, is a natural transformation $\omega\colon u \to v\colon \wedge \to \mathbf{C}$ of cospans, or equivalently a commutative diagram of **C** as in the right diagram below

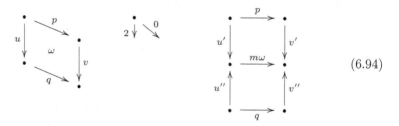

$$(6.94)$$

Their 0-composition $\omega'\omega$ is obvious again, and gives the category $\mathrm{tv}_2(\mathsf{A})$. Their 2-composition $\omega +_2 \zeta$ is computed by two pushouts in **C**.

(c″) A 12-cell $\pi \in A_{12}$ is a functor $\pi\colon \vee \times \wedge \to \mathbf{C}$, or equivalently a commutative diagram of **C**, as at the right below, with three spans in direction 1 and three cospans in direction 2

$$(6.95)$$

Their 1-composition, written as $\pi +_1 \pi'$, is computed by three pullbacks in **C**; their 2-composition, written as $\pi +_2 \rho$, by three pushouts in **C**.

Each set A_{ij} (for $0 \leqslant i < j \leqslant 2$) has two obvious degeneracies and four obvious faces (already used in the composition laws described above)

$$
\begin{aligned}
e_i &\colon A_j \to A_{ij}, & e_j &\colon A_i \to A_{ij}, \\
\partial_i^\alpha &\colon A_{ij} \to A_j, & \partial_j^\alpha &\colon A_{ij} \to A_i.
\end{aligned}
$$

$$(6.96)$$

(d) Finally an element of A_{012}, i.e. a triple cell of A, is a natural transformation $\Pi\colon \pi \to \rho\colon \vee \times \wedge \to \mathbf{C}$, or a functor $\Pi\colon \vee \times \wedge \times \mathbf{2} \to \mathbf{C}$

$$(6.97)$$

Its boundary consists of two 12-cells π, ρ (its 0-faces), two 01-cells φ, ψ and two 02-cells ω, ζ with consistent boundaries. Moreover Π has an additional transversal arrow $m\Pi\colon m\pi \to m\rho$ between the central objects of π

and ρ, forming a commutative diagram in $\mathbf{C}$ (four cubes, joining the four squares of π, in (6.95), to those of ω).

The set A_{012} has the six faces described above and three obvious degeneracies

$$\partial_i^\alpha\colon A_3 \to A_{3|i}, \qquad e_i\colon A_{3|i} \to A_3. \qquad (6.98)$$

The 0-composition of such cells, written as $\Pi'\Pi$, is that of natural transformations and gives the category $\mathrm{tv}_{12}(A)$. Their 1-composition is computed by six pullbacks in $\mathbf{C}$ and written as $\Pi +_1 \Pi'$; their 2-composition, computed by six pushouts in $\mathbf{C}$, is written as $\Pi +_2 \Pi'$.

(e) These sets A_i, with the faces and degeneracies considered above, form a triple set (i.e. a 3-dimensional truncated multiple set), equipped with composition laws organised by its faces.

6.6.3 Comparisons

We have already remarked that the 0-directed composition in each set A_{0i} is categorical, with units given by $e_0\colon A_i \to A_{0i}$. It is also easy to see that the 0-directed composition has a strict interchange with all the other compositions. Because of our assumption on the choice of pushouts and pullbacks, all 1- or 2-directed composition laws are strictly unitary.

To complete the construction of the chiral triple category $\mathsf{SC}(\mathbf{C})$, there are *invertible* comparisons for the associativity of 1- and 2-directed composition, and a *directed* comparison for their interchange. All these comparisons are constructed in [GP7], Section 6, where their coherence is proved.

The comparisons for the 1-directed (resp. 2-directed) composition are constructed as in the weak double category $\mathbb{S}\mathrm{pan}\mathbf{C}$ (resp. $\mathbb{C}\mathrm{osp}\mathbf{C}$). The interchanger $\chi = \chi_{12}$ is defined for a consistent matrix $\left(\begin{smallmatrix}\pi & \pi' \\ \rho & \rho'\end{smallmatrix}\right)$ of 12-cells, and is a 12-special map, natural under 0-composition (*and not invertible*)

$$\chi(\pi,\pi',\rho,\rho')\colon (\pi +_1 \pi') +_2 (\rho +_1 \rho') \to_0 (\pi +_2 \rho) +_1 (\pi' +_2 \rho'). \qquad (6.99)$$

In fact in our hypotheses the weak double category $\mathbb{C}\mathrm{osp}\mathbf{C}$ (with vertical composition $\otimes$ computed by pushout) has double pullbacks, forming a *lax* double functor $F\colon (\mathbb{C}\mathrm{osp}\mathbf{C})^\wedge \to \mathbb{C}\mathrm{osp}\mathbf{C}$, as any double limit (see 3.3.6 for this exponential). Its comparison for composition gives our interchanger:

$$F(\pi,\pi') \otimes F(\rho,\rho') \to F((\pi,\pi') \otimes (\rho,\rho')) = F(\pi \otimes \rho, \pi' \otimes \rho'). \qquad (6.100)$$

As in Section 6.1, a good way of exploring the chiral triple category $\mathsf{SC}(\mathbf{C})$ is to show that it has all five kinds of tabulators and cotabulators: this will be the subject of Exercise 6.7.3.

6.6.4 Higher dimensional structures

More generally, one can form a chiral n-tuple category $S_p C_q(\mathbf{C})$ for $p, q > 0$ and $n = p + q + 1$.

Its i-directed arrows are morphisms of $\mathbf{C}$ for $i = 0$, spans for $1 \leqslant i \leqslant p$ and cospans for $p + 1 \leqslant i \leqslant p + q$. More generally an $\mathbf{i}$-cube is a functor

$$x \colon \vee^{\mathbf{j}} \times \wedge^{\mathbf{k}} \to \mathbf{C}, \qquad \mathbf{j} \subset \{1, ..., p\}, \quad \mathbf{k} \subset \{p + 1, ..., p + q\}, \qquad (6.101)$$

where $\mathbf{i} = \mathbf{j} \cup \mathbf{k}$, and

$$\mathrm{tv}_{\mathbf{i}}(S_p C_q(\mathbf{C})) = \mathbf{Cat}(\vee^{\mathbf{j}} \times \wedge^{\mathbf{k}}, \mathbf{C}). \qquad (6.102)$$

Again, the ij-interchanger χ_{ij} is not invertible whenever it involves both spans and cospans, i.e. for $i \leqslant p < j$. Allowing the cases $p = 0$ or $q = 0$ we would get the weak n-tuple categories

$$S_0 C_{n-1}(\mathbf{C}) = \mathsf{Cosp_n}(\mathbf{C}), \quad S_{n-1} C_0(\mathbf{C}) = \mathsf{Span_n}(\mathbf{C}). \qquad (6.103)$$

We also have the infinite-dimensional structure $S_p C_\infty(\mathbf{C})$, but it is more interesting to consider the 'unbounded' chiral multiple category $S_{-\infty} C_\infty(\mathbf{C})$ indexed by the ordered set of *integers*, pointed at 0, where i-arrows are spans for $i < 0$, ordinary arrows for $i = 0$ and cospans for $i > 0$.

Similarly we have a *left* chiral n-tuple category $C_p S_q(\mathbf{C})$ (see 6.4.8) whose geometric i-arrows are cospans of $\mathbf{C}$ for $1 \leqslant i \leqslant p$ and spans of $\mathbf{C}$ for $p + 1 \leqslant i \leqslant p + q$. It is transversally dual to $S_p C_q(\mathbf{C}^{\mathrm{op}})$.

We also have the left chiral multiple category $C_{-\infty} S_\infty(\mathbf{C})$, transversally dual to $S_{-\infty} C_\infty(\mathbf{C}^{\mathrm{op}})$.

6.6.5 Multiple categories of partial mappings and partial bijections

(a) We have defined in 3.1.9 the double category $\mathbb{P}\mathrm{map}(\mathbf{C})$ of morphisms and partial maps, over a category $\mathbf{C}$ with pullbacks of monomorphisms along arbitrary mappings. (The category $\mathbf{C}$ is supposed to have a choice of subobjects, stable under composition; counterimages of subobjects are determined by this choice.)

This construction has an obvious extension to a strict multiple category $\mathsf{Pmap}(\mathbf{C})$ of *cubical partial map*. The n-cubes are the n-cubical spans formed by the spans of $\mathbf{C}$ whose first arrow is a subobject, as shown below for $n = 1, 2$

$$(6.104)$$

The transversal maps are arbitrary morphisms of these cubical spans.

If $\mathbf{C}$ has all pullbacks, $\mathsf{Pmap}(\mathbf{C})$ has a transversally full pseudo embedding in $\mathsf{Span}(\mathbf{C})$.

(b) Similarly, if the category $\mathbf{C}$ has pullbacks of monos along monos (in $\mathbf{C}$), and a choice of subobjects stable under composition, there is a strict multiple category $\mathsf{Piso}(\mathbf{C})$ *of cubical partial isomorphisms.*

6.6.6 A premultiple category of cubical relations

One can form a structure RelSet of cubical relations, which is a flat *premultiple category with compositions*: it satisfies all the axioms of a strict symmetric cubical category *except interchange*, and has no interchanger, either of left- or right-chiral type. (This corrects an error in [G2].)

Nevertheless RelSet is of interest here, as it contains various more 'regular' substructures, like the multiple categories $\mathsf{Piso}(\mathbf{Set})$ and $\mathsf{Pmap}(\mathbf{Set})$ of 6.6.5. *RelSet should perhaps be reformulated in an 'unbiased' framework, with multiple operations instead of binary ones; this is far out of our current goals.*

The cubes of RelSet will be indexed by the three-element set $\{0, \iota, 1\}$ and its cartesian powers. A 1-cubical relation is an ordinary relation $a \colon a_0 \nrightarrow a_1$ of sets, viewed as a subset $a_\iota \subset a_0 \times a_1$; their composition will be written in additive notation.

A 2-cubical relation a is indexed by $\{0, \iota, 1\}^2$ and consists of (see the diagram below):

- a four-tuple of vertices $(a_{ij}) \colon 2 \times 2 \to \mathbf{Set}$ (where $2 \times 2 = \{0, 1\}^2$ is a discrete category on four objects),

- four (binary) relations on the sides of a square: $a_{\iota j} \subset a_{0j} \times a_{1j}$ and $a_{i\iota} \subset a_{i0} \times a_{i1}$,

- and one *quaternary relation* $a_{\iota\iota} \subset \prod a_{ij}$ whose projection on each side is contained in the corresponding binary relation

$$a_{00} \xrightarrow{a_{\iota 0}} a_{10}$$

$$\begin{matrix} a_{00} & \xrightarrow{a_{\iota 0}} & a_{10} \\ a_{0\iota} \downarrow & a_{\iota\iota} & \downarrow a_{1\iota} \\ a_{01} & \xrightarrow{a_{\iota 1}} & a_{11} \end{matrix} \qquad \begin{matrix} \bullet & \xrightarrow{1} & \\ & \downarrow 2 & \end{matrix}$$

$$(p_{0j}, p_{1j})(a_{\iota\iota}) \subset a_{\iota j}, \qquad (6.105)$$

$$(p_{i0}, p_{i1})(a_{\iota\iota}) \subset a_{i\iota},$$

where $p_{ij} \colon a_{00} \times a_{01} \times a_{10} \times a_{11} \to a_{ij}$ denotes a cartesian projection.

The 1-concatenation $c = a +_1 b$ is (of course) defined when the 2-cubes a, b are consecutive in direction 1, i.e. $a_{1\iota} = b_{0\iota}$, and is computed below, at the right

$$\begin{matrix} a_{00} & \xrightarrow{a_{\iota 0}} \bullet & \xrightarrow{b_{\iota 0}} b_{10} \\ a_{0\iota} \downarrow & a_{\iota\iota} & b_{\iota\iota} & \downarrow b_{1\iota} \\ a_{01} & \xrightarrow{a_{\iota 1}} \bullet & \xrightarrow{b_{\iota 1}} b_{11} \end{matrix} \quad = \quad \begin{matrix} a_{00} & \xrightarrow{c_{\iota 0}} b_{10} \\ a_{0\iota} \downarrow & c_{\iota\iota} & \downarrow b_{1\iota} \\ a_{01} & \xrightarrow{c_{\iota 1}} b_{11} \end{matrix} \qquad (6.106)$$

where the subset

$$c_{\iota\iota} = a_{\iota\iota} +_1 b_{\iota\iota} \subset a_{00} \times a_{01} \times b_{10} \times b_{11}, \qquad (6.107)$$

is an ordinary composition of relations, provided we view $a_{\iota\iota}$ and $b_{\iota\iota}$ as *binary* relations 'in direction 1'

$$a_{\iota\iota} \colon a_{00} \times a_{01} \dashrightarrow a_{10} \times a_{11}, \qquad b_{\iota\iota} \colon b_{00} \times b_{01} \dashrightarrow b_{10} \times b_{11}. \qquad (6.108)$$

This proves that 1-concatenation is strictly categorical, i.e. strictly associative, with strict units; the degeneracy $e_1(a)$ of an ordinary relation $a \colon a_0 \dashrightarrow a_1$ is

$$\begin{matrix} a_0 & \xrightarrow{\text{id}} & a_0 \\ a_\iota \downarrow & (e_1 a)_{\iota\iota} & \downarrow a_\iota \\ a_1 & \xrightarrow{\text{id}} & a_1 \end{matrix} \qquad (e_1 a)_{\iota\iota} = a_\iota \times a_\iota. \qquad (6.109)$$

2-concatenation can be defined in the symmetric way, or by the transposition s_1 (that permutes directions 1 and 2) and the previous operation:

$$a +_2 a' = s_1(s_1 a +_1 s_1 a'). \qquad (6.110)$$

We proceed analogously in higher dimension: an n-cube is a family $a = (a_t)$, indexed by the n-tuples $t = (t_1, ..., t_n) \in \{0, \iota, 1\}^n$ and satisfying the following conditions (i)–(iii).

(i) If $t \in 2^n$, then a_t is a set.

(ii) Otherwise let $w = |t|$ be its *weight*, defined as the number of $\iota's$ in the n-tuple; then a_t is a 2^w-ary relation

$$a_t \subset A_t = \prod a_{i_1...i_n}, \qquad (6.111)$$

the cartesian product being indexed by those n-tuples $i = (i_1, ..., i_n) \in 2^n$ where i_j coincides with t_j when the latter is 0 or 1 (there are 2^w such n-tuples).

(iii) Finally, if t' is a multi-index obtained by replacing *one* occurrence of ι in t with 0 or 1 (of weight $w' = w - 1$), the corresponding projection $p_{tt'}$ must send the 2^w-ary relation a_t into the $2^{w'}$-ary relation $a_{t'}$

$$p_{tt'} \colon A_t \to A_{t'}, \qquad p_{tt'}(a_t) \subset a_{t'}. \qquad (6.112)$$

We define now a transversal map $f \colon a \to b$ as a natural transformation on the discrete category 2^n

$$f = (f_i) \colon a \to b \colon 2^n \to \mathbf{Set}, \qquad f_i \colon a_i \to b_i \quad (i \in 2^n), \qquad (6.113)$$

which is 'coherent' with the 'multiple' relations inside a and b:

(*) for every n-tuple $t \in \{0, \iota, 1\}^n$, the mapping $f_t \colon A_t \to B_t$ defined by the cartesian product of the components f_i singled out in (6.111) carries the subset a_t into b_t.

Faces are defined using the cofaces

$$\partial_i^\alpha \colon \{0, \iota, 1\}^{n-1} \to \{0, \iota, 1\}^n,$$
$$\partial_i^\alpha(t_1, ..., t_{n-1}) = (t_1, ..., \alpha, t_i, ..., t_{n-1}) \qquad (\alpha = 0, 1). \qquad (6.114)$$

The n-map $f \colon a \to b$ is determined by its 0-faces a, b together with its 2^n total faces, so that the structure RelSet is flat, according to condition 6.4.9(ii).

The definition of degeneracies extends (6.109).

There is no interchanger, as we show below: already in dimension three, the truncated structure $\mathsf{Rel_3}(\mathbf{Set})$ of 2-cubical relations and their maps is neither left-chiral nor right-chiral.

6.6.7 *A counterexample*

The computations below will be used in various points. We begin by working in the weak triple category $\mathsf{Span_3}(\mathbf{Set})$, where we have an invertible interchanger χ. The cubes of $\mathsf{Rel_3}(\mathbf{Set})$ will be viewed *inside this structure*, as determined by jointly monic spans, and we write as $x \otimes_i y$ their i-concatenation, namely the 12-cubical relation associated to the 12-cubical span $x +_i y$ (for $i = 1, 2$).

A consistent matrix $\left(\begin{smallmatrix} x & y \\ z & u \end{smallmatrix}\right)$ of 12-cubes of $\mathsf{Rel}_3(\mathbf{Set})$ is shown in the left diagram below, where $0, 1, 2$ are cardinal sets and the mappings $1 \to 2 = \{0, 1\}$ are labelled by their image. First we compute the operations in $\mathsf{Span}_3(\mathbf{Set})$, where interchange works

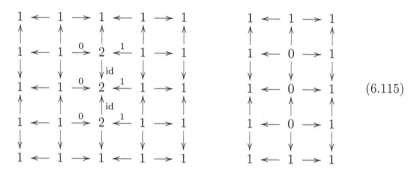

(6.115)

In the right diagram above we compute $a = x +_1 y$ and $b = z +_1 u$. In the left diagram below we have $c = x +_2 z$ and $d = y +_2 u$

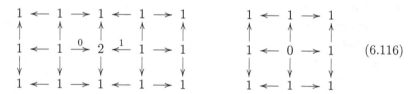

(6.116)

Finally the right diagram above is the 12-cube $a +_2 b = c +_1 d$.

Note now that a, b and $a +_2 b$ already are 12-cubical relations, while c and d (in the left diagram above) are not: the common vertical span of c and d is not jointly monic. The associated 12-cubical relations $c' = x \otimes_2 z$ and $d' = y \otimes_2 u$ are shown in the left diagram below, and their composition $c' \otimes_1 d'$ at the right

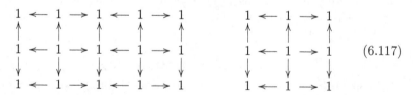

(6.117)

Therefore there is no transversal map $c' \otimes_1 d' \to a \otimes_2 b$ and it is not possible to define an interchanger

$$(x \otimes_2 z) \otimes_1 (y \otimes_2 u) \to (x \otimes_1 y) \otimes_2 (z \otimes_1 u).$$

The other way round there can be no general interchanger either, by symmetry.

6.7 Multiple tabulators and span representations

Tabulators in the strict triple category Stcⅅbl have already been examined in Section 6.1. We study here these important multiple limits, which often clarify the structure we are examining, and can lead to representing it by cubical spans. This extends the span representation of weak double categories, in Section 3.6.

We work in a chiral multiple category A. (The extension to intercategories will be briefly considered in Section 6.9.)

6.7.1 Tabulators

Let $\mathbf{i}$ be a positive multi-index and $\mathbf{j} \subset \mathbf{i}$, with complement $\mathbf{k} = \mathbf{i} \setminus \mathbf{j}$. Letting $\mathbf{j} = \{j_1, ..., j_r\}$, we consider the iterated degeneracy of A

$$e_{\mathbf{j}} \colon \mathrm{tv}_{\mathbf{k}}(\mathsf{A}) \to \mathrm{tv}_{\mathbf{i}}(\mathsf{A}), \qquad e_{\mathbf{j}} = e_{j_1} ... e_{j_r}. \qquad (6.118)$$

The $e_{\mathbf{j}}$-*tabulator* of an $\mathbf{i}$-cube $x \in A_{\mathbf{i}}$ is a $\mathbf{k}$-cube $T = \mathsf{T}_{\mathbf{j}} x \in A_{\mathbf{k}}$ equipped with a universal $\mathbf{i}$-map $t_x \colon e_{\mathbf{j}}(\mathsf{T}_{\mathbf{j}} x) \to_0 x$ from the functor $e_{\mathbf{j}}$ to the object x. The existence of all of them amounts to an ordinary functor $\mathsf{T}_{\mathbf{j}} \colon \mathrm{tv}_{\mathbf{i}}(\mathsf{A}) \to \mathrm{tv}_{\mathbf{k}}(\mathsf{A})$ right adjoint to $e_{\mathbf{j}}$.

The universal property of the pair $(\mathsf{T}_{\mathbf{j}} x, t_x)$ says that, for every $\mathbf{k}$-cube A and every $\mathbf{i}$-map $h \colon e_{\mathbf{j}}(A) \to_0 x$ there is a unique $\mathbf{k}$-map u such that

$$e_{\mathbf{j}}(A) \xrightarrow{e_{\mathbf{j}}(u)} e_{\mathbf{j}}(T) \qquad\qquad u \colon A \to_0 T,$$
$$\searrow^{h} \quad \downarrow^{t_x} \qquad\qquad\qquad\qquad (6.119)$$
$$x \qquad\qquad\qquad t_x . e_{\mathbf{j}}(u) = h.$$

An $\mathbf{i}$-cube x of degree n can thus have $\binom{n}{k}$ tabulators of degree k, and globally 2^n of them, including the trivial case $\mathsf{T}_{\emptyset} x = x$, the tabulator of degree n. On the other hand, the $e_{\mathbf{i}}$-tabulator $\mathsf{T} x = \mathsf{T}_{\mathbf{i}} x$ is an object of A (if it exists), and will be called a *total tabulator*, or a *tabulator of degree zero*.

We say that the chiral multiple category A *has all tabulators*, or *tabulators of all degrees*, if every $\mathbf{i}$-cube $x \in A_{\mathbf{i}}$ has all $e_{\mathbf{j}}$-tabulators $\mathsf{T}_{\mathbf{j}} x$ (for $\mathbf{j} \subset \mathbf{i}$). We say that A *has multiple tabulators* if it has tabulators of all degrees, preserved by faces and degeneracies.

In this case, if A is transversally invariant, one can always make a choice of multiple tabulators such that this preservation is strict (see a similar proof in 7.6.4):

$$\partial_i^\alpha(\mathsf{T_j}x) = \mathsf{T_j}(\partial_i^\alpha x), \qquad \mathsf{T_j}(e_i(y)) = e_i(\mathsf{T_j}(y)), \tag{6.120}$$

for $\mathbf{j} \subset \mathbf{i}$, $i \in \mathbf{i} \setminus \mathbf{j}$, $x \in A_\mathbf{i}$ and $y \in A_{\mathbf{i}|i}$.

Note that the conditions (6.120) are trivial if $\mathbf{j} = \emptyset$ or $\mathbf{j} = \mathbf{i}$, *whence for all weak double categories*, where there is only one positive index; this fact is commented in 6.7.8(d).

6.7.2 Lemma (Basic properties of tabulators)

Let A *be a chiral multiple category.*

(a) For an $\mathbf{i}$*-cube* x *and a disjoint union* $\mathbf{j} = \mathbf{j}' \cup \mathbf{j}'' \subset \mathbf{i}$ *we have*

$$\mathsf{T_j}x = \mathsf{T_{j''}}\mathsf{T_{j'}}(x), \tag{6.121}$$

provided that $\mathsf{T_{j'}}(x)$ *and* $\mathsf{T_{j''}}\mathsf{T_{j'}}(x)$ *exist.*

(b) A *has all tabulators if and only it has all* elementary tabulators $\mathsf{T}_i x$ *(for every positive multi-index* $\mathbf{i}$*, every* $i \in \mathbf{i}$ *and every* $\mathbf{i}$*-cube* x*).*

(c) For a fixed $i \in \mathbf{i}$*, the existence in* A *of all* e_i*-tabulators of* $\mathbf{i}$*-cubes amounts to an ordinary adjunction*

$$e_i \colon \mathrm{tv}_{\mathbf{i}|i}A \rightleftarrows \mathrm{tv}_\mathbf{i}A \colon \mathsf{T}_i, \qquad e_i \dashv \mathsf{T}_i. \tag{6.122}$$

Then the degeneracy $e_i \colon \mathrm{tv}_{\mathbf{i}|i}A \to \mathrm{tv}_\mathbf{i}A$ *preserves colimits.*

Proof Obvious, composing universal arrows of ordinary functors. □

6.7.3 Exercises and complements, I

Prove that the chiral triple category $\mathsf{SC}(\mathbf{C})$, over a category $\mathbf{C}$ with pullbacks and pushouts (see Section 6.6), has multiple tabulators and cotabulators.

6.7.4 Exercises and complements, II

(a) Study tabulators and cotabulators in the weak multiple categories $\mathsf{Cub}(\mathbf{C})$, $\mathsf{Cosp}(\mathbf{C})$, $\mathsf{Span}(\mathbf{C})$ and $\mathsf{Bisp}(\mathbf{C})$ of 6.3.3 and 6.4.6.

(b) Same exercises in the chiral multiple categories $\mathsf{S}_p\mathsf{C}_q(\mathbf{C})$ and $\mathsf{S}_{-\infty}\mathsf{C}_\infty(\mathbf{C})$ of 6.6.4.

(c) Note that all these structures are transversally invariant – a property whose interest was remarked in 6.7.1.

6.7.5 Maps of tabulators

We now examine the relationship of tabulators of $\mathbf{i}$-cubes with faces, degeneracies and concatenation, in direction $i \in \mathbf{i}$.

(a) If the cube $x \in A_{\mathbf{i}}$ and its face $z = \partial_i^\alpha x$ have total tabulators, the projection $p = p_i^\alpha x$ of $\mathsf{T}x$ $(= \mathsf{T}_i x)$ will be the $*$-map of A determined by the following condition

$$e_{\mathbf{i}|i}\mathsf{T}x \xrightarrow{e_{\mathbf{i}|i}(p)} e_{\mathbf{i}|i}\mathsf{T}(\partial_i^\alpha x) \qquad p_i^\alpha x\colon \mathsf{T}x \to_0 \mathsf{T}(\partial_i^\alpha x),$$

$$\searrow_{\partial_i^\alpha(t_x)} \quad \downarrow_{t_z} \tag{6.123}$$

$$z = \partial_i^\alpha x \qquad\qquad t_z.e_{\mathbf{i}|i}(p_i^\alpha x) = \partial_i^\alpha(t_x).$$

(b) If the degenerate $\mathbf{i}$-cube $x = e_i z$ and the $\mathbf{i}|i$-cube z have total tabulators in A, they are linked by a *diagonal* transversal $*$-map $d_i z$, defined as follows

$$e_{\mathbf{i}}(\mathsf{T}z) \xrightarrow{e_{\mathbf{i}}(d_i z)} e_{\mathbf{i}}(\mathsf{T}(e_i z)) \qquad d_i z\colon \mathsf{T}z \to_0 \mathsf{T}(e_i z),$$

$$\searrow_{e_i t_z} \quad \downarrow_{t_x} \tag{6.124}$$

$$x = e_i z \qquad\qquad t_x.e_{\mathbf{i}}(d_i z) = e_{\mathbf{i}}(t_z).$$

This $*$-map $d_i z$ is a section of both projections $p_i^\alpha x$ (defined above) because

$$t_z.e_{\mathbf{i}|i}(p_i^\alpha x.d_i z) = \partial_i^\alpha(t_x).e_{\mathbf{i}|i}(d_i z) = \partial_i^\alpha(t_x.e_{\mathbf{i}}(d_i z)) = \partial_i^\alpha(e_{\mathbf{i}}(t_z)) = t_z.$$

(c) For a concatenation $z = x +_i y$ of $\mathbf{i}$-cubes, the three total tabulators of x, y, z are also related. The link goes through the ordinary pullback $\mathsf{T}_i(x, y)$ of the objects $\mathsf{T}x$ and $\mathsf{T}y$, over the tabulator $\mathsf{T}w$ of the $\mathbf{i}|i$-cube $w = \partial_i^+ x = \partial_i^- y$ (provided all these tabulators and this pullback exist)

$$
\begin{array}{c}
\mathsf{T}x \\
p_i(x,y) \nearrow \qquad \searrow p_i^+ x \\
\mathsf{T}_i(x, y) \quad \big) \qquad \mathsf{T}w \\
q_i(x,y) \searrow \qquad \nearrow p_i^- x \\
\mathsf{T}y
\end{array}
\qquad
\begin{array}{c}
t_w.e_{\mathbf{i}|i}(p_i^+ x) = \partial_i^+(t_x), \\[2mm]
\\
t_w.e_{\mathbf{i}|i}(p_i^- y) = \partial_i^-(t_y).
\end{array}
\tag{6.125}
$$

We now have a *diagonal* transversal $*$-map $d_i(x, y)$ given by the universal property of $\mathsf{T}z$

$$d_i(x, y)\colon \mathsf{T}_i(x, y) \to_0 \mathsf{T}z,$$

$$t_z.e_{\mathbf{i}}(d_i(x, y)) = t_x.e_{\mathbf{i}}p_i(x, y) +_i t_y.e_{\mathbf{i}}q_i(x, y). \tag{6.126}$$

The i-composition above is legitimate, by construction

$$\partial_i^+(t_x.e_ip_i(x,y)) = \partial_i^+(t_x).e_{i|i}p_i(x,y)$$
$$= t_w.e_{i|i}(p_i^+ x).e_{i|i}p_i(x,y) = t_w.e_{i|i}(p_i^- y).e_{i|i}q_i(x,y)$$
$$= \partial_i^-(t_y).e_{i|i}q_i(x,y) = \partial_i^-(t_y.e_iq_i(x,y)).$$

Defining general multiple limits as in [GP9], $\mathsf{T}_i(x,y)$ is the limit of the diagram formed by the i-consecutive $\mathbf{i}$-cubes x, y.

6.7.6 Theorem (Tabulators and cubical spans)

Let A *be a chiral multiple category, and suppose that:*

(a) A *has all total tabulators,*

(b) the ordinary category $\mathbf{C} = \mathrm{tv}_* A$ *of objects and transversal arrows has pullbacks.*

Then there is a canonical lax functor

$$T\colon A \to \mathrm{Span}(\mathbf{C}), \qquad \mathrm{tv}_*(T) = \mathrm{id}\mathbf{C}, \qquad (6.127)$$

which is trivial in degree zero and takes:

(i) an $\mathbf{i}$-cube x of A *of degree n to an n-cubical span $Tx\colon \vee^n \to \mathbf{C}$ whose central object is the total tabulator T_ix,*

(ii) an n-map $f\colon x \to y$ in A *to an n-map $Tf\colon Tx \to Ty$ whose central component is T_if.*

Proof For $n = 0$ we take $Tx = \mathsf{T}x = x$. Having defined T up to degree $n-1$, the faces of Tx are determined: $\partial_i^\alpha(Tx) = T(\partial_i^\alpha x)$. To complete the n-cubical span Tx we add the central object T_ix equipped with a family of $\mathbf{C}$-maps (p_i^α) defined as in (6.123)

$$p_i^\alpha\colon \mathsf{T}_ix \to \mathsf{T}_{i|i}(z),$$
$$t_z.e_{i|i}(p_i^\alpha x) = \partial_i^\alpha t_x\colon e_{i|i}\mathsf{T}_ix \to z \qquad (z = \partial_i^\alpha x). \qquad (6.128)$$

In the 1-dimensional case, $t_z = 1_z$ and the maps $p_i^\alpha = \partial_i^\alpha t_x$ are the two projections of the tabulator T_ix of the i-arrow x (as in 3.6.4)

$$\partial_i^- x \xleftarrow{\;p_i^-\;} \mathsf{T}_ix \xrightarrow{\;p_i^+\;} \partial_i^+ x.$$

The 2-dimensional case is shown in this diagram

$$
\begin{array}{ccccc}
\partial_{ij}^{--}x & \xleftarrow{\ p_i^-\ } & \mathsf{T}_i\partial_j^- x & \xrightarrow{\ p_i^+\ } & \partial_{ij}^{+-}x \\[4pt]
{\scriptstyle p_j^-}\big\uparrow & & {\scriptstyle p_j^-}\big\uparrow & & {\scriptstyle p_j^-}\big\uparrow \\[4pt]
\mathsf{T}_j\partial_i^- x & \xleftarrow{\ p_i^-\ } & \mathsf{T}_{ij}x & \xrightarrow{\ p_i^+\ } & \mathsf{T}_j\partial_i^+ x \\[4pt]
{\scriptstyle p_j^+}\big\downarrow & & {\scriptstyle p_j^+}\big\downarrow & & {\scriptstyle p_j^+}\big\downarrow \\[4pt]
\partial_{ij}^{-+}x & \xleftarrow[\ p_i^-\ ]{} & \mathsf{T}_i\partial_j^+ x & \xrightarrow[\ p_i^+\ ]{} & \partial_{ij}^{++}x
\end{array}
\qquad (6.129)
$$

The diagram commutes (and is a 2-cubical span), because

$$
p_j^\beta p_i^\alpha = \partial_j^\beta t_z . p_i^\alpha = \partial_j^\beta (t_z . e_j(p_i^\alpha)) = \partial_j^\beta \partial_i^\alpha (t_x)
$$
$$
= \partial_i^\alpha \partial_j^\beta (t_x) = \ldots = p_i^\alpha p_j^\beta.
$$

Similarly, all the n-cubical spans Tx are well defined.

Using inductively the ordinary functors $\mathsf{T}_i \colon \mathbf{A}_i \to \mathbf{C}$, there is precisely one extension of T to transversal maps that satisfies condition (ii). The comparisons of T are then constructed by an inductive procedure that extends what we have seen in Theorem 3.6.4 for weak double categories. $\qquad\square$

6.7.7 Representability

Let A be a chiral multiple category.

(a) We say that A is *span representable*, or *representable by cubical spans*, if it has total tabulators, the ordinary category $\mathbf{C} = \mathrm{tv}_*(\mathsf{A})$ has pullbacks and moreover the lax functor $T \colon \mathsf{A} \to \mathrm{Span}(\mathbf{C})$ constructed above is transversally faithful.

(a*) By transversal duality, assuming that A has total cotabulators and $\mathbf{C} = \mathrm{tv}_*(\mathsf{A})$ has pushouts, we form a canonical colax functor

$$
C \colon \mathsf{A} \to \mathrm{Cosp}(\mathbf{C}), \qquad \mathrm{tv}_*(C) = \mathrm{id}\,\mathbf{C}, \qquad (6.130)
$$

that takes an $\mathbf{i}$-cube x of A of degree n to an n-cubical cospan $Cx \colon \wedge^n \to \mathbf{C}$ whose central object is the total cotabulator $\perp_{\mathbf{i}} x$.

In this situation we say that A is *cospan representable* if this colax functor is transversally faithful.

6.7.8 Exercises and complements

(a) Examine the span representability of $\mathrm{Span}(\mathbf{C})$.

(b) Same question for the cubical categories Pmap($\mathbf{C}$) and Piso($\mathbf{C}$) of 6.6.5.

(c) Same question for the chiral triple category SC($\mathbf{Set}$).

(d) Let $\mathbb{A}$ be a weak double category and let A be the associated weak multiple category $sk_2(\mathbb{A})$, obtained by adding degenerate items of all the missing types (see 6.2.8).

In A the only non-trivial tabulators are the total tabulators of 1-cubes, i.e. vertical arrows; we have already remarked in 6.7.1 that they are bound by no conditions of coherence with faces and degeneracies. Therefore the weak multiple category A has multiple tabulators if and only if $\mathbb{A}$ has 1-dimensional tabulators. We are not considering in A any higher-dimensional universal property.

6.8 A sketch of intercategories

Three-dimensional *intercategories*, introduced and studied in [GP6, GP7], generalise the notion of chiral triple category by replacing all *strict* or *weak* interchangers with *lax interchangers* of four types $(\tau, \mu, \delta, \chi)$, which deal with the four possible cases of zero-ary or binary composition in the *positive* directions 1, 2.

The difference can be better appreciated noting that a 3-dimensional intercategory is a pseudo category in the 2-category of weak double categories, lax double functors and horizontal transformations (see [GP6], Section 2), while a chiral triple category is a *unitary* pseudo category in the 2-category of weak double categories, *unitary* lax double functors and horizontal transformations.

Three-dimensional intercategories are extensively studied in [GP7], showing that this notion encompasses a wide variety of structures from the literature, like duoidal categories [AM, BhCZ, BkS], monoidal double categories [Shu], cubical bicategories [GaG], Verity's double bicategories [Ve] and Gray categories [Gra1]. The general framework, besides providing an effective unification of these three-dimensional structures, makes also possible to consider morphisms between them and study how they relate to each other.

Here we present infinite-dimensional intercategories, introduced in [GP8], and give a short synopsis of the results recalled above. Other examples of 3-dimensional intercategories can be found in Section 6.9 and [GP7].

6.8.1 Intercategories

An (infinite-dimensional, right) *intercategory* is a kind of lax multiple category more general than a chiral multiple category: with respect to the latter we replace the three strict interchangers listed in (6.70) with lax interchangers.

For any two *positive* directions $i < j$ we have thus the following families of ij-special transversal maps (including χ_{ij}, already present in the chiral case):

(a) $\tau_{ij}(x)$: $e_j e_i(x) \to_0 e_i e_j(x)$ (*zeroary ij-interchanger*),

(b) $\mu_{ij}(x, y)$: $e_i(x) +_j e_i(y) \to_0 e_i(x +_j y)$
$\qquad\qquad\qquad$ (ij-*interchanger for i-identities and j-composition*),

(c) $\delta_{ij}(x, y)$: $e_j(x +_i y) \to_0 e_j(x) +_i e_j(y)$
$\qquad\qquad\qquad$ (ij-*interchanger for i-composition and j-identities*),

(d) $\chi_{ij}(x, y, z, u)$: $(x +_i y) +_j (z +_i u) \to_0 (x +_j z) +_i (y +_j u)$
$\qquad\qquad\qquad$ (*binary ij-interchanger*),

where all compositions are assumed to be legitimate.

All these maps must be natural for transversal maps and commute with the faces ∂_k^α, for $k \neq i, j$. The coherence axioms stated in 6.4.3 are required. Furthermore there are coherence conditions for the interchangers, stated below in 6.8.2 and 6.8.3.

The transversally dual notion of a *left intercategory* has interchangers in the opposite direction.

With respect to the chiral case, let us note that an intercategory with $e_i e_j \neq e_j e_i$ is no longer a multiple set. Moreover a degeneracy e_i (for $i > 0$) is now *colax* with respect to every lower j-composition (for $0 < j < i$, via τ_{ji} and δ_{ji}) and *lax* with respect to every higher j-composition (for $j > i$, via τ_{ij} and μ_{ij}).

6.8.2 Lower coherence axioms for the interchangers

We list here and in 6.8.3 the conditions involving the interchangers.

The following diagrams of transversal maps must commute for $0 < i < j$, assuming that all compositions are legitimate. We let $\tau = \tau_{ij}$, $\mu = \mu_{ij}$, $\delta = \delta_{ij}$, $\chi = \chi_{ij}$.

(i) *Coherence hexagon of $\chi = \chi_{ij}$ and κ_i:* see (wmc.9.iii) in 6.4.4.

(ii) *Coherence hexagon of $\chi = \chi_{ij}$ and κ_j:* see (wmc.9.iv).

(iii) *Coherence hexagon of* $\delta = \delta_{ij}$ *and* κ_i

$$
\begin{array}{ccc}
e_j(x +_i (y +_i z)) & \xrightarrow{\;e_j\kappa_i\;} & e_j((x +_i y) +_i z) \\
{\scriptstyle \delta}\downarrow & & \downarrow{\scriptstyle \delta} \\
e_j(x) +_i e_j(y +_i z) & & e_j(x +_i y) +_i e_j(z) \\
{\scriptstyle 1 +_i \delta}\downarrow & & \downarrow{\scriptstyle \delta +_i 1} \\
e_j(x) +_i (e_j(y) +_i e_j(z)) & \xrightarrow[\;\kappa_i e_j\;]{} & (e_j(x) +_i e_j(y)) +_i e_j(z)
\end{array}
$$

(iv) *Coherence hexagon of* $\mu = \mu_{ij}$ *and* κ_j

$$
\begin{array}{ccc}
e_i(x) +_j (e_i(y) +_j e_i(z)) & \xrightarrow{\;\kappa_j e_i\;} & (e_i(x) +_j e_i(y)) +_j e_i(z) \\
{\scriptstyle 1 +_j \mu}\downarrow & & \downarrow{\scriptstyle \mu +_j 1} \\
e_i(x) +_j e_i(y +_j z) & & e_i(x +_j y) +_j e_i(z) \\
{\scriptstyle \mu}\downarrow & & \downarrow{\scriptstyle \mu} \\
e_i(x +_j (y +_j z)) & \xrightarrow[\;e_i\kappa_j\;]{} & e_i((x +_j y) +_j z)
\end{array}
$$

(v) *Coherence laws of* χ, μ *with* λ_i *and* ρ_i

$$
\begin{array}{ccccc}
(e_i\partial_i^- x +_i x) +_j (e_i\partial_i^- y +_i y) & \xrightarrow{\;\lambda_i +_j \lambda_i\;} x +_j y \xleftarrow{\;\rho_i +_j \rho_i\;} & (x +_i e_i\partial_i^+ x) +_j (y +_i e_i\partial_i^+ y) \\
{\scriptstyle \chi}\downarrow & \Big\| & \downarrow{\scriptstyle \chi} \\
(e_i\partial_i^- x +_j e_i\partial_i^- y) +_i (x +_j y) & & (x +_j y) +_i (e_i\partial_i^+ x +_j e_i\partial_i^+ y) \\
{\scriptstyle \mu +_i 1}\downarrow & & \downarrow{\scriptstyle 1 +_i \mu} \\
e_i\partial_i^- (x +_j y) +_i (x +_j y) & \xrightarrow[\;\lambda_i\;]{} x +_j y \xleftarrow[\;\rho_i\;]{} & (x +_j y) +_i e_i\partial_i^+ (x +_j y)
\end{array}
$$

(vi) *Coherence laws of* χ, δ *with* λ_j *and* ρ_j

$$
\begin{array}{ccccc}
e_j\partial_j^- (x +_i y) +_j (x +_i y) & \xrightarrow{\;\lambda_j\;} x +_i y \xleftarrow{\;\rho_j\;} & (x +_i y) +_j e_j\partial_j^+ (x +_i y) \\
{\scriptstyle \delta +_j 1}\downarrow & \Big\| & \downarrow{\scriptstyle 1 +_j \delta} \\
(e_j\partial_j^- x +_i e_j\partial_j^- y) +_j (x +_i y) & & (x +_i y) +_j (e_j\partial_j^+ x +_i e_j\partial_j^+ y) \\
{\scriptstyle \chi}\downarrow & & \downarrow{\scriptstyle \chi} \\
(e_j\partial_j^- x +_j x) +_i (e_j\partial_j^- y +_j y) & \xrightarrow[\;\lambda_j +_i \lambda_j\;]{} x +_i y \xleftarrow[\;\rho_j +_i \rho_j\;]{} & (x +_j e_j\partial_j^+ x) +_i (y +_j e_j\partial_j^+ y)
\end{array}
$$

(vii) *Coherence laws of* δ, τ *with* λ_i *and* ρ_i

$$
\begin{array}{ccccc}
e_j((e_i\partial_i^- x) +_i x) & \xrightarrow{\;e_j\lambda_i\;} e_j(x) \xleftarrow{\;e_j\rho_i\;} & e_j(x +_i (e_i\partial_i^- x)) \\
{\scriptstyle \delta}\downarrow & \Big\| & \downarrow{\scriptstyle \delta} \\
e_j e_i\partial_i^- (x) +_i e_j(x) & & e_j(x) +_i e_j e_i\partial_i^- (x) \\
{\scriptstyle \tau +_i 1}\downarrow & & \downarrow{\scriptstyle 1 +_i \tau} \\
e_i\partial_i^- e_j(x) +_i e_j(x) & \xrightarrow[\;\lambda_i e_j\;]{} e_j(x) \xleftarrow[\;\rho_i e_j\;]{} & e_j(x) +_i e_i\partial_i^- e_j(x)
\end{array}
$$

(viii) *Coherence laws of μ, τ with λ_j and ρ_j*

$$
\begin{array}{ccccc}
e_j \partial_j^- e_i(x) +_j e_i(x) & \xrightarrow{\lambda_j e_i} & e_i(x) & \xleftarrow{\rho_j e_i} & e_i(x) +_j e_j \partial_j^- e_i(x) \\[2pt]
\tau +_j 1 \downarrow & & \Big\| & & \downarrow 1 +_j \tau \\[2pt]
e_i e_j \partial_j^-(x) +_j e_i(x) & & \Big\| & & e_i(x) +_j e_i e_j \partial_j^-(x) \\[2pt]
\mu \downarrow & & \Big\| & & \downarrow \mu \\[2pt]
e_i((e_j \partial_j^- x) +_j x) & \xrightarrow{e_i \lambda_j} & e_i(x) & \xleftarrow{e_i \rho_j} & e_i(x +_j (e_j \partial_j^- x))
\end{array}
$$

6.8.3 Higher coherence axioms

Finally we list the coherence conditions involving three interchangers and three positive directions $i < j < k$ at a time. These axioms vanish in the 3-dimensional case, where we only have two positive indices. The first condition is axiom (wmc.9.vii) of weak (and chiral) multiple categories, in 6.4.4; here it forms a guideline for the others.

The following diagrams of transversal maps must commute, assuming that all the positive compositions make sense.

(i) (Case 2×2×2) *Coherence hexagon of the interchangers χ_{ij}, χ_{jk} and χ_{ik},* for a 2×2×2 matrix of **i**-cubes

$$
\begin{array}{c}
((x +_i y) +_j (z +_i u)) +_k ((x' +_i y') +_j (z' +_i u')) \\[2pt]
\chi_{jk} \downarrow \qquad\qquad\qquad\qquad\qquad\qquad \Big| \\[2pt]
((x +_i y) +_k (x' +_i y')) +_j ((z +_i u) +_k (z' +_i u')) \quad \Big| \; \chi_{ij} +_k \chi_{ij} \\[2pt]
\\[2pt]
\chi_{ik} +_j \chi_{ik} \Big| \qquad ((x +_j z) +_i (y +_j u)) +_k ((x' +_j z') +_i (y' +_j u')) \\[2pt]
\\[2pt]
((x +_k x') +_i (y +_k y')) +_j ((z +_k z') +_i (u +_k u')) \quad \Big| \chi_{ik} \\[2pt]
\\[2pt]
\chi_{ij} \Big| \qquad ((x +_j z) +_k (x' +_j z')) +_i ((y +_j u) +_k (y' +_j u')) \\[2pt]
\qquad\qquad\qquad\qquad\qquad\qquad \downarrow \chi_{jk} +_i \chi_{jk} \\[2pt]
((x +_k x') +_j (z +_k z')) +_i ((y +_k y') +_j (u +_k u'))
\end{array}
$$

(ii) (Case 0×2×2) *Coherence hexagon of the interchangers μ_{ij}, χ_{jk}, and*

μ_{ik}, for a 2×2 matrix of $\mathbf{i}|i$-cubes

$$
\begin{array}{ccc}
(e_i x +_j e_i y) +_k (e_i z +_j e_i u) & \xrightarrow{\chi_{jk} e_i} & (e_i x +_k e_i z) +_j (e_i y +_k e_i u) \\
{\scriptstyle \mu_{ij} +_k \mu_{ij}} \downarrow & & \downarrow {\scriptstyle \mu_{ik} +_j \mu_{ik}} \\
e_i(x +_j y) +_k e_i(z +_j u) & & e_i(x +_k z) +_j e_i(y +_k u) \\
{\scriptstyle \mu_{ik}} \downarrow & & \downarrow {\scriptstyle \mu_{ij}} \\
e_i((x +_j y) +_k (z +_j u)) & \xrightarrow{e_i \chi_{jk}} & e_i((x +_k z) +_j (y +_k u))
\end{array}
$$

(iii) (Case $2\times 0\times 2$) *Coherence hexagon of the interchangers* δ_{ij}, μ_{jk} *and* χ_{ik}, *for a* 2×2 *matrix of* $\mathbf{i}|j$-*cubes*

$$
\begin{array}{ccc}
e_j(x +_i y) +_k e_j(z +_i u) & \xrightarrow{\delta_{ij} +_k \delta_{ij}} & (e_j x +_i e_j y) +_k (e_j z +_i e_j u) \\
{\scriptstyle \mu_{jk}} \downarrow & & \downarrow {\scriptstyle \chi_{ik} e_j} \\
e_j((x +_i y) +_k (z +_i u)) & & (e_j x +_k e_j z) +_i (e_j y +_k e_j u) \\
{\scriptstyle e_j \chi_{ik}} \downarrow & & \downarrow {\scriptstyle \mu_{jk} +_i \mu_{jk}} \\
e_j((x +_k z) +_i (y +_j u)) & \xrightarrow{\delta_{ij}} & e_j(x +_k z) +_i e_j(y +_j u)
\end{array}
$$

(iv) (Case $2\times 2\times 0$) *Coherence hexagon of the interchangers* χ_{ij}, δ_{jk} *and* δ_{ik} *for a* 2×2 *matrix of* $\mathbf{i}|k$-*cubes*

$$
\begin{array}{ccc}
e_k((x +_i y) +_j (z +_i u)) & \xrightarrow{e_k \chi_{ij}} & e_k((x +_j z) +_i (y +_j u)) \\
{\scriptstyle \delta_{jk}} \downarrow & & \downarrow {\scriptstyle \delta_{ik}} \\
e_k(x +_i y) +_j e_k(z +_i u) & & e_k(x +_j z) +_i e_k(y +_j u) \\
{\scriptstyle \delta_{ik} +_j \delta_{ik}} \downarrow & & \downarrow {\scriptstyle \delta_{jk} +_i \delta_{jk}} \\
(e_k x +_i e_k y) +_j (e_k z +_i e_k u) & \xrightarrow{\chi_{ij} e_k} & (e_k x +_j e_k z) +_i (e_k y +_j e_k u)
\end{array}
$$

(v) (Case $0\times 0\times 2$) *Coherence hexagon of the interchangers* τ_{ij}, μ_{jk} *and* μ_{ik} *for a pair of cubes indexed by* $\mathbf{i} \setminus \{i, j\}$

$$
\begin{array}{ccc}
e_j e_i x +_k e_j e_i y & \xrightarrow{\tau_{ij} +_k \tau_{ij}} & e_i e_j x +_k e_i e_j y \\
{\scriptstyle \mu_{jk} e_i} \downarrow & & \downarrow {\scriptstyle \mu_{ik} e_j} \\
e_j(e_i x +_k e_i y) & & e_i(e_j x +_k e_j y) \\
{\scriptstyle e_j \mu_{ik}} \downarrow & & \downarrow {\scriptstyle e_i \mu_{jk}} \\
e_j e_i(x +_k y) & \xrightarrow{\tau_{ij}} & e_i e_j(x +_k y)
\end{array}
$$

(vi) (Case 0×2×0) *Coherence hexagon of the interchangers* μ_{ij}, δ_{jk} *and* τ_{ik}

$$
\begin{array}{ccc}
e_k(e_i x +_j e_i y) & \xrightarrow{\;\delta_{jk} e_i\;} & e_k e_i x +_j e_k e_i y \\
{\scriptstyle e_k \mu_{ij}} \downarrow & & \downarrow {\scriptstyle \tau_{ik} +_j \tau_{ik}} \\
e_k e_i(x +_j y) & & e_i e_k x +_j e_i e_k y \\
{\scriptstyle \tau_{ik}} \downarrow & & \downarrow {\scriptstyle \mu_{ij} e_k} \\
e_i e_k(x +_j y) & \xrightarrow{\;e_i \delta_{jk}\;} & e_i(e_k x +_j e_k y)
\end{array}
$$

(vii) (Case 2×0×0) *Coherence hexagon of the interchangers* δ_{ij}, τ_{jk} *and* δ_{ik}

$$
\begin{array}{ccc}
e_k e_j(x +_i y) & \xrightarrow{\;\tau_{jk}\;} & e_j e_k(x +_i y) \\
{\scriptstyle e_k \delta_{ij}} \downarrow & & \downarrow {\scriptstyle e_j \delta_{ik}} \\
e_k(e_j x +_i e_j y) & & e_j(e_k x +_i e_k y) \\
{\scriptstyle \delta_{ik} e_j} \downarrow & & \downarrow {\scriptstyle \delta_{ij} e_k} \\
e_k e_j x +_i e_k e_j y & \xrightarrow{\;\tau_{jk} +_i \tau_{jk}\;} & e_j e_k x +_i e_j e_k y
\end{array}
$$

(viii) (Case 0×0×0) *Coherence hexagon of the interchangers* τ_{ij}, τ_{jk} *and* τ_{ik}

$$
\begin{array}{ccc}
e_k e_j e_i x & \xrightarrow{\;\tau_{jk} e_i\;} & e_j e_k e_i x \\
{\scriptstyle e_k \tau_{ij}} \downarrow & & \downarrow {\scriptstyle e_j \tau_{ik}} \\
e_k e_i e_j x & & e_j e_i e_k x \\
{\scriptstyle \tau_{ik} e_j} \downarrow & & \downarrow {\scriptstyle \tau_{ij} e_k} \\
e_i e_k e_j x & \xrightarrow{\;e_i \tau_{jk}\;} & e_i e_j e_k x
\end{array}
$$

(The latter condition can be viewed as a Moore relation for transpositions, in the symmetric group S_3; see (B.26).)

6.8.4 Duoidal categories as intercategories

We begin now to show how intercategories encompass a wide variety of three-dimensional categorical structures. Full details and proofs for this part can be found in [GP7].

Loosely speaking, a *duoidal category* A is a category equipped with two tensor products related by interchange morphisms; it can be defined as a 'pseudo monoid' in the 2-category of monoidal categories and lax monoidal functors. The interchangers express the fact that the second tensor is a lax functor with respect to the first, or equivalently that the first tensor is colax with respect to the second (as in the structure of a bialgebra).

More explicitly, A is a category equipped with two monoidal structures $(+_i, E_i, \kappa_i, \lambda_i, \rho_i)$, for $i = 1, 2$, that are linked by four 12-interchangers (for A, B, C, D objects of A)

(a) $\tau \colon E_2 \to E_1$ (*zeroary interchanger*),

(b) $\mu \colon E_1 +_2 E_1 \to E_1$ (*interchanger for 1-identities and 2-product*),

(c) $\delta \colon E_2 \to E_2 +_1 E_2$ (*interchanger for 1-product and 2-identities*),

(d) $\chi(A, B, C, D) \colon (A +_1 B) +_2 (C +_1 D) \to (A +_2 C) +_1 (B +_2 D)$
                (*binary interchanger*).

These interchangers must be natural and satisfy the coherence axioms 6.8.2, for $i = 1, j = 2$.

This is the same as a three-dimensional intercategory on a single object (written as a dot, and called the *vertex*), with trivial arrows in all directions and trivial 01- and 02-cells. Its 12-cubes and 12-maps are the objects and morphisms of A, respectively. The 12-map corresponding to the morphism $f \colon A \to B$ will thus inhabit the following 'general' cube:

$$(6.131)$$

Let us note that there is precisely one 1-arrow and one 2-arrow, namely $e_1(\bullet)$ and $e_2(\bullet)$, which in the diagrams above are only distinguished by their direction. There are two degenerate 12-cubes, namely $e_1 e_2(\bullet)$ and $e_2 e_1(\bullet)$, which respectively correspond to the objects E_1 and E_2 of **C** (possibly equal).

6.8.5 Some examples

(a) A category **C** with finite products and sums has a structure of duoidal category $\mathsf{Fps}(\mathbf{C}) = (\mathbf{C}, \times, \top, +, \bot)$ given by (a choice of) these operations, in this order. This gives an intercategory whose only non trivial cells are the 12-cubes and their transversal maps, respectively corresponding to the objects and morphisms of **C**.

The interchangers τ, μ, δ are the following morphisms of **C**, uniquely determined

$$\tau \colon \bot \to \top, \qquad \mu \colon \top + \top \to \top, \qquad \delta \colon \bot \to \bot \times \bot, \qquad (6.132)$$

and χ has the following components as a morphism with values in a product

$(p_i, q_i$ denote cartesian projections)

$$\chi_{ABCD} \colon (A \times B) + (C \times D) \to (A + C) \times (B + D),$$
$$p_1 + q_1 \colon (A \times B) + (C \times D) \to A + C, \tag{6.133}$$
$$p_2 + q_2 \colon (A \times B) + (C \times D) \to B + D.$$

The morphism $\tau \colon \bot \to \top$ is invertible if and only if $\mathbf{C}$ has a zero object, a case examined below.

We can also note that in the (non-chiral) duoidal category $\mathsf{Fps}(\mathbf{Set})$ the interchanger $\delta \colon \emptyset \to \emptyset \times \emptyset$ is trivial while the other interchangers are not invertible: $\tau \colon \emptyset \to 1$, $\mu \colon 1 + 1 \to 1$, and χ.

In $\mathsf{Fps}(\mathbf{Set}^{\mathrm{op}})$ the interchanger $\delta \colon \bot \twoheadrightarrow \bot \times \bot$ is the previous (non invertible) mapping $\mu \colon 1 + 1 \to 1$. In $\mathsf{Fps}(\mathbf{Set} \times \mathbf{Set}^{\mathrm{op}})$ all the four interchangers are non invertible.

Assuming that $\mathbf{C}$ has finite limits and colimits, there are a lax and a colax inclusion $F, G \colon \mathsf{Fps}(\mathbf{C}) \to \mathsf{SC}(\mathbf{C})$ of intercategories (see Section 6.6). F (resp. G) takes the 12-cube at A, displayed in the left diagram below, to the span-cospan cube represented in the central (resp. right) diagram

$$F(A) \qquad\qquad G(A) \tag{6.134}$$

Furthermore F is horizontally pseudo, while G is vertically pseudo: both are thus *colax-lax* functors (see 6.5.7).

If $\mathbf{C}$ has a zero object $\top = 0 = \bot$, these inclusions coincide and $\mathsf{Fps}(\mathbf{C})$ becomes a substructure of the chiral triple category $\mathsf{SC}(\mathbf{C})$; in this case $\mathsf{Fps}(\mathbf{C})$ is chiral itself: τ, μ, δ are identities. If $\mathbf{C}$ is semiadditive, finite products and sums coincide and we simply have a monoidal structure on $\mathbf{C}$.

Tabulators and cotabulators of $\mathsf{Fps}(\mathbf{C})$ as an intercategory are examined in 6.9.4.

(b) More generally a monoidal category $(\mathbf{C}, \otimes, E)$ with finite products has a structure of duoidal category $\mathsf{Fpt}(\mathbf{C}) = (\mathbf{C}, \times, \top, \otimes, E)$, as in Example 6.19 of [AM].

Here the comparisons $\tau \colon E \to \top$ and $\mu \colon \top \otimes \top \to \top$ are determined by the terminal object; $\delta \colon E \to E \times E$ is the diagonal and χ has the following components

$$\chi_{ABCD}\colon (\overset{\ast}{A}\times B)\otimes(C\times D)\to (A\otimes C)\times(B\otimes D),$$

$$p_1\otimes q_1\colon (A\times B)\otimes(C\times D)\to A\otimes C, \tag{6.135}$$

$$p_2\otimes q_2\colon (A\times B)\otimes(C\times D)\to B\otimes D.$$

(c) Dually if the monoidal category $\mathbf{C}$ has finite sums, we have a duoidal category $\mathsf{Fts}(\mathbf{C}) = (\mathbf{C},\otimes,E,+,\bot)$ with obvious $\tau\colon \bot\to E$ and $\delta\colon \bot\to \bot\otimes\bot$; then $\mu\colon E+E\to E$ is the codiagonal and

$$\chi_{ABCD}\colon (A\otimes B)+(C\otimes D)\to (A+C)\otimes(B+D), \tag{6.136}$$

is determined by obvious cocomponents.

6.8.6 Trioidal categories

Putting the previous structures together, a monoidal category $(\mathbf{C},\otimes,E)$ with finite products and sums can be said to have a structure of 'trioidal category'

$$\mathsf{A} = \mathsf{Fpts}(\mathbf{C}) = (\mathbf{C},\times,\top,\otimes,E,+,\bot), \tag{6.137}$$

with $\mathrm{tv}_{123}(\mathsf{A}) = \mathbf{C}$.

By a *trioidal category* we mean a 4-dimensional intercategory whose only non-trivial items are the 123-cubes (the highest dimensional) and their transversal maps. Note that now, after the coherence axioms of 6.8.2, we also have to retain those of 6.8.3 for the positive multi-index $\{i,j,k\} = \{1,2,3\}$.

The interchangers come from the partial duoidal structures

- the 12-interchangers from $\mathsf{Fpt}(\mathbf{C}) = (\mathbf{C},\times,\top,\otimes,E))$,

- the 23-interchangers from $\mathsf{Fts}(\mathbf{C}) = (\mathbf{C},\otimes,E,+,\bot)$,

- the 13-interchangers from $\mathsf{Fps}(\mathbf{C}) = (\mathbf{C},\times,\top,+,\bot)$.

In particular:

$$\tau_{12}\colon E\to\top,\qquad \tau_{23}\colon \bot\to E,\qquad \tau_{13} = \tau_{12}.\tau_{23}\colon \bot\to\top, \tag{6.138}$$

and the relation $\tau_{13} = \tau_{12}.\tau_{23}$ is precisely axiom 6.8.3(viii), in the present degenerate form.

(a) As a particular case one can take the monoidal category $\mathbf{Set}_{\bullet}$ of pointed sets. Binary products and sums are constructed as follows:

$$(X,x_0)\times(Y,y_0) = (X\times Y,(x_0,y_0)),$$

$$(X,x_0)+(Y,y_0) = (X\times\{y_0\}\cup\{x_0\}\times Y,(x_0,y_0)), \tag{6.139}$$

and there is a canonical morphism, given by the inclusion

$$u\colon (X, x_0) + (Y, y_0) \to (X, x_0) \times (Y, y_0), \qquad (6.140)$$

(that can be easily extended to any pointed category with binary products and sums). The *smash* product defining our monoidal structure is the cokernel of u (see 1.8.2), and is usually denoted as follows

$$(X, x_0) \wedge (Y, y_0) = ((X, x_0) \times (Y, y_0))/((X, x_0) + (Y, y_0)). \qquad (6.141)$$

Concretely, it is a quotient of the cartesian product $X \times Y$, where all equivalence-classes are singletons except (generally) the following, which gives the base-point of the smash product:

$$[(x_0, y_0)] = X \times \{y_0\} \cup \{x_0\} \times Y.$$

The identity is the cardinal $2 = \{0, 1\}$, pointed at 0.

For the monoidal category **Top.** of pointed spaces one proceeds in the same way, putting on the smash product the quotient topology of the cartesian product.

(b) As another example we have Fpts(**Ord**), where the tensor product $X \otimes Y$ is the ordinal sum of ordered sets: it extends the order relations of $X + Y$ by letting $x < y$ for all $x \in X$ and $y \in Y$.

(c) Finally we also have a trioidal category Fpts(**A**) for every monoidal semiadditive category **A** (see A3.4). Here the product and sum coincide and are written $A \oplus B$; the 13-interchanger

$$(A \oplus B) \oplus (C \oplus D) \to (A \oplus C) \oplus (B \oplus D)$$

is invertible.

6.8.7 *Monoidal double categories and locally cubical bicategories

The rest of this chapter is more technical, and can be of interest for a reader already acquainted with higher dimensional category theory.

(a) A *monoidal double category* of Shulman [Shu] can be defined as a pseudo monoid in the 2-category Ps**Dbl** of weak double categories, pseudo functors and horizontal transformations.

Shulman uses this notion to construct monoidal bicategories. The new notion is effective and simpler, because its coherence morphisms are isomorphisms rather than equivalences and the coherence conditions are much easier. Many examples are given.

As proved in [GP7], Subsection 3.1, this is the same as a 3-dimensional intercategory with one object, one 0-arrow, one 2-arrow and one 02-cell (all identities of course); furthermore all interchangers are invertible.

Simple examples come from any weak double category with pseudo functorial finite products.

(b) More generally, a *locally cubical bicategory* of Garner and Gurski [GaG] is a multiobject version of the previous structure. It can be defined as a category weakly enriched on the monoidal (cartesian) 2-category Ps**Dbl**.

As proved in [GP7], Subsection 3.5, this is the same as an intercategory in which all transversal arrows, 2-arrows and 02-cells are identities, with invertible interchangers.

For two objects A, B the general 3-dimensional cells $\Pi\colon e_{02}(A) \to_1 e_{02}(B)$ are cells in a weak double category $[A, B]$, where Π has horizontal arrows φ, ψ in direction 0 and vertical arrows π, ρ in direction 2

$$(6.142)$$

6.8.8 *Verity double bicategories and Gray categories*

(a) *Verity double bicategories* have been introduced in [Ve], and used in a study of higher cobordism, in [Mo].

As proved in [GP7], Theorem 4.1.1, they can be viewed as weak triple categories whose transversal arrows and all interchangers are identities, under an additional condition (*) too technical to be recalled here.

(b) Finally *Gray categories* [Gra1] have been generalised in [GP7], Section 5, under the name of 'true Gray categories'; it is proved that the latter can be viewed as intercategories (in three ways).

6.8.9 *Comments on lax multiple categories*

We end this section by remarking that the term 'lax multiple category' can cover various 'kinds' of laxity, where – with respect to a weak multiple category – some comparisons are still invertible while others (even some strict ones!) acquire a particular direction depending on the kind we are considering.

Thus, an intercategory is a particular type of 'interchange-lax' triple category, which is not even a triple set: the positive degeneracies need not commute.

As another example, we can recall the 'symmetric quasi cubical category' COSP(**Top**) of higher cospans of topological spaces, constructed in [G4], where the positive compositions work with homotopy pushouts. This is of interest for higher cobordism, because homotopy pushouts are homotopy invariant, while ordinary pushouts are not.

To give an idea of this structure, let us replace **Top** with a more regular 2-dimensional structure: a 2-category **C** with iso-pushouts (i.e. cocomma squares with an invertible cell). Then we can modify the weak symmetric cubical category Cosp(**C**) recalled in 6.4.6 *by composing cubical cospans with iso-pushouts.* We obtain a kind of lax symmetric cubical category COSP(**C**) where all comparisons are invertible except the unitors, directed as

$$\lambda_i(x)\colon e_i(\partial_i^- x) +_i x \to x, \qquad \rho_i(x)\colon x +_i e_i(\partial_i^+ x) \to x. \qquad (6.143)$$

Dually, if the 2-category **C** has (a fixed choice of) iso-pullbacks, we can form a structure SPAN(**C**) by composing cubical spans with iso-pullbacks; we get a different kind of laxity, where unitors are directed the other way round with respect to (6.143).

6.9 *Tabulators in intercategories

Tabulators can be easily extended to intercategories, *and even acquire richer forms* when degeneracies do not commute.

Here we construct a 3-dimensional intercategory B with $e_1 e_2 \neq e_2 e_1$, where the $e_1 e_2$-tabulator and the $e_2 e_1$-tabulator of a 12-cube always exist and are different, generally. This example is rather artificial, but intercategories having $e_1 e_2 \neq e_2 e_1$ *and* tabulators are not easy to build – seemingly.

Finally, in 6.9.4, we give a simpler case where $e_1 e_2$- and $e_2 e_1$-tabulators must still be distinguished, but only the second kind always exists.

6.9.1 Definition

In an intercategory A any composed degeneracy

$$e = e_{j_1} e_{j_2} \dots e_{j_r} \colon \mathrm{tv}_{\mathbf{k}}(\mathsf{A}) \to \mathrm{tv}_{\mathbf{i}}(\mathsf{A}) \qquad (\mathbf{k} = \mathbf{i} \setminus \{j_1, \dots, j_r\}), \qquad (6.144)$$

depends on the order of the r-tuple $(j_1, \dots, j_r)$, and gives a distinct notion of *e-tabulator* (of degree $k = n - r$) of an **i**-cube x of degree n: it is an object $T = \mathsf{T}_{j_1, \dots, j_r}(x)$ with a universal transversal map $t_x\colon e(T) \to_0 x$ from the functor e to the object x.

In degree $n - r$ there are thus $\binom{n}{r}.r!$ of them, that fall down to $\binom{n}{r}$ in the chiral case.

Below we only consider 3-dimensional intercategories, where we have *two* total degeneracies $\mathrm{tv}_*(\mathsf{A}) \to \mathrm{tv}_{12}(\mathsf{A})$. The $e_1 e_2$-*tabulator* of a 12-cube x is thus defined as an object $T = \mathsf{T}_{12}x$ with a universal transversal map $t_x \colon e_1 e_2(\mathsf{T}x) \to_0 x$ from the functor $e_1 e_2$ to the object x. Symmetrically, we (can) have the $e_2 e_1$-*tabulator* $\mathsf{T}_{21}x$.

Globally, adding the tabulators $\mathsf{T}_1 x$ (of 1-arrows and 12-cubes) and the tabulators $\mathsf{T}_2 x$ (of 2-arrows and 12-cubes), there are six kinds of non-trivial tabulators instead of the previous five.

6.9.2 An intercategory

We start from the chiral triple category *of spans and cospans* $\mathsf{A} = \mathsf{SC}(\mathbf{C})$, studied in Section 6.6, where the category $\mathbf{C}$ is equipped with a choice of pullbacks and pushouts that preserves identities. We also assume that $\mathbf{C}$ has a (chosen) initial object 0, and therefore *all finite colimits* (by 1.3.5(i)); furthermore, we assume that every morphism $u \colon 0 \to X$ is *mono* (which holds in many concrete categories but fails in **Rng**, for instance) and the chosen pullback of (u, u) is precisely 0.

We now restrict the items of A, so that the only remaining 1-arrows are the 'null spans' $X \leftarrow 0 \to Y$. We can thus form an intercategory B that is not a substructure of A and is no longer chiral: it has a different e_1 and its interchangers τ, μ are directed – while δ is still trivial.

(a) The sets B_*, B_0, B_2 and B_{02} coincide, respectively, with A_*, A_0, A_2, A_{02}, and have the same composition laws in direction 0 and 2.

(b) The subset $B_1 \subset A_1$ consists of the *null spans* $(X \leftarrow 0 \to Y)$ of $\mathbf{C}$, also written as $0_{XY} \colon X \to_1 Y$. These new 1-cells compose as in A (by our assumptions on the initial object) but have different identities (as the old ones do not belong to B_1)

$$0_{XY} +_1 0_{YZ} = 0_{XZ}, \qquad e_1(X) = 0_{XX} = (X \leftarrow 0 \to X).$$

This forms a category, isomorphic to the indiscrete category on the objects of $\mathbf{C}$.

(c) A 01-cell $(p, q) \in A_{01}$ amounts to an arbitrary pair of morphisms of $\mathbf{C}$

$$\tag{6.145}$$

Their 0-composition is obvious and gives a category isomorphic to $\mathbf{C} \times \mathbf{C}$.

Their 1-composition is that of the indiscrete category on the morphisms of
C: $(p,q) +_1 (q,r) = (p,r)$.

(c') A 12-cell $\pi \in B_{12} \subset A_{12}$

$$
\begin{array}{ccc}
X & 0 & X' \\
u\!\!\downarrow & \pi & \downarrow v \\
Y & 0 & Y'
\end{array}
\qquad
\begin{array}{c}
\bullet \;\xrightarrow{\;1\;} \\
2 \downarrow
\end{array}
\tag{6.146}
$$

is a cell of A_{12} whose boundary are null spans and arbitrary cospans, as in
the left diagram below (automatically commutative in **C**)

$$
\begin{array}{ccccc}
X & \xleftarrow{\;0\;} & Y & \qquad & X \qquad\qquad Y \\
u'\downarrow & \downarrow & \downarrow v' & & u'\downarrow \qquad\qquad \downarrow v' \\
A & \xleftarrow{f'} P \xrightarrow{f''} & B & & A \xleftarrow{f'} P \xrightarrow{f''} B \\
u''\uparrow & \uparrow & \uparrow v'' & & u''\uparrow \qquad\qquad \uparrow v'' \\
Z & \xleftarrow{\;0\;} & U & & Z \qquad\qquad U
\end{array}
$$

It amounts to a triple (u, f, v) containing two cospans, u and v, and a
span f as in the right diagram above. The 1- and 2-composition of these
12-cells are as in **A** (computed by pullbacks and pushouts, respectively),
with the same associators. In particular

$$
(u, f, v) +_1 (v, g, w) = (u, f +_1 g, w). \tag{6.147}
$$

The degeneracies $e_1 \colon B_2 \to B_{12}$ and $e_2 \colon B_1 \to B_{12}$ work as follows

$$
\begin{aligned}
e_1(u) &= e_1(X \to A \leftarrow Y) = (u, (1_A, 1_A), u), \\
e_2(0_{XY}) &= e_2(X \leftarrow 0 \to Y) = ((1_X, 1_X), 0_{XY}, (1_Y, 1_Y)),
\end{aligned}
\tag{6.148}
$$

$$
\begin{array}{ccc}
X & & X \\
u'\downarrow & & \downarrow u' \\
A \xleftarrow{\;1\;} A \xrightarrow{\;1\;} A \\
u''\uparrow & & \uparrow u'' \\
Y & & Y
\end{array}
\qquad\qquad
\begin{array}{ccc}
X & & Y \\
1\downarrow & & \downarrow 1 \\
X \xleftarrow{\;0\;} Y \\
1\uparrow & & \uparrow 1 \\
X & & Y
\end{array}
$$

$$
\qquad\qquad e_1(u) \qquad\qquad\qquad\qquad e_2(0_{XY})
$$

In particular, the second is the restriction of that of **A**, and sends a null
span 0_{XY} to the obvious 12-cell with three degenerate cospans (on X, 0
and Y).

The binary interchanger χ is the (non-invertible) restriction of that of **A**.
The lower interchangers are defined as follows. Firstly, δ is trivial

$$
\delta(0_{XY}, 0_{YZ}) \colon e_2(0_{XY} +_1 0_{YZ}) \to e_2(0_{XY}) +_1 e_2(0_{YZ}), \tag{6.149}
$$

namely the identity of the 12-cell $((1_X, 1_X), 0_{XZ}(1_X, 1_X))$.

Secondly, μ amounts to the canonical morphism $h: A + B \to C$, where A, B and $C = A +_Y B$ are the central objects of the cospans u, v and $u +_2 v$, respectively:

$$\mu(u, v): e_1(u) +_2 e_1(v) \to e_1(u +_2 v),$$

$$e_1(u) +_2 e_1(v) = (u +_2 v, (h, h), u +_2 v), \tag{6.150}$$

$$e_1(u +_2 v) = (u +_2 v, (1_C, 1_C), u +_2 v).$$

Thirdly, also τ is not invertible (in general)

$$\tau(X): e_2 e_1(X) \to e_1 e_2(X), \tag{6.151}$$

$$e_2 e_1(X) \qquad\qquad\qquad e_1 e_2(X)$$

(d) Finally B_{012} is the set of triple cells of A whose boundary 1-arrows are null spans. They compose as in A.

6.9.3 Exercises

Let us suppose that, in the previous construction, the category **C** also has a terminal object (and therefore all finite limits). Then one can prove that the intercategory B has all (six kinds of) tabulators: the tabulator of a 1-arrow, the tabulator of a 2-arrow and four tabulators for the 12-cube $\pi = (u, f, v)$ of (6.146). These limits are not always preserved by faces and degeneracies.

6.9.4 Tabulators and cotabulators for a duoidal category

We end with another case where $e_1 e_2$- and $e_2 e_1$-tabulators must be distinguished.

We have seen in 6.8.5 that a category **C** with finite products and sums can be seen as a duoidal category $\mathsf{Fps}(\mathbf{C}) = (\mathbf{C}, \times, \top, +, \bot)$, and as an intercategory F whose only non trivial cells are the 12-cubes and their transversal maps, respectively corresponding to the objects and morphisms of **C**.

F has precisely one object, the formal vertex •. Its totally degenerate 12-cells $e_1e_2(\bullet)$ and $e_2e_1(\bullet)$ correspond, respectively, to the unit-objects $E_1 = \top$ and $E_2 = \bot$ of **C** (as remarked at the end of 6.8.4)

$$\begin{array}{cc} \boxed{\top} & \boxed{\bot} \\ e_1e_2(\bullet) & e_2e_1(\bullet) \end{array} \tag{6.152}$$

(a) Therefore every object A (as a 12-cube of F) admits the vertex • as its e_2e_1-tabulator, via the unique transversal map $e_2e_1(\bullet) \to A$ (given by the unique morphism $\bot \to A$ of **C**). Dually every object A admits the vertex • as its e_1e_2-cotabulator, via the unique transversal map $A \to e_1e_2(\bullet) = \top$.

The objects $\bot$ and $\top$ of **C** are thus determined as follows, as 12-cubes of F (for every A)

$$\bot = e_2e_1(\bullet) = e_2e_1(\top_{21}A), \quad \top = e_1e_2(\bullet) = e_1e_2(\bot_{12}A). \tag{6.153}$$

(b) The symmetric cases, namely e_1e_2-tabulators and e_2e_1-cotabulators, behave differently. In fact, an object A admits the vertex • as its e_1e_2-tabulator (resp. e_2e_1-cotabulator) if and only if there is precisely one morphism $\top \to A$ (resp. $A \to \bot$) in **C**.

This is always true if **C** is pointed. On the other hand, in Fps(**Set**) only a singleton has e_1e_2-tabulator and only the empty set has e_2e_1-cotabulator.

7

Multiple adjunctions

This chapter is devoted to adjunctions for chiral multiple categories, and is based on [GP10].

Section 7.1 is an informal introduction to the subject of colax-lax adjunctions between chiral multiple categories.

In Section 7.2 we introduce the strict *double* category $\mathbb{C}mc$ of chiral multiple categories (or cm-categories), lax and colax functors, with suitable double cells. Comma cm-categories are also considered. Both topics extend notions of weak double categories developed in Section 4.2.

Sections 7.3 and 7.4 introduce and study multiple colax-lax adjunctions, as adjoint arrows in the double category $\mathbb{C}mc$.

In Section 7.5 the double category $\mathbb{C}mc$ is extended to a multiple category Cmc where the arrows in each direction are mixed laxity functors (defined in 6.5.7), varying from the lax functors to the colax ones.

In Section 7.6, we briefly examine *level limits*, a particular case of multiple limits. The exponential A^{X} of (strict) multiple categories is constructed in Section 7.7.

7.1 Some basic examples of adjunctions

As an informal introduction to multiple adjunctions, we begin by recalling examples of adjunctions for *weak double categories*, and explore their extension to *weak multiple categories* of cubical type. We derive from the latter some instances of adjunctions between *chiral multiple* categories. Other examples are given in 7.1.5–7.1.7.

7.1.1 A double adjunction

Let $\mathbf{C}$ be a category with (a choice of) pullbacks and pushouts. For the sake of simplicity we still assume that the distinguished pullback or pushout of an identity along any map is an identity.

The weak double categories $\mathbb{Span}(\mathbf{C})$ and $\mathbb{Cosp}(\mathbf{C})$ of spans and cospans of $\mathbf{C}$ are linked by an (idempotent) colax-lax adjunction described in 4.5.6 and called the pushout-pullback adjunction of $\mathbf{C}$

$$F \colon \mathbb{Span}(\mathbf{C}) \rightleftarrows \mathbb{Cosp}(\mathbf{C}) \colon G, \qquad (\eta, \varepsilon) \colon F \dashv G, \qquad (7.1)$$

which is trivial in degree zero; in degree 1, F acts by pushout on spans and G acts by pullback on cospans.

7.1.2 The pushout-pullback multiple adjunction

The unitary colax double functor $F \colon \mathbb{Span}(\mathbf{C}) \to \mathbb{Cosp}(\mathbf{C})$ can be extended to a unitary colax multiple functor $F \colon \mathsf{Span}(\mathbf{C}) \to \mathsf{Cosp}(\mathbf{C})$, between the weak multiple categories of cubical type introduced in 6.4.6.

For instance, let us take a 2-dimensional span $x \in \mathsf{Span}_\mathbf{i}(\mathbf{C})$ indexed by $\mathbf{i} = \{i, j\}$. In the notation of 6.4.6, this is a diagram $x \colon \mathsf{V}^\mathbf{i} \to \mathbf{C}$

$$
\begin{array}{ccccc}
x_{00} & \leftarrow & x_{\iota 0} & \rightarrow & x_{10} \\
\uparrow & & \uparrow & & \uparrow \\
x_{0\iota} & \leftarrow & x_{\iota\iota} & \rightarrow & x_{1\iota} \\
\downarrow & & \downarrow & & \downarrow \\
x_{01} & \leftarrow & x_{\iota 1} & \rightarrow & x_{11}
\end{array}
\qquad
\begin{array}{c}
\bullet \overset{i}{\longrightarrow} \\[2pt]
\Big\downarrow j \\[2pt]
{}
\end{array}
\qquad (7.2)
$$

The 2-dimensional cospan $F(x) = F_\mathbf{i}(x)$ is constructed in the following diagram

$$
\begin{array}{ccccc}
x_{00} & \longrightarrow & F(\partial_j^- x)(\iota) & \longleftarrow & x_{10} \\
\downarrow & & \downarrow & & \downarrow \\
F(\partial_i^- x)(\iota) & \longrightarrow & \mathrm{Colim}\,(x) & \longleftarrow & F(\partial_i^+ x)(\iota) \\
\uparrow & & \uparrow & & \uparrow \\
x_{01} & \longrightarrow & F(\partial_j^+ x)(\iota) & \longleftarrow & x_{11}
\end{array}
\qquad
\begin{array}{c}
\bullet \overset{i}{\longrightarrow} \\[2pt]
\Big\downarrow j \\[2pt]
{}
\end{array}
\qquad (7.3)
$$

with the pushouts $F(\partial_i^\alpha x)$, $F(\partial_j^\alpha x)$ of the four faces and, in the central vertex, the colimit of the whole diagram $x \colon \mathsf{V}^\mathbf{i} \to \mathbf{C}$; the latter exists in $\mathbf{C}$ and can be constructed as a pushout of pushouts. (A general characterisation of the dual topic, limits 'generated' by pullbacks, can be found in [Pa2].)

One proceeds by induction, defining $F_{\mathbf{i}}$ for a positive multi-index $\mathbf{i} = \{i_1, ..., i_n\}$ of degree n, after all instances of degree $n - 1$

$$\partial_i^\alpha(F_{\mathbf{i}}(x)) = F_{\mathbf{i}|i}(\partial_i^\alpha x) \qquad \text{(for } \alpha = \pm \text{ and } i \in \mathbf{i}\text{)},$$
$$F_{\mathbf{i}}(x)(\underline{\iota}) = \mathrm{Colim}\,(x),$$

(7.4)

where $\underline{\iota} = (\iota, ..., \iota)$ is the central vertex of $\vee^{\mathbf{i}}$.

The definition of F on transversal $\mathbf{i}$-maps is obvious. The comparison cell for the i-directed concatenation $F_i(x, y)\colon F(x +_i y) \to Fx +_i Fy$ derives from the universal properties of the colimits involved in the construction of $F(x +_i y)$.

The unitary lax double functor $G\colon \mathbb{C}\mathrm{osp}(\mathbf{C}) \to \mathbb{S}\mathrm{pan}(\mathbf{C})$ is similarly extended, using limits instead of colimits, and gives a unitary lax multiple functor $G\colon \mathsf{Cosp}(\mathbf{C}) \to \mathsf{Span}(\mathbf{C})$.

One extends the unit $\eta\colon 1 \dashrightarrow GF$ by a similar inductive procedure:

$$\partial_i^\alpha(\eta_{\mathbf{i}}(x)) = \eta_{\mathbf{i}|i}(\partial_i^\alpha x) \qquad \text{(for } \alpha = \pm \text{ and } i \in \mathbf{i}\text{)},$$
$$(\eta_{\mathbf{i}} x)(\underline{\iota})\colon x(\underline{\iota}) \to (G_{\mathbf{i}} F_{\mathbf{i}} x)(\underline{\iota}) = \mathrm{Lim}\,(F_{\mathbf{i}} x),$$

(7.5)

where the map $(\eta_{\mathbf{i}} x)(\underline{\iota})$ is given by the universal property of $\mathrm{Lim}\,(F_{\mathbf{i}} x)$ as the limit of the cubical cospan $F_{\mathbf{i}} x\colon \wedge^{\mathbf{i}} \to \mathbf{C}$.

Analogously for the counit $\varepsilon\colon FG \dashrightarrow 1$. The triangular equations hold.

Extending what we have already seen in 4.5.6, for ordinary spans and cospans, one can prove that $F_{\mathbf{i}} \eta_{\mathbf{i}} x$ is invertible (Exercise 7.1.3). We have thus an *idempotent* colax-lax multiple adjunction $F \dashv G$ (as defined in 8.4.1).

7.1.3 Exercise

Prove that, in the previous adjunction, $F_{\mathbf{i}} \eta_{\mathbf{i}} x$ is always invertible.

7.1.4 Chiral examples

The colax-lax adjunction of weak triple categories

$$F\colon \mathsf{Span}_3(\mathbf{C}) \rightleftarrows \mathsf{Cosp}_3(\mathbf{C}) : G,$$

(7.6)

can be factorised through the chiral triple category $\mathsf{SC}(\mathbf{C})$ of spans and cospans of $\mathbf{C}$, obtaining two colax-lax adjunctions of chiral triple categories (no longer of cubical type)

$$\mathsf{Span}_3(\mathbf{C}) \underset{G'}{\overset{F'}{\rightleftarrows}} \mathsf{SC}(\mathbf{C}) \underset{G''}{\overset{F''}{\rightleftarrows}} \mathsf{Cosp}_3(\mathbf{C})$$

(7.7)

Here the functor F': $\mathsf{Span}_3(\mathbf{C}) \to \mathsf{SC}(\mathbf{C})$ acts on a 12-cube x by pushout on the three 2-directed spans of x, as the identity on the 1-directed boundary spans $\partial_2^\alpha(x)$ and by induced morphisms on the middle 1-directed span. On the other hand the functor G': $\mathsf{SC}(\mathbf{C}) \to \mathsf{Span}_3(\mathbf{C})$ acts by pullback on the three (2-directed) cospans of x, as the identity on the (1-directed) boundary spans $\partial_2^\alpha(x)$ and by induced morphisms on the middle span. Similarly for F'' and G''.

One can also factorise the adjunction (7.6) through the left chiral triple category $\mathsf{CS}(\mathbf{C})$ *of cospans and spans*, obtaining two colax-lax adjunctions of left chiral triple categories.

Similarly, the multiple adjunction constructed in 7.1.2 can be factorised through any right chiral multiple category $\mathsf{S}_p\mathsf{C}_\infty(\mathbf{C})$, *or* through any left chiral multiple category $\mathsf{C}_p\mathsf{S}_\infty(\mathbf{C})$.

However, in infinite dimension, one may prefer to consider a more symmetric situation, starting from a colax-lax adjunction of weak multiple categories *indexed by the ordered set* $\mathbb{Z}$ of integers (pointed at 0, see 6.2.3)

$$F: \mathsf{Span}_\mathbb{Z}(\mathbf{C}) \rightleftarrows \mathsf{Cosp}_\mathbb{Z}(\mathbf{C}) : G. \tag{7.8}$$

This can be factorised through the chiral multiple category $\mathsf{S}_{-\infty}\mathsf{C}_\infty(\mathbf{C})$ defined in 6.4.8, obtaining two colax-lax adjunctions of 'unbounded' chiral multiple categories

$$\mathsf{Span}_\mathbb{Z}(\mathbf{C}) \rightleftarrows \mathsf{S}_{-\infty}\mathsf{C}_\infty(\mathbf{C}) \rightleftarrows \mathsf{Cosp}_\mathbb{Z}(\mathbf{C}). \tag{7.9}$$

7.1.5 Extending ordinary adjunctions to cubical spans

As in the 2-dimensional extensions of 4.5.8, we start from an adjunction $F \dashv G$ between ordinary categories

$$F: \mathbf{X} \rightleftarrows \mathbf{A} : G, \qquad \eta: 1 \to GF, \quad \varepsilon: FG \to 1. \tag{7.10}$$

If $\mathbf{X}$ and $\mathbf{A}$ have (a fixed choice of) pullbacks, this adjunction can be extended in a natural way to a unitary *colax-pseudo* adjunction between the weak multiple categories of higher spans

$$\mathsf{Span}(F): \mathsf{Span}(\mathbf{X}) \rightleftarrows \mathsf{Span}(\mathbf{A}) : \mathsf{Span}(G). \tag{7.11}$$

In fact there is an obvious 2-functor

$$\mathsf{Span}: \mathbf{Cat}_{\mathrm{pb}} \to \mathrm{CxCmc}, \tag{7.12}$$

defined on the full sub-2-category of $\mathbf{Cat}$ containing all categories with (a choice of) pullbacks, with values in the 2-category of chiral multiple categories, colax functors and their transversal transformations (see 6.5.3).

It sends a category $\mathbf{C}$ with pullbacks to the chiral multiple category $\mathsf{Span}(\mathbf{C})$ (actually a weak multiple category of symmetric cubical type). For a functor $F\colon \mathbf{X} \to \mathbf{A}$ (between categories with pullbacks) $\mathsf{Span}(F)$, also written as F for brevity, simply acts by computing F over the diagrams of $\mathbf{X}$ that form **i**-cubes and **i**-maps; formally, over an **i**-map $f\colon x \to y\colon \vee^n \to \mathbf{X}$, $F(f)\colon F(x) \to F(y)$ is the composite $F.f\colon F.x \to F.y\colon \vee^n \to \mathbf{A}$.

This extension is, in a natural way, a unitary *colax* functor, since identities of $\mathbf{X}$ go to identities of $\mathbf{A}$ and a composition $x +_i y$ of two *spans* $x, y\colon \vee \to \mathbf{X}$ (in any direction $i > 0$) gives rise to a diagram in $\mathbf{X}$ and a diagram in $\mathbf{A}$

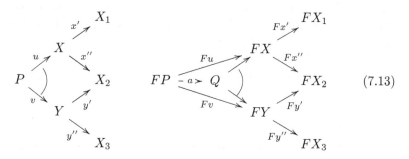

(7.13)

where the comparison $\underline{F}_i(x, y)\colon F(x +_i y) \to F(x) +_i F(y)$ is an i-special transversal map given by the $\mathbf{A}$-morphism $a\colon FP \to Q$ determined by the universal property of the pullback Q. Similarly we define $\underline{F}_i(x, y)$ for every i-composition of **i**-cubes. Note that $\mathsf{Span}(F)$ is pseudo if (and only if) $F\colon \mathbf{X} \to \mathbf{A}$ preserves pullbacks.

A natural transformation $\varphi\colon F \to F'\colon \mathbf{X} \to \mathbf{A}$ yields a transversal transformation

$$\mathsf{Span}(\varphi)\colon \mathsf{Span}(F) \to \mathsf{Span}(F')\colon \mathsf{Span}(\mathbf{X}) \to \mathsf{Span}(\mathbf{A}), \qquad (7.14)$$

that again will often be written as φ. On an **i**-cube $x\colon \vee^n \to \mathbf{X}$, the transversal map $\varphi x\colon F(x) \to_0 F'(x)$ is the whisker composite of the functor $x\colon \vee^n \to \mathbf{X}$ with the natural transformation $\varphi\colon F \to F'\colon \mathbf{X} \to \mathbf{A}$. Concretely, the transversal **i**-map $\varphi x\colon F(x) \to_0 F'(x)$ has components $\varphi(x(t))$, for every vertex t of $\vee^n$.

Now, letting the 2-functor $\mathsf{Span}\colon \mathbf{Cat}_{\mathrm{pb}} \to \mathrm{Cx}\mathbf{Cmc}$ act on the adjunction (7.10), we get a *colax-pseudo* adjunction of weak multiple categories

$$F\colon \mathsf{Span}(\mathbf{X}) \rightleftarrows \mathsf{Span}(\mathbf{A}) :G. \qquad (7.15)$$

This gives a colax endofunctor GF on $\mathsf{Span}(\mathbf{X})$ (and a colax multiple monad, see Section 8.2).

7.1.6 Extending ordinary adjunctions to cubical cospans

Dually, we consider now the full sub-2-category $\mathbf{Cat}_{\mathrm{po}}$ of $\mathbf{Cat}$ containing all categories with (a choice of) pushouts. There is now a 2-functor

$$\mathsf{Cosp}\colon \mathbf{Cat}_{\mathrm{po}} \to \mathrm{Lx}\mathbf{Cmc}, \quad \mathsf{Cosp}(\mathbf{C}) = (\mathsf{Span}(\mathbf{C}^{\mathrm{op}}))^{\mathrm{tr}}, \tag{7.16}$$

with values in the 2-category of chiral multiple categories, lax functors and their transversal transformations.

The adjunction (7.10) yields thus a *pseudo-lax* adjunction of weak multiple categories

$$F\colon \mathsf{Cosp}(\mathbf{X}) \rightleftarrows \mathsf{Cosp}(\mathbf{A}) : G. \tag{7.17}$$

This gives a lax endofunctor GF on $\mathsf{Span}(\mathbf{X})$ (and a lax multiple monad, see Section 8.3).

7.1.7 Mixed extensions

Finally, we have a 2-functor

$$\mathsf{SC}\colon \mathbf{Cat}_{\mathrm{pbpo}} \to \mathrm{Mx}_2(\mathbf{Cmc_3}), \tag{7.18}$$

defined on the 2-category of small categories with a fixed choice of pullbacks and pushouts, arbitrary functors and natural transformations. It takes values in the 2-category of chiral triple categories, colax-lax functors (or 2-mixed functors) and transversal transformations, defined in 6.5.8.

The adjunction (7.10) yields now an adjunction of chiral triple categories

$$F'\colon \mathsf{SC}(\mathbf{X}) \rightleftarrows \mathsf{SC}(\mathbf{A}) : G', \tag{7.19}$$

where $F' = \mathsf{SC}(F)$ is a colax-pseudo functor (because F preserves pushouts), while $G' = \mathsf{SC}(G)$ is a pseudo-lax functor (because G preserves pullbacks).

Their composites $G'F' = \mathsf{SC}(GF)$ and $F'G' = \mathsf{SC}(FG)$ are colax-lax functors (and we have a colax-lax monad $G'F'$ on $\mathsf{SC}(\mathbf{X})$).

Similarly, we have a 2-functor

$$\mathsf{S}_p\mathsf{C}_\infty\colon \mathbf{Cat}_{\mathrm{pbpo}} \to \mathrm{Mx}_{p+1}(\mathbf{Cmc}), \tag{7.20}$$

and variations taking into account the cases $\mathsf{S}_p\mathsf{C}_q$ and $\mathsf{S}_{-\infty}\mathsf{C}_\infty$.

7.2 The double category of lax and colax multiple functors

In the 2-dimensional case, weak double categories, with lax and colax double functors and suitable double cells, form the strict double category $\mathbb{D}\mathrm{bl}$ (see Section 4.2). This construction is now extended, forming the strict *double* category $\mathbb{C}\mathrm{mc}$ of chiral multiple categories, lax and colax multiple functors, that will be used below to define colax-lax adjunctions between chiral multiple categories.

Comma chiral multiple categories are also considered, extending again the case of double categories.

In $\mathbb{C}$mc we follow the notation for double categories used in Chapter 3: the horizontal and vertical compositions of cells are written as $\pi \mid \rho$ and $\frac{\pi}{\sigma} = \pi \otimes \sigma$. Horizontal identities, of an object or a vertical arrow, are written as 1_A and 1_U; vertical identities, of an object or a horizontal arrow, as e_A and e_F.

7.2.1 The double category $\mathbb{C}$mc

As usual, we do not want to compose lax and colax multiple functors: they give the horizontal and vertical arrows of a *strict* double category $\mathbb{C}$mc.

The objects of $\mathbb{C}$mc are the chiral multiple categories, or cm-categories, $A, B, C, ...$; its horizontal arrows are the *lax* functors $F, G...$; its vertical arrows are the *colax* functors $U, V...$ A cell π

$$
\begin{array}{ccc}
A & \xrightarrow{F} & B \\
U \downarrow & \pi & \downarrow V \\
C & \xrightarrow{G} & D
\end{array}
\qquad (7.21)
$$

is – roughly speaking – a 'transformation' $\pi \colon VF \dashrightarrow GU$. Again, the composites VF and GU are neither lax nor colax, and the coherence conditions of π require the individual knowledge of the four 'functors', including the comparison cells of each of them.

Precisely, the double cell π consists of the following data:

(a) two lax multiple functors $F \colon A \to B$, $G \colon C \to D$ with comparisons

$$\underline{F}_i(x) \colon e_i(Fx) \to_0 F(e_i x), \qquad \underline{F}_i(x, y) \colon Fx +_i Fy \to_0 F(x +_i y),$$
$$\underline{G}_i(x) \colon e_i(Gx) \to_0 G(e_i x), \qquad \underline{G}_i(x, y) \colon Gx +_i Gy \to_0 G(x +_i y),$$

(b) two colax multiple functors $U \colon A \to C$, $V \colon B \to D$ with comparisons

$$\underline{U}_i(x) \colon U(e_i x) \to_0 e_i(Ux), \qquad \underline{U}_i(x, y) \colon U(x +_i y) \to_0 Ux +_i Uy,$$
$$\underline{V}_i(x) \colon V(e_i x) \to_0 e_i(Vx), \qquad \underline{V}_i(x, y) \colon V(x +_i y) \to_0 Vx +_i Vy,$$

(c) a family of **i**-maps $\pi x \colon VFx \to_0 GUx$ of D, for every **i**-cube x in A, consistent with faces

$$\pi(\partial_i^\alpha x) = \partial_i^\alpha(\pi x). \qquad (7.22)$$

These data have to satisfy the naturality condition (dc.1) and the coherence conditions (dc.2), (dc.3) (with respect to i-degeneracies and i-composition, respectively)

(dc.1) $GUf.\pi x = \pi y.VFf \colon VFx \to_0 GUy$ $\qquad$ (for $f \colon x \to_0 y$ in $\mathbf{A}_i$),

(dc.2) $G\underline{U}_i x.\pi e_i x.V\underline{F}_i x = \underline{G}_i Ux.e_i \pi x.\underline{V}_i Fx$ $\qquad$ (for x in $\mathbf{A}_{i|i}$),

$$
\begin{array}{ccc}
Ve_i(Fx) & \xrightarrow{\;V\underline{F}_i x\;} & VF(e_ix) & \xrightarrow{\;\pi e_i x\;} & GU(e_ix) \\
{\scriptstyle \underline{V}_i Fx}\Big\downarrow & & & & \Big\downarrow{\scriptstyle GU_i x} \\
e_iVFx & \xrightarrow[\;e_i \pi x\;]{} & e_iGUx & \xrightarrow[\;\underline{G}_i Ux\;]{} & Ge_iUx
\end{array}
$$

(dc.3) $G\underline{U}_i(x,y).\pi z.V\underline{F}_i(x,y) = \underline{G}_i(Ux,Uy).(\pi x +_i \pi y).\underline{V}_i(Fx,Fy)$
$\qquad$ (for $z = x +_i y$ in $\mathbf{A}_i$),

$$
\begin{array}{ccc}
V(Fx +_i Fy) & \xrightarrow{\;V\underline{F}_i(x,y)\;} & VFz & \xrightarrow{\;\pi z\;} & GUz \\
{\scriptstyle \underline{V}_i(Fx,Fy)}\Big\downarrow & & & & \Big\downarrow{\scriptstyle G\underline{U}_i(x,y)} \\
VFx +_i VFy & \xrightarrow[\;\pi x +_i \pi y\;]{} & GUx +_i GUy & \xrightarrow[\;\underline{G}_i(Ux,Uy)\;]{} & G(Ux +_i Uy)
\end{array}
$$

The horizontal composition $\pi \mid \rho$ and the vertical composition $\pi \otimes \sigma$ of double cells are both defined via the composition of transversal maps (for x in $\mathbf{A}$)

$$
\begin{array}{ccccc}
\mathbf{A} & \xrightarrow{\;F\;} & \bullet & \xrightarrow{\;F'\;} & \bullet \\
{\scriptstyle U}\Big\downarrow & \pi & {\scriptstyle V}\Big\downarrow & \rho & {\scriptstyle W}\Big\downarrow \\
\bullet & \xrightarrow{\;-\,G\,\succ\;} & \bullet & \xrightarrow{\;-\,G'\,\succ\;} & \bullet \\
{\scriptstyle U'}\Big\downarrow & \sigma & {\scriptstyle V'}\Big\downarrow \; \tau & & {\scriptstyle W'}\Big\downarrow \\
\bullet & \xrightarrow{\;H\;} & \bullet & \xrightarrow{\;H'\;} & \bullet
\end{array}
\tag{7.23}
$$

$$
\begin{aligned}
(\pi \mid \rho)(x) &= G'\pi x.\rho Fx \colon WF'Fx \to_0 G'VFx \to_0 G'GUx, \\
(\pi \otimes \sigma)(x) &= \sigma Ux.V'\pi x \colon V'VFx \to_0 V'GUx \to_0 HU'Ux.
\end{aligned}
\tag{7.24}
$$

The fact that these compositions are well-defined, and satisfy the axioms of a double category, will follow from Theorem 7.5.6 about the multiple category Cmc: in fact Cmc amounts to the restriction of Cmc to the indices 1 and ∞. (Alternatively, the reader can see a direct proof for Cmc in [GP10], Theorem 2.3.)

Within Cmc, we have the strict 2-category **LxCmc** of *cm-categories, lax functors and transversal transformations*: namely, **LxCmc** is the restriction of Cmc to trivial vertical arrows. Symmetrically, Cmc also contains the strict 2-category **CxCmc**, whose arrows are the colax functors.

As in 4.2.3 (for weak double categories), we note that a double cell $\pi\colon (U \overset{F}{\underset{1}{}} 1)$ gives a notion of *transversal transformation* $\pi\colon F \dashrightarrow U\colon \mathsf{A} \to \mathsf{B}$ *from a lax to a colax functor,* while a double cell $\pi\colon (1 \overset{1}{\underset{G}{}} V)$ gives a notion of *transversal transformation* $\pi\colon V \dashrightarrow G\colon \mathsf{A} \to \mathsf{B}$ *from a colax to a lax functor.* Therefore, for a fixed pair A, B of chiral multiple categories, the four possible kinds of transversal transformations between lax and colax functors $\mathsf{A} \to \mathsf{B}$ *compose* forming the morphisms of a category $\mathrm{HV}(\mathsf{A}, \mathsf{B})$. (It is a consequence of Exercise 4.2.8(a), taking $\mathbb{D} = \mathbb{C}\mathrm{mc}$.)

7.2.2 Comma cm-categories

Comma double categories (see 4.2.4) also have a natural extension to chiral multiple categories.

Given a *colax* functor $U\colon \mathsf{A} \to \mathsf{C}$ and a *lax* functor $F\colon \mathsf{X} \to \mathsf{C}$ with the same codomain, we can construct the *comma cm-category* $U \downarrow F$, where the projections P and Q are strict functors and π is a cell of $\mathbb{C}\mathrm{mc}$

$$
\begin{array}{ccc}
U \downarrow F & \overset{P}{\longrightarrow} & \mathsf{A} \\
{\scriptstyle Q}\downarrow & \pi & \downarrow{\scriptstyle U} \\
\mathsf{X} & \underset{F}{\longrightarrow} & \mathsf{C}
\end{array}
\tag{7.25}
$$

An **i**-cube of $U \downarrow F$ is a triple $(a, x; c\colon Ua \to_0 Fx)$ where a is an **i**-cube of A, x is an **i**-cube of X and c is an **i**-map of C. An **i**-map

$$(h, f)\colon (a, x; c) \to_0 (a', x'; c')$$

comes from a pair of **i**-maps $h\colon a \to_0 a'$ (in A) and $f\colon x \to_0 x'$ (in X) that give in C a commutative square of transversal maps

$$
\begin{array}{ccc}
Ua & \overset{c}{\longrightarrow} & Fx \\
{\scriptstyle Uh}\downarrow & & \downarrow{\scriptstyle Ff} \qquad Ff.c = c'.Uh. \\
Ua' & \underset{c'}{\longrightarrow} & Fx'
\end{array}
\tag{7.26}
$$

Faces and transversal composition are obvious. The degeneracies are defined using the fact that U is *colax* and F is *lax*:

$$e_i(a, x; c\colon Ua \to_0 Fx) = (e_i a, e_i x; \underline{F}_i(x).e_i c.\underline{U}_i(a)). \tag{7.27}$$

Similarly the i-concatenation is defined as follows

$$
\begin{aligned}
(a, x; c\colon Ua \to Fx) +_i (b, y; d\colon Ub \to Fy) &= (a +_i b, x +_i y; u), \\
u = \underline{F}_i(x, y).(c +_i d).\underline{U}_i(a, b)\colon U(a +_i b) &\to_0 F(x +_i y),
\end{aligned}
\tag{7.28}
$$

$$U(a +_i b) \xrightarrow{U_i(a,b)} Ua +_i Ub \xrightarrow{c+_id} Fx +_i Fy \xrightarrow{E_i(x,y)} F(x +_i y).$$

The associativity comparison for the i-composition of three i-consecutive $\mathbf{i}$-cubes of $U \downarrow F$

$$(a, x; c), \qquad (a', x'; c'), \qquad (a'', x''; c''),$$

is given by the pair $(\kappa_i(\underline{a}), \kappa_i(\underline{x}))$ of associativity i-isomorphisms for our two triples of i-cubes, namely $\underline{a} = (a, a', a'')$ in A and $\underline{x} = (x, x', x'')$ in X

$$(\kappa_i(\underline{a}), \kappa_i(\underline{x})): \; (a, x; c) +_i ((a', x'; c') +_i (a'', x''; c'')) \to_0 \atop ((a, x; c) +_i (a', x'; c')) +_i (a'', x''; c''). \tag{7.29}$$

The coherence of this $\mathbf{i}$-map, as in diagram (7.26) above, is deferred to Exercise 7.2.3.

Similarly one constructs the unitors λ_i, ρ_i and the interchangers χ_{ii} of $U \downarrow F$, using those of A and X.

Finally, the strict functors P and Q are projections, while the component of the 'transformation' π on the $\mathbf{i}$-cube $(a, x; c)$ of $U \downarrow F$ is the transversal map

$$\pi(a, x; c) = c \colon Ua \to_0 Fx. \tag{7.30}$$

7.2.3 Exercise

The pair $(\kappa_i(\underline{a}), \kappa_i(\underline{x}))$ in (7.29) is indeed an $\mathbf{i}$-map of $U \downarrow F$.

7.2.4 Theorem (Universal properties of commas)

(a) For a pair of lax functors G, H and a cell φ as below (in $\mathbb{C}mc$) there is a unique lax functor $L \colon Z \to U \downarrow F$ such that $G = PL$, $H = QL$ and $\varphi = \psi \mid \pi$ where the cell ψ is defined by the identity $1 \colon QL \to H$ (a horizontal transformation of lax functors)

$$\tag{7.31}$$

Moreover, L is pseudo or strict if and only if both G and H are.

(b) A similar property holds for a pair of colax functors $V \colon Z \to X$, $W \colon Z \to A$ and a double cell $\varphi' \colon (V \overset{1}{F} UW)$.

Proof (a) L is defined as follows on an **i**-cube x and an **i**-map $f\colon x \to x'$ of Z

$$L(x) = (Gx, Hx; \varphi x\colon UGx \to FHx), \qquad L(f) = (Gf, Hf). \qquad (7.32)$$

The comparison transversal maps $\underline{L}_i$ are constructed with the laxity transversal maps of G and H (for t in $Z_{i|i}$ and $z = x +_i y$ in Z_i), and are invertible or degenerate if and only if the latter are

$$\underline{L}_i t = (\underline{G}_i t, \underline{H}_i t)\colon e_i(Lt) \to Le_i(t),$$
$$\underline{L}_i(x, y) = (\underline{G}_i(x, y), \underline{H}_i(x, y))\colon Lx +_i Ly \to Lz. \qquad (7.33)$$

Here $Lx +_i Ly$ is the **i**-cube defined as follows, applying (7.27)–(7.28)

$$\begin{aligned} Lx +_i Ly &= (Gx, Hx; \varphi x) +_i (Gy, Hy; \varphi y) \\ &= (Gx +_i Gy, Hx +_i Hy;\ u), \\ u = \underline{F}_i(Hx, Hy).(\varphi x &+_i \varphi y).\underline{U}_i(Gx, Gy)\colon \\ U(Gx &+_i Gy) \to F(Hx +_i Hy), \end{aligned} \qquad (7.34)$$

$$U(Gx +_i Gy) \longrightarrow UGx +_i UGy \longrightarrow FHx +_i FHy \longrightarrow F(Hx +_i Hy).$$

The coherence condition (7.26) on the transversal map $\underline{L}_i(x, y)$ of $U \downarrow F$, i.e. the commutativity of the square

$$\begin{array}{ccc} U(Gx +_i Gy) & \xrightarrow{\ u\ } & F(Hx +_i Hy) \\ {\scriptstyle U\underline{G}_i(x,y)}\big\downarrow & & \big\downarrow{\scriptstyle F\underline{H}_i(x,y)} \\ UGz & \xrightarrow[\ \varphi z\ ]{} & FHz \end{array} \qquad (7.35)$$

follows from the coherence condition (dc.3) of $\varphi\colon (1 \ {}^{G}_{FH}\, U)$ as a double cell in $\mathbb{C}\mathrm{mc}$. In fact, this condition gives the commutative diagram

where the lower row is the composed comparison of FH.

The uniqueness of L is obvious. $\qquad\qquad\qquad\qquad\qquad\qquad\qquad\square$

7.3 Main definitions

We now define a colax-lax adjunction between chiral multiple categories, a notion that occurs naturally in various situations – as we have already seen in Section 7.1.

7.3.1 Multiple adjunctions

A *multiple adjunction* $(\eta, \varepsilon)\colon F \dashv G$, or a colax-lax adjunction between chiral multiple categories, will be an orthogonal adjunction in the double category $\mathbb{C}$mc (as defined in 4.1.2).

The data amount thus to:

- a colax functor $F\colon \mathsf{X} \to \mathsf{A}$, the left adjoint,
- a lax functor $G\colon \mathsf{A} \to \mathsf{X}$, the right adjoint,
- two $\mathbb{C}$mc-cells $\eta\colon 1_\mathsf{X} \dashrightarrow GF$ and $\varepsilon\colon FG \dashrightarrow 1_\mathsf{A}$ (unit and counit) that satisfy the triangle equalities:

$$
\begin{array}{ccc}
\mathsf{X} =\!=\!= \mathsf{X} & \mathsf{A} \xrightarrow{\;G\;} \mathsf{X} & \eta \otimes \varepsilon = 1_F, \\[2pt]
\left\downarrow{\scriptstyle F}\quad \eta \quad\right\| & \left\|\quad \varepsilon \quad\right\downarrow{\scriptstyle F} & \\[2pt]
\mathsf{A} \xrightarrow[\;G\;]{} \mathsf{X} & \mathsf{A} =\!=\!= \mathsf{A} & \varepsilon \mid \eta = e_G.
\end{array}
\tag{7.36}
$$

We speak of a *pseudo-lax* (resp. a *colax-pseudo*) adjunction when the left (resp. right) adjoint is pseudo, and of a *pseudo* (or *strict*) adjunction when both adjoints are pseudo (or strict).

From general properties (see 4.1.2), we already know that the left adjoint of a lax functor G is determined up to transversal isomorphism (which amounts to a special invertible cell between vertical arrows in $\mathbb{C}$mc), and that adjunctions can be composed.

As in 7.2.1, the arrow of a colax functor is marked with a dot *when displayed vertically*, in a double cell of $\mathbb{C}$mc. Again, we may write unit and counit as $\eta\colon 1 \dashrightarrow GF$ and $\varepsilon\colon FG \dashrightarrow 1$, but we know that the coherence conditions of these 'transformations' involve the comparison cells of F and G. As for double categories, in Chapter 4, a general colax-lax adjunction cannot be seen as an adjunction in some 2-category; but this is possible for a *pseudo-lax* or a *colax-pseudo* adjunction, as we shall see in the next section.

7.3.2 A description

A colax-lax adjunction $(\eta, \varepsilon)\colon F \dashv G$ between the cm-categories X and A consists thus of the following items.

(a) A colax functor $F: \mathsf{X} \to \mathsf{A}$, with comparison transversal maps

$$\underline{F}_i(x): F(e_i x) \to_0 e_i(Fx), \qquad \underline{F}_i(x, y): F(x +_i y) \to_0 Fx +_i Fy.$$

(b) A lax functor $G: \mathsf{A} \to \mathsf{X}$, with comparison transversal maps

$$\underline{G}_i(a): e_i(Ga) \to_0 G(e_i a), \qquad \underline{G}_i(a, b): Ga +_i Gb \to_0 G(a +_i b).$$

(c) An ordinary adjunction $F_{\mathbf{i}} \dashv G_{\mathbf{i}}$ of the transversal components, for every positive multi-index $\mathbf{i}$

$$\eta_{\mathbf{i}}: 1 \to G_{\mathbf{i}} F_{\mathbf{i}}: \mathsf{X}_{\mathbf{i}} \to \mathsf{X}_{\mathbf{i}}, \qquad \varepsilon_{\mathbf{i}}: F_{\mathbf{i}} G_{\mathbf{i}} \to 1: \mathsf{A}_{\mathbf{i}} \to \mathsf{A}_{\mathbf{i}},$$

$$\varepsilon_{\mathbf{i}} F_{\mathbf{i}} . F_{\mathbf{i}} \eta_{\mathbf{i}} = 1_{F\mathbf{i}}, \qquad\qquad G_{\mathbf{i}} \varepsilon_{\mathbf{i}} . \eta_{\mathbf{i}} G_{\mathbf{i}} = 1_{G\mathbf{i}}.$$

Explicitly, point (c) means that we are assigning:

- transversal maps $\eta_{\mathbf{i}} x: x \to_0 G_{\mathbf{i}} F_{\mathbf{i}} x$ in X (for x in $\mathsf{X}_{\mathbf{i}}$), also written as $\eta x: x \to_0 GFx$,

- transversal maps $\varepsilon_{\mathbf{i}} a: F_{\mathbf{i}} G_{\mathbf{i}} a \to_0 a$ in A (for a in $\mathsf{A}_{\mathbf{i}}$), also written as $\varepsilon a: FGa \to_0 a$,

satisfying the naturality conditions (ad.1), for $f: x \to_0 y$ in X and $h: a \to_0 b$ in A, and the triangle equations (ad.2), for x in X and a in A

(ad.1) $\eta y . f = GFf . \eta x, \qquad\qquad \varepsilon b . FGh = h . \varepsilon a,$

(ad.2) $\varepsilon Fx . F\eta x = 1_{Fx}, \qquad\qquad G\varepsilon a . \eta Ga = 1_{Ga}.$

(d) These families $\eta = (\eta_{\mathbf{i}})$ and $\varepsilon = (\varepsilon_{\mathbf{i}})$ must be consistent with faces

$$\eta(\partial_i^\alpha x) = \partial_i^\alpha(\eta x), \qquad\qquad \varepsilon(\partial_i^\alpha a) = \partial_i^\alpha(\varepsilon a), \qquad (7.37)$$

and with the geometric compositions (with respect to the comparison maps of F and G):

(ad.3) (*coherence of η and ε with i-identities*) for x in X and a in A:

$$G\underline{F}_i(x) . \eta(e_i x) = \underline{G}_i(Fx) . e_i(\eta x), \qquad (7.38)$$

$$\varepsilon(e_i a) . F\underline{G}_i(a) = e_i(\varepsilon a) . \underline{F}_i(Ga), \qquad (7.39)$$

(ad.4) (*coherence of η and ε with i-composition*) for $z = x +_i y$ in X and $c = a +_1 b$ in A:

$$G\underline{F}_i(x, y) . \eta z = \underline{G}_i(Fx, Fy) . (\eta x +_i \eta y), \qquad (7.40)$$

$$\varepsilon c . F\underline{G}_i(a, b) = (\varepsilon a +_i \varepsilon b) . \underline{F}_i(Ga, Gb), \qquad (7.41)$$

$$z \xrightarrow{\ \eta z\ } GFz \qquad F(Ga +_i Gb) \xrightarrow{\ F\underline{G}_i(a,b)\ } FGc$$

with vertical maps $\eta x +_i \eta y$, $G\underline{F}_i(x,y)$ on the left diagram and $\underline{F}_i(Ga,Gb)$, εc on the right, and bottom rows

$$GFx +_i GFy \xrightarrow{\ \underline{G}_i(Fx,Fy)\ } G(Fx +_i Fy) \qquad FGa +_i FGb \xrightarrow{\ \varepsilon a +_i \varepsilon b\ } c$$

When F and G are unitary, (ad.3) amounts to: $\eta(e_i x) = e_i(\eta x)$ and $\varepsilon(e_i a) = e_i(\varepsilon a)$.

7.3.3 Lemma

(a) In a colax-lax cm-adjunction $(\eta, \varepsilon)\colon F \dashv G$ the comparison maps of G determine the comparison maps of F, through the ordinary adjunctions $F_\mathbf{i} \dashv G_\mathbf{i}$, as

$$\underline{F}_i(x) = \varepsilon e_i(Fx).F\underline{G}_i(Fx).Fe_i(\eta x)\colon Fe_i(x) \to e_i Fx, \tag{7.42}$$

$$\underline{F}_i(x,y) = \varepsilon(Fx +_i Fy).F\underline{G}_i(Fx,Fy).F(\eta x +_i \eta y)\colon \tag{7.43}$$
$$F(x +_i y) \to Fx +_i Fy.$$

Dually, the comparison maps of F determine the comparison maps of G, through the ordinary adjunctions.

(b) If all the components of η, ε are invertible, then G is pseudo if and only if F is.

Note. Loosely speaking, point (a) says that a lax multiple functor *can only have* a colax left adjoint (if any), and symmetrically; this fact will be made precise in Corollary 7.3.7. Moreover Theorem 7.4.3 will show that if a lax functor has a *lax* left adjoint, the latter is necessarily pseudo.

Proof (a) The first equation of (ad.3) says that the adjoint map of $\underline{F}_i(x)$, i.e. $(\underline{F}_i(x))' = G\underline{F}_i(x).\eta(e_i x)$, must be equal to $f = \underline{G}_i(Fx).e_i(\eta x)$. The adjoint map of the latter gives $\underline{F}_i(x) = f' = \varepsilon e_i(Fx).F(f)$. Similarly for $\underline{F}_i(x,y)$, from (ad.4).

(b) A straightforward consequence of the previous point and its dual. □

7.3.4 Theorem (Characterisation by transversal hom-sets)

A multiple adjunction $(\eta, \varepsilon)\colon F \dashv G$ can equivalently be given by a colax functor $F\colon \mathsf{X} \to \mathsf{A}$, a lax functor $G\colon \mathsf{A} \to \mathsf{X}$ and a family $(L_\mathbf{i})$ of isomorphisms of ordinary functors, indexed by the positive multi-indices $\mathbf{i} \subset \mathbb{N}$

$$L_\mathbf{i}\colon \mathbf{A_i}(F_\mathbf{i}-, =) \to \mathbf{X_i}(-, G_\mathbf{i} =)\colon (\mathbf{X_i})^{\mathrm{op}} \times \mathbf{A_i} \to \mathbf{Set}, \tag{7.44}$$

$$L_\mathbf{i}(x,a)\colon \mathbf{A_i}(Fx, a) \to \mathbf{X_i}(x, Ga).$$

The components $L_i(x, a)$, *also written as* $L(x, a)$ *or just* L, *must be consistent with faces and the positive operations (through the comparison maps of* F *and* G), *i.e. must satisfy the following conditions:*

(adh.1) $\quad L_i(\partial_i^\alpha x, \partial_i^\alpha a) = \partial_i^\alpha(L_i(x, a))$,

(adh.2) $\quad L(e_i x, e_i a)(e_i h.\underline{F}_i x) = \underline{G}_i a.e_i(Lh) \qquad$ (*for* $h\colon Fx \to_0 a$ *in* $\mathbf{A}_{i|i}$),

$$Fe_i x \xrightarrow{\ F_i\ } e_i Fx \xrightarrow{\ e_i h\ } e_i a \qquad\qquad e_i x \xrightarrow{\ e_i(Lh)\ } e_i Ga \xrightarrow{\ G_i\ } Ge_i a,$$

(adh.3) $\quad L(x +_i y, a +_i b)((h +_i k).\underline{F}_i(x, y)) = \underline{G}_i(a, b).(Lh +_i Lk)$

$$\qquad\qquad\qquad\qquad (\textit{for } h\colon Fx \to_0 a,\ k\colon Fy \to_0 b \textit{ in } \mathbf{A}_i),$$

$$F(x +_i y) \xrightarrow{\ F_i\ } Fx +_i Fy \xrightarrow{\ h +_i k\ } a +_i b$$

$$x +_i y \xrightarrow{\ Lh +_i Lk\ } Ga +_i Gb \xrightarrow{\ G_i\ } G(a +_i b).$$

In this equivalence $L_i(x, a)$ *is defined by the unit* η *as*

$$L_i(x, a)(h) = Gh.\eta_i x\colon x \to_0 Ga \quad (\textit{for } h\colon Fx \to_0 a \textit{ in } \mathbf{A}_i). \qquad (7.45)$$

The other way round, the component $\eta_i\colon 1 \to G_i F_i\colon \mathbf{X}_i \to \mathbf{X}_i$ *of the unit is defined by* L *as*

$$\eta_i(x) = L_i(x, Fx)(\mathrm{id}Fx)\colon x \to_0 GF(x) \qquad\qquad (\textit{for } x \textit{ in } \mathbf{X}_i). \qquad (7.46)$$

Proof We only have to verify the equivalence of the conditions (7.37)–(7.41) with the conditions above.

This is straightforward. For instance, to show that (7.40) implies (adh.3), let $h\colon Fx \to a$ and $k\colon Fy \to b$ be i-consecutive i-maps in A, and apply $L = L(x +_i y, a +_i a)$ as defined above, in (7.45):

$$L((h +_i k).\underline{F}_i(x, y))$$
$$= G(h +_i k).G\underline{F}_i(x, y).\eta(x +_i y)$$
$$= G(h +_i k).\underline{G}_i(Fx, Fy).(\eta x +_i \eta y) \qquad\qquad\qquad (\text{by } (7.40))$$
$$= \underline{G}_i(a, b).(Gh +_i Gk).(\eta x +_i \eta y) \qquad (\text{by axiom (lmf.2) in } 6.5.2)$$
$$= \underline{G}_i(a, b).(Lh +_i Lk).$$

$\square$

7.3.5 Corollary (Characterisation by multiple commas)

With the previous notation, a multiple adjunction amounts to an isomorphism of chiral multiple categories $L\colon F \downarrow A \to X \downarrow G$ *over the product* $X \times A$

$$F \downarrow A \xrightarrow{\quad L \quad} X \downarrow G \qquad \qquad (7.47)$$
$$\searrow \qquad \swarrow$$
$$X \times A$$

Proof It is a straightforward consequence of the previous theorem. $\square$

7.3.6 Theorem (Right adjoint by universal properties)

Let a colax functor $F\colon X \to A$ *be given.*

The existence and choice of a right adjoint lax functor G *amounts to a family (rad.***i***) of conditions and choices, indexed by the positive multi-indices* **i***:*

(rad.**i**) *for every* **i***-cube* a *in* A *there is a universal arrow*

$$(Ga, \varepsilon_{\mathbf{i}} a\colon F(Ga) \to_0 a)$$

from the functor $F_{\mathbf{i}}\colon X_{\mathbf{i}} \to A_{\mathbf{i}}$ *to the object* x*, and we choose one, consistently with faces.*

Explicitly, the universal property means that, for each **i***-cube* x *in* X *and* **i***-map* $h\colon Fx \to_0 a$ *in* A *there is a unique* $f\colon x \to_0 Ga$ *such that* $h = \varepsilon a.Ff\colon Fx \to_0 F(Ga) \to_0 a.$

The comparison **i***-maps of* G

$$\underline{G}_i(a)\colon e_i(Ga) \to_0 G(e_i a), \qquad \underline{G}_i(a,b)\colon Ga +_i Gb \to_0 G(a +_i b), \quad (7.48)$$

are then given by the universal property of ε*, as the unique solution of the equations* (7.39)*,* (7.41)*, respectively.*

Proof The conditions (rad.**i**) are plainly necessary.

Conversely, each of them provides an ordinary adjunction $(\eta_{\mathbf{i}}, \varepsilon_{\mathbf{i}})\colon F_{\mathbf{i}} \dashv G_{\mathbf{i}}$ between the categories $X_{\mathbf{i}}$ and $A_{\mathbf{i}}$, so that G, η and ε are correctly defined – as far as cubes, transversal maps, faces, transversal composition and transversal identities are concerned.

Now, we define the comparison maps $\underline{G}_i$ as specified in the statement, so that the coherence properties of ε, in (7.39) and (7.41), are satisfied. One verifies easily the axioms of naturality and coherence for these comparisons (see 6.5.2).

Finally, we have to prove that $\eta\colon 1 \dashrightarrow GF$ satisfies the coherence property (7.40)

$$GF_i(x,y).\eta z = G_i(Fx, Fy).(\eta x +_i \eta y), \qquad (7.49)$$

for a composite $z = x +_i y$ of **i**-cubes in X (similarly one proves property (7.38)).

Applying the universal property of ε, it is sufficient to show that the composite $\varepsilon(Fx +_i Fy).F(-)$ takes the same value on both terms of (7.49). In fact we get $F_i(x,y)$ in both cases:

$$\varepsilon(Fx +_i Fy).FGF_i(x,y).F\eta z = F_i(x,y).\varepsilon Fz.F\eta z = F_i(x,y),$$

$$\begin{aligned}
&\varepsilon(Fx +_i Fy).FG_i(Fx, Fy).F(\eta x +_i \eta y) \\
&= (\varepsilon Fx +_i \varepsilon Fy).F_i(GFx, GFy).F(\eta x +_i \eta y) &\text{(by (7.41))} \\
&= (\varepsilon Fx +_i \varepsilon Fy).(F\eta x +_i F\eta y).F_i(x,y) &\text{(by naturality of } F_i) \\
&= (\varepsilon Fx.F\eta x +_i \varepsilon Fy.F\eta y).F_i(x,y) &\text{(by 0}i\text{-interchange)} \\
&= (1_{Fx} +_i 1_{Fy}).F_i(x,y) = F_i(x,y).
\end{aligned}$$

$$\square$$

7.3.7 Corollary (Right adjoint by transversal components)

A colax functor $F\colon \mathsf{X} \to \mathsf{A}$ has a lax right adjoint if and only if each component $F_i\colon \mathbf{X_i} \to \mathbf{A_i}$ has an (ordinary) right adjoint $G_i\colon \mathbf{A_i} \to \mathbf{X_i}$, consistently with faces

$$\partial_i^\alpha G_i = G_{i|i}\partial_i^\alpha \qquad (i \in \mathbf{i}). \qquad (7.50)$$

Proof A straightforward consequence of the previous theorem. $\square$

7.3.8 Theorem (Factorisation of adjunctions)

Let $F \dashv G$ be a colax-lax adjunction between X and A. Using the isomorphism of cm-categories $L\colon F\downarrow\mathsf{A} \to \mathsf{X}\downarrow G$ (of Corollary 7.3.5), we can factorise the adjunction

$$\mathsf{X} \underset{P}{\overset{F'}{\rightleftarrows}} F\downarrow\mathsf{A} \underset{L^{-1}}{\overset{L}{\rightleftarrows}} \mathsf{X}\downarrow G \underset{G'}{\overset{Q}{\rightleftarrows}} \mathsf{A} \qquad (7.51)$$

as a composite of:

- a coreflective colax-strict adjunction $F' \dashv P$ (with unit $PF' = 1$),

- *an isomorphism $L \dashv L^{-1}$,*

- *a reflective strict-lax adjunction $Q \dashv G'$ (with counit $QG' = 1$),*

where the strict functors P and Q are comma-projections.

Proof We define the lax functor $G' \colon \mathsf{A} \to \mathsf{X} \downarrow G$ by the universal property of commas 7.2.4(a), applied to $G \colon \mathsf{A} \to \mathsf{X}$, $1 \colon \mathsf{A} \to \mathsf{A}$ and $\varphi = e_G$, as in the diagram below

$$
\begin{array}{ccc}
\mathsf{A} \xrightarrow{\ G\ } \mathsf{X} & \quad & \mathsf{A} \xrightarrow{G'} \mathsf{X} \downarrow G \xrightarrow{\ P\ } \mathsf{X} \\
\downarrow^{1} \quad \varphi \quad \downarrow^{1} & = & \downarrow^{1} \quad \psi \quad \downarrow^{Q} \quad \pi \quad \downarrow^{1} \\
\mathsf{A} \xrightarrow[1]{} \mathsf{A} \xrightarrow[G]{} \mathsf{X} & \quad & \mathsf{A} \xrightarrow[\ \ 1\ \ ]{} \mathsf{A} \xrightarrow[\ \ G\ \ ]{} \mathsf{X}
\end{array}
\qquad (7.52)
$$

$$
\begin{aligned}
& G'(a) = (Ga, a; 1 \colon Ga \to Ga), \\
& \underline{G}'_i(a) = (\underline{G}_i(a), 1) \colon (e_i Ga, e_i a; \underline{G}_i a) \to (G(e_i a), e_i a; 1), \\
& \underline{G}'_i(a, b) = (\underline{G}_i(a, b), 1) \colon \\
& \qquad (Ga +_i Gb, a +_i b; \underline{G}_i(a, b)) \to (G(a +_i b), a +_i b; 1).
\end{aligned}
$$

The colax functor $F' \colon \mathsf{X} \to F \downarrow \mathsf{A}$ is defined by the dual property 7.2.4(b)

$$
\begin{aligned}
& F'(x) = (x, Fx; 1 \colon Fx \to Fx), \\
& \underline{F}'_i(x) = (1, \underline{F}_i(x)) \colon (e_i x, F(e_i x); 1) \to (e_i x, Fe_i x; \underline{F}_i(x)), \\
& \underline{F}'_i(x, y) = (1, \underline{F}_i(x, y)) \colon \\
& \qquad (x +_i y, F(x +_i y); 1) \to (x +_i y, Fx +_i Fy; \underline{F}_i(x, y)).
\end{aligned}
\qquad (7.53)
$$

The coreflective adjunction $F' \dashv P$ is obvious

$$
\begin{aligned}
& \eta' x = 1_x \colon x \to PF' x, \\
& \varepsilon'(x, a; c \colon Fx \to a) = (1_x, c) \colon (x, Fx; 1_{Fx}) \to (x, a; c),
\end{aligned}
\qquad (7.54)
$$

as well as the reflective adjunction $Q \dashv G'$, and the factorisation (7.51). $\qquad \square$

7.4 Multiple adjunctions and pseudo functors

We consider now cm-adjunctions where the left or right adjoint is a pseudo functor, and we introduce adjoint equivalences of chiral multiple categories.

7.4.1 Comments

Let us recall, from 7.3.1, that a *pseudo-lax* cm-adjunction $F \dashv G$ is a multiple adjunction between cm-categories where the left adjoint F is pseudo.

Here the comparison cells of F are horizontally invertible and the composites GF and FG are lax functors; it follows (from 7.2.1) that the unit and counit are horizontal transformations of lax functors. Therefore a pseudo-lax cm-adjunction gives an adjunction in the 2-category LxCmc of cm-categories, lax functors and their transversal transformations (introduced in 7.2.1); we shall prove that these two facts are actually equivalent (Theorem 7.4.3).

7.4.2 Theorem (Companions in ℂmc)

A lax functor G has an orthogonal companion F in the double category $\mathbb{C}$mc if and only if it is pseudo; then one can define $F = G_*$ as the colax functor which coincides with G except for comparison maps, that are transversally inverse to those of G.

Proof We restrict to unitary cm-categories, for simplicity. If G is pseudo, it is obvious that G_*, as defined above, is an orthogonal companion.

Conversely, suppose that the lax functor $G\colon \mathsf{A} \to \mathsf{X}$ has an orthogonal companion F (colax). There are thus two double cells η, ε in ℂmc

$$
\begin{array}{ccc}
\mathsf{A} = \mathsf{A} & \qquad & \mathsf{A} \xrightarrow{\;G\;} \mathsf{X} \\
\| \quad \eta \quad \downarrow F & & F \downarrow \quad \varepsilon \quad \| \\
\mathsf{A} \xrightarrow[G]{} \mathsf{X} & & \mathsf{X} = \mathsf{X}
\end{array}
\qquad (7.55)
$$

which satisfy the equations $\eta \,|\, \varepsilon = e_G,\ \eta \otimes \varepsilon = 1_F$.

This means two 'transformations' $\eta\colon F \dashrightarrow G$ and $\varepsilon\colon G \dashrightarrow F$, as defined in 7.2.1; for every **i**-cube a in A, we have two transversal maps ηa and εa in X

$$
\eta a\colon Fa \to Ga, \qquad\qquad \varepsilon a\colon Ga \to Fa, \qquad (7.56)
$$

consistently with faces. These maps are transversally inverse, because of the previous equations (see (7.24))

$$
\eta a.\varepsilon a = (\eta \,|\, \varepsilon)(a) = 1_{Ga}, \qquad \varepsilon a.\eta a = (\eta \otimes \varepsilon)(a) = 1_{Fa}. \qquad (7.57)
$$

Applying now the coherence condition (dc.3) (of 7.2.1) for the transformations η, ε and a concatenation $c = a +_i b$ in A we find

$$\eta c = \underline{G}_i(a,b).(\eta a +_i \eta b).\underline{F}_i(a,b)\colon Fc \to Gc,$$
$$\varepsilon a +_i \varepsilon b = \underline{F}_i(a,b).\varepsilon c.\underline{G}_i(a,b)\colon Ga +_i Gb \to Fa +_i Fb. \tag{7.58}$$

Since all the components of η and ε are transversally invertible, this proves that $\underline{G}_i(a,b)$ has a right inverse and a left inverse transversal map, whence it is invertible. Similarly for the comparisons of degeneracies.

Therefore G is pseudo and F is transversally isomorphic to G_*. □

7.4.3 Theorem

(a) (Pseudo-lax adjunctions) For every adjunction $F \dashv G$ in the 2-category **LxCmc**, *the functor F is pseudo and the adjunction is pseudo-lax, in the sense of 7.3.1.*

(b) (Colax-pseudo adjunctions) For every adjunction $F \dashv G$ in the 2-category **CxCmc**, *the functor G is pseudo and the adjunction is colax-pseudo.*

Proof It suffices to prove (a); again, we only deal with the comparisons of a composition.

Let the lax structures of $F\colon \mathsf{X} \to \mathsf{A}$ and $G\colon \mathsf{A} \to \mathsf{X}$ be given by the following comparison maps, where $z = x +_i y$ and $c = a +_i b$

$$\lambda_i(x,y)\colon Fx +_i Fy \to Fz, \qquad \underline{G}_i(a,b)\colon Ga +_i Gb \to Gc,$$

so that we have:

$$\eta z = G\lambda_i(x,y).\underline{G}_i(Fx,Fy).(\eta x +_i \eta y)\colon z \to GFz,$$
$$\varepsilon a +_i \varepsilon b = \varepsilon c.F\underline{G}_i(a,b).\lambda_i(Ga,Gb)\colon FGa +_i FGb \to c. \tag{7.59}$$

We construct a colax structure $\underline{F}$ for F, letting

$$\underline{F}_i(x,y) = \varepsilon(Fx +_i Fy).F\underline{G}_i(Fx,Fy).F(\eta x +_i \eta y)\colon Fz \to Fx +_i Fy.$$

Now it is sufficient to verify that $\underline{F}_i(x,y)$ and $\lambda_i(x,y)$ are transversally inverse:

$$\lambda_i(x,y).\underline{F}_i(x,y)$$
$$= \lambda_i(x,y).\varepsilon(Fx +_i Fy).F\underline{G}_i(Fx,Fy).F(\eta x +_i \eta y)$$
$$= \varepsilon Fz.FG\lambda_i(x,y).F\underline{G}_i(Fx,Fy).F(\eta x +_i \eta y) \qquad \text{(by naturality of } \varepsilon\text{)}$$
$$= \varepsilon Fz.F(\eta z) = 1_{Fz} \qquad \text{(by (7.59)),}$$

$\underline{F}_i(x,y).\lambda_i(x,y)$

$\quad = \varepsilon(Fx +_i Fy).F\underline{G}_i(Fx, Fy).F(\eta x +_i \eta y).\lambda_i(x, y)$

$\quad = \varepsilon(Fx +_i Fy).F\underline{G}_i(Fx, Fy).\lambda_i(GFx, GFy).(F\eta x +_i F\eta y)$

$\qquad\qquad\qquad\qquad\qquad\qquad\qquad\qquad$ (by naturality of λ)

$\quad = (\varepsilon Fx +_i \varepsilon Fy).(F\eta x +_i F\eta y)$ $\qquad\qquad\qquad$ (by (7.59))

$\quad = (\varepsilon Fx.F\eta x) +_i (\varepsilon Fy.F\eta y)$ $\qquad\qquad$ (by $0i$-interchange)

$\quad = 1_{Fx} +_i 1_{Fy} = 1_{Fx +_i Fy}.$

More formally, (a) can be rewritten saying that, in the double category $\mathbb{C}$mc, if the horizontal arrow G has a 'horizontal left adjoint' H (within the horizontal 2-category $\mathbf{Hor}\mathbb{C}$mc = $\mathbf{LxCmc}$), then it also has an orthogonal adjoint F (colax). After this is proved as above, we apply Proposition 4.1.4 showing that H and F are companions, whence H is pseudo, by Theorem 7.4.2, and isomorphic to F. $\qquad\qquad\qquad\qquad\qquad\qquad\square$

7.4.4 Equivalences of cm-categories

An *equivalence* between two cm-categories X and A will be a pseudo functor $F\colon \mathsf{X} \to \mathsf{A}$ which has a weak inverse $G\colon \mathsf{A} \to \mathsf{X}$, with $GF \cong 1$ and $FG \cong 1$.

An *adjoint equivalence* will be a pseudo cm-adjunction $(\eta, \varepsilon)\colon F \dashv G$ where the transversal transformations $\eta\colon 1_{\mathsf{X}} \to GF$ and $\varepsilon\colon FG \to 1_{\mathsf{A}}$ are invertible.

The following properties of a lax or colax functor $F\colon \mathsf{X} \to \mathsf{A}$ will allow us (in the next theorem) to characterise the equivalence of cm-categories in the usual way, under the mild restriction of *transversal invariance*.

(a) We say that F is (transversally) *faithful* if all the ordinary functors $F_{\mathbf{i}}\colon \mathbf{X_i} \to \mathbf{A_i}$ (between the categories of **i**-cubes and their transversal maps) are faithful: given two **i**-maps $f, g\colon x \to_0 y$ of X between the same **i**-cubes, $F(f) = F(g)$ implies $f = g$.

(b) Similarly, we say that F is (transversally) *full* if all the ordinary functors $F_{\mathbf{i}}\colon \mathbf{X_i} \to \mathbf{A_i}$ are full: for every **i**-map $h\colon Fx \to_0 Fy$ in A there is an **i**-map $f\colon x \to_0 y$ in X such that $F(f) = h$.

(c) Finally, we say that F is *essentially surjective on cubes*, or on extended objects, if every functor $F_{\mathbf{i}}$ is essentially surjective on objects: for every **i**-cube a in A there is some **i**-cube x in X and some invertible **i**-map $h\colon F(x) \to_0 a$ in A.

7.4.5 Theorem (Characterisations of equivalences)

Let $F\colon \mathsf{X} \to \mathsf{A}$ be a pseudo functor between two transversally invariant cm-categories (see 6.5.5). The following conditions are equivalent:

(i) $F\colon \mathsf{X} \to \mathsf{A}$ belongs to an adjoint equivalence of cm-categories,

(ii) $F\colon \mathsf{X} \to \mathsf{A}$ is an equivalence of cm-categories,

(iii) F is faithful, full and essentially surjective on cubes (see 7.4.4),

(iv) every ordinary functor $F_{\mathbf{i}}\colon \mathsf{X}_{\mathbf{i}} \to \mathsf{A}_{\mathbf{i}}$ is an equivalence of categories.

Moreover, if F is unitary, one can make its 'quasi-inverse' unitary as well.

Proof (Of course the proof depends on the axiom of choice.) By our previous definitions in 7.4.4, conditions (iii) and (iv) are about the family of ordinary functors $(F_{\mathbf{i}})$ and are well known to be equivalent, see 1.2.5. The implications (i) $\Rightarrow$ (ii) $\Rightarrow$ (iv) are obvious.

Conversely, let us assume that every $F_{\mathbf{i}}$ is an equivalence of ordinary categories and let us extend the pseudo functor F to an adjoint equivalence, proceeding by induction on the degree $n \geqslant 0$ of the positive multi-index $\mathbf{i}$.

First, $F_*\colon \mathrm{tv}_*(\mathsf{X}) \to \mathrm{tv}_*(\mathsf{A})$ is an equivalence of categories and we begin by choosing an adjoint quasi-inverse $G_*\colon \mathrm{tv}_*(\mathsf{A}) \to \mathrm{tv}_*(\mathsf{X})$.

In other words, we choose for every object a in A some $G(a)$ in X and some isomorphism $\varepsilon a\colon FG(a) \to a$ in A; then a transversal map $h\colon a \to b$ in A is sent to the unique X-map $G(h)\colon G(a) \to G(b)$ coherent with the previous choices (since F_* is full and faithful). Finally the isomorphism $\eta x\colon x \to GF(x)$ is determined by the triangle equations (for every $*$-cube x of X).

We assume now that the components of G, ε and η have been defined up to degree $n - 1 \geqslant 0$, and define them for a multi-index $\mathbf{i}$ of degree n, taking care that the new choices be consistent with the previous ones.

For every $\mathbf{i}$-cube a in A we want to choose some $\mathbf{i}$-cube $G(a)$ in X and some $\mathbf{i}$-isomorphism $\varepsilon a\colon FG(a) \to a$ in A, consistently with all faces ∂_i^α ($i \in \mathbf{i}$). We begin by choosing an $\mathbf{i}$-cube x and an $\mathbf{i}$-isomorphism $u\colon F(x) \to a$. By the inductive hypothesis, we have a family of $2n$ transversal isomorphisms of A

$$v_i^\alpha = \partial_i^\alpha(u^{-1}).\varepsilon(\partial_i^\alpha a)\colon FG(\partial_i^\alpha a) \to \partial_i^\alpha a \to F(\partial_i^\alpha x) \quad (i \in \mathbf{i},\ \alpha = \pm),$$

which can be uniquely lifted as transversal isomorphisms t_i^α of X, since F is full and faithful

$$t_i^\alpha\colon G(\partial_i^\alpha a) \to \partial_i^\alpha x, \qquad v_i^\alpha = F(t_i^\alpha).$$

The family (v_i^α) has consistent positive faces (see 6.5.5), because this is true of the family $(\partial_i^\alpha u^{-1})_{i,\alpha}$, by commuting faces, and of the family $(\varepsilon(\partial_i^\alpha a))_{i,\alpha}$, by inductive assumption. It follows that also the family (t_i^α) has consistent positive faces.

By transversal invariance in X we can fill this family (t_i^α) with a (chosen) transversal i-isomorphism $t\colon y \to x$, and we define the i-cube $G(a)$ and the i-isomorphism εa as follows:

$$G(a) = y, \qquad \varepsilon a = u.Ft\colon FG(a) \to F(x) \to a.$$

This choice is consistent with faces:

$$\partial_i^\alpha(\varepsilon a) = (\partial_i^\alpha u).Ft_i^\alpha = (\partial_i^\alpha u).v_i^\alpha = \varepsilon(\partial_i^\alpha a).$$

Now, since F_i is full and faithful, a transversal i-map $h\colon a \to b$ in A is sent to the unique X-map $G(h)\colon G(a) \to G(b)$ satisfying the condition $\varepsilon b.F(Gh) = h.\varepsilon a$ (naturality of ε).

Again, the i-isomorphism $\eta x\colon x \to GF(x)$ is determined by the triangle equations, for every i-cube x of X.

The comparison i-maps $\underline{G}_i$ are uniquely determined by their coherence conditions (see 7.3.2)

$$\varepsilon e_i a.F\underline{G}_i(a) = e_i(\varepsilon a).\underline{F}_i(Ga) \qquad\qquad \text{(for } a \in \mathbf{A}_{i|i}),$$

$$\varepsilon c.F\underline{G}_i(a,b) = (\varepsilon a +_i \varepsilon b).\underline{F}_i(Ga, Gb) \qquad \text{(for } c = a +_i b \text{ in } \mathbf{A}_i).$$

Moreover $\underline{G}_i(a)$ and $\underline{G}_i(a,b)$ are invertible, because so are their images by F, which is full and faithful.

The construction of G, ε and η is now achieved. One ends by proving that G is indeed a pseudo functor, and that ε, η are coherent with the comparison cells of F and G.

Finally, let us assume that F is unitary: $\underline{F}_i(x)\colon F(e_i x) \to e_i(Fx)$ is always an identity. To make G unitary we assume that – in the previous inductive construction – the following constraint has been followed: for a j-degenerate i-cube $a = e_j c$ we always choose the transversal isomorphism $u = e_j(\varepsilon c)\colon F(e_j(Gc)) \to e_j c$. It follows that each

$$v_i^\alpha\colon FG(\partial_i^\alpha e_j c) \to F(\partial_i^\alpha e_j Gc)$$

is the identity; then $t_i^\alpha\colon G(\partial_i^\alpha e_j c) \to \partial_i^\alpha e_j Gc$ is the identity as well. We choose to fill their family with the identity $t\colon e_j Gc \to e_j Gc$, which gives

$$G(e_j c) = e_j Gc, \qquad \varepsilon(e_j c) = u.Ft = e_j(\varepsilon c).$$

If a is also j'-degenerate, the commutativity of degeneracies ensures that both constructions give the same result. $\qquad\square$

7.5 The multiple category of chiral multiple categories

In Section 7.2 we constructed the double category $\mathbb{C}$mc of chiral multiple categories, with lax and colax multiple functors, and suitable 'quintets' for double cells.

We extend now this structure, forming a (strict) multiple category Cmc of chiral multiple categories, indexed by 'extended' multi-indices

$$\mathbf{p} = \{p, q, r...\} \subset \boldsymbol{\omega} + \mathbf{1} = \{0, 1, 2, ..., \infty\},$$

where the p-directed arrows are the p-mixed functors of 6.5.7. The strict functors give the 0-directed arrows, and are also called 0-*mixed functors*. The (small) chiral multiple categories A, B... forming the objects of Cmc are still indexed by finite multi-indices $\mathbf{i} = \{i, j, k...\} \subset \boldsymbol{\omega} = \mathbb{N}$.

Similarly, the $(n + 1)$-dimensional multiple category $\mathsf{Cmc_n}$ of chiral n-tuple categories will be indexed by multi-indices $\mathbf{p} \subset \mathbf{n} + \mathbf{1} = \{0, 1, ..., n\}$.

Let us note that, if we restrict Cmc to the weak multiple categories of cubical type (see 6.4.5), we still have a multiple category of *non*-cubical type, with different kinds of arrows in each direction.

All this can be generalised to a multiple category Inc of intercategories (indexed by $\boldsymbol{\omega} + \mathbf{1}$), and an $(n + 1)$-dimensional multiple category $\mathsf{Inc_n}$ of n-intercategories.

This section is based on [GP12], but we use here a wider structure with a simpler construction, where all cells – including the transversal maps – are given by 'quintets': see 7.5.5(d).

7.5.1 Low dimensional cells

We begin the construction of the multiple category Cmc.

In dimension 0, the objects of Cmc are the (small) chiral multiple categories.

In dimension 1 and direction p (for $0 \leqslant p \leqslant \infty$), a p-*morphism* $F \colon \mathsf{A} \to_p \mathsf{B}$ between chiral multiple categories is a strict functor if $p = 0$, and a p-mixed functor, as defined in 6.5.7, for $p > 0$: colax in all positive directions $i < p$ and lax in all directions $i \geqslant p$. In particular, it is a lax functor when $p = 1$ and a colax functor when $p = \infty$.

7.5.2 Two-dimensional cells

To define a pq-cell (for $0 \leqslant p < q \leqslant \infty$) we have to adapt the axioms of transversal transformation in 6.5.8.

A *pq-cell* $\varphi \colon (U \; {}^F_G \; V)$ is a 'generalised quintet' consisting of two p-morphisms F, G, two q-morphisms U, V, together with, roughly speaking, a 'transversal transformation' $\varphi \colon VF \dashrightarrow GU$

$$
\begin{array}{ccc}
A \xrightarrow{\ F\ } \bullet & & \bullet \xrightarrow{\ p\ } \\[2pt]
U \downarrow \quad \varphi \quad \downarrow V & & q \downarrow \\[2pt]
\bullet \xrightarrow[\ G\ ]{} B & &
\end{array}
\qquad\qquad (7.60)
$$

Again this is an abuse of notation since, generally, there are no composites VF and GU in our structure: the coherence conditions of φ are based on the four mixed functors F, G, U, V and all their comparison maps. Precisely, the cell φ consists of a face-consistent family of transversal maps in B

$$
\begin{aligned}
\varphi(x) &= \varphi_{\mathbf{i}}(x) \colon VF(x) \to_0 GU(x), & \text{(for } x \text{ in } \mathbf{A_i}\text{)}, \\
\partial_i^\alpha . \varphi_{\mathbf{i}} &= \varphi_{\mathbf{i}|i} . \partial_i^\alpha & \text{(for } i \in \mathbf{i}\text{)},
\end{aligned}
\qquad (7.61)
$$

so that each component $\varphi_{\mathbf{i}} \colon V_{\mathbf{i}} F_{\mathbf{i}} \to G_{\mathbf{i}} U_{\mathbf{i}} \colon \mathbf{A_i} \to \mathbf{B_i}$ is a natural transformation of ordinary functors:

(nat) for all $f \colon x \to_0 y$ in A, we have a commutative diagram of transversal maps in B

$$
\begin{array}{ccc}
VFx & \xrightarrow{\ \varphi x\ } & GUx \\[2pt]
{\scriptstyle VFf}\downarrow & & \downarrow{\scriptstyle GUf} \\[2pt]
VFy & \xrightarrow[\ \varphi y\ ]{} & GUy
\end{array}
\qquad (7.62)
$$

Moreover φ has to satisfy the following coherence conditions (coh.a) – (coh.c) with the comparisons of F, G, U, V for a degenerated cube $e_i(t)$, with t in $\mathbf{A_{i|i}}$, and a composite $z = x +_i y$ in $\mathbf{A_i}$. *(Of course, if $p = 0$ the comparisons of F, G are identities.)*

(coh.a) If $p < q \leqslant i$, so that F, G, U, V are lax in direction i, we have commutative diagrams:

$$
\begin{array}{ccc}
e_i VF(t) & \xrightarrow{\ e_i(\varphi t)\ } & e_i GU(t) \\[2pt]
{\scriptstyle \underline{V_i}F}\downarrow & & \downarrow{\scriptstyle \underline{G_i}U} \\[2pt]
V(e_i Ft) & & Ge_i U(t) \\[2pt]
{\scriptstyle V\underline{F_i}}\downarrow & & \downarrow{\scriptstyle \underline{GU_i}} \\[2pt]
VF(e_i t) & \xrightarrow[\ \varphi(e_i t)\ ]{} & GU(e_i t)
\end{array}
\qquad
\begin{array}{ccc}
VFx +_i VFy & \xrightarrow{\ \varphi x +_i \varphi y\ } & GUx +_i GUy \\[2pt]
{\scriptstyle \underline{V_i}F}\downarrow & & \downarrow{\scriptstyle \underline{G_i}U} \\[2pt]
V(Fx +_i Fy) & & G(Ux +_i Uy) \\[2pt]
{\scriptstyle V\underline{F_i}}\downarrow & & \downarrow{\scriptstyle \underline{GU_i}} \\[2pt]
VFz & \xrightarrow[\ \varphi z\ ]{} & GUz
\end{array}
$$

If $p \leqslant i < q$, so that F, G are lax and U, V are colax in direction i, we have commutative diagrams (where the comparisons of U, V are reversed):

$$
\begin{array}{ccc}
e_i VF(t) \xrightarrow{e_i(\varphi t)} e_i GU(t) \\
\end{array}
\qquad
\begin{array}{ccc}
VFx +_i VFy \xrightarrow{\varphi x +_i \varphi y} GUx +_i GUy \\
\end{array}
$$

$$
\begin{array}{ccc}
\underline{V}_i F \uparrow & & \downarrow \underline{G}_i U \\
V(e_i Ft) & & Ge_i U(t) \\
V\underline{F}_i \downarrow & & \uparrow GU_i \\
VF(e_i t) \xrightarrow{\varphi(e_i t)} GU(e_i t)
\end{array}
\qquad
\begin{array}{ccc}
\underline{V}_i F \uparrow & & \downarrow \underline{G}_i U \\
V(Fx +_i Fy) & & G(Ux +_i Uy) \\
V\underline{F}_i \downarrow & & \uparrow GU_i \\
VFz \xrightarrow{\varphi z} GUz
\end{array}
$$

(coh.c) If $i < p < q$, so that F, G, U, V are colax in direction i, we have commutative diagrams:

$$
\begin{array}{ccc}
e_i VF(t) \xrightarrow{e_i(\varphi t)} e_i GU(t) \\
\end{array}
\qquad
\begin{array}{ccc}
VFx +_i VFy \xrightarrow{\varphi x +_i \varphi y} GUx +_i GUy \\
\end{array}
$$

$$
\begin{array}{ccc}
\underline{V}_i F \uparrow & & \uparrow \underline{G}_i U \\
V(e_i Ft) & & Ge_i U(t) \\
V\underline{F}_i \uparrow & & \uparrow GU_i \\
VF(e_i t) \xrightarrow{\varphi(e_i t)} GU(e_i t)
\end{array}
\qquad
\begin{array}{ccc}
\underline{V}_i F \uparrow & & \uparrow \underline{G}_i U \\
V(Fx +_i Fy) & & G(Ux +_i Uy) \\
V\underline{F}_i \uparrow & & \uparrow GU_i \\
VFz \xrightarrow{\varphi z} GUz
\end{array}
$$

The p- and q-composition of pq-cells are both defined using component-wise the transversal composition of a chiral multiple category. Namely, for a consistent matrix of pq-cells

$$(7.63)$$

$$(\varphi +_p \psi)(x) = \psi(Fx) +_0 G'(\varphi x) \colon WF'Fx \to G'VFx \to G'GUx, \quad (7.64)$$

$$(\varphi +_q \sigma)(x) = V'(\varphi x) +_0 \sigma(Ux) \colon V'VFx \to V'GUx \to HU'Ux. \quad (7.65)$$

We prove below, in Theorem 7.5.6, that these composition laws are well-defined, i.e. the cells above do satisfy the previous coherence conditions, and satisfy the middle-four interchange law.

Moreover, they have been defined via the composition of transversal maps, and therefore are strictly unitary and associative.

7.5.3 Three-dimensional cells

A *pqr-cell* (for $0 \leqslant p < q < r \leqslant \infty$) will be a 'commutative cell' Π determined by its six faces:

- two *pq*-cells φ, ψ (the faces $\partial_r^\alpha \Pi$),

- two *pr*-cells π, ρ (the faces $\partial_q^\alpha \Pi$),

- two *qr*-cells ω, ζ (the faces $\partial_p^\alpha \Pi$),

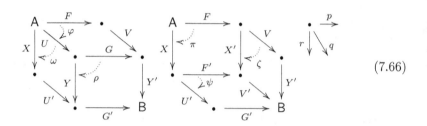

$$(7.66)$$

They must satisfy the following commutativity relation in B, for every **i**-cube x of A

$$G'\omega x.\rho Ux.Y'\varphi x = \psi Xx.V'\pi x.\zeta Fx:$$
$$Y'VFx \to_0 G'U'Xx.$$

$$(7.67)$$

These cells are composed in direction p, q, or r, by pasting 2-dimensional cells. Also here, these operations are associative, unitary and satisfy the middle-four interchange by pairs.

7.5.4 Higher cells

Finally, we have the multiple category Cmc (indexed by the ordinal $\omega + 1$), where each cell of dimension $\geqslant 4$ is 'coskeletally' determined by a face-consistent family of all its iterated faces of dimension 3. Compositions work accordingly.

In the truncated case we have the $(n+1)$-multiple category $\mathsf{Cmc_n}$ of (small) chiral n-tuple categories, where the objects are indexed by the ordinal $\mathbf{n} = \{0, ..., n-1\}$, while $\mathsf{Cmc_n}$ is indexed by $\mathbf{n}+1$ (the previous index ∞ being replaced by n). Let us note that $\mathsf{Cmc_n}$ is not an ordinary truncation of Cmc, as its objects too are truncated.

Cmc is a substructure of the – similarly defined – multiple category Inc of small infinite-dimensional intercategories.

7.5.5 Comments

These multiple categories are related to various double or triple categories previously constructed.

(a) A chiral 1-dimensional multiple category is just a category, and $\mathsf{Cmc_1}$ is the double category of $\mathbb{Q}(\mathbf{Cat})$ of quintets on the 2-category $\mathbf{Cat}$.

(b) A chiral 2-dimensional multiple category is a weak double category, and $\mathsf{Cmc_2}$ is 'similar' to the triple category $\mathsf{Stc}\mathbb{D}\mathrm{bl}$ of weak double categories (see 6.1.1): the two structures have the same arrows, but the quintets of the 01-cells are reversed.

(c) The quadruple category $\mathsf{Inc_3}$ of 3-dimensional intercategories is an extension of the triple category ICat of [GP6], Section 6, obtained by adding strict functors in the new transversal direction.

(d) The structure Cmc studied in [GP12] is a substructure of the current one, with the same cubes and restricted transversal maps.

7.5.6 Theorem

In Cmc *the composition law* $\varphi +_p \psi$ *of pq-cells is well-defined by the following formula, in* (7.64)

$$(\varphi +_p \psi)(x) = \psi(Fx) +_0 G'(\varphi x):$$
$$WF'Fx \to G'GUx, \tag{7.68}$$

in the sense that this family of transversal maps does satisfy the conditions (coh.a)–(coh.c) of 7.5.2.

   Similarly the composition $\varphi +_q \sigma$ *is well defined in* (7.65). *Moreover these laws satisfy the middle-four interchange law.*

Proof The argument is an extension of a similar one for the double category $\mathbb{D}\mathrm{bl}$ in Section 4.2, taking into account the mixed-laxity of the present 'functors'. We prove the three coherence axioms with respect to a composed cube $z = x +_i y$ in $\mathbf{A_i}$; one would work in a similar way for a degenerate cube $e_i(t)$, with t in $\mathbf{A_{i|i}}$.

   First we prove (coh.a), letting $p < q \leqslant i$, so that all our functors F, F', G, G', U, V, W are lax in direction i. This amounts to the commutativity of the outer diagram below, formed of transversal maps (the index i

is omitted in $+_i$, in all comparisons $\underline{F}_i$ etc.)

$$
\begin{array}{ccccc}
WF'Fz & \xrightarrow{\ \psi Fz\ } & G'VFz & \xrightarrow{\ G'\varphi z\ } & G'GUz \\[2pt]
\uparrow\scriptstyle{WF'\underline{F}} & & \uparrow\scriptstyle{G'V\underline{F}} & & \uparrow\scriptstyle{G'G\underline{U}} \\[2pt]
WF'(Fx+Fy) & \xrightarrow{\ \psi(Fx+Fy)\ } & G'V(Fx+Fy) & & G'G(Ux+Uy) \\[2pt]
\uparrow\scriptstyle{W\underline{F}'F} & & \uparrow\scriptstyle{G'\underline{V}F} & & \uparrow\scriptstyle{G'\underline{GU}} \\[2pt]
W(F'Fx+F'Fy) & & G'(VFx+VFy) & \xrightarrow{\ G'(\varphi x+\varphi y)\ } & G'(GUx+GUy) \\[2pt]
\uparrow\scriptstyle{\underline{W}F'F} & & \uparrow\scriptstyle{\underline{G'}VF} & & \uparrow\scriptstyle{\underline{G'}GU} \\[2pt]
WF'Fx+WF'Fy & \xrightarrow[\psi Fx+\psi Fy]{} & G'VFx+G'VFy & \xrightarrow[G'\varphi x+G'\varphi y]{} & G'GUx+G'GUy
\end{array}
$$

Indeed, the two hexagons commute by (coh.a), applied to φ and ψ, respectively. The upper square commutes by naturality of ψ on $\underline{F}_i(x,y)$. The lower square commutes by axiom (lmf.2) of 6.5.2 on the lax functor G', with respect to the transversal **i**-maps $\varphi x\colon VF(x) \to_0 GU(x)$ and $\varphi y\colon VF(y) \to_0 GU(y)$

$$
G'(\varphi x+_i\varphi y).\underline{G}_i'(VFx,VFy) = \underline{G}_i'(GUx,GUy).(G'(\varphi x)+_iG'(\varphi y)). \quad (7.69)
$$

The proof of (coh.c) is transversally dual to the previous one.

To prove (coh.b) we let $p \leqslant i < q$, so that F, F', G, G' are lax and U, V, W are colax in direction i. After reversing the comparisons of U, V, W in the diagram above, the two hexagons still commute, by (coh.b) on φ and ψ, and the two squares are unchanged. (This part corresponds to part (a) of the proof of Theorem 4.2.2.)

Finally, to verify the middle-four interchange law on the four double cells of diagram (7.63), we compute the compositions $(\varphi +_p \psi) +_q (\sigma +_p \tau)$ and $(\varphi +_q \sigma) +_p (\psi +_q \tau)$ on an **i**-cube x. This gives the two transversal maps $W'WF'Fx \to_0 H'HU'Ux$ of the upper or lower path in the following diagram

$$
\begin{array}{ccccc}
W'WF'Fx & \xrightarrow{\ W'\psi Fx\ } & W'G'VFx & \xrightarrow{\ W'G'\varphi x\ } & W'G'GUx \\[2pt]
\downarrow\scriptstyle{\tau VFx} & & \downarrow\scriptstyle{\tau GUx} & & \\[2pt]
H'V'VFx & \xrightarrow[\ H'V'\varphi x\ ]{} & H'V'GUx & \xrightarrow[\ H'\sigma Ux\ ]{} & H'HU'Ux
\end{array}
$$

The square commutes, by the naturality of the double cell τ (with respect to the transversal map $\varphi x\colon VFx \to_0 GUx$), so that these two composites coincide. $\qquad\square$

7.6 Multiple level limits

The general theory of multiple limits in a chiral multiple category A was studied in [GP9]; we do not cover here the whole topic, which does not completely agree with our theory of double limits and might perhaps be improved.

Tabulators have already been considered in Sections 6.1, 6.7 and 6.9; their relationship with the 2-dimensional case is discussed in 6.7.8(d). We add here the elementary case of *level limits*, which are ordinary limits in a transversal category $tv_i A$.

Level limits can be extended to intercategories with the same definitions, and minor modifications in some results (for instance in 7.6.4).

7.6.1 Products

We begin by examining various kinds of products in the chiral triple category $A = SC(C)$.

Supposing that C has products, the same is true of its functor categories (by 1.5.4(f)), and (using the categories $\vee$ and $\wedge$ recalled in 6.4.6), we have four types of products in A (indexed by a small set X):

- of objects, in $C = tv_*(A)$: $C = \prod_x C_x,$ $p_x : C \to_0 C_x,$

- of 1-arrows, in $C^\vee = tv_1(A)$: $f = \prod_x f_x,$ $p_x : f \to_0 f_x,$

- of 2-arrows, in $C^\wedge = tv_2(A)$: $u = \prod_x u_x,$ $p_x : u \to_0 u_x,$

- of 12-cells, in $C^{\vee \times \wedge} = tv_{12}(A)$: $\pi = \prod_x \pi_x,$ $p_x : \pi \to_0 \pi_x.$

Faces and degeneracies preserve these products. Saying that the triple category $SC(C)$ *has triple products* we mean all this. This is a global condition: we shall *not* define when, in a chiral triple category, a *single* product of items should be called 'a triple product'.

It is now simpler and clearer to work in a chiral *multiple* category A, rather than in a truncated case, as above.

Let $n \geqslant 0$ and let i be a positive multi-index (possibly empty). An i-product $a = \prod_x a_x$ will be an ordinary product in the transversal category A_i of i-cubes of A. It comes with a family $p_x : a \to_0 a_x$ of i-maps (i.e. cells of A_{0i}) that satisfies the usual universal property.

We say that A:

(i) *has i-products*, or products of type i, if all these products exist (for an arbitrary small set X),

(ii) *has products* if it has i-products for all positive multi-indices i,

(iii) *has multiple products* if it has all products, and these are preserved by faces and degeneracies, as ordinary functors

$$\partial_i^\alpha \colon \mathbf{A_i} \to \mathbf{A_{i|i}}, \quad e_i \colon \mathbf{A_{i|i}} \to \mathbf{A_i} \qquad (i \in \mathbf{i}, \, \alpha = \pm). \qquad (7.70)$$

This preservation is meant in the usual sense, up to isomorphism (i.e. invertible transversal maps); however, if this holds and A is transversally invariant (see 6.5.5), one can construct a choice of products that is strictly preserved by faces and degeneracies, starting from ∗-products and going up. This will be proved in Theorem 7.6.4, for all level limits.

A ∗-product is also called a *product of degree 0.*

7.6.2 Level limits

We now let $\mathbf{X}$ be a small category and consider the ordinary functors $F \colon \mathbf{X} \to \mathbf{A_i}$ with values in the transversal category $\mathbf{A_i} = \mathrm{tv_i}(\mathsf{A})$ of $\mathbf{i}$-cubes of A, for a fixed positive multi-index $\mathbf{i}$.

The limit of the functor F, called an $\mathbf{i}$-*level limit* in A, is an $\mathbf{i}$-cube $L \in \mathbf{A_i}$ equipped with a universal natural transformation $t \colon DL \to F \colon \mathbf{X} \to \mathbf{A_i}$, where $DL \colon \mathbf{X} \to \mathbf{A_i}$ is the constant functor at L. It is an $\mathbf{i}$-product if $\mathbf{X}$ is discrete and an $\mathbf{i}$-*equaliser* if $\mathbf{X}$ is the category $0 \rightrightarrows 1$.

We say that A:

(i) *has $\mathbf{i}$-level limits on* $\mathbf{X}$ if all the functors $\mathbf{X} \to \mathbf{A_i}$ have a limit,

(ii) *has level limits on* $\mathbf{X}$ if it has such limits for all positive multi-indices $\mathbf{i}$,

(iii) *has level multiple limits on* $\mathbf{X}$ if it has such level limits, and these are preserved by faces and degeneracies (as limits),

(iv) *has level multiple limits* if the previous property holds for every small category $\mathbf{X}$.

In particular, a ∗-*level limit* is a limit in the transversal category $\mathrm{tv}_*(\mathsf{A})$, associated to the multi-index $\emptyset$ of degree 0; it will also be called a *level limit of degree 0.*

All this can be extended to intercategories, in the same way.

7.6.3 Remarks

(a) Obviously, the cm-category A has level multiple limits if and only if it has multiple products and multiple equalisers. *Finite level limits* work similarly, with finite multiple products.

(b) Given a colax-lax cm-adjunction $F \dashv G$, the lax functor $G \colon \mathsf{B} \to \mathsf{A}$

preserves all the (existing) level limits of functors $\mathbf{X} \to tv_\mathbf{i}A$, while the colax functor $F \colon A \to B$ preserves all level colimits.

This simply depends on the ordinary adjunction $F_\mathbf{i} \dashv G_\mathbf{i}$, as for weak double categories in Exercise 5.5.7(a).

(c) If the category $\mathbf{C}$ is complete, so are its functor categories. Therefore the chiral triple category $SC(\mathbf{C})$ has all level triple limits.

(d) Extending limit pairs in weak double categories (see 5.2.4(b)), one can consider *multiple limit pairs*, where preservation by degeneracies can fail. This cannot happen when our cm-category has all cotabulators – as was proved in Lemma 6.7.2(c) for the dual case.

7.6.4 *Theorem* (Level limits and invariance)

Let $\mathbf{X}$ *be a category and* A *a* transversally invariant *chiral multiple category (see 6.5.5). If* A *has level multiple limits on* $\mathbf{X}$, *one can find a* strictly consistent *choice of such limits. More precisely, one can fix for every positive multi-index* $\mathbf{i}$ *and every functor* $F \colon \mathbf{X} \to A_\mathbf{i}$ *a limit of* F

$$L(F) \in A_\mathbf{i}, \qquad t(F) \colon DL(F) \to F \colon \mathbf{X} \to A_\mathbf{i}, \qquad (7.71)$$

so that these choices are strictly preserved by faces and degeneracies:

$$\begin{aligned} \partial_i^\alpha(L(F)) = L(\partial_i^\alpha F), \qquad & \partial_i^\alpha(t(F)) = t(\partial_i^\alpha F) && (i \in \mathbf{i}), \\ e_i(L(F)) = L(e_i F), \qquad & e_i(t(F)) = t(e_i F) && (i \notin \mathbf{i}). \end{aligned} \qquad (7.72)$$

Note. If A is a transversally invariant *intercategory*, the statement still holds, omitting everything about degeneracies.

Proof We proceed by induction on the degree n of positive multi-indices, following the same pattern of the proof of Proposition 6.5.6.

For $n = 0$ we just fix a choice $(L(F), t(F))$ of $*$-level limits on $\mathbf{X}$, for all $F \colon \mathbf{X} \to tv_*(A)$. For $n \geqslant 1$ we suppose to have a consistent choice for all positive multi-indices of degree up to $n-1$ and extend this choice to degree n, as follows.

For a functor $F \colon \mathbf{X} \to A_\mathbf{i}$ of degree n, we already have a choice

$$(L(\partial_i^\alpha F), t(\partial_i^\alpha F))$$

of the limit of each of its faces. Let (L, t) be an arbitrary limit of F; since faces are supposed to preserve limits (in the usual, non-strict sense), there is a unique family of transversal isomorphisms h_i^α coherent with the limit cones (of degree $n-1$)

$$h_i^\alpha \colon L(\partial_i^\alpha F) \to_0 \partial_i^\alpha L, \qquad t(\partial_i^\alpha F) = (\partial_i^\alpha t).h_i^\alpha \qquad (i \in \mathbf{i}, \, \alpha = \pm), \qquad (7.73)$$

and this family has consistent faces (see (6.82)), as follows easily from their coherence with the limit cones of a lower degree (when $n \geqslant 2$, otherwise the consistency condition is void).

Now, because of transversal invariance, this family can be filled with a transversal isomorphism h, *yielding a choice* for $L(F)$ and $t(F)$

$$h\colon L(F) \to_0 L, \qquad t(F) = t.Dh\colon DL(F) \to F. \qquad (7.74)$$

By construction this extension of L is strictly preserved by all faces. To make it also consistent with degeneracies, we assume that – in the previous construction – the following constraint has been followed: for an i-degenerate functor $F = e_i G\colon \mathbf{X} \to \mathbf{A_i}$ we always choose the pair $(e_i L(G), e_i t(G))$ as its limit (L, t). This allows us to take $h_i^\alpha = \mathrm{id}(L(G))$ (for all $i \in \mathbf{i}$ and $\alpha = 0, 1$), and finally $h = \mathrm{id}(L)$, that is

$$L(F) = e_i L(G), \qquad t(F) = e_i t(G)\colon DL(F) \to F. \qquad (7.75)$$

If F is also j-degenerate, then $F = e_i e_j H = e_j e_i H$; therefore, by the inductive assumption, both procedures give the same result:

$$e_i L(G) = e_i e_j L(H) = e_j e_i L(H) = e_j L(e_i H).$$

In an intercategory with $e_i e_j \neq e_j e_i$, the last point fails – and only that. $\qquad \square$

7.6.5 Exercises and complements

(a) $\mathbf{X}$ is a small category. Construct the exponential $\mathsf{A}^\mathbf{X}$ of a chiral multiple category A.

(b) Construct the diagonal functor $D\colon \mathsf{A} \to \mathsf{A}^\mathbf{X}$.

*(c) Define the product $-\times \mathbf{X}\colon \mathbf{StCmc} \to \mathbf{StCmc}$ as an endofunctor of the category of chiral multiple categories and strict functors.

*(d) Define a strict evaluation functor $\mathrm{ev}\colon \mathsf{A}^\mathbf{X} \times \mathbf{X} \to \mathsf{A}$, so that $(\mathsf{A}^\mathbf{X}, \mathrm{ev})$ is a universal arrow from the functor $-\times \mathbf{X}$ to the object A.

7.6.6 Level limits as unitary lax functors

Taking into account this exponential $\mathsf{A}^\mathbf{X}$, Theorem 7.6.4 shows that, if the chiral multiple category A is transversally invariant and has level multiple limits on the small category $\mathbf{X}$, we can form a *unitary* lax functor L and a transversal transformation t

$$L\colon \mathsf{A}^\mathbf{X} \to \mathsf{A}, \qquad t\colon DL \to 1\colon \mathsf{A}^\mathbf{X} \to \mathsf{A}^\mathbf{X}, \qquad (7.76)$$

such that, on every **i**-cube F, the pair $(L(F), t(F))$ is the level limit of the functor F, as in (7.71). Then L is right adjoint to the diagonal functor $D\colon \mathsf{A} \to \mathsf{A}^{\mathbf{X}}$, with counit t.

Indeed, after defining L and t on **i**-cubes *by a consistent choice* (which is possible by Theorem 7.6.4), we define $L(\varphi)$ for every natural transformation $\varphi\colon F \to G\colon \mathbf{X} \to \mathsf{A_i}$. By the universal property of limits, there is precisely one **i**-map $L(\varphi)$ such that

$$L(\varphi)\colon L(F) \to_0 L(G), \qquad \varphi.t(F) = t(G).DL(\varphi). \qquad (7.77)$$

This extension on **i**-maps is obviously the only one that makes the family $t(F)\colon DL(F) \to F$ into a transversal transformation $DL \to 1$. The lax comparison for i-composition (with $i \in \mathbf{i}$)

$$\underline{L}(F, G)\colon LF +_i LG \to L(F +_i G), \qquad t(F +_i G).D\underline{L}(F, G) = tF +_i tG,$$

comes from the universal property of $L(F +_i G)$ as a limit.

In the hypotheses above we say that A *has lax functorial level limits* on $\mathbf{X}$. We say that A *has pseudo* (resp. *strict*) *functorial level limits* on $\mathbf{X}$ if L is a pseudo functor (resp. can be chosen as a strict functor).

For limit pairs one adds a family of comparisons $\underline{L}_i F\colon e_i LF \to L(e_i F)$, for $i \notin \mathbf{i}$.

7.6.7 Level limits in weak double categories

Let $\mathbb{D}$ be a weak double category, and let A be the associated weak multiple category $\mathrm{sk}_2(\mathbb{D})$, obtained by adding degenerate items of all the missing types (see 6.2.8).

Level limits in A present slight differences in terminology with the level limits in $\mathbb{D}$, as defined in 5.3.1, essentially because the 2-dimensional universal property of double limits here is not required from the start but comes out of a condition of preservation by degeneracies.

As a particular case of the definitions in 7.6.2, we have the following cases for A, and a small category $\mathbf{X}$

(i) A *has ∗-level limits on* $\mathbf{X}$ if all the functors $\mathbf{X} \to \mathrm{tv}_*(\mathsf{A})$ have a limit. All of them can be constructed from:

- small products $\prod A_x$ of objects,

- equalisers of pairs $f, g\colon A \to B$ of parallel horizontal arrows.

(i′) A *has 1-level limits on* $\mathbf{X}$ if all the functors $\mathbf{X} \to \mathrm{tv}_1(\mathsf{A})$ have a limit. All of them can be constructed from:

- small products $\prod u_x$ of vertical arrows,

- equalisers of pairs $a, b \colon u \to v$ of double cells (between the same vertical arrows).

(ii) A *has level limits on* **X** if it has $*$- and 1-level limits on **X**.

(iii) A *has level multiple limits on* **X** if it has such level limits, preserved by faces and degeneracies.

(iv) A *has level multiple limits* if the previous property holds for every small category **X**; this is equivalent to the existence of small *multiple products* and *multiple equalisers*.

In Chapter 5, case (i) is expressed saying that $\mathbb{D}$ has 1-*dimensional* limits of horizontal functors on **X**, case (iii) saying that $\mathbb{D}$ has lax functorial level limits on **X**, case (iv) saying that $\mathbb{D}$ has lax functorial level limits.

7.6.8 Exercises and complements

These exercises are entirely left to the reader. Many results can be found in [GP9].

(a) Prove that the strict triple category Stcⅅbl, defined in 6.1.1, has products.

*(b) Study level limits and colimits in the chiral multiple categories of Sections 6.4 and 6.6.

7.7 *The exponential of strict multiple categories

The exponential A^X of strict multiple categories can be constructed as an extension of the exponential of double categories. This construction will not be used elsewhere; it is shown here because of its interesting, non-obvious machinery.

7.7.1 Reviewing the exponential of double categories

The exponential $\mathbb{F} = \mathbb{A}^{\mathbb{X}}$ of strict double categories, where $\mathbb{X}$ is small, has been defined in 3.2.7. We rewrite the definition in terms adapted to its extension to multiple categories: in particular the horizontal and vertical directions are now denoted by the index $i = 0$ or 1.

(A) An object of $\mathbb{F}$ is a (double) functor $F \colon \mathbb{X} \to \mathbb{A}$.

(B) A 0-directed arrow $h \colon F \to_0 F' \colon \mathbb{X} \to \mathbb{A}$ is a horizontal transformation. It assigns

(a) to each object x of $\mathbb{X}$ a 0-arrow $hx \colon Fx \to_0 F'x$ in $\mathbb{A}$,

(b) to each 1-arrow $u: x \to_1 y$ of $\mathbb{X}$ a 01-cell $hu: (Fu \; {}^{hx}_{hy} \; F'u)$, as below in (7.79), preserving vertical composition and identities and satisfying two naturality conditions:

(i) $h(u +_1 v) = hu +_1 hv, \qquad h(e_1 x) = e_1(hx),$

(ii) for each 0-arrow $f: x \to_0 x'$ of $\mathbb{X}$ we have a commutative square of 0-arrows in $\mathbb{A}$, whose diagonal will be written as $h(f)$

$$
\begin{array}{ccc}
Fx & \xrightarrow{\;hx\;} & F'x \\
{\scriptstyle Ff}\downarrow & & \downarrow{\scriptstyle F'f} \\
Fx' & \xrightarrow[\;hx'\;]{} & F'x'
\end{array}
\qquad h(f) = hx +_0 F'f = Ff +_0 hx', \qquad (7.78)
$$

(iii) for each 01-cell $\xi: (u \; {}^f_g \; v)$ of $\mathbb{X}$ we have a commutative square of 01-cells in $\mathbb{A}$, composed in direction 0

$$ hu +_0 F'\xi = F\xi +_0 hv, \qquad (7.79) $$

$$
\begin{array}{ccccc}
Fx & \xrightarrow{\;hx\;} & F'x & \xrightarrow{\;F'f\;} & F'x' \\
{\scriptstyle Fu}\downarrow & hu & {\scriptstyle F'u}\downarrow & F'\xi & \downarrow{\scriptstyle F'v} \\
Fy & \xrightarrow[\;hy\;]{} & F'y & \xrightarrow[\;F'g\;]{} & F'y'
\end{array}
\;=\;
\begin{array}{ccccc}
Fx & \xrightarrow{\;Ff\;} & Fx' & \xrightarrow{\;hx'\;} & F'x' \\
{\scriptstyle Fu}\downarrow & F\xi & \downarrow{\scriptstyle Fv} & hv & \downarrow{\scriptstyle F'v} \\
Fy & \xrightarrow[\;Fg\;]{} & Fy' & \xrightarrow[\;hy'\;]{} & F'y'
\end{array}
$$

(B') Symmetrically, a 1-directed arrow $r: F \to_1 G: \mathbb{X} \to \mathbb{A}$ is a vertical transformation, with data

$$ rx: Fx \to_1 Gx, \qquad rf: (rx \; {}^{Ff}_{Gf} \; rx'), \qquad (7.80) $$

for each object x and each 0-arrow $f: x \to_0 x'$ of $\mathbb{X}$, satisfying symmetric conditions. In particular, for a 1-arrow $u: x \to_1 y$ of **X**, we have a commutative square of 1-arrows in $\mathbb{A}$, whose diagonal will be written as

$$ r(u) = rx +_1 Gu = Fu +_1 ry. \qquad (7.81) $$

(C) A double cell $\Phi: (r \; {}^h_k \; s)$ is a modification. It assigns to each object x of $\mathbb{X}$ a double cell $\Phi x: (rx \; {}^{hx}_{kx} \; sx)$ in $\mathbb{A}$, which is natural for each $f: x \to_0 x'$ and each $u: x \to_1 y$ of $\mathbb{X}$

$$ \Phi x +_0 sf = rx +_0 \Phi x', \qquad \Phi x +_1 ku = hx +_1 \Phi y. \qquad (7.82) $$

(D) One defines, in a pointwise way, the horizontal and vertical operations (compositions and identities), and $\mathbb{A}^{\mathbb{X}}$ is made into a double category.

7.7.2 Evaluation and cartesian closedness

We prove now the statement of Exercise 3.2.8(a). The double functor of evaluation is defined in the obvious way, for:

$$F, \quad h\colon F \to F', \quad r\colon F \to G, \quad \Phi\colon (r \overset{h}{\underset{k}{}} s) \quad \text{in } \mathbb{A}^{\mathbb{X}},$$

$$x, \quad f\colon x \to_0 x', \quad u\colon x \to_1 y, \quad \xi\colon (u \overset{f}{\underset{g}{}} v) \quad \text{in } \mathbb{X},$$

$$\mathrm{ev}\colon \mathbb{A}^{\mathbb{X}} \times \mathbb{X} \to \mathbb{A}, \qquad \mathrm{ev}(F, x) = F(x),$$

$$\mathrm{ev}(h, f) = h(f)\colon Fx \to F'x',$$

$$\mathrm{ev}(r, u) = r(u)\colon Fx \to Gy, \tag{7.83}$$

$$\mathrm{ev}(\Phi, \xi) = \frac{\Phi x \mid sf}{ku \mid G'\xi} \colon \left(r(u) \overset{h(f)}{\underset{k(g)}{}} s(v) \right),$$

$$\begin{array}{ccccc}
Fx & \xrightarrow{\ hx\ } & F'x & \xrightarrow{\ F'f\ } & F'y \\[2pt]
{\scriptstyle rx}\downarrow & \Phi x & \downarrow{\scriptstyle sx} \quad sf & & \downarrow{\scriptstyle sy} \\[2pt]
Gx & \xrightarrow{\ kx\ } & G'x & \xrightarrow{\ G'f\ } & G'y \\[2pt]
{\scriptstyle Gu}\downarrow & ku & \downarrow{\scriptstyle G'u} \ G'\xi & & \downarrow{\scriptstyle G'v} \\[2pt]
Gy & \xrightarrow[\ ky\ ]{} & G'y & \xrightarrow[\ G'g\ ]{} & G'y'
\end{array} \tag{7.84}$$

Now the pair $(\mathbb{A}^{\mathbb{X}}, \mathrm{ev})$ satisfies the universal property of exponentiation: for every double functor $L\colon \mathbb{B} \times \mathbb{X} \to \mathbb{A}$ there is precisely one double functor $M\colon \mathbb{B} \to \mathbb{A}^{\mathbb{X}}$ such that $\mathrm{ev}.(M \times \mathbb{X}) = L$. Given L, the double functor M is defined as follows

$$M(b) = L(b, -)\colon \mathbb{X} \to \mathbb{A}, \qquad Mb(x) = L(b, x),$$

$$Mb(f) = L(1_b, f), \quad Mb(u) = L(e_b, u), \quad Mb(\xi) = L(\square b, \xi), \tag{7.85}$$

$$M(h\colon b \to_0 b')\colon L(b, -) \to_0 L(b', -),$$

$$Mh(x) = L(h, 1_x)\colon L(b, x) \to_0 L(b', x), \tag{7.86}$$

$$Mh(u\colon x \to_1 y) = L(e_h, 1_u)\colon \left(Mbu \overset{Mhx}{\underset{Mhy}{}} Mb'u \right),$$

$$M(v\colon b \to_1 c)\colon L(b, -) \to_1 L(c, -),$$

$$Mv(x) = L(v, e_x)\colon L(b, x) \to_1 L(c, x), \tag{7.87}$$

$$Mv(f\colon x \to_0 x') = L(1_v, e_f)\colon \left(Mvx \overset{Mbf}{\underset{Mcf}{}} Mvx' \right),$$

$$M(\beta)\colon (Mv \overset{Mh}{\underset{Mh'}{}} Mv'), \qquad M\beta(x) = L(\beta, \square x). \tag{7.88}$$

For the following extension, it is important to note that M is computed by the values of L on all pairs of $\mathbb{B} \times \mathbb{X}$ whose items have *disjoint supports* $\mathbf{i}, \mathbf{j} \subset \{0, 1\}$. These pairs fall in the following 9 cases, listed in the same order as above, in (7.85)–(7.88):

$$(b, x) \in B_* \times X_*, \ (b, f) \in B_* \times X_0, \ (b, u) \in B_* \times X_1, \ (b, \xi) \in B_* \times X_{01},$$
$$(h, x) \in B_0 \times X_*, \qquad (h, u) \in B_0 \times X_1,$$
$$(v, x) \in B_1 \times X_*, \qquad (v, f) \in B_1 \times X_0,$$
$$(\beta, x) \in B_{01} \times X_*.$$

Conversely, given M one reconstructs L decomposing each pair of $\mathbb{B} \times \mathbb{X}$ by pairs of this kind:

$$(h, f) = (h, 1_x) \,|\, (1_{b'}, f) = (1_b, f) \,|\, (h, 1_{x'}) \quad \text{for } h \colon b \to_0 b', \ f \colon x \to_0 x',$$
$$(v, u) = (v, e_x) \otimes (e_c, u) = (e_b, u) \otimes (v, e_y) \quad \text{for } v \colon b \to_1 c, \ u \colon x \to_1 y,$$

$$(\beta, \xi) \quad = \quad
\begin{array}{ccccc}
bx & \xrightarrow{\ hx\ } & b'x & \xrightarrow{\ b'f\ } & b'x' \\
{\scriptstyle vx}\downarrow & (\beta, \Box x) & {\scriptstyle v'x}\downarrow\ (1_{v'}, e_f) & & \downarrow{\scriptstyle v'x'} \\
cx & \xrightarrow{\ h'x\ } & c'x & \xrightarrow{\ c'f\ } & c'x' \\
{\scriptstyle cu}\downarrow & (e_{h'}, 1_u) & {\scriptstyle c'u}\downarrow\ (\Box c', \xi) & & \downarrow{\scriptstyle c'u'} \\
cy & \xrightarrow{\ h'y\ } & c'y & \xrightarrow{\ c'f'\ } & c'y'
\end{array}$$

for $\beta \colon (v \, {\overset{h}{\underset{h'}{}}} \, v') \colon \binom{b\ b'}{c\ c'}$ in $\mathbb{B}$ and $\xi \colon (u \, {\overset{f}{\underset{f'}{}}} \, u') \colon \binom{x\ x'}{y\ y'}$ in $\mathbb{X}$.

7.7.3 The infinite-dimensional extension

Inspired by the previous remarks, we extend this definition to strict multiple categories, building the exponential $\mathsf{F} = \mathsf{A}^{\mathsf{X}}$ where X is small. The 2-dimensional case can be recovered restricting all items to the multi-indices contained in $\mathbf{2} = \{0, 1\}$.

An $\mathbf{i}$-cell F of A^{X} (for $\mathbf{i} \subset \mathbb{N}$) consists of a family of cells $Fx \in A_{\mathbf{i} \cup \mathbf{j}}$, where the multi-index $\mathbf{j}$ is disjoint from $\mathbf{i}$ and $x \in X_{\mathbf{j}}$. These data must satisfy the following conditions, whenever $\mathbf{i} \cap \mathbf{j} = \emptyset$

(i) $\quad F(\partial_j^\alpha x) = \partial_j^\alpha (Fx)$ $\hfill$ (for $j \in \mathbf{j}$, $x \in X_{\mathbf{j}}$),

(ii) $\quad F(e_j x) = e_j (Fx)$ $\hfill$ (for $j \in \mathbf{j}$, $x \in X_{\mathbf{j}|j}$),

(iii) $\quad F(x +_j y) = Fx +_j Fy$ $\hfill$ (for $j \in \mathbf{j}$, $x, y \in X_{\mathbf{j}}$),

(iv) $\quad Fx +_i (\partial_i^- F)(f) = (\partial_i^+ F)(f) +_i Fy$ $\hfill$ (for $i \in \mathbf{i}$, $f \colon x \to_i y \in X_{\mathbf{j}i}$).

(This includes multiple functors $X \to A$, their transformations, modifications and so on, as examined in the exercises below.) Our cells form a multiple set, with the following faces and degeneracies:

$$(\partial_i^\alpha F)(x) = \partial_i^\alpha(Fx) \qquad (\text{for } i \in \mathbf{i}, \, x \in X_{\mathbf{j}}, \, \mathbf{i} \cap \mathbf{j} = \emptyset),$$

$$(e_i F)(x) = e_i(Fx) \qquad (\text{for } i \notin \mathbf{i}, \, x \in X_{\mathbf{j}}, \, \mathbf{i}i \cap \mathbf{j} = \emptyset).$$

(7.89)

A^X is made into a multiple category by defining the i-directed operations of $\mathbf{i}$-cells in a pointwise way:

$$(F +_i G)(x) = Fx +_i Gx \qquad (\text{for } x \in X_{\mathbf{j}}, \, \mathbf{i} \cap \mathbf{j} = \emptyset). \qquad (7.90)$$

Finally one verifies that the pair (A^X, ev) satisfies the universal property of exponentiation: for every double functor $L \colon B \times X \to A$ there is precisely one double functor $M \colon B \to A^X$ such that $\mathrm{ev}.(M \times X) = L$. This can be computed as follows

$$(Mb)(a) = L(e_{\mathbf{j}} b, e_{\mathbf{i}} x) \qquad (\text{for } b \in B_{\mathbf{i}}, \, x \in X_{\mathbf{j}}, \, \mathbf{i} \cap \mathbf{j} = \emptyset). \qquad (7.91)$$

7.7.4 Exercises and complements

Characterise the low dimensional cases of $\mathbf{i}$-cells of the multiple category A^X, for $\mathbf{i} = \emptyset$, or $\{i\}$, or $\{i, j\}$ (with $i \neq j$).

Note that, in this context of strict multiple categories, the index 0 plays no special role (as it was already the case in Section 6.2, up to 6.2.4): there is no real reason of distinguishing transversal transformations among the i-directed ones.

8

Monads and algebras for multiple categories

Here we study lax and colax monads on chiral multiple categories and their relationship with multiple adjunctions. Non-standard results deal with 'strong' idempotent lax monads and their algebras. A more detailed outline will be given in 8.1.2.

This chapter is mostly new. Unless otherwise specified, an 'algebra' is an Eilenberg–Moore algebra for a monad.

8.1 Adjunctions, monads and comonads

This section is a brief presentation of the problems and results of the present chapter.

8.1.1 The problem

Studying the monadicity of multiple adjunctions is not straightforward.

As we have seen in Chapter 7, a general adjunction

$$(F, G, \eta, \varepsilon) \colon \mathsf{X} \rightarrowtail \mathsf{A}, \tag{8.1}$$

between chiral multiple categories consists of a colax functor $F \colon \mathsf{X} \to \mathsf{A}$, a lax functor $G \colon \mathsf{A} \to \mathsf{X}$ and two double cells η, ε in the double category $\mathbb{C}\mathrm{mc}$.

We often write the unit as $\eta \colon 1 \dashrightarrow GF$ (because it has components $\eta x \colon x \to GFx$), but – in general – there are no arrows of $\mathbb{C}\mathrm{mc}$ that can be written as GF (or FG), and it is not evident how one should define an associated monad (or comonad).

The next sections will deal with particular situations where one can form a colax *or* a lax monad $T = GF$, or more generally a p-mixed monad.

True, one can always view (8.1) as an adjunction in the 2-category p**Mlt** of premultiple functors (see 6.2.7), and $T = GF$ as a premultiple monad on X, which leads to a notion of 'transversal monadicity', defined in 8.7.4. This – at best – can allow us to reconstruct the underlying premultiple category |A|, deprived of its 'geometric' structure. Finally, even though this shortcut can give a general framework for the whole topic, the real interest lies in the different ways of filling up the premultiple framework.

8.1.2 An outline of particular cases

We shall examine various particular cases of multiple adjunctions, with forms of monadicity more effective than the transversal one.

(a) As we know, a *colax-pseudo adjunction* (where the right adjoint G is a pseudo functor) is the same as an adjunction in the 2-category CxCmc of chiral multiple categories, colax functors and horizontal transformations; it produces a *colax monad* $T = GF$ and a *colax comonad* $S = FG$ by the usual 2-categorical procedure. On the other hand, given a colax monad T over a chiral multiple category X, one can construct – in a standard way – a chiral multiple category X^T of Eilenberg–Moore algebras, with an associated *colax-strict* adjunction (see Section 8.2).

(b) Similarly, a *pseudo-lax adjunction* (where the left adjoint F is a pseudo functor) is the same as an adjunction in the 2-category LxCmc of chiral multiple categories, lax functors and horizontal transformations; it produces a *lax monad* T and a *lax comonad* S. Now, a lax monad has a standard construction of Kleisli algebras and an associated *strict-lax* adjunction (in the Exercises of 8.3.3).

(c) As an extension of case (b), we introduce in 8.3.4 a 'left-strong colax-lax adjunction' which still produces a lax monad.

In Section 8.4 we begin the less standard part of our analysis, about strong *idempotent* lax monads and their links with left-strong idempotent adjunctions. The interest of idempotence lies in the fact that it allows the construction of a chiral multiple category of 'algebraic cubes'. Actually, we perform the construction in the (equivalent) *strictly* idempotent case, to avoid complicated computations.

There are interesting examples of this kind, in dimension two. In Section 8.5 we study the 'jointly-monic monad' on the weak double category $\mathbb{S}\text{pan}(\mathbf{C})$ of spans over a regular category. Its algebras form the weak double category $\mathbb{R}\text{el}'(\mathbf{C})$ of 'jointly-monic relations' (see 2.5.7), which is equivalent to the double category $\mathbb{R}\text{el}(\mathbf{C})$ of 'cartesian relations' – as we have seen in Section 2.5.

We also consider the idempotent pushout-pullback adjunction $F \dashv G$: $\mathbb{S}\text{pan}(\mathbf{C}) \twoheadrightarrow \mathbb{C}\text{osp}(\mathbf{C})$. If $\mathbf{C}$ is abelian, this adjunction is strong and gives the jointly-monic monad on $\mathbb{S}\mathrm{pan}(\mathbf{C})$; in this case the relations on $\mathbf{C}$ can be equivalently presented as algebras (jointly-monic spans) or coalgebras (jointly-epic cospans). On the other hand, for **Set** this adjunction is *not* strong, and does not yield the jointly-monic monad on $\mathbb{S}\mathrm{pan}(\mathbf{Set})$.

In Section 8.6 we briefly consider idempotent comonads, dualising Section 8.4; then we deal with the idempotent adjunction

$$C \dashv P\colon \mathbb{C}\mathrm{at} \rightarrowtail \mathbb{C}\mathrm{osp}\mathbf{Cat}$$

of 4.5.5, showing that it is comonadic: profunctors are the coalgebraic cospans of the comonad CP.

Higher dimensional examples are considered in Section 8.7, together with the notion of transversal monadicity mentioned above.

8.1.3 Elementary examples

Let us take the embedding $U\colon \mathbf{Ab} \to \mathbf{Gp}$ and its left adjoint, the abelianisation functor $F(X) = X^{\mathrm{ab}} = X/[X, X]$

$$F\colon \mathbf{Gp} \rightleftarrows \mathbf{Ab}\colon U, \qquad\qquad F \dashv U. \tag{8.2}$$

It gives the abelianisation monad $T = UF\colon \mathbf{Gp} \to \mathbf{Gp}$, which is idempotent (see 1.7.5). Its algebraic objects are the commutative groups.

It is easy to see that U does not preserve pushouts while F does not preserve pullbacks.

(For instance, the biproduct $\mathbb{Z} \oplus \mathbb{Z}$ in **Ab** is not a sum of groups. For the second point, one can take a simple group X and the embedding $f\colon A \rightarrowtail X$ of a non-trivial commutative subgroup, e.g. a cyclic one; then the kernel pair of f is A, while the kernel pair of $F(f)\colon A \to 0$ is $A \oplus A$.)

(a) Applying to this adjunction the 2-functor $\mathsf{Span}\colon \mathbf{Cat}_{\mathrm{pb}} \to \mathrm{Cx}\mathbf{Cmc}$ of 7.1.5, we get a *colax-pseudo* adjunction $F' \dashv U'$, which is not a pseudo adjunction, because of the previous remark. On $\mathsf{Span}(\mathbf{Gp})$ we have an idempotent colax monad $U'F'$, whose algebraic cubes (i.e. the cubical spans of groups with invertible unit-component) are the cubical spans of commutative groups. This case will fit in Section 8.2, about the algebras of colax monads.

(b) Applying the 2-functor $\mathsf{Cosp}\colon \mathbf{Cat}_{\mathrm{po}} \to \mathrm{Lx}\mathbf{Cmc}$ (of 7.1.6) we get a *pseudo-lax* adjunction $F' \dashv U'$ and an idempotent lax monad $U'F'$ on $\mathsf{Cosp}(\mathbf{Gp})$, whose algebraic cubes are the cubical cospans of commutative groups. This case will fit in Section 8.4, about the algebras of *idempotent* lax monads.

(c) Applying the 2-functor $\mathsf{SC}\colon \mathbf{Cat}_{\mathrm{pbpo}} \to \mathrm{Mx}_2(\mathbf{Cmc_3})$ (of 7.1.7) we get an adjunction where F' is a colax-pseudo functor and U' is a pseudo-lax functor. We have an idempotent colax-lax triple monad $U'F'$ on $\mathsf{SC}(\mathbf{Gp})$, whose algebraic cubes $\vee \times \wedge \to \mathbf{Gp}$ are the 'spans of cospans' of commutative groups.

8.2 Colax adjunctions and monads

We begin by working in the 2-category $\mathrm{Cx}\mathbf{Cmc}$ of chiral multiple categories, colax functors and transversal transformations (see 6.5.3), studying multiple adjunctions and monads of a colax type, correlated by Eilenberg–Moore algebras.

X and A are always chiral multiple categories; i denotes a positive multi-index of dimension $n \geqslant 0$ (possibly empty).

8.2.1 Colax monads

A *colax monad* $T = (T, \eta, \mu)$ on the chiral multiple category X is defined as a monad in the 2-category $\mathrm{Cx}\mathbf{Cmc}$. Thus $T \colon \mathsf{X} \to \mathsf{X}$ is a colax endofunctor, with comparison maps

$$\underline{T}_i(x) \colon T e_i(x) \to e_i(Tx), \qquad \underline{T}_i(x, y) \colon T(x +_i y) \to Tx +_i Ty. \qquad (8.3)$$

It is equipped with transversal transformations of colax functors $\eta \colon 1 \to T$ and $\mu \colon T^2 \to T$, that satisfy the usual axioms: $\mu.\eta T = 1 = \mu.T\eta$ and $\mu.\mu T = \mu.T\mu$.

These axioms only concern the family of transversal categories $\mathbf{X_i} = \mathrm{tv_i}\mathsf{X}$, where we have ordinary monads

$$T_i \colon \mathbf{X_i} \to \mathbf{X_i}, \qquad \eta_i \colon 1 \to T_i, \qquad \mu_i \colon T_i^2 \to T_i. \qquad (8.4)$$

On the other hand the coherence conditions of T, η, μ (see 6.5.2, 6.5.3) cannot be expressed within the transversal components: one needs all the weak double and chiral triple categories 'contained' in X.

8.2.2 Adjunctions and monads, the colax case

As we have seen in Theorem 7.4.3, a colax-pseudo adjunction

$$(F, G, \eta, \varepsilon) \colon \mathsf{X} \nrightarrow \mathsf{A}$$

between chiral multiple categories is the same as an internal adjunction in the 2-category $\mathrm{Cx}\mathbf{Cmc}$, consisting of two colax functors $F \colon \mathsf{X} \to \mathsf{A}$ and $G \colon \mathsf{A} \to \mathsf{X}$ with transversal transformations $\eta \colon 1 \to GF$ and $\varepsilon \colon FG \to 1$ that satisfy the triangular equations: then G is automatically pseudo.

By the usual 2-categorical procedure (in Cx**Cmc**), this colax-pseudo adjunction produces a colax monad $T = GF$ on the domain X of F, with the same unit η and multiplication $\mu = G\varepsilon F$.

8.2.3 Eilenberg–Moore algebras for colax monads

Given a colax monad $T = (T, \eta, \mu)$ on X, the chiral multiple category X^T of (Eilenberg–Moore) T-*algebras* can be constructed in a standard way, as a substructure of the comma chiral multiple category $T \downarrow 1_X$ (which is legitimate, because T is colax, see 7.2.2): the **i**-cubes of the latter are triples $(x, x'; h\colon Tx \to x')$, and we take those for which $x = x'$ and the axioms of algebras are satisfied; we proceed in the same way for transversal maps.

More analytically, we have the following description.

(a) First, the transversal category $\mathrm{tv_i}(X^T)$ is constructed as the category of algebras $(\mathbf{X_i})^{T_i}$; in other words

- an **i**-cube $(x, h\colon Tx \to x)$ consists of an **i**-cube x and an **i**-map h in X (the *algebraic structure*) satisfying the usual axioms:

$$h.\eta x = 1_x, \qquad h.Th = h.\mu x\colon T^2 x \to x,$$

- an **i**-map $f\colon (x, h) \to (y, k)$ comes from an **i**-map $f\colon x \to y$ in X such that $f.h = k.Tf$,
- their transversal composition is that of X; let us note that $f\colon (x, h) \to (y, k)$ is invertible if and only if $f\colon x \to y$ is.

When convenient the symbol $\hat{x}$ may denote an algebra (x, h) of X^T over the **i**-cube x of X.

(b) The faces are obvious (for $i \in \mathbf{i}$):

$$\partial_i^\alpha (x, h\colon Tx \to x) = (\partial_i^\alpha x, \partial_i^\alpha h\colon T\partial_i^\alpha x \to x), \quad \partial_i^\alpha(f) = \partial_i^\alpha f. \qquad (8.5)$$

(c) Degeneracies and compositions in a positive direction i are constructed with the colaxity comparisons of T

$$e_i(x, h\colon Tx \to x) = (e_i x, e_i h.\underline{T}_i(x)\colon Te_i x \to e_i x),$$
$$e_i(f\colon (x, h) \to (x', h')) = e_i f\colon e_i(x, h) \to e_i(x', h'), \qquad (8.6)$$

$$(x, h) +_i (y, k) = (x +_i y, (h +_i k).\underline{T}_i(x, y)\colon T(x +_i y) \to x +_i y),$$
$$(f\colon (x, h) \to (x', h')) +_i (g\colon (y, k) \to (y', k')) = f +_i g. \qquad (8.7)$$

To verify that (8.6) is well defined it suffices to apply the definition and axiom (cmf.1) of 6.5.2 for T

$$e_i f.e_i h.\underline{T}_i(x) = e_i(fh).\underline{T}_i(x) = e_i h'.e_i Tf.\underline{T}_i(x) = e_i h'.\underline{T}_i(x').Te_i f,$$

while for (8.7) one applies (cmf.2).

(d) The i-associator of three i-consecutive **i**-cubes $\hat{x}, \hat{y}, \hat{z}$ of X^T is given by the associator $\kappa_i(x, y, z)$ of the underlying cubes x, y, z of X, viewed as a transversal map in X^T

$$\kappa_i(\hat{x}, \hat{y}, \hat{z}) = \kappa_i(x, y, z) : \hat{x} +_i (\hat{y} +_i \hat{z}) \to (\hat{x} +_i \hat{y}) +_i \hat{z}. \qquad (8.8)$$

Its coherence as a transversal map of X^T comes from Lemma 7.2.3, about comma multiple categories. Its invertibility is proved below, in point (e).

Similarly the unitors and interchangers of X^T come from those of X

$$\lambda_i(\hat{x}) = \lambda_i x : e_i \partial_i^\alpha(\hat{x}) +_i \hat{x} \to \hat{x},$$
$$\rho_i(\hat{x}) = \rho_i x : \hat{x} +_i e_i \partial_i^\alpha(\hat{x}) \to \hat{x}, \qquad (8.9)$$
$$\chi_{ij}(\hat{x}, \hat{y}, \hat{z}, \hat{u}) = \chi_{ij}(x, y, z, u).$$

Note that if X is a weak (or strict) multiple category, so is X^T.

(e) Finally the coherence axioms for X^T trivially hold, because (under some abuse of language) the forgetful (strict) multiple functor

$$G^T : X^T \to X, \qquad G^T(x, h) = x,$$
$$G^T(f : (x, h) \to (y, k)) = f : x \to y, \qquad (8.10)$$

is faithful (which means that it gives faithful ordinary functors on the transversal components, see 7.4.4).

We have already remarked above that it reflects invertible transversal maps: this proves that the new unitors and associators are invertible, as well as the interchangers in the weak case. G^T also reflects identities, which accounts for the strict case.

8.2.4 *From colax monads to adjunctions and back*

By extending the usual procedure (in 1.6.4), a colax monad $T = (T, \eta, \mu)$ on X produces a *colax-strict* adjunction

$$F^T : X \rightleftarrows X^T : G^T, \qquad (\eta^T, \varepsilon^T) : F^T \dashv G^T, \qquad (8.11)$$

whose associated monad is the given one.

As we have already seen, G^T is the obvious (strict) forgetful functor $G^T(x, h) = x$.

The free-algebra functor $F^T : X \to X^T$ is colax, with comparisons derived from those of T

$$F^T(x) = (Tx, \mu x \colon T^2 x \to Tx),$$

$$F^T(f \colon x \to y) = Tf \colon F^T x \to F^T y,$$

$$\underline{F}_i^T(x) = \underline{T}_i(x) \colon F^T(e_i x) \to e_i F^T x,$$

$$\underline{F}_i^T(x,y) = \underline{T}_i(x,y) \colon F^T(x +_i y) \to F^T x +_i F^T y.$$

(8.12)

Its coherence is proved by the fact that $G^T F^T = T$ is coherent and G^T is faithful.

Unit and counit are defined as usual

$$\eta^T = \eta \colon 1 \to G^T F^T = T \colon \mathsf{X} \to \mathsf{X},$$

$$\varepsilon^T \colon F^T G^T \to 1 \colon \mathsf{X}^T \to \mathsf{X}^T,$$

(8.13)

$$\varepsilon^T(x,h) = h \colon F^T G^T(x,h) = (Tx, \mu x) \to (x,h),$$

and the triangular equations hold, because they hold on transversal categories.

We have already seen that $G^T F^T = T$ and $\eta^T = \eta$. The new multiplication also coincides with the old one: $G^T \varepsilon^T F^T(x) = G^T \varepsilon^T(Tx, \mu x) = \mu x$.

8.2.5 The comparison

The other way round, a colax-pseudo adjunction $(\eta, \varepsilon) \colon F \dashv G$ (with $F \colon \mathsf{X} \to \mathsf{A}$) has an associated colax monad $T = GF \colon \mathsf{X} \to \mathsf{X}$, which determines a chiral multiple category of algebras X^T and its adjunction $F^T \dashv G^T$ (as for ordinary categories, in 1.6.5).

The two adjunctions are linked by a *comparison* pseudo functor K, defined in the usual way on cubes and transversal maps

$$K \colon \mathsf{A} \to \mathsf{X}^T, \quad Ka = (Ga, G\varepsilon a \colon TGa \to Ga),$$

$$K(f \colon a \to b) = Gf \colon Ka \to Kb,$$

$$\underline{K}_i(a) = \underline{G}_i(a) \colon e_i Ka \to K(e_i a),$$

$$\underline{K}_i(a,b) = \underline{G}_i(a,b) \colon Ka +_i Kb \to K(a +_i b),$$

(8.14)

$$
\begin{array}{ccc}
\mathsf{X} \xrightarrow{\ F\ } \mathsf{A} & & KF = F^T, \\
\Big\| \quad G \quad \Big\downarrow K & & G^T K = G, \\
\mathsf{X} \xrightarrow[G^T]{F^T} \mathsf{X}^T & & \varepsilon^T K = K\varepsilon, \quad \eta^T = \eta.
\end{array}
$$

The coherence and invertibility of the comparisons of K comes from the relation $G^T K = G$ and the faithfulness of G^T.

We say that the pseudo functor $G\colon \mathsf{A} \to \mathsf{X}$ is *pseudo monadic* if it has a colax left adjoint and the comparison K is a pseudo isomorphism of chiral multiple categories (see 6.5.4): this last point is equivalent to saying that each transversal component $G_i\colon \mathbf{A}_i \to \mathbf{X}_i$ is monadic. More particularly, we say that G is *monadic, or algebraic,* if K is an isomorphism, which means that (moreover) its comparisons $\underline{K}_i$, for units and compositions, are identities.

8.2.6 Examples

As in 7.1.5, we start from an adjunction $(\eta, \varepsilon)\colon F \dashv G$ between ordinary categories with pullbacks, to which we add now the associated monad (T, η, μ)

$$F\colon \mathbf{X} \rightleftarrows \mathbf{A} \colon G, \qquad \eta\colon 1 \to GF, \quad \varepsilon\colon FG \to 1,$$
$$T = GF\colon \mathbf{X} \to \mathbf{X}, \quad \eta\colon 1 \to T, \quad \mu = G\varepsilon F\colon T^2 \to T. \tag{8.15}$$

As we have already seen in 8.1.3(a), all this can be extended to the weak multiple categories of cubical spans over $\mathbf{X}$ and $\mathbf{A}$, using the 2-functor constructed in 7.1.5

$$\mathsf{Span}\colon \mathbf{Cat}_{\mathrm{pb}} \to \mathbf{CxCmc}, \qquad \mathbf{C} \mapsto \mathsf{Span}(\mathbf{C}), \tag{8.16}$$

and defined on the full sub-2-category of **Cat** containing all categories with (a fixed choice of) pullbacks.

In this way we obtain the unitary *colax-pseudo* adjunction $\mathsf{Span}(F) \dashv \mathsf{Span}(G)$ already considered in 7.1.5, together with the associated unitary *colax* monad $\mathsf{Span}(T)$

$$\mathsf{Span}(F)\colon \mathsf{Span}(\mathbf{X}) \rightleftarrows \mathsf{Span}(\mathbf{A}) \colon \mathsf{Span}(G),$$
$$\mathsf{Span}(T)\colon \mathsf{Span}(\mathbf{X}) \to \mathsf{Span}(\mathbf{X}). \tag{8.17}$$

We already know, from Exercise 1.6.6(c), that $\mathbf{X}^T$ inherits from $\mathbf{X}$ a choice of pullbacks strictly preserved by the forgetful functor $G^T\colon \mathbf{X}^T \to \mathbf{X}$.

We prove below, in the next theorem, that the weak multiple category of algebras of $\mathsf{Span}(T)$ can be identified with $\mathsf{Span}(\mathbf{X}^T)$; this identification is coherent with the comparisons of the ordinary adjunction and the associated multiple adjunction.

8.2.7 Theorem

An ordinary adjunction $(\eta, \varepsilon)\colon F \dashv G$ *is given between categories with a fixed choice of pullbacks, as in* (8.15).

(a) There is a canonical isomorphism

$$J\colon \mathsf{Span}(\mathbf{X})^{\mathsf{Span}(T)} \to \mathsf{Span}(\mathbf{X}^T) \tag{8.18}$$

which allows us to identify these two weak multiple categories.

(b) Under this identification the pseudo multiple functor

$$K\colon \mathsf{Span}(\mathbf{A}) \to \mathsf{Span}(\mathbf{X})^{\mathsf{Span}(T)}$$

defined in 8.2.5 is identified with

$$\mathsf{Span}(K_*)\colon \mathsf{Span}(\mathbf{A}) \to \mathsf{Span}(\mathbf{X}^T),$$

where $K_*\colon \mathbf{A} \to \mathbf{X}^T$ is the comparison of the ordinary adjunction.

Thus, if the latter is monadic, the adjunction $\mathsf{Span}(F) \dashv \mathsf{Span}(G)$ is pseudo monadic; it is monadic if the given functor $G\colon \mathbf{A} \to \mathbf{X}$ strictly preserves the choices of pullbacks.

Proof (a) We have already recalled, in 8.2.6, that $\mathbf{X}^T$ has a choice of pullbacks strictly preserved by the forgetful functor $G^T\colon \mathbf{X}^T \to \mathbf{X}$. The colax endofunctor $T' = \mathsf{Span}(T)$ acts on an n-cube $x\colon \vee^n \to \mathbf{X}$ of $\mathsf{Span}(\mathbf{X})$, by composition: $T'x = T.x\colon \vee^n \to \mathbf{X}$.

As defined in 8.2.3, an n-cube $(x, h\colon T'x \to x)$ of $\mathsf{Span}(\mathbf{X})^{T'}$ is a functor $x\colon \vee^n \to \mathbf{X}$ equipped with a natural transformation $h\colon T.x \to x$ satisfying the usual axioms

$$
\begin{array}{lll}
\begin{array}{ccc}
\vee^n & \xrightarrow{x} & \mathbf{X} \\
\| & \overset{h}{\nearrow} \, T\!\downarrow \uparrow^{\eta} \downarrow 1 \\
\vee^n & \xrightarrow{x} & \mathbf{X}
\end{array}
&
\begin{array}{ccc}
\vee^n & \xrightarrow{x} & \mathbf{X} \\
\| & \overset{h}{\nearrow} \, T\!\downarrow \uparrow^{\mu} \downarrow T^2 \\
\vee^n & \xrightarrow{x} & \mathbf{X}
\end{array}
&
\begin{array}{l}
h.\eta x = 1_x, \\[6pt]
\\
h.\mu x = h.Th.
\end{array}
\end{array}
\tag{8.19}
$$

On the other hand the category $\mathbf{X}^T$ is the full subcategory of the comma category $T \downarrow \mathbf{X}$ determined by the objects $(x, h\colon Tx \to x)$ satisfying the 'same' axioms in degree zero, with $x\colon \vee^0 \to \mathbf{X}$. An n-cube y of $\mathsf{Span}(\mathbf{X}^T)$ is a functor $y\colon \vee^n \to \mathbf{X}^T$, or equivalently a functor $y\colon \vee^n \to (T \downarrow \mathbf{X})$ that satisfies the following axioms

$$
\begin{array}{ll}
\begin{array}{ccc}
 & \mathbf{X} & \\
 & \overset{P}{\nearrow} \quad \overset{T}{\searrow} & \\
\vee^n \xrightarrow{y} T\!\downarrow\!\mathbf{X} & \Big)\, \pi & \mathbf{X} \\
 & \overset{Q}{\searrow} \quad \overset{1}{\nearrow} & \\
 & \mathbf{X} &
\end{array}
&
\begin{array}{l}
Py = Qy, \\[6pt]
\pi y.\eta Py = 1_{Py}, \\[6pt]
\pi y.\mu Py = \mu Py.T(\mu Py).
\end{array}
\end{array}
\tag{8.20}
$$

The canonical isomorphism $\mathsf{Span}(\mathbf{X})^{T'} \cong \mathsf{Span}(\mathbf{X}^T)$ is now obvious: a functor $x\colon \vee^n \to \mathbf{X}$ equipped with a natural transformation $h\colon T.x \to x$

is 'the same' as a functor $y: \vee^n \to (T \downarrow \mathbf{X})$ such that $Py = Qy$: the correspondence is given by the equations

$$Py = Qy = x, \qquad \pi y = h: T.x \to x, \qquad (8.21)$$

under which the axioms of (8.19) correspond to the remaining axioms of (8.20). The rest is obvious. $\qquad\qquad\square$

8.2.8 An example

We start from the free-group adjunction and the associated monad, with comparison isomorphism K

$$F: \mathbf{Set} \rightleftarrows \mathbf{Gp} : U, \qquad \eta: 1 \to UF, \quad \varepsilon: FU \to 1,$$

$$T = UF: \mathbf{Set} \to \mathbf{Set}, \qquad \eta: 1 \to T, \quad \mu = U\varepsilon F: T^2 \to T, \qquad (8.22)$$

$$K: \mathbf{Gp} \to \mathbf{Set}^T, \qquad K(A) = (U(A), U\varepsilon A: TUA \to UA).$$

Applying the 2-functor Span, as in 8.2.6, all this is transformed into a colax-pseudo adjunction, the associated colax monad and a comparison isomorphism of weak multiple categories.

Note that $\mathsf{Span}(U)$ is even strict, if we let the relevant limits of groups be constructed 'as in' **Set**; or, more precisely, created by the underlying-set functor U and the choice in **Set**, see 1.3.7.

The same can be done replacing **Gp** with any variety of algebras (see 1.6.6(b)).

8.3 Lax adjunctions and monads

We now work in the 2-category Lx**Cmc** of chiral multiple categories, lax functors and transversal transformations (see 6.5.3), studying multiple monads and adjunctions of a lax type.

Here the construction of Kleisli algebras is standard (see 8.3.3), but we are more interested in Eilenberg–Moore algebras, which will be constructed in the idempotent 'strong' case, in Section 8.4.

Again X and A are chiral multiple categories; **i** denotes a positive multi-index of dimension $n \geqslant 0$.

8.3.1 Adjunctions and monads, the lax case

A *lax monad* $T = (T, \eta, \mu)$ over the chiral multiple category X is defined as a monad in the 2-category Lx**Cmc**.

Thus $T\colon \mathsf{X} \to \mathsf{X}$ is a lax endofunctor, with comparison maps

$$\underline{T}_i(x)\colon e_i(Tx) \to T(e_i x),$$
$$\underline{T}_i(x,y)\colon Tx +_i Ty \to T(x +_i y). \tag{8.23}$$

Furthermore we have two transversal transformations of lax functors $\eta\colon 1 \to T$ and $\mu\colon T^2 \to T$ that satisfy the usual axioms:

$$\mu.\eta T = 1 = \mu.T\eta, \qquad \mu.\mu T = \mu.T\mu.$$

Again these axioms only concern the family of transversal categories $\mathbf{X}_i$, where we have ordinary monads (T_i, η_i, μ_i), while the coherence conditions of T, η, μ (see 6.5.2, 6.5.3) cannot be expressed within the transversal components.

As we have seen in Theorem 7.4.3, a *pseudo-lax adjunction*

$$(F, G, \eta, \varepsilon)\colon \mathsf{X} \nrightarrow \mathsf{A}$$

between chiral multiple categories is the same as an internal adjunction in this 2-category. By the usual 2-categorical procedure it produces a lax monad $T = GF$ on the domain of F, with the same unit η and $\mu = G\varepsilon F$.

8.3.2 Examples

Let us assume that the adjunction $(\eta, \varepsilon)\colon F \dashv G$ and the associated monad (T, η, μ) considered above, in (8.15), involve two categories $\mathbf{X}$ and $\mathbf{A}$ that have pushouts.

We apply to these data the 2-functor

$$\mathsf{Cosp}\colon \mathbf{Cat}_{\mathrm{po}} \to \mathrm{Lx}\mathbf{Cmc}, \qquad \mathsf{Cosp}(\mathbf{C}) = (\mathsf{Span}(\mathbf{C}^{\mathrm{op}}))^{\mathrm{tr}}, \tag{8.24}$$

defined in 7.1.6.

We get now a unitary pseudo-lax adjunction (already considered in 7.1.6) and a unitary lax monad of weak multiple categories that coincides with the lax monad associated to the multiple adjunction (letting F stand for $\mathsf{Cosp}(F)$, etc.)

$$F\colon \mathsf{Cosp}(\mathbf{X}) \rightleftarrows \mathsf{Cosp}(\mathbf{A})\colon G,$$
$$T = GF\colon \mathsf{Cosp}(\mathbf{X}) \to \mathsf{Cosp}(\mathbf{X}). \tag{8.25}$$

The idempotent case will be studied in Section 8.7.

8.3.3 *Exercises and complements* (Kleisli algebras)

(a) Construct the chiral multiple category X_T of *Kleisli T-algebras*, for a lax monad (T, η, μ) on a chiral multiple category X.

(b) Define the associated strict-lax adjunction $F_T \dashv U_T$.

8.3.4 Definition (Left-strong adjunctions)

As a generalisation of a pseudo-lax adjunction, we say that a colax-lax adjunction $(F, G, \eta, \varepsilon) \colon \mathsf{X} \rightarrowtail \mathsf{A}$ between chiral multiple categories is *left-strong* if:

(i) the comparison cells of F are made invertible by applying G, i.e. all the following transversal maps are invertible

$$GF_i(x) \colon GF(e_i x) \to G(e_i F x),$$

$$GF_i(x, y) \colon GF(x +_i y) \to G(Fx +_i Fy), \tag{8.26}$$

(ii) the premultiple functor $T = GF$ becomes a lax functor, when equipped with the following comparison maps, derived from those of F and G:

$$T_i(x) = (GF_i(x))^{-1}.G_i(Fx) \colon e_i(Tx) \to T(e_i x),$$

$$T_i(x, y) = (GF_i(x, y))^{-1}.G_i(Fx, Fy) \colon Tx +_i Ty \to T(x +_i y). \tag{8.27}$$

We prove below that, in these hypotheses, we still have an *associated lax monad* (T, η, μ) over X, whose transversal components form a face-consistent family of ordinary monads

$$T_\mathbf{i} = G_\mathbf{i} F_\mathbf{i} \colon \mathbf{X}_\mathbf{i} \to \mathbf{X}_\mathbf{i},$$

$$\eta_\mathbf{i} \colon 1 \to T_\mathbf{i}, \qquad \mu_\mathbf{i} = G_\mathbf{i} \varepsilon_\mathbf{i} F_\mathbf{i} \colon T_\mathbf{i} T_\mathbf{i} \to T_\mathbf{i}, \tag{8.28}$$

constructed in the usual way from the transversal components of the adjunction. We shall often write the terms $T_\mathbf{i} x$, $\eta_\mathbf{i} x$ and $\mu_\mathbf{i} x$ omitting the multi-index $\mathbf{i}$.

By transversal duality we say that our adjunction is *right-strong* if the comparison cells of G are made invertible by applying F, and the premultiple endofunctor $S = FG$ of A is colax, via the obvious comparison maps derived from those of F and G.

A *strong* colax-lax adjunction is left- and right-strong, by definition.

8.3.5 Theorem (From left-strong adjunctions to lax monads)

Let $(F, G, \eta, \varepsilon) \colon \mathsf{X} \rightarrowtail \mathsf{A}$ be a left-strong colax-lax adjunction between chiral multiple categories. There is an associated lax monad (T, η, μ) over X, constructed as specified above (in 8.3.4).

Proof T is a lax endofunctor of X, by hypothesis.

To prove that η is a transversal transformation $1 \to T$ of lax functors it is now sufficient to consider the coherence conditions (7.38) and (7.40) of $\eta\colon 1 \dashrightarrow GF$ (as a double cell in $\mathbb{C}$mc), and invert the transversal maps $G\underline{F}_i(x)$ and $G\underline{F}_i(x,y)$.

As to multiplication, the counit ε (which is *not* a transversal transformation of lax functors, generally) does produce, under condition (i) in 8.3.4, a transversal transformation of lax functors $\mu\colon T^2 \to T$ with components $\mu(x) = G(\varepsilon(F(x)))$ (even though $G\varepsilon F$ makes no sense, generally).

A direct proof would be complicated, but the machinery of the double category $\mathbb{C}$mc makes it easy. We consider the double cell ι of $\mathbb{C}$mc, with components $\iota x = 1_{Tx}\colon Tx \to GFx$, which is coherent precisely by definition (8.27) (this cell is a sort of 'co-operative flipping' of F over G)

$$\tag{8.29}$$

Now we paste two copies of ι with ε, as in the right diagram above, getting a cell $\overline{\mu}\colon (F\,{}^{TT}_{\ G}\,X)$; informally $\overline{\mu}\colon 1.T^2 \dashrightarrow GF$.

The coherence properties of this cell show that the family of transversal arrows $\mu(x) = G\varepsilon F(x)$ is indeed a transversal transformation of lax functors $T^2 \to T$. More formally, there is precisely one double cell $\mu\colon (X\,{}^{TT}_{\ T}\,X)$ such that the vertical composite $\mu \otimes \iota$ coincides with $\overline{\mu}$.

Finally the monad axioms 'live' in the transversal components of X and are satisfied. $\qquad\square$

8.4 Strong idempotent lax monads

We define *left-strong idempotent adjunctions* and *strong idempotent lax monads* for chiral multiple categories. Most of the theory *will be developed in the strict version* (extending the 1-dimensional case of 1.7.8), since the geometric operations of the algebras of the monad have a simpler construction in the strict case.

We shall consider examples that yield double categories of relations as algebraic spans, in Section 8.5, and higher dimensional cases in Section 8.7.

The index i will often be omitted in large diagrams and long formulas.

8.4.1 Definition (Left-strong idempotent adjunctions)

Let us start from a general colax-lax adjunction

$$(F, G, \eta, \varepsilon) \colon \mathsf{X} \rightarrowtail \mathsf{A}, \qquad (8.30)$$

between chiral multiple categories.

We say that the adjunction is *idempotent* if all the ordinary adjunctions

$$F_i \colon \mathbf{X}_i \rightleftarrows \mathbf{A}_i \colon G_i,$$

$$\eta_i \colon 1 \to G_i F_i, \qquad \varepsilon_i \colon F_i G_i \to 1, \qquad (8.31)$$

are idempotent, or equivalently all the natural transformations $F_i \eta_i$, $\varepsilon_i F_i$, $\eta_i G_i$ and $G_i \varepsilon_i$ are invertible, or equivalently this holds for one of these four families.

We say that (8.30) is a *left-strong idempotent adjunction* if moreover:

(i) the adjunction is left-strong (see 8.3.4),

(ii) all the transversal maps $F\underline{G}_i(Fx)$ and $F\underline{G}_i(Fx, Fy)$ are invertible,

(ii') all the transversal maps $Fe_i(\eta x)$ and $F(\eta x +_i \eta y)$ are invertible.

The conditions (ii) and (ii') are *here* equivalent, as follows immediately from the coherence conditions (7.38) and (7.41). In fact, from (7.38) we have

$$FG\underline{F}_i(x).F\eta(e_i x) = F\underline{G}_i(Fx).Fe_i(\eta x),$$

where the left term is invertible, by (i) and idempotence; therefore $F\underline{G}_i(Fx)$ is invertible if and only if $Fe_i(\eta x)$ is.

This notion is adequate to get a (strong) idempotent lax monad GF on X, as we prove below (Theorem 8.4.4). By transversal duality, a *right-strong idempotent adjunction* gives an idempotent colax comonad FG on A.

A pseudo-lax idempotent adjunction is always left-strong idempotent. Indeed, condition (i) is obviously satisfied. To prove that $F\underline{G}_i(Fx, Fy)$ is always invertible it is sufficient to note that the functor $FGF \colon \mathsf{X} \to \mathsf{A}$ is isomorphic to F (via $F\eta$) and is pseudo as well; moreover its (invertible) comparison factorises as

$$FG\underline{F}_i(x, y).F\underline{G}_i(Fx, Fy).\underline{F}_i(GFx, GFy) \colon$$

$$FGFx +_i FGFy \to F(GFx +_i GFy) \to FG(Fx +_i Fy) \to FGF(x +_i y),$$

where the first and third arrow are known to be invertible. Similarly $F\underline{G}_i(Fx)$ is always invertible.

We say that (8.30) is a *strong idempotent adjunction* if it is idempotent (as defined above) and a strong adjunction (as defined in 8.3.4). The last condition implies that all the transversal maps of type GF_i and FG_i are invertible, which also implies condition (ii) above, and its dual. In other words, a colax-lax adjunction is strong idempotent if and only if it is both left- and right-strong idempotent.

Finally we say that (8.30) is a *left-strict idempotent adjunction* if it is left-strong and $\eta G = 1_G$, or equivalently $G\varepsilon = 1_G$; it follows that $GFG = G$.

8.4.2 *Definition* (Strong idempotent lax monads)

Let us recall that a lax monad (T, η, μ) over a chiral multiple category X is a monad in the 2-category $\mathrm{Lx}\mathbf{Cmc}$ (see 8.3.1), and induces *transversal monads* of ordinary categories

$$T_i \colon \mathbf{X}_i \to \mathbf{X}_i, \qquad \eta_i \colon 1 \to T_i, \qquad \mu_i \colon T_i^2 \to T_i. \qquad (8.32)$$

Our monad is said to be *idempotent* if it satisfies the following conditions, equivalent because a transversal transformation is invertible if and only if all its components are

(idm.1) all the ordinary monads in (8.32) are idempotent,

(idm.1′) the transversal transformation $\mu \colon T^2 \to T$ is invertible, with inverse $T\eta = \eta T \colon T \to T^2$,

(idm.1″) $T\eta$ is invertible,

(idm.1‴) ηT is invertible.

Equivalently we shall write an idempotent lax monad over X as a pair (T, η) where $T \colon \mathsf{X} \to \mathsf{X}$ is a lax endofunctor, $\eta \colon 1 \to T$ is a transversal transformation and $T\eta = \eta T \colon T \to T^2$ is invertible (again the last condition lives on transversal categories). One recovers μ as the inverse of $T\eta$.

We also note that the components $\eta x \colon x \to TX$ give commutative diagrams:

$$(8.33)$$

where the slanting arrows are invertible.

We say that the idempotent lax monad (T, η) is *strong* if it satisfies the following conditions, equivalent by the previous remark:

(idm.2) all the transversal maps $T\underline{T}_i(x)$ and $T\underline{T}_i(x, y)$ are invertible,

(idm.2′) all the following transversal maps are invertible

$$Te_i(\eta x) \colon T(e_i x) \to T(e_i Tx), \qquad T(\eta x +_i \eta y) \colon T(x +_i y) \to T(Tx +_i Ty).$$

In this situation we say that an **i**-cube x is *algebraic* if $\eta x\colon x \to Tx$ is invertible, or equivalently if x is transversally isomorphic to Tx in X (by Theorem 1.7.3).

More particularly, a *strictly idempotent lax monad* is a strong idempotent lax monad (T, η) where $T = T^2$ and $T\eta = \eta T = 1_T$. The cube x is said to be *strictly algebraic* if $Tx = x$, or equivalently $\eta x = 1_x$ (from $Tx = x$ we get $\eta x = \eta Tx = 1_{Tx} = 1_x$). We shall see that, in this case, one can define a chiral multiple category of algebras in a simple way.

Extending the 1-dimensional case of 1.7.9, and using Proposition 6.5.6, one proves that every strong idempotent lax monad (T, η) over a transversally invariant chiral multiple category can be replaced by an isomorphic strictly idempotent one $(U, \varphi\eta)$, so that the algebraic objects of T are precisely the strictly algebraic objects of U.

8.4.3 *Proposition* (Basic properties)

*Let (T, η) be a strong idempotent lax monad on the chiral multiple category X; let x, y, z, u be **i**-consecutive **i**-cubes of X; $+$ denotes $+_i$.*

(a) The following transversal maps are invertible

$$T(\eta + (\eta + \eta))\colon T(x + (y + (z + u))) \to T(Tx + (Ty + T(z + u))), \quad (8.34)$$

$$T((\eta + \eta) + \eta)\colon T(((x + y) + z) + u) \to T((T(x + y) + Tz) + Tu), \quad (8.35)$$

$$T(\eta + (\eta(y + z) + \eta)), \qquad T((\eta + \eta(y + z)) + \eta). \qquad (8.36)$$

(b) The laxity comparisons of T form the dashed arrows of these commutative diagrams

$$(8.37)$$

and are therefore determined by the components of $\eta\colon 1 \to T$ as follows:

$$\underline{T}_i(x) = (T(e_i\eta x))^{-1}.\eta e_i(Tx)\colon e_i(Tx) \to T(e_i x),$$
$$\underline{T}_i(x, y) = (T(\eta x + \eta y))^{-1}.\eta(Tx + Ty)\colon Tx + Ty \to T(x + y). \qquad (8.38)$$

Proof (a) Property (8.34) follows from the following commutative square, where the other three arrows are invertible

$$
\begin{array}{ccc}
T(x + (y + (z + u))) & \xrightarrow{T(\eta + (\eta + \eta))} & T(Tx + (Ty + T(z + u))) \\
{\scriptstyle T(\eta + \eta)}\downarrow & & \downarrow{\scriptstyle T(\eta + \eta)} \\
T(Tx + T(y + (z + u))) & \xrightarrow[T(1 + T(\eta + \eta))]{} & T(Tx + T(y + T(z + u)))
\end{array}
$$

One verifies (8.35) and (8.36) in a similar way.

(b) The two upper triangles in (8.37) commute by axiom (trt.2L) of 6.5.3 on η. For the lower triangles we verify the right-hand one.

This follows from the commutative diagram below, where the square commutes by the naturality of η, the triangle by (8.34) (since $\eta T = T\eta$), and $T\underline{T}_i(x, y)$ is invertible

$$\square$$

8.4.4 *Theorem* (From left-strong idempotent adjunctions to monads)

Let $(F, G, \eta, \varepsilon): \mathsf{X} \to \mathsf{A}$ be a left-strong idempotent adjunction *between chiral multiple categories, as defined in 8.4.1.*

Then the lax monad (T, η, μ) on X constructed in 8.3.4 is a strong idempotent lax monad, as defined in 8.4.2.

If $F \dashv G$ is a left-strict idempotent adjunction, then (T, η, μ) is a strictly idempotent lax monad.

Proof Applying Theorem 8.3.5 we know that (T, η, μ) is indeed a lax monad.

Now the condition that ηT be invertible lives in the transversal components of X and is satisfied. The invertibility of the transversal maps $T\underline{T}_i(x)$ and $T\underline{T}_i(x, y)$ follows immediately from the definition of the comparisons of T in (8.27), using condition 8.4.1(ii) on the adjunction.

The strict case is obvious, because $\mu = G\varepsilon F$ is now an identity. $\square$

8.4.5 Algebras of strictly idempotent lax monads

We go on examining the strict case, because working in the strong one would make the geometric operations $x \circ_i y$ of algebraic cubes complicated, and their comparisons even more.

Let (T, η) be a strictly idempotent lax monad over the chiral multiple category X. We want to construct the chiral multiple category $\mathsf{Alg}_*(T)$ *of strictly algebraic cubes of* X, extending the construction we have seen in 1.7.8 for ordinary categories. Note that $\mathsf{Alg}_*(T)$ will have its own geometric structure, with a *lax* embedding in X.

(a) For the transversal part of the construction we simply apply the ordinary procedure (of 1.7.8) to each strictly idempotent monad $(T_\mathbf{i}, \eta_\mathbf{i})$ on the category $\mathbf{X}_\mathbf{i} = \mathrm{tv_i} \mathsf{X}$, getting a full subcategory $\mathbf{Alg}_*(T_\mathbf{i}) \subset \mathbf{X}_\mathbf{i}$.

An object of $\mathbf{Alg}_*(T_\mathbf{i})$ is thus a strictly algebraic $\mathbf{i}$-cube x of X (as defined in 8.4.2): $Tx = x$, or equivalently $\eta x = 1_x$. The algebra $\hat{x} = (x, 1_x)$ coincides with the free $T_\mathbf{i}$-algebra $(Tx, \mu x)$ on x.

A morphism $f \colon x \to y$ of $\mathbf{Alg}_*(T_\mathbf{i})$ is any transversal map of X between strictly algebraic $\mathbf{i}$-cubes. It satisfies $Tf = f$ and gives an arrow of $T_\mathbf{i}$-algebras $f \colon \hat{x} \to \hat{y}$.

All this is consistent with faces: $\partial_i^\alpha(\mathbf{Alg}_*(T_\mathbf{i})) \subset \mathbf{Alg}_*(T_{\mathbf{i}|i})$ for $i \in \mathbf{i}$, because the monads $(T_\mathbf{i}, \eta_\mathbf{i})$ are face-consistent.

(b) Now we extend this family $\mathbf{Alg}_*(T_\mathbf{i})$ to a chiral multiple category $\mathsf{Alg}_*(T)$, writing $T_\mathbf{i} x$ and $\eta_\mathbf{i} x$ as Tx and ηx.

First we define the new degeneracies and concatenations of strictly algebraic $\mathbf{i}$-cubes and their maps as follows

$$
\begin{aligned}
e_i^\circ(x) = Te_i(x), &\qquad e_i^\circ(f) = Te_i(f) &\quad (i \notin \mathbf{i}), \\
x \circ_i y = T(x +_i y), &\qquad f \circ_i g = T(f +_i g) &\quad (i \in \mathbf{i}).
\end{aligned}
\tag{8.39}
$$

It should be noted that these cubes have the correct i-faces because x and y are strictly algebraic and T does not modify the i-faces of $e_i(x)$ and $x +_i y$

$$
\partial_i^\alpha(e_i^\circ(x)) = T(\partial_i^\alpha e_i(x)) = Tx = x,
$$
$$
\partial_i^-(x \circ_i y) = T(\partial_i^-(x +_i y)) = T(\partial_i^- x) = \partial_i^- Tx, = \partial_i^- x,
$$
$$
\partial_i^+(x \circ_i y) = \ldots = \partial_i^+ y.
$$

(Working with algebraic cubes one should 'correct' these definitions, assuming that X is transversally invariant; this can be done, but would lead to complicated computations below.)

(c) Secondly we define unitors, associators and interchangers for cubes x, y, z, u in $\mathsf{Alg}_*(T)$ so that they form the following commutative diagrams, where the upper maps are T-transformed comparisons of X, the vertical maps are transversal isomorphisms by 8.4.2, and the lower (dashed) arrows are the new comparisons we are defining for strict algebras (where $\eta_x = 1_x \colon x \to x$, and so on)

$$
\begin{array}{ccc}
T(e_i \partial_i^- x +_i x) & \xrightarrow{\ T\lambda\ } Tx \xleftarrow{\ T\rho\ } & T(x +_i (e_i \partial_i^+ x)) \\[2pt]
\big\downarrow{\scriptstyle T(\eta+1)} & \big\uparrow{\scriptstyle \eta} & \big\downarrow{\scriptstyle T(1+\eta)} \\[2pt]
T(Te_i(\partial_i^- x) +_i x) \;-\;\underset{\lambda}{\dashrightarrow}\; & x \;\underset{\rho}{\dashleftarrow}\;-\; & T(x +_i T(e_i \partial_i^+ x)) \\[2pt]
\| & & \| \\[2pt]
e_i^\circ(\partial_i^- x) \circ_i x & & x \circ_i e_i^\circ(\partial_i^+ x)
\end{array}
\tag{8.40}
$$

$$
\begin{array}{ccc}
T(x +_i (y +_i z)) & \xrightarrow{\ T\kappa\ } & T((x +_i y) +_i z) \\[2pt]
\big\downarrow{\scriptstyle T(1+\eta)} & & \big\downarrow{\scriptstyle T(\eta+1)} \\[2pt]
T(x +_i T(y +_i z)) & \dashrightarrow_{\ \kappa\ } & T(T(x +_i y) +_i z) \\[2pt]
\| & & \| \\[2pt]
x \circ_i (y \circ_i z) & & (x \circ_i y) \circ_i z
\end{array}
\tag{8.41}
$$

$$
\begin{array}{ccc}
T((x +_i y) +_j (z +_i u)) & \xrightarrow{\ T\chi_{ij}\ } & T((x +_j z) +_i (y +_j u)) \\[2pt]
\big\downarrow{\scriptstyle T(\eta +_j \eta)} & & \big\downarrow{\scriptstyle T(\eta +_j \eta)} \\[2pt]
T(T(x +_i y) +_j T(z +_i u)) & \dashrightarrow_{\ \chi_{ij}\ } & T(T(x +_j z) +_i T(y +_j u)) \\[2pt]
\| & & \| \\[2pt]
(x \circ_i y) \circ_j (z \circ_i u) & & (x \circ_j z) \circ_i (y \circ_j u)
\end{array}
\tag{8.42}
$$

The coherence axioms are verified below, in Theorem 8.4.6.

Let us note that if X is a weak multiple category, its interchangers are invertible and so are those of $\mathsf{Alg}_*(T)$; more precisely, the latter is a weak multiple category if and only if $T(\chi_{ij}(x, y, z, u))$ is invertible, for every consistent matrix of strictly algebraic cubes.

8.4.6 Theorem (Algebras of strictly idempotent lax monads)

If (T, η) is a strictly idempotent lax monad over X, the previous construction gives a chiral multiple category $\mathsf{Alg}_*(T)$.

Proof The verification of the coherence axioms is long. We write down a complete proof for the pentagon-diagram of the i-associator κ_i, i.e. axiom (wmc.9.i) of 6.4.3. The proof of the other axioms follows the same pattern.

We start from the commutative pentagon of κ_i in X (omitting the index

i, as we often do)

$$
\begin{array}{ccc}
& (x+y)+(z+u) & \\
& \nearrow^{\kappa} \qquad \searrow^{\kappa} & \\
x+(y+(z+u)) & & ((x+y)+z)+u \qquad (8.43) \\
{}_{1+\kappa}\searrow & & \nearrow_{\kappa+1} \\
x+((y+z)+u) & \xrightarrow[\kappa]{} & (x+(y+z))+u
\end{array}
$$

Let x, y, z, u be strictly algebraic cubes. We want to show that the (commutative) T-image of this diagram is related to the corresponding pentagon for the operation $\circ = \circ_i$ by a coherent family of transversal isomorphisms (even though we just need to cancel one of them, at the left).

For the T-image of the upper path of (8.43) we form the following commutative diagram (recalling again that $\eta_x = 1_x \colon x \to x$, etc.). The solid vertical arrows are invertible by Proposition 8.4.3; the dashed diagonal $T(\eta +_i \eta)$ of the diamond is invertible by definition (while the upper slanting arrows need not be)

$$
\begin{array}{ccccc}
T(x+(y+(z+u))) & \xrightarrow{T\kappa} & T((x+y)+(z+u)) & \xrightarrow{T\kappa} & T(((x+y)+z)+u) \\
\downarrow{\scriptstyle T(1+(1+\eta))} & {\scriptstyle T(1+\eta)\nearrow} & \Big\downarrow & {\scriptstyle\searrow T(\eta+1)} & {\scriptstyle T((\eta+1)+1)}\Big\downarrow \\
X_1 & \xrightarrow{T\kappa} & X_2 \quad {\scriptstyle T(\eta+\eta)} & X_3 & \xrightarrow{T\kappa} \quad X_4 \\
\downarrow{\scriptstyle T(1+\eta)} & {\scriptstyle T(\eta+1)\searrow} & \Big\downarrow & {\scriptstyle\nearrow T(1+\eta)} & {\scriptstyle T(\eta+1)}\Big\downarrow \\
T(x+T(y+T(z+u))) & \dashrightarrow[\kappa] & T(T(x+y)+T(z+u)) & \dashrightarrow[\kappa] & T(T(T(x+y)+z)+u)
\end{array}
$$

$$X_1 = T(x +_i (y +_i T(z +_i u))), \qquad X_2 = T((x +_i y) +_i T(z +_i u)),$$
$$X_3 = T(T(x +_i y) +_i (z +_i u)), \qquad X_4 = T((T(x +_i y) +_i z) +_i u).$$

For the T-image of the lower path of (8.43) we have a commutative diagram (that should be symmetrically completed, at the right) where all the vertical and slanting arrows are invertible

$$
\begin{array}{ccccc}
T(x+(y+(z+u))) & \xrightarrow{T(1+\kappa)} T(x+((y+z)+u)) & \xrightarrow{T\kappa} & T((x+(y+z))+u) \\
{\scriptstyle T(1+\eta)}\Big\downarrow & {\scriptstyle T(1+\eta)\nearrow \quad \searrow T(1+(\eta+1))} & & {\scriptstyle\nearrow T((1+\eta)+1)} \\
Y_1 & \xrightarrow{T(1+T\kappa)} Y_2 & Y_3 \xrightarrow{T\kappa} Y_4 \\
{\scriptstyle T(1+T(1+\eta))}\Big\downarrow & {\scriptstyle T(1+T(\eta+1))\searrow \quad \nearrow T(1+\eta)} & & {\scriptstyle\searrow T(\eta+1)} \\
T(x+T(y+T(z+u))) & \xrightarrow[1\circ\kappa]{} T(x+T(T(y+z)+u)) & \dashrightarrow[\kappa] & T(T(x+T(y+z))+u)
\end{array}
$$

$$Y_1 = T(x +_i T(y +_i (z +_i u))), \qquad Y_2 = T(x +_i T((y +_i z) +_i u)),$$
$$Y_3 = T(x +_i (T(y +_i z) +_i u)), \qquad Y_4 = T((x +_i T(y +_i z)) +_i u).$$

Since the left-most vertical arrow (through Y_1) coincides with the previous one (through X_1), and – symmetrically – the same happens for the right-most vertical arrows, the conclusion follows. $\qquad\square$

8.4.7 Theorem (From strictly idempotent monads to adjunctions)

Let (T, η) be again a strictly idempotent lax monad over X, with $\mathsf{Alg}_*(T)$ constructed as above. There is an associated strictly idempotent adjunction

$$(F', G', \eta, \varepsilon') \colon \mathsf{X} \rightharpoondown \mathsf{Alg}_*(T),$$

which is pseudo-lax and gives back the monad; moreover $F'G' = 1$ and $\varepsilon' = 1_{F'G'}$. It is constructed as follows.

(a) The inclusion $G' \colon \mathsf{Alg}_*(T) \to \mathsf{X}$ is a full lax functor (where full means transversally full, see 7.4.4), with comparisons derived from η

$$\underline{G}'_i x \colon e_i(G'x) \to G'e_i^\circ(x), \qquad \underline{G}'_i x = \eta e_i x \colon e_i x \to T e_i x,$$
$$\underline{G}'_i(x, y) \colon G'x +_i G'y \to G'(x \circ_i y), \tag{8.44}$$
$$\underline{G}'_i(x, y) = \eta(x +_i y) \colon x +_i y \to T(x +_i y).$$

(b) The pseudo functor $F' \colon \mathsf{X} \to \mathsf{Alg}_*(T)$ is a codomain-restriction of T, with comparisons derived again from η (written below in the colax direction, and invertible)

$$F'x = Tx, \qquad F'f = Tf,$$
$$\underline{F}'_i x \colon F'(e_i x) \to e_i^\circ(F'x),$$
$$\underline{F}'_i x = T(e_i \eta x) \colon T(e_i x) \to T e_i(Tx), \tag{8.45}$$
$$\underline{F}'_i(x, y) \colon F'(x +_i y) \to F'x \circ_i F'y,$$
$$\underline{F}'_i(x, y) = T(\eta x +_i \eta y) \colon T(x +_i y) \to T(Tx +_i Ty).$$

(c) The composite $G'F'$ coincides with T (including comparisons, of course); η is a transversal transformation of lax functors $1 \to G'F'$ such that $G'F'\eta = \eta G'F'$ is the identity.

(d) The composite $F'G' \colon \mathsf{Alg}_*(T) \to \mathsf{Alg}_*(T)$ is the identity, as well as the counit $\varepsilon' \colon F'G' \to 1$.

(e) The pseudo-lax adjunction $F' \dashv G'$ is left-strict idempotent; its associated lax idempotent monad is precisely (T, η).

Proof (a) For G' we only verify condition (lmf.1) on a transversal map $f \colon x \to y$ in $\mathsf{Alg}_*(T)$

$$G'e_i^\circ(f).\underline{G}'_i(x) = T e_i(f).\eta e_i(x) = \eta e_i(y).e_i(f) = \underline{G}'_i(y).e_i(G'f),$$

and the coherence condition (lmf.4) on three i-consecutive **i**-cubes x, y, z of $\mathsf{Alg}_*(T)$

where all the quadrilaterals commute by naturality of η or by definition of the new comparisons κ_i in (8.41).

(b) Similarly for F' we verify condition (cmf.1) on a transversal map $f\colon x \to y$ in X

$$e_i^\circ(F'f).\underline{F}'_i(x) = Te_i(Tf.\eta x) = Te_i(\eta y.f) = \underline{F}'_i(y).F'e_i(f),$$

and (cmf.4) on three i-consecutive **i**-cubes x, y, z of X

Again all the quadrilaterals commute by naturality of η or by definition of the new comparisons κ_i in (8.41).

(c) The composite $G'F'\colon \mathsf{X} \to \mathsf{X}$ coincides with T on cubes and transversal maps. Its comparisons are derived from those of F' and G'

$$(T(e_i\eta x))^{-1}.\eta e_i(Tx)\colon e_i(Tx) \to T(e_i x),$$

$$(T(\eta x +_i \eta y))^{-1}.\eta(Tx +_i Ty)\colon Tx +_i Ty \to T(x +_i y),$$

and coincide with those of T by Proposition 8.4.3(b).

(d) The composite $S = F'G'\colon \mathsf{Alg}_*(T) \to \mathsf{Alg}_*(T)$ is computed as T on cubes and maps, which means the identity. This is also true of its comparisons: for instance for the i-degeneracy of the strictly algebraic cube x we have:

$$\underline{S}_i x = F' \underline{G}'_i x . (\underline{F}'_i G' x)^{-1} = T\eta e_i x . (T(e_i \eta x))^{-1} = 1 \colon e_i x \to e_i x.$$

(e) We already know that the adjunction is strictly idempotent, since ε' is the identity. Moreover F is a pseudo functor and the transversal maps $F' \underline{G}'_i(x) = T\eta e_i(x)$ are invertible (and similarly for i-composition). Finally $T = GF$ has been proved above. □

8.4.8 Theorem (The comparison functor of algebras)

On the other hand, let us start from a left-strict idempotent adjunction

$$(F, G, \eta, \varepsilon) \colon \mathsf{X} \to \mathsf{A}.$$

We have the associated strictly idempotent lax monad (T, η) (by Theorem 8.4.4), its chiral multiple category $\mathsf{Alg}_*(T)$ of strictly algebraic cubes (by 8.4.5) and the associated left-strict pseudo-lax adjunction $(\eta, \varepsilon') \colon F' \dashv G'$ constructed above (in 8.4.7), with $G'F' = T$ and $\varepsilon' = 1_{F'G'}$.

There is a comparison lax functor $K \colon \mathsf{A} \to \mathsf{Alg}_*(T)$, defined as a restriction of $G \colon \mathsf{A} \to \mathsf{X}$ on codomain; its comparison-maps are modified to be coherent with the modified degeneracies and operations of $\mathsf{Alg}_*(T)$, using the fact that $\eta G = 1$ (and $GFG = G$)

$$K \colon \mathsf{A} \to \mathsf{Alg}_*(T), \qquad Ka = Ga, \qquad K(f \colon a \to b) = Gf,$$

$$\underline{K}_i a = (\eta G e_i a)^{-1} . GF\underline{G}_i a \colon e_i^{\circ} Ga \to G e_i a, \tag{8.46}$$

$$\underline{K}_i(a, b) = (\eta G(a +_i b))^{-1} . GF\underline{G}_i(a, b) \colon Ga \circ_i Gb \to G(a +_i b).$$

Moreover we have

$$
\begin{array}{ccc}
\mathsf{X} \underset{G}{\overset{F}{\rightleftarrows}} \mathsf{A} & & KF = F', \\
\| \qquad \downarrow K & & G'K = G, \\
\mathsf{X} \underset{G'}{\overset{F'}{\rightleftarrows}} \mathsf{X}^T & & K\varepsilon = 1_K = \varepsilon'K.
\end{array}
\tag{8.47}
$$

The relation $KF = F'$ must be interpreted: the composite of F (colax) with K (lax) is 'naturally' a pseudo functor (as F'), because all maps $\underline{K}_i(Fx, Fy)$ and $K\underline{F}_i(x, y)$ are invertible. (The same holds for the comparisons of degeneracies.)

Proof The definition of $K \colon \mathsf{A} \to \mathsf{Alg}_*(T)$ on cubes and transversal maps is legitimate because all components $\eta G(a)$ are identities.

We write down the verification of axiom (lmf.4), of coherence with associators, following the same pattern as in many previous verifications.

This axiom, applied to the lax functor G, says that for three i-consecutive i-cubes a, b, c in A we have a commutative diagram of i-maps:

$$
\begin{array}{ccc}
Ga +_i (Gb +_i Gc) & \xrightarrow{\;\kappa_i G\;} & (Ga +_i Gb) +_i Gc \\
{\scriptstyle 1+_i\underline{G}_i}\downarrow & & \downarrow{\scriptstyle \underline{G}_i+_i 1} \\
Ga +_i G(b +_i c) & & G(a +_i b) +_i Gc \\
{\scriptstyle \underline{G}_i}\downarrow & & \downarrow{\scriptstyle \underline{G}_i} \\
G(a +_i (b +_i c)) & \xrightarrow[\;G\kappa_i\;]{} & G((a +_i b) +_i c)
\end{array}
$$

The T-image of the upper-right composite is linked to the corresponding composite for K by a commutative diagram in which the vertical maps are invertible (here $A = (a +_i b) +_i c$)

$$
\begin{array}{ccccccc}
T(Ga+(Gb+Gc)) & \xrightarrow{T\kappa_i G} & T((Ga+Gb)+Gc) & \xrightarrow{T(\underline{G}+1)} & T(G(a+b)+Gc) & \xrightarrow{T\underline{G}} & TGA \\
{\scriptstyle T(1+\eta)}\downarrow & & {\scriptstyle T(\eta+1)}\downarrow & & \| & & \downarrow{\scriptstyle (\eta G)^{-1}} \\
T(Ga+T(Gb+Gc)) & \underset{\kappa_i K}{\to} & T(T(Ga+Gb)+Gc) & \underset{\underline{K}\circ_i 1}{\to} & T(G(a+b)+Gc) & \underset{\underline{K}}{\to} & GA \\
\| & & \| & & \| & & \| \\
Ka \circ_i (Kb \circ_i Kc) & & (Ka \circ_i Kb) \circ_i Kc & & K(a+b) \circ_i Kc & & KA
\end{array}
$$

In fact the left square commutes by definition of κ in $\mathsf{Alg}_*(T)$ and the right one by definition of $\underline{K}$. Finally the central square commutes by definition of $\underline{K}$ and by naturality of η:

$$
\begin{aligned}
(\underline{K}_i(a,b) &\circ_i 1_{Kc}) \,.\, T\left(\eta(Ga +_i Gb) +_i 1_{Gc}\right) \\
&= T\left(\underline{K}_i(a,b) +_i 1_{Gc}\right) .\, T\left(\eta(Ga +_i Gb) +_i 1_{Gc}\right) \\
&= T\left((\eta G(a +_i b))^{-1}.GF\underline{G}_i(a,b).\eta(Ga +_i Gb) +_i 1_{Gc}\right) \\
&= T\left(\underline{G}_i(a,b) +_i 1_{Gc}\right).
\end{aligned}
$$

(Here $\eta G = 1$, but we only need its invertibility.) One deals similarly with the left-lower path.

Finally the relations (8.47) are obvious on cubes and transversal maps. In particular the third amounts to the relation $G\varepsilon(a) = (\eta Ga)^{-1}$, which holds in every idempotent adjunction.

We check that the first of these relations, namely $KF = F'$, is also true for the comparisons of i-concatenation (as said in the statement)

$$\underline{KF}_i(x,y) = (\underline{K}_i(Fx, Fy))^{-1}.\underline{KF}_i(x,y)$$
$$= (T\underline{G}_i(Fx, Fy))^{-1}.\eta G(Fx +_i Fy).\underline{GF}_i(x,y)$$
$$= (T\underline{G}_i(Fx, Fy))^{-1}.TG\underline{F}_i(x,y).\eta T(x +_i y) \qquad \text{(by naturality of } \eta\text{)}$$
$$= (T\underline{G}_i(Fx, Fy))^{-1}.T\underline{G}_i(Fx, Fy).T(\eta x +_i \eta y) \quad \text{(by } T\eta = \eta T \text{ and (7.38))}$$
$$= T(\eta x +_i \eta y) = \underline{F}'_i(x,y).$$

$\square$

8.4.9 Theorem (Strong idempotent lax monads by basic data)

In order to assign a strong idempotent lax monad over a chiral multiple category X it suffices to give the following data:

(i) a face-consistent family of idempotent ordinary monads, indexed by the positive multi-indices $\mathbf{i}$

$$T_\mathbf{i} \colon \mathbf{X_i} \to \mathbf{X_i}, \qquad \eta_\mathbf{i} \colon 1 \to T_\mathbf{i},$$
$$\partial_i^\alpha(Tx) = T(\partial_i^\alpha x), \qquad \partial_i^\alpha(\eta x) = \eta(\partial_i^\alpha x), \qquad (8.48)$$
$$T_\mathbf{i}\eta_\mathbf{i} = \eta_\mathbf{i}T_\mathbf{i} \text{ is invertible,}$$

(ii) where $Te_i(\eta x)$ and $T(\eta x +_i \eta y)$ are invertible transversal maps (whenever this makes sense).

All this can be uniquely completed, forming a lax functor $T \colon \mathsf{X} \to \mathsf{X}$ and a transversal transformation $\eta \colon 1 \to T$ of lax functors with $T\eta = \eta T$ invertible.

For the strict case one modifies (i) requiring each $(T_\mathbf{i}, \eta_\mathbf{i})$ to be a strictly idempotent ordinary monad: $T_\mathbf{i}\eta_\mathbf{i} = \eta_\mathbf{i}T_\mathbf{i}$ is an identity.

Proof We (must) define the comparisons of T as specified in (8.38)

$$\underline{T}_i(x) = (T(e_i\eta x))^{-1}.\eta(e_iTx) \colon e_i(Tx) \to Te_i(x),$$
$$\underline{T}_i(x,y) = (T(\eta x +_i \eta y))^{-1}.\eta(Tx +_i Ty) \colon Tx +_i Ty \to T(x +_i y). \qquad (8.49)$$

Before checking that these comparisons make T a lax functor, it will be useful to verify the following properties, corresponding to axiom (trt.2L) for $\eta \colon 1 \to T$

$$\eta(e_i x) = \underline{T}_i(x).e_i(\eta x),$$
$$\eta(x +_i y) = \underline{T}_i(x,y).(\eta x +_i \eta y). \qquad (8.50)$$

This follows from the following identities, cancelling the invertible maps $T(e_i \eta x)$ and $T(\eta x +_i \eta y)$

$$T(e_i \eta x).\eta(e_i x) = \eta(e_i T x).e_i(\eta x) = T(e_i \eta x).\underline{T}_i(x).e_i(\eta x),$$
$$T(\eta x +_i \eta y).\eta(x +_i y) = \eta(T x +_i T y).(\eta x +_i \eta y)$$
$$= T(\eta x +_i \eta y).\underline{T}_i(x, y).(\eta x +_i \eta y).$$

We now verify that T is lax. First, the axioms (lmf.1, 2) on naturality of comparisons say that:

$$T e_i f.\underline{T}_i x = \underline{T}_i y.e_i T f \colon e_i T x \to T e_i y,$$
$$T(f +_i g).\underline{T}_i(x, y) = \underline{T}_i(x', y').(T f +_i T g). \tag{8.51}$$

To prove the first identity it is sufficient to cancel the invertible map $T(e_i \eta y)$ from the relation:

$$T(e_i \eta y).T e_i f.\underline{T}_i x = T e_i T f.T(e_i \eta x).\underline{T}_i x = T(e_i T f).\eta(e_i T x)$$
$$= \eta(e_i T y).e_i(T f) = T(e_i \eta y).\underline{T}_i(y).e_i(T f),$$

that comes from applying (twice) the naturality of η and the definition of $\underline{T}$.

The definition of the comparisons $\underline{T}$ and their naturality give the following relations, that will be repeatedly used below, together with (8.50):

$$\eta(e_i T x) = T(e_i \eta x).\underline{T}_i(x) = \underline{T}_i(T x).e_i(T \eta x),$$
$$\eta(T x +_i T y) = T(\eta x +_i \eta y).\underline{T}_i(x, y) = \underline{T}_i(T x, T y).(T \eta x +_i T \eta y). \tag{8.52}$$

(lmf.3) (*coherence with unitors*) For an **i**-cube x with $\partial_i^- x = z$ we have to prove the commutativity of this diagram of **i**-maps (and similarly for the right unitor ρ):

$$
\begin{array}{ccc}
e_i(T z) +_i T x & \xrightarrow{\ \lambda_i(T x)\ } & T x \\
{\scriptstyle \underline{T}_i(z) +_i 1} \downarrow & & \downarrow {\scriptstyle (T \lambda_i x)^{-1}} \\
T(e_i z) +_i T x & \xrightarrow[\ \underline{T}_i(e_i z, x)\ ]{} & T(e_i z +_i x)
\end{array}
\tag{8.53}
$$

Consider the following transversal isomorphism (the equality comes from the first relation of (8.50)):

$$T(\eta(e_i z) +_i \eta x)) = T(\underline{T}_i(z) +_i 1_{T x}).T(e_i \eta z +_i \eta x).$$

Our goal follows from cancelling it:

$$T(\eta(e_i z) +_i \eta x).\underline{T}_i(e_i z, x).(\underline{T}_i z +_i 1_{Tx})$$
$$= \eta(T(e_i z) +_i Tx).(\underline{T}_i z +_i 1_{Tx})$$
$$= T(\underline{T}_i z +_i 1_{Tx}).\eta(T(e_i z) +_i Tx),$$

$$T(\underline{T}_i z +_i 1_{Tx}).T(e_i \eta z +_i \eta x).(T\lambda_i x)^{-1}.\lambda_i(Tx)$$
$$= T(\underline{T}_i z +_i 1_{Tx}).(T\lambda_i Tx)^{-1}.T\eta x.\lambda_i(Tx)$$
$$= T(\underline{T}_i z +_i 1_{Tx}).(T\lambda_i Tx)^{-1}.(T\lambda_i Tx).\eta(T(e_i z) +_i Tx)$$
$$= T(\underline{T}_i z +_i 1_{Tx}).\eta(T(e_i z) +_i Tx).$$

(lmf.4) (*Coherence with associators*) For three *i*-consecutive **i**-cubes x, y, z in X we have to prove the commutativity of the hexagon marked with a question mark:

Consider now the transversal isomorphism appearing in the triangle (which is commutative by (8.50))

$$T(\eta(x +_i y) +_i \eta z) = T(\underline{T}_i(x, y) +_i 1_{Tz}).T((\eta x +_i \eta y) +_i \eta z).$$

Again, the goal follows from cancelling it (writing $\kappa' = \kappa_i(x', y', z')$),

$$T(\eta(x +_i y) +_i \eta z).\underline{T}_i(x +_i y, z).(\underline{T}_i(x, y) +_i 1).\kappa'$$
$$= \eta(T(x +_i y) +_i Tz).(\underline{T}_i(x, y) +_i 1).\kappa'$$
$$= T(\underline{T}_i(x, y) +_i 1).\eta((Tx +_i Ty) +_i Tz).\kappa'$$
$$= T(\underline{T}_i(x, y) +_i 1).T\kappa'.\eta(Tx +_i (Ty +_i Tz)),$$

$$T(\underline{T}_i(x,y)+1).T((\eta x+\eta y)+\eta z).T\kappa_i(x,y,z).\underline{T}_i(x,y+z).(1+\underline{T}_i(y,z))$$
$$= T(\underline{T}_i(x,y)+1).T\kappa'.T(\eta x+(\eta y+\eta z)).\underline{T}_i(x,y+z).(1+\underline{T}_i(y,z))$$
$$= T(\underline{T}_i(x,y)+1).T\kappa'.\underline{T}_i(Tx,T(y+z)).(T\eta x+T(\eta y+\eta z)).(1+\underline{T}_i(y,z))$$
$$= T(\underline{T}_i(x,y)+1).T\kappa'.\underline{T}_i(Tx,T(y+z)).(T\eta x+\eta(Ty+Tz))$$
$$= T(\underline{T}_i(x,y)+1).T\kappa'.\underline{T}_i(Tx,T(y+z)).(\eta Tx+\eta(Ty+Tz))$$
$$= T(\underline{T}_i(x,y)+1).T\kappa'.\underline{T}_i(Tx,T(y+z)).(\eta Tx+\eta(Ty+Tz))$$
$$= T(\underline{T}_i(x,y)+1).T\kappa'.\eta(Tx+(Ty+Tz)).$$

(lmf.5) (*Coherence with interchangers*) For $0 < i < j$ we have to prove the commutativity of this diagram of transversal maps

$$
\begin{array}{ccc}
(Tx +_i Ty) +_j (Tz +_i Tu) & \xrightarrow{\;\chi_{ij}T\;} & (Tx +_j Tz) +_i (Ty +_j Tu) \\
{\scriptstyle \underline{T}_i+_j\underline{T}_i}\big\downarrow & & \big\downarrow{\scriptstyle \underline{T}_j+_i\underline{T}_j} \\
T(x +_i y) +_j T(z +_i u) & & T(x +_j y) +_i T(z +_j u) \qquad (8.54) \\
{\scriptstyle \underline{T}_j}\big\downarrow & & \big\downarrow{\scriptstyle \underline{T}_i} \\
T((x +_i y) +_j (z +_i u)) & \xrightarrow[\;T\chi_{ij}\;]{} & T((x +_j z) +_i (y +_j u))
\end{array}
$$

As above this is proved by cancelling the following transversal isomorphism (and again, the equality follows from (8.50))

$$T(\eta(x +_j y) +_i \eta(z +_j u))$$
$$= T(\underline{T}_j(x,y) +_j \underline{T}_j(z,u)).T((\eta x +_j \eta y) +_i \eta(\eta z +_j \eta u)).$$

□

8.5 The jointly-monic monad of relations

This section is about weak *double* categories, viewed as truncated weak multiple categories, in order to apply the theory developed in the previous section. Thus an n-cube is an object (for $n = 0$) or a vertical arrow (for $n = 1$), and a transversal n-map is a horizontal arrow (for $n = 0$) or a double cell (for $n = 1$); the transversal direction is the horizontal one.

For a *regular* category $\mathbf{C}$, the double category $\mathbb{Rel}(\mathbf{C})$ has been introduced in 3.1.3. It can now be obtained from the previous theory of algebras applied to a strong idempotent lax monad T over the weak double category $\mathbb{Span}(\mathbf{C})$.

This can be done in essentially two ways, obtaining a *weak* double category $\mathbb{Rel}'(\mathbf{C})$ of *jointly-monic relations* (as in 8.5.3) or a *strict* double category $\mathbb{Rel}(\mathbf{C})$ of *cartesian relations* (as in 8.5.4). The monad T is trivial in degree zero, so that – in this truncated 2-dimensional case – there would be no advantage in making T strictly idempotent (see 8.5.3).

Extending this approach to higher dimension only works in a defective way (see 8.5.5), related to the *premultiple* category of cubical relations examined in 6.6.6.

8.5.1 Constructing a 2-dimensional monad

We construct a lax monad $T \colon \operatorname{Span}(\mathbf{C}) \to \operatorname{Span}(\mathbf{C})$, by unconditioned choices.

Everything is trivial in degree zero: $\operatorname{Hor}_0(T) = \operatorname{id}\mathbf{C}$ and $\operatorname{Hor}_0(\eta) = 1$.

Working in degree 1 we note that a double cell $y \to x$ *with values in a jointly monic span* x and boundary $(y \overset{f}{\underset{g}{}} x)$ is determined by its horizontal mappings f, g; thus every span y can have at most one *special* cell $y \to x$.

Now an arbitrary span $x \colon \vee \to \mathbf{C}$ amounts to a pair of maps $u_i \colon x_\iota \to x_i$ $(i = 0, 1)$ and determines a map $u \colon x_\iota \to x_0 \times x_1$; choosing a canonical factorisation $u = wp$ in $\mathbf{C}$, by a regular epi and a monomorphism, we get a 'factorisation' $u_i = w_i p$ of the given span (u_i), by a regular epi p and a jointly monic span $Tx = (w_0, w_1)$

$$
\begin{array}{ccc}
x_0 & \!\!=\!\!\!=\!\!\!=\!\! & x_0 \\
\uparrow{\scriptstyle u_0} & & \uparrow{\scriptstyle w_0} \\
x_\iota & \xrightarrow{\ p\ } & x_\iota \\
\downarrow{\scriptstyle u_1} & & \downarrow{\scriptstyle w_1} \\
x_1 & \!\!=\!\!\!=\!\!\!=\!\! & x_1
\end{array}
\qquad
\begin{array}{c}
\bullet \xrightarrow{\ 0\ } \\
\ \ \downarrow{\scriptstyle 1}
\end{array}
\tag{8.55}
$$

The special cell $\eta x \colon x \to Tx$ is already shown above, with central map p. Moreover, for every map $f \colon x \to y$ of spans there is a unique $Tf \colon Tx \to Ty$ consistent with η. We have thus defined a face-consistent family of ordinary functors and a face-consistent family of natural transformations

$$
T_n \colon \operatorname{Hor}_n(\operatorname{Span}(\mathbf{C})) \to \operatorname{Hor}_n(\operatorname{Span}(\mathbf{C})), \quad \eta_n \colon 1 \to T_n \quad (n = 0, 1).
$$

T could be easily made strictly idempotent, by a simple constraint on the previous choice: if the span x is jointly monic we take $p = 1$ and $Tx = x$. Later on, in 8.5.4, we shall be interested in stronger constraints.

8.5.2 Comparisons

It is now sufficient to verify the hypotheses of Theorem 8.4.9, to conclude that these data can be completed to a unique strong idempotent lax monad (T, η), with comparisons for T defined as in (8.49).

First the special cells $T(\eta x), \eta(Tx)\colon Tx \to T^2x$ coincide because the span T^2x is jointly monic.

We now remark that a special cell $f\colon x \to y\colon \vee \to \mathbf{C}$ whose central component is a regular epi of $\mathbf{C}$ gives an *invertible* cell $Tf\colon Tx \to Ty$. Indeed the special map $\eta y.f\colon x \to Ty$ has a central component of this kind and takes values in a jointly monic span; therefore Ty is isomorphic to Tx and the unique map $Tx \to Ty$ is an isomorphism.

Since the components of the special cells ηx and $e(\eta x)$ are regular epis, we already have that $T(\eta x)$ and $T(e(\eta x))$ are invertible. (In fact the latter is trivial, because x must be of degree 0.)

We further remark that a concatenation $f \otimes g$ of special cells whose central components f_ι, g_ι are regular epis is again of this type. Indeed this concatenation is computed by the pullbacks of the two cospans represented in the commutative diagram below, and the map induced between them is a regular epi, by Lemma 2.5.9 applied in two steps

$$(8.56)$$

Thus the central component of the special cell $\eta x \otimes \eta y\colon x \otimes y \to Tx \otimes Ty$ is a regular epi and $T(\eta x \otimes \eta y)$ is invertible.

8.5.3 Algebraic spans

A T-algebra of degree 1 is thus any span x isomorphic to Tx, i.e. any jointly monic span. The weak double category $\mathbb{A}\mathrm{lg}(T)$ has vertical composition defined as in (8.39)

$$x \circ y = T(x \otimes y), \qquad\qquad f \circ g = T(f \otimes g). \qquad (8.57)$$

This is automatically consistent with faces (without making T strict) because the faces of our spans (and cells) have degree zero, where T is trivial:

$$\partial_1^-(x \circ y) = T(\partial_1^-(x \otimes y)) = T(\partial_1^- x) = \partial_1^- x,$$
$$\partial_1^+(x \circ y) = \ldots = \partial_1^+ y.$$

For the same reason we can take $e^\circ = e$. We interpret $\mathbb{A}\mathrm{lg}(T)$ as the weak double category $\mathbb{R}\mathrm{el}'(\mathbf{C})$ of *jointly-monic relations*.

8.5.4 Cartesian choice and strict algebras

We now assume, as in 2.5.2, that $\mathbf{C}$ is equipped with a choice of binary products and subobjects.

Pullbacks in $\mathbf{C}$ will now be chosen as subobjects of a binary product (and the unit constraint of 1.3.5 is not followed). The lax endofunctor T is also constructed by a conditioned choice: in the canonical factorisation $u = wp$ used in (8.55) the monomorphism w is required to be a subobject of $x_0 \times x_1$. In this way T is strictly idempotent, with $T\eta = \eta T = 1_T$.

Its strictly algebraic spans, of the form Tx, are the *cartesian relations* and form the substructure $\mathbb{A}\mathrm{lg}_*(T) = \mathbb{R}\mathrm{el}(\mathbf{C})$; it is a 'special horizontal skeleton' of $\mathbb{A}\mathrm{lg}(T)$, i.e. an equivalent structure (see 4.4.4) which has precisely one span in any class of special isomorphism. Therefore it is a *strict* double category.

8.5.5 A higher dimensional premultiple monad

The 'obvious' higher dimensional extension of the idempotent lax monad T of $\mathbb{S}\mathrm{pan}(\mathbf{Set})$ studied above only works in a defective way.

In fact we get a family of strictly idempotent monads

$$T_i \colon \mathrm{tv}_i(\mathsf{Span}(\mathbf{Set})) \to \mathrm{tv}_i(\mathsf{Span}(\mathbf{Set})), \qquad \eta_i \colon 1 \to T_i, \qquad (8.58)$$

that associates to any n-cubical span of sets an (obvious) n-cubical relation, as defined in the premultiple category $\mathsf{Rel}(\mathbf{Set})$ described in 6.6.6.

Globally, the family (T_i) preserves degeneracies but – in dimension three or higher – can*not* be made into a lax endofunctor of $\mathsf{Span}(\mathbf{Set})$, because of the counterexample exposed in 6.6.7. (With the notation used there, the 12-cube $c +_1 d$ of $\mathsf{Span}(\mathbf{Set})$ has an empty central object, while $Tc +_1 Td = c' +_1 d'$ has a singleton in the central position; therefore there is no transversal map $Tc +_1 Td \to T(c +_1 d) = c +_1 d$.)

Here we only get a strictly idempotent premultiple monad on the underlying premultiple category $|\mathsf{Span}(\mathbf{Set})|$; its premultiple category of strictly algebraic cubes is $\mathsf{Rel}(\mathbf{Set})$.

8.5.6 The pushout-pullback adjunction

We end this section coming back to the idempotent pushout-pullback adjunction, studied in 4.5.6

$$(F, G, \eta, \varepsilon) \colon \mathbb{S}\mathrm{pan}(\mathbf{C}) \rightharpoondown \mathbb{C}\mathrm{osp}(\mathbf{C}), \qquad (8.59)$$

between the weak double categories of spans and cospans, over a category

C with pullbacks and pushouts. The colax functor F is computed by pushouts, the lax functor G by pullbacks. The adjunction reduces to the identity in degree zero.

Depending on **C**, *the adjunction can be strong or not.*

8.5.7 Exercises and complements

(a) For $\mathbf{C} = \mathbf{Set}$ the pushout-pullback adjunction is not strong.

(b) If **C** is an abelian category, the adjunction is strong and its algebras are the relations on **C**.

(c) The following facts can be found in [Me].

For a regular category **C**, all *pullback* relations (determined by a span which 'is a pullback') are *regular* (in the sense of von Neumann, also called difunctional), i.e. satisfy the condition $u \circ u^\sharp \circ u = u$, where $u^\sharp$ is the opposite relation.

For **Set**, a relation is regular if and only if it is a pullback relation, but – as we know – not all relations are regular.

For **Gp**, all relations are regular (as we have seen in 2.1.5), but not all of them are pullback relations.

8.6 Idempotent colax comonads and profunctors

We briefly dualise Section 8.4, without writing down the strict case. In 8.6.6 we extend to infinite dimensional strong idempotent adjunctions the equivalence between algebras and coalgebras proved in 1.7.7 for ordinary categories.

Finally we show in 8.6.7 that $\mathbb{C}$at is comonadic over $\mathbb{C}$osp**Cat**, for an idempotent comonad: profunctors can be seen as coalgebraic cospans of categories.

8.6.1 Definition (Strong idempotent colax comonads)

A colax comonad (S, ε, δ) over a chiral multiple category **A** is a comonad in the 2-category Cx**Cmc**, and induces *transversal comonads* of ordinary categories

$$S_\mathbf{i} \colon \mathbf{A_i} \to \mathbf{A_i}, \qquad \varepsilon_\mathbf{i} \colon S_\mathbf{i} \to 1, \quad \delta_\mathbf{i} \colon S_\mathbf{i} \to S_\mathbf{i}^2. \qquad (8.60)$$

Our comonad is said to be *idempotent* if it satisfies the following equivalent conditions:

(idc.1) all the ordinary comonads (8.60) are idempotent,

(idc.1′) the transversal transformation $\delta\colon S^2 \to S$ is invertible, with inverse $S\varepsilon = \varepsilon S\colon S \to S^2$,

(idc.1″) $S\varepsilon$ is invertible,

(idc.1‴) εS is invertible.

An idempotent colax comonad over A will be written as a pair (S, ε) where $S\colon \mathsf{A} \to \mathsf{A}$ is a colax endofunctor, $\varepsilon\colon 1 \to S$ is a transversal transformation and $S\varepsilon = \varepsilon S\colon S \to S^2$ is invertible. The comultiplication δ is recovered as the inverse of $S\varepsilon$.

We say that (S, ε) is a *strong* idempotent colax comonad if it also satisfies the following equivalent conditions:

(idc.2) the transversal maps $S\underline{S}_i(a)$ and $S\underline{S}_i(a, b)$ are invertible,

(idc.2′) the following transversal maps are invertible

$$Se_i(\varepsilon a)\colon S(e_i Sa) \to S(e_i a), \quad S(\varepsilon a +_i \varepsilon a)\colon S(Sa +_i Sb) \to S(a +_i b).$$

In this situation we say that an **i**-cube a is *coalgebraic* if $\varepsilon a\colon Sa \to a$ is invertible, or equivalently if a is transversally isomorphic to Sa in A.

8.6.2 From right-strong idempotent adjunctions to comonads

Right-strong idempotent adjunctions between chiral multiple categories have been defined in 8.4.1.

Dualising Theorem 8.4.4, if $(F, G, \eta, \varepsilon)\colon \mathsf{X} \rightharpoonup \mathsf{A}$ is of this kind, then the associated colax comonad (S, ε) over A defined in 8.3.4 is strong idempotent. Its transversal components are

$$S_i = F_i G_i\colon \mathbf{A}_i \to \mathbf{A}_i, \quad \varepsilon_i\colon S_i \to 1, \quad \delta_i = F_i \eta_i G_i\colon S_i \to S_i S_i. \tag{8.61}$$

8.6.3 Coalgebras of idempotent comonads

Dualising 8.4.5 and Theorem 8.4.6, if (S, ε) is a strong idempotent colax comonad over A we can form a chiral multiple category $\mathsf{Coalg}(S)$ of coalgebraic cubes.

First, an **i**-cube a of A is *coalgebraic* for (S, ε) if it satisfies the following equivalent conditions:

(i) εa is invertible,

(ii) there exists a transversal map $k\colon a \to Sa$ such that $\varepsilon a.k = 1_a$,

(iii) there exists in $\mathbf{A}_i$ an S_i-coalgebra (a, k) over a,

(iv) εA is invertible and $\hat{a} = (a, (\varepsilon a)^{-1})$ is the unique S_i-coalgebra over a,

(v) a is transversally isomorphic to Sa.

In this case $\hat{a} = (a, (\varepsilon a)^{-1})$ is transversally isomorphic to the cofree S-coalgebra $(Sa, \delta a)$ on a.

Now $\mathsf{Coalg}(S)$ consists of all these cubes of A and all transversal maps between them. Faces are inherited from A, but positive degeneracies and compositions are redefined as follows in $\mathsf{Coalg}(S)$

$$e_i^\circ(a) = Se_i(a), \qquad e_i^\circ(f) = Se_i(f),$$
$$a \circ_i b = S(a +_i b), \qquad f \circ_i g = S(f +_i g). \tag{8.62}$$

8.6.4 From idempotent comonads to adjunctions and back

Dualising Theorem 8.4.7, let (S, ε) be a strong idempotent colax comonad over A, with $\mathsf{Coalg}(S)$ constructed as above. There is an associated strong idempotent adjunction $(F'', G'', \eta'', \varepsilon) \colon \mathsf{Coalg}(S) \rightharpoonup$ A, which is colax-pseudo and gives back the comonad. The construction is only sketched here.

(a) The inclusion $F'' \colon \mathsf{Coalg}(S) \to$ A is a full colax functor, with comparisons constructed with ε

$$\underline{F}_i''(a) = \varepsilon e_i(a) \colon Se_i(a) \to e_i(a),$$
$$\underline{F}_i''(a, b) = \varepsilon(a +_i b) \colon S(a +_i b) \to a +_i b, \tag{8.63}$$

since $Se_i(a) = e_i^\circ(a) = F''e_i^\circ(a)$, $e_i(a) = e_i(F''a)$, etc.

(b) The pseudo functor $G'' \colon$ A $\to \mathsf{Coalg}(S)$ is a restriction of S, with comparisons constructed with ε (and written below in the lax direction)

$$G''a = Sa, \qquad G''f = Sf,$$
$$\underline{G}_i''a = S(e_i \varepsilon a) \colon Se_i(Sa) \to S(e_i a), \tag{8.64}$$
$$\underline{G}_i''(a, b) = S(\varepsilon a +_i \varepsilon b) \colon S(Sa +_i Sb) \to S(a +_i b).$$

Now $F''G'' = S$ (including comparisons).

The colax-pseudo adjunction is completed with the given transversal transformation $\varepsilon \colon S \to 1$ and the invertible transversal transformation

$$\eta'' \colon 1 \to G''F'' \colon \mathsf{Coalg}(S) \to \mathsf{Coalg}(S), \quad \eta''a = (\varepsilon a)^{-1} \colon a \to SA. \tag{8.65}$$

8.6.5 The comparison functor of coalgebras

Dualising Theorem 8.4.8 we now start from a right-strong idempotent adjunction $(F, G, \eta, \varepsilon) \colon$ X $\rightharpoonup$ A. We have the associated strong idempotent colax comonad (S, ε) recalled in 8.6.2, its chiral multiple category $\mathsf{Coalg}(S)$ of coalgebraic cubes (see 8.6.3) and the associated strong colax-pseudo adjunction $(\eta'', \varepsilon) \colon F'' \dashv G''$ constructed above (in 8.6.4), with $F''G'' = S$ and η'' invertible.

The two adjunctions are correlated by a *comparison colax functor H*

$$H: X \to \text{Coalg}(S), \quad Hx = Fx, \quad H(f: x \to y) = Ff,$$

$$\underline{H}_i x = FG\underline{F}_i(x).(\varepsilon Fe_i x)^{-1}: F(e_i x) \to FGe_i(Fx) = e_i^\circ(Fx),$$

$$\underline{H}_i(x, y) = FG\underline{F}_i(x, y).(\varepsilon F(x +_i y))^{-1}: \qquad (8.66)$$

$$F(x +_i y) \to FG(Fx +_i Fy) = Fx \circ_i Fy.$$

Moreover we have

$$HG = G'', \qquad F''H = F, \qquad H\eta = \eta'H. \qquad (8.67)$$

The first relation, where G'' is a pseudo functor, must – again – be interpreted: the composite of G (lax) with H (colax) is 'naturally' a pseudo functor, because all maps $\underline{H}_i(Gx, Gy)$ and $H\underline{G}_i(x, y)$ are invertible (similarly for the comparisons of degeneracies).

8.6.6 Strong idempotent adjunctions, algebras and coalgebras

Finally we start from a strong idempotent multiple adjunction

$$(F, G, \eta, \varepsilon): X \rightharpoonup A,$$

with associated strong idempotent colax monad (T, η) and strong idempotent lax comonad (S, ε). We want to show that the chiral multiple categories $\text{Alg}(T)$ (of algebraic cubes of X, defined in 8.4.5) and $\text{Coalg}(S)$ (of coalgebraic cubes of A, defined in 8.6.3) are equivalent, extending a similar result for ordinary categories (see 1.7.7).

More precisely, we can factorise the given multiple adjunction as follows, up to isomorphism

$$X \underset{G'}{\overset{F'}{\rightleftarrows}} \text{Alg}(T) \underset{G^\sharp}{\overset{F^\sharp}{\rightleftarrows}} \text{Coalg}(S) \underset{G''}{\overset{F''}{\rightleftarrows}} A. \qquad (8.68)$$

(a) The reflective pseudo-lax adjunction $(\eta, \varepsilon'): F' \dashv G'$ is described in Theorem 8.4.7; G' is the full embedding of algebraic objects, with reflector $F'x = Tx$ and invertible counit $\varepsilon'x = (\eta x)^{-1}$. The comparison lax functor $K: A \to \text{Alg}(T)$ is defined in (8.46).

(b) The coreflective colax-pseudo adjunction $(\eta'', \varepsilon): F'' \dashv G''$ is described in 8.6.4; F'' is the full embedding of coalgebraic objects, with coreflector $G''a = Sa$ and invertible unit $\eta''a = (\varepsilon a)^{-1}$. The comparison colax functor $H: X \to \text{Coalg}(S)$ is described in (8.66).

(c) There is a pseudo adjunction $F^\sharp \dashv G^\sharp$. We write the comparisons of

$F^\sharp$ in the colax direction and those of $G^\sharp$ in the lax direction, because this is useful below

$$(F^\sharp, G^\sharp, \eta^\sharp, \varepsilon^\sharp) : \mathsf{Alg}(T) \rightsquigarrow \mathsf{Coalg}(S), \qquad (8.69)$$

$$F^\sharp : \mathsf{Alg}(T) \to \mathsf{Coalg}(S), \quad F^\sharp x = Fx, \quad F^\sharp f = Ff,$$

$$\underline{F}_i^\sharp(x, y) = FG\underline{F}_i(x, y) : F(GF(x +_i y)) \to FG(Fx +_i Fy),$$

$$G^\sharp : \mathsf{Coalg}(S) \to \mathsf{Alg}(T), \quad G^\sharp a = Ga, \quad G^\sharp f = Gf,$$

$$\underline{G}_i^\sharp(a, b) = GF\underline{G}_i(a, b) : GF(Ga +_i Gb) \to GFG(a +_i b),$$

$$\eta^\sharp : 1 \to G^\sharp F^\sharp, \quad \eta^\sharp(x) = \eta x,$$

$$\varepsilon^\sharp : F^\sharp G^\sharp \to 1, \quad \varepsilon^\sharp(a) = \varepsilon a.$$

(The comparison maps $\underline{F}_i^\sharp(x)$ and $\underline{G}_i^\sharp(a)$ for unitarity are similar.) We do not write down the long verifications of coherence. This adjunction is an equivalence, because η and ε are invertible on algebraic and coalgebraic cubes, respectively.

(d) Composing the three adjunctions above we get $F'' F^\sharp F' = SF$ and $G' G^\sharp G'' = TG$, which are isomorphic to F and G, respectively (because of the idempotence of $F \dashv G$). We only verify that the equality of lax functors $G' G^\sharp G'' = TG$ does hold for comparisons of i-concatenation:

$$(G' G^\sharp G'')_i(a, b) = G\underline{G}_i''(a, b).\underline{G}_i^\sharp(Sa, Sb).\underline{G}_i'(GFGa, GFGb)$$

$$= GFG(\varepsilon a +_i \varepsilon b).GF\underline{G}_i(Sa, Sb).\eta(GFGa +_i GFGb)$$

$$= GF\underline{G}_i(a, b).GF(G\varepsilon a +_i G\varepsilon b).\eta(GFGa +_i GFGb) \text{ (by naturality of } \underline{G}_i)$$

$$= GF\underline{G}_i(a, b).\eta(Ga +_i Gb).(G\varepsilon a +_i G\varepsilon b) \qquad \text{(by naturality of } \eta)$$

$$= GF\underline{G}_i(a, b).(G\underline{F}_i(Ga, Gb))^{-1}.G\underline{F}_i(Ga, Gb).\eta(Ga +_i Gb).(G\varepsilon a +_i G\varepsilon b)$$

$$= GF\underline{G}_i(a, b).(G\underline{F}_i(Ga, Gb))^{-1}.\underline{G}_i(FGa, FGb).(\eta Ga + \eta Gb).(G\varepsilon a + G\varepsilon b)$$

$$\text{(by (7.38))}$$

$$= GF\underline{G}_i(a, b).(G\underline{F}_i(Ga, Gb))^{-1}.\underline{G}_i(FGa, FGb) \qquad \text{(by idempotence)}$$

$$= T\underline{G}_i(a, b).\underline{T}_i(Ga, Gb) = \underline{TG}_i(a, b).$$

8.6.7 Profunctors as coalgebraic cospans

There is an important example in dimension 2. We have already seen in 4.5.5 the colax-lax adjunction

$$C \dashv P : \mathbb{C}\mathsf{at} \rightsquigarrow \mathbb{C}\mathsf{osp}\mathbf{Cat}, \qquad (8.70)$$

where the colax functor C is the cospan representation of profunctors, by cotabulators. This adjunction is trivial in degree zero and has unit $\eta = \mathrm{id}\colon 1 \to PC\colon \mathbb{C}\mathrm{at} \to \mathbb{C}\mathrm{at}$.

The adjunction gives a strictly idempotent colax comonad

$$S = CP\colon \mathbb{C}\mathrm{osp}\mathbf{Cat} \to \mathbb{C}\mathrm{osp}\mathbf{Cat}, \quad \varepsilon\colon S \to 1, \qquad (8.71)$$

whose counit ε is computed in (4.48).

The left adjoint C is comonadic and $\mathbb{C}\mathrm{at}$ can be identified with the weak double category $\mathbb{C}\mathrm{oalg}_*(CP)$ formed by the strictly coalgebraic cospans in $\mathbb{C}\mathrm{osp}\mathbf{Cat}$, namely those of the form Cp for some profunctor p: this means a cospan of categories satisfying the conditions (i), (ii) of 4.5.5. They must be composed as in (8.62): $Cp \circ Cq = C(Cp \otimes Cq)$, where $\otimes$ is the composition of cospans.

8.7 *Complements

In 8.2.6 we have extended an ordinary monad to a colax monad of cubical spans. One can similarly obtain a lax monad of cubical cospans. We restrict to the idempotent case, to apply the previous theory of algebras as algebraic cubes.

We end with a notion of transversal monadicity, already discussed in 8.1.1.

8.7.1 Idempotent lax monads of cubical cospans

We start from a strictly idempotent monad (T, η) over an ordinary category $\mathbf{C}$ with pushouts; e.g. the abelianisation monad over the category of groups, as in 8.1.3.

The associated idempotent adjunction will be written in the following form, where G is a full embedding with reflector F (a codomain-restriction of T) and trivial counit (see 1.7.3(c))

$$F\colon \mathbf{C} \rightleftarrows \mathbf{Alg}_*(T)\colon G, \qquad \eta\colon 1 \to GF = T, \quad \varepsilon = 1\colon FG \to 1,$$
$$FX = TX, \qquad F(f\colon X \to Y) = Tf\colon TX \to TY. \qquad (8.72)$$

The category $\mathbf{Alg}_*(T)$ has pushouts (since a span in $\mathbf{Alg}_*(T)$ has a pushout in $\mathbf{C}$, which is preserved by F), and we can apply the 2-functor $\mathsf{Cosp}\colon \mathbf{Cat}_{\mathrm{po}} \to \mathrm{Lx}\mathbf{Cmc}$, obtaining a pseudo-lax idempotent adjunction

$$F\colon \mathsf{Cosp}(\mathbf{C}) \rightleftarrows \mathsf{Cosp}(\mathbf{Alg}_*(T))\colon G, \qquad \eta\colon 1 \to GF, \quad \varepsilon = 1\colon FG \to 1,$$

and the associated idempotent lax monad

$$T\colon \mathsf{Cosp}(\mathbf{C}) \to \mathsf{Cosp}(\mathbf{C}), \quad \eta\colon 1 \to T \quad (T\eta = \eta T = 1). \tag{8.73}$$

The new adjunction is left-strong idempotent, as proved in 8.4.1, actually left-strict because $\varepsilon = 1$; its monad T is strictly idempotent.

The strictly algebraic n-cubes of this monad are the cubical spans $\mathsf{V}^n \to \mathbf{C}$ whose vertices are T-invariant, i.e. the cubical spans of $\mathbf{Alg}_*(T)$, and we can identify their weak multiple categories

$$\mathsf{Alg}_*(T) = \mathsf{Cosp}(\mathbf{Alg}_*(T)). \tag{8.74}$$

8.7.2 A chiral triple category of spans and relations

There is a chiral triple category R with:

$$\mathrm{tv}_{01}\mathsf{R} = \mathbb{Span}(\mathbf{Set}), \qquad \mathrm{tv}_{02}\mathsf{R} = \mathbb{Rel}'(\mathbf{Set}),$$

that can be constructed as the structure of algebraic cubes of a 'semi-strong' lax idempotent monad T over the weak triple category $\mathsf{Span}_3(\mathbf{Set})$.

Here we only sketch this topic, without proofs. A 12-cube of R is a 2-cubical span $x\colon \mathsf{V}^2 \to \mathbf{Set}$ where the three 2-*directed* spans are jointly monic. (We use the notation of (6.67).)

The lax endofunctor T of $\mathsf{Span}_3(\mathbf{Set})$ is the identity on 1-directed arrows and sends:

- a 2-directed arrow $x\colon \mathsf{V} \to \mathbf{Set}$ to the associated jointly monic pair, the image of the mapping $x_\iota \to x_0 \times x_1$,

- a 12-cube $x\colon \mathsf{V}^2 \to \mathbf{Set}$ to the associated 12-cube of R, whose vertical spans are the images of the mappings $x_{tt} \to x_{t0} \times x_{t1}$ (for $t = 0, \iota, 1$).

The unit $\eta x\colon x \to Tx$ has canonical projections in each component. The operations of the new 12-cubes are deduced from those of $\mathsf{Span}_3(\mathbf{Set})$

$$x \circ_1 y = T(x +_1 y), \qquad x \circ_2 y = T(x +_2 y).$$

The interchanger

$$\chi(x, y, z, u)\colon (x \circ_1 y) \circ_2 (z \circ_1 u) \to (x \circ_2 z) \circ_1 (y \circ_2 u)$$

is not invertible, generally. This can be seen on the consistent matrix of 12-cubes $\left(\begin{smallmatrix} x & y \\ z & u \end{smallmatrix}\right)$ shown in 6.6.7.

Note also that the central component of the transversal map

$$\eta c +_1 \eta d\colon c +_1 d \to Tc +_1 Td$$

is not surjective (being $\emptyset \to 1$), so that $T(\eta c +_1 \eta d)$ is not invertible.

8.7.3 Premultiple adjunctions and monads

Forgetting everything about geometric compositions, let us consider a *pre-multiple adjunction*

$$(F, G, \eta, \varepsilon) \colon X \rightsquigarrow A, \tag{8.75}$$

i.e. an adjunction in the 2-category $\mathbf{C} = \mathbf{pMlt}$ of premultiple categories, or multiple objects in $\mathbf{Cat}$ (see 6.2.7). For every positive multi-index $\mathbf{i}$, its component is an ordinary adjunction

$$(F_\mathbf{i}, G_\mathbf{i}, \eta_\mathbf{i}, \varepsilon_\mathbf{i}) \colon X_\mathbf{i} \rightsquigarrow A_\mathbf{i}, \tag{8.76}$$

and the adjunction (8.75) is equivalent to giving the family of its components, consistent with faces and degeneracies.

We have thus an associated *premultiple monad* $T = GF \colon X \to X$, i.e. a monad in $\mathbf{pMlt}$ (as already considered in 8.5.5). Again, this amounts to a family $T = (T_\mathbf{i})$ of ordinary monads (for $\mathbf{i} \subset \mathbb{N}^*$)

$$T_\mathbf{i} = G_\mathbf{i} F_\mathbf{i} \colon X_\mathbf{i} \to X_\mathbf{i}, \quad \eta_\mathbf{i} \colon 1 \to T_\mathbf{i}, \quad \mu_\mathbf{i} = G_\mathbf{i} \varepsilon_\mathbf{i} F_\mathbf{i} \colon T_\mathbf{i}^2 \to T_\mathbf{i}, \tag{8.77}$$

consistent with faces and degeneracies:

$$\partial_i^\alpha . T_\mathbf{i} = T_{\mathbf{i}|i} . \partial_i^\alpha \qquad e_i . T_{\mathbf{i}|i} = T_\mathbf{i} . e_i \qquad (i \in \mathbf{i}), \tag{8.78}$$

and similarly for $\eta = (\eta_\mathbf{i}) \colon 1 \to T$ and $\mu = (\mu_\mathbf{i}) \colon T^2 \to T$.

It is easy to see that the 2-category $\mathbf{C} = \mathbf{pMlt}$ of premultiple categories admits the construction of algebras (in the sense of 2.3.8): for every premultiple monad (T, η, μ) on X, the premultiple category of algebras X^T is obtained as the family of categories $(X_\mathbf{i})^{T_\mathbf{i}}$, with obvious faces and degeneracies (for x in $X_\mathbf{i}$, $i \in \mathbf{i}$ and $j \notin \mathbf{i}$)

$$\partial_i^\alpha(x, h \colon T_\mathbf{i} x \to x) = (\partial_i^\alpha x, \partial_i^\alpha h \colon T_{\mathbf{i}|i}(\partial_i^\alpha x) \to \partial_i^\alpha x),$$
$$e_j(x, h \colon T_\mathbf{i} x \to x) = (e_j x, e_j h \colon T_{\mathbf{i}j}(e_j x) \to e_j x). \tag{8.79}$$

If the premultiple monad T derives from the premultiple adjunction (8.75), the premultiple comparison functor $K \colon A \to X^T$ has components given by the comparisons of the ordinary adjunctions (8.76)

$$K_\mathbf{i} \colon A_\mathbf{i} \to (X_\mathbf{i})^{T_\mathbf{i}}, \quad K_\mathbf{i} A = (G_\mathbf{i} A, G_\mathbf{i} \varepsilon_\mathbf{i} A \colon G_\mathbf{i} F_\mathbf{i} G_\mathbf{i} A \to G_\mathbf{i} A). \tag{8.80}$$

Finally, the premultiple adjunction is monadic (i.e. K is a premultiple isomorphism) if and only if all its components (8.76) are ordinary monadic adjunctions.

8.7.4 Transversal monadicity

A unitary multiple adjunction $(F, G, \eta, \varepsilon) \colon \mathsf{X} \rightarrowtail \mathsf{A}$ between chiral multiple categories can be said to be *transversally monadic* if the associated premultiple adjunction $|\mathsf{X}| \rightarrowtail |\mathsf{A}|$ is monadic, and gives an isomorphism $K \colon |\mathsf{A}| \to |\mathsf{X}|^T$ of premultiple categories. As we have seen, this simply means that all the ordinary adjunctions $F_{\mathsf{i}} \dashv G_{\mathsf{i}}$ are monadic.

Of course, if this is the case, there is precisely one structure of chiral multiple category X^T on $|\mathsf{X}|^T$ that makes K an isomorphism $\mathsf{A} \to \mathsf{X}^T$, *but this is of little help,* since we need A for this construction, and we can simply keep A as it is.

Appendix A

Applications in homological algebra and algebraic topology

We give here some applications of double categories.

We begin, in Section A1, by working in categories of modules $R\mathbf{Mod}$, with the aid of suitable double categories of modular lattices or relations. This approach can be extended to abelian categories, but the natural framework is more general: Puppe-exact categories. In fact we need from the very beginning a *non-abelian* category $\mathbf{Mlc}$ of this kind, and a related double category $\mathbb{Mlhc}$, to describe direct and inverse images of subobjects in $R\mathbf{Mod}$ (or in an abelian category, or more generally in a Puppe-exact one).

Puppe-exact and abelian categories are briefly dealt with in Sections A2–A4, after the book [G8] devoted to homological algebra in Puppe-exact categories. Generalisations of these subjects can be found in [G9] and references therein.

Section A5, the last, deals with constructions based on the weak double categories $\mathbb{Span}\mathbf{Top}$ and $\mathbb{Cosp}\mathbf{Top}$ of topological spans or cospans. This can be of interest in studying tangles, ribbons and cobordisms between manifolds, including topological quantum field theories: see [KaRT, Kk, Ye, G2, G3, G4] and their references.

For the sake of simplicity we often follow the common abuses of notation for subobjects. Thus a subobject $m\colon M \rightarrowtail A$ can be simply denoted by its domain M; if $M \leqslant N$ in $\mathrm{Sub}A$ (i.e. we have a subobject $n\colon N \rightarrowtail A$ with $m \leqslant n$), M can also denote the corresponding subobject of N, equivalent to *the* monomorphism $u\colon M \rightarrowtail N$ such that $m = nu$.

A1 Double categories in basic homological algebra

This section is about Homological Algebra in an elementary context, the category $R\mathbf{Mod}$ of (left) modules on a (unitary) ring R. We show, in an informal way, how double categories of lattices and double categories of relations can be useful.

This section is also a preparation for the next: the introduction of the non-abelian category **Mlc** of modular lattices and modular connections, together with the analysis of its exactness properties, will prepare the definition of Puppe-exact categories in A2.2.

A1.1 An outline

Direct and inverse images of subsets have been dealt with in 1.4.5. For the category $R\mathbf{Mod}$, images and preimages of submodules are analysed below, according to the following outline.

We already know that, for every object A in $R\mathbf{Mod}$, the ordered set $\mathrm{Sub}A$ of its subobjects (i.e. submodules) is a modular lattice (see 1.4.2), with maximum $1_A\colon A \to A$ and minimum $0_A\colon 0 \to A$. For submodules $X, X' \subset A$ the meet is $X \wedge X' = X \cap X'$ and the join is

$$X \vee X' = X + X' = \{x + x' \mid x \in X,\ x' \in X'\}.$$

A homomorphism $f\colon A \to B$ gives two monotone mappings, called *direct* and *inverse image*

$$f_*\colon \mathrm{Sub}A \rightleftarrows \mathrm{Sub}B : f^*,$$

$$f_*(X) = f(X), \quad f^*(Y) = f^{-1}(Y) \qquad (X \subset A,\ Y \subset B). \tag{A.1}$$

These mappings form a Galois connection $f_* \dashv f^*$, as for sets in 1.4.5, but here we have a stronger property

$$f^* f_*(X) = X \vee f^*(0_B) \supset X, \qquad f_* f^*(Y) = Y \wedge f_*(1_A) \subset Y, \tag{A.2}$$

which we describe saying that the pair $(f_*, f^*)\colon \mathrm{Sub}A \rightarrowtail \mathrm{Sub}B$ is a 'modular connection'. It is thus a morphism of the category **Mlc** *of modular lattices and modular connections*, a subcategory of $\mathbf{AdjOrd}$ that will be studied in A1.2 and A1.4. We recall that f_* preserves joins and the minimum, while f^* preserve meets and maximum.

All this defines a *transfer functor* for subobjects of $R\mathbf{Mod}$

$$\mathrm{Sub}_R\colon R\mathbf{Mod} \to \mathbf{Mlc}, \tag{A.3}$$

which will be seen to be exact in A1.4, i.e. a functor which preserves kernels and cokernels (or equivalently exact sequences, as defined in A2.4).

Now let $F\colon R\mathbf{Mod} \to S\mathbf{Mod}$ be an exact functor, in the same sense. We shall see (in A2.6) that it also preserves subobjects, their meets and joins. For every R-module A we have thus a *homomorphism of lattices*, which belongs to the category **Mlh** of 1.4.2 (*not* to **Mlc**)

$$(\mathrm{Sub}F)_A\colon \mathrm{Sub}_R(A) \to \mathrm{Sub}_S(FA), \qquad X \mapsto F(X). \tag{A.4}$$

We shall also see that F preserves direct and inverse images along morphisms: for a homomorphism $f \colon A \to B$ in $R\mathbf{Mod}$ we have two commutative squares (of increasing mappings)

$$\begin{array}{ccc}
\mathrm{Sub}_R(A) & \xrightarrow{\ (\mathrm{Sub}F)_A\ } & \mathrm{Sub}_S(FA) \\[4pt]
f_* \Big\downarrow\Big\uparrow f^* & & Ff_* \Big\downarrow\Big\uparrow Ff^* \\[4pt]
\mathrm{Sub}_R(B) & \xrightarrow[\ (\mathrm{Sub}F)_B\ ]{} & \mathrm{Sub}_S(FB)
\end{array} \qquad (\mathrm{A.5})$$

$$F(f_*X) = (Ff)_*(FX), \qquad F(f^*Y) = (Ff)^*(FY).$$

This diagram is a cell in the double category $\mathbb{A}\mathrm{dj}_0\mathbf{Ord}$ of ordered sets, increasing mappings (as horizontal arrows), Galois connections (as vertical arrows) and bicommutative cells, defined in 3.1.7(b).

To express this interaction between lattice-homomorphisms and modular connections we shall use – more precisely – a double subcategory $\mathbb{M}\mathrm{lhc} \subset \mathbb{A}\mathrm{dj}_0\mathbf{Ord}$ introduced below (in A1.5) and consisting of: modular lattices, homomorphisms, modular connections and bicommutative cells.

The transfer functor (A.3) can now be seen as a *vertical functor*

$$\mathrm{Sub}_R \colon R\mathbf{Mod} \to \mathbb{M}\mathrm{lhc}, \qquad (\mathrm{A.6})$$

i.e. a double functor defined on the *vertical* double category $R\mathbf{Mod}$ (with trivial horizontal arrows). For every exact functor $F \colon R\mathbf{Mod} \to S\mathbf{Mod}$ we have a *horizontal transformation of vertical functors*

$$\mathrm{Sub}_F \colon \mathrm{Sub}_R \to \mathrm{Sub}_S.F \colon R\mathbf{Mod} \to \mathbb{M}\mathrm{lhc}, \qquad (\mathrm{A.7})$$

where $\mathrm{Sub}_S.F$ is the composite vertical functor

$$R\mathbf{Mod} \to S\mathbf{Mod} \to \mathbb{M}\mathrm{lhc}.$$

A1.2 Modular lattices and modular connections

We begin now to study the framework of *modular lattices and modular connections*, that we have seen to abstract the properties of direct and inverse images for $R\mathbf{Mod}$ (and will similarly work for a wide class of categories).

An object is a modular lattice (with 0 and 1). A morphism

$$f = (f_\bullet, f^\bullet) \colon X \to Y$$

is a pair where

(i) $f_\bullet \colon X \to Y$ and $f^\bullet \colon Y \to X$ are increasing mappings,

(ii) $f^\bullet f_\bullet(x) = x \vee f^\bullet 0, \quad f_\bullet f^\bullet(y) = y \wedge f_\bullet 1 \quad$ (for $x \in X$, $y \in Y$).

As a consequence $f^{\bullet}f_{\bullet} \geqslant \mathrm{id}X$ and $f_{\bullet}f^{\bullet} \leqslant \mathrm{id}Y$, and we have an adjunction $f_{\bullet} \dashv f^{\bullet}$ between ordered sets, i.e. a covariant Galois connection (see 1.4.3). As we have seen $f_{\bullet}$ determines $f^{\bullet}$, and conversely

$$f^{\bullet}(y) = \max\{x \in X \mid f_{\bullet}(x) \leqslant y\},$$
$$f_{\bullet}(x) = \min\{y \in Y \mid f^{\bullet}(y) \geqslant x\}. \tag{A.8}$$

Moreover $f_{\bullet}$ preserves all the existing joins (including $0 = \vee\emptyset$), $f^{\bullet}$ preserves all the existing meets (including $1 = \wedge\emptyset$), and

$$f_{\bullet}f^{\bullet}f_{\bullet} = f_{\bullet}, \qquad f^{\bullet}f_{\bullet}f^{\bullet} = f^{\bullet}. \tag{A.9}$$

Condition (ii) can be equivalently rewritten in a seemingly stronger form:

(ii') $f^{\bullet}(f_{\bullet}x \vee y) = x \vee f^{\bullet}y, \quad f_{\bullet}(f^{\bullet}y \wedge x) = y \wedge f_{\bullet}x \quad (x \in X, y \in Y),$

as proved in Exercise A1.3(a).

This rewriting shows that modular connections *are closed under composition* in the category **AdjOrd** of ordered sets and Galois connections (introduced in 1.4.4), and form a subcategory **Mlc** of the latter

$$(g_{\bullet}, g^{\bullet}).(f_{\bullet}, f^{\bullet}) = (g_{\bullet}f_{\bullet}, f^{\bullet}g^{\bullet}), \tag{A.10}$$

whose isomorphisms are characterised in Exercise A1.3(b).

Mlc is selfdual: it inherits a (contravariant) involution from **AdjOrd**, together with an order consistent with the involution (see (1.39))

$$X \mapsto X^{\mathrm{op}}, \quad ((f_{\bullet}, f^{\bullet}) \colon X \twoheadrightarrow Y) \mapsto ((f^{\bullet}, f_{\bullet}) \colon Y^{\mathrm{op}} \twoheadrightarrow X^{\mathrm{op}}),$$
$$f \leqslant g \quad \Leftrightarrow \quad f_{\bullet} \leqslant g_{\bullet} \quad \Leftrightarrow \quad f^{\bullet} \geqslant g^{\bullet}. \tag{A.11}$$

We are also interested in the full subcategory **Dlc** of **Mlc** determined by distributive lattices. Its morphisms are analysed in Exercise A1.3(d).

A1.3 Exercises and complements

(a) Prove that, in A1.2, property (ii) implies (ii'). Deduce that modular connections are closed under composition in **AdjOrd**.

(b) (*Isomorphisms of lattices*) Prove that the morphism $(f_{\bullet}, f^{\bullet}) \colon X \twoheadrightarrow Y$ is an isomorphism in **Mlc** if and only if the following equivalent conditions hold:

(i) $f^\bullet f_\bullet = 1_X$ and $f_\bullet f^\bullet = 1_Y$,

(ii) $f^\bullet 0 = 0$ and $f_\bullet 1 = 1$,

(iii) $f_\bullet$ is a bijective mapping,

(iii*) $f^\bullet$ is a bijective mapping,

(iv) $f_\bullet$ is an isomorphism of ordered sets (hence of **Mlh**),

(iv*) $f^\bullet$ is an isomorphism of ordered sets (hence of **Mlh**).

(c) As we have seen, *the isomorphisms of the category* **Mlc** *correspond to the isomorphisms of* **Mlh**, i.e. the bijective homomorphisms of modular lattices. Furthermore, the following (faithful) forgetful functor reflects the isomorphisms ($|X|$ denotes the underlying set of X)

$$U\colon \mathbf{Mlc} \to \mathbf{Set}, \quad X \mapsto |X|, \quad (f_\bullet, f^\bullet) \mapsto f_\bullet, \tag{A.12}$$

as well as the functor **Mlc** $\to$ **Set**^{op} where $(f_\bullet, f^\bullet)$ is sent to $f^\bullet$.

(d) A morphism $(f_\bullet, f^\bullet)\colon X \to Y$ of **Dlc** is a modular connection between distributive lattices. Prove that the mappings $f_\bullet$ and $f^\bullet$ are homomorphisms of *quasi* lattices (in the sense of 1.4.1): in other words, $f_\bullet$ also preserves binary meets (but need not preserve the maximum, of course), while $f^\bullet$ also preserves binary joins (but need not preserve the minimum).

A1.4 Exactness properties

We prove now that the category **Mlc** of modular lattices and modular connections satisfies exactness conditions that will make it Puppe-exact (according to a definition made precise below, in A2.2), or p-exact for short. The reader is warned that **Mlc** is not abelian, and cannot even be exactly embedded in an abelian category (this is proved in [G8], Lemma 2.3.5).

First there is a *zero object* (both terminal and initial, see 1.3.2), namely the one-point lattice $\{*\}$, since every object X has unique morphisms t and s:

$$X \overset{t}{\dashrightarrow} \{*\} \overset{s}{\dashrightarrow} X \qquad (t^\bullet(*) = 1, \; s_\bullet(*) = 0). \tag{A.13}$$

The null morphism $0_{XY}\colon X \twoheadrightarrow \{*\} \twoheadrightarrow Y$ is defined by $x \mapsto 0$ and $y \mapsto 1$. Every morphism $f = (f_\bullet, f^\bullet)\colon X \twoheadrightarrow Y$ has a *kernel* ker f and a *cokernel*

$\operatorname{cok} f$ (see 1.8.2)

$$m = \ker f \colon\ \downarrow f^\bullet 0 \rightarrowtail X, \qquad m_\bullet(x) = x, \quad m^\bullet(x) = x \wedge f^\bullet 0,$$
$$p = \operatorname{cok} f \colon Y \twoheadrightarrow \uparrow f_\bullet 1, \qquad p_\bullet(y) = y \vee f_\bullet 1, \quad p^\bullet(y) = y. \tag{A.14}$$

The verification of the universal property of m as an equaliser of f and 0_{XY} is a straightforward computation: essentially, if $h = (h_\bullet, h^\bullet)\colon Z \rightarrow X$ is annihilated by f, then for every $z \in Z$ we have:

$$f^\bullet 0 = f^\bullet f_\bullet h_\bullet(z) = h_\bullet(z) \vee f^\bullet 0 \geqslant h_\bullet(z),$$

so that we can restrict the codomain of $h_\bullet$ to $\downarrow f^\bullet 0$. Dually for p.

As a consequence of these properties, the morphism f has a canonical factorisation $f = ngq$ through $q = \operatorname{cok}(\ker f)$ (the *coimage* of f) and $n = \ker(\operatorname{cok} f)$ (the *image* of f), that determines a precise morphism g

$$\begin{array}{ccccccc}
\downarrow f^\bullet 0 & \overset{m}{\rightarrowtail} & X & \overset{f}{\longrightarrow} & Y & \overset{p}{\twoheadrightarrow} & \uparrow f_\bullet 1 \\
& & q \downarrow & & \uparrow n & & \\
& & \uparrow f^\bullet 0 & \underset{g}{\longrightarrow} & \downarrow f_\bullet 1 & &
\end{array} \tag{A.15}$$

$$q_\bullet(x) = x \vee f^\bullet 0, \quad q^\bullet(x) = x, \qquad n_\bullet(y) = y, \quad n^\bullet(y) = y \wedge f_\bullet 1,$$
$$g_\bullet(x) = f_\bullet(x), \qquad\qquad\qquad g^\bullet(y) = f^\bullet(y).$$

Moreover g is an *isomorphism* of **Mlc** (as characterised in A1.3), because

$$g^\bullet g_\bullet(x) = f^\bullet f_\bullet(x) = x \qquad \text{(for } x \geqslant f^\bullet 0\text{),}$$
$$g_\bullet g^\bullet(y) = f_\bullet f^\bullet(y) = y \qquad \text{(for } y \leqslant f_\bullet 1\text{).}$$

(This will mean that the category **Mlc** is Puppe-exact.)

We also note that every element $a \in X$ determines a *subobject* and a *quotient* of X

$$m \colon\ \downarrow a \rightarrowtail X, \qquad m_\bullet(x') = x', \quad m^\bullet(x) = x \wedge a, \tag{A.16}$$

$$p \colon X \twoheadrightarrow \uparrow a, \qquad p_\bullet(x) = x \vee a, \quad p^\bullet(x') = x'. \tag{A.17}$$

The correspondence $a \mapsto m$ establishes an isomorphism between the lattice X and the ordered set $\operatorname{Sub} X$ of subobjects of X in **Mlc**, while the correspondence $a \mapsto p$ gives an anti-isomorphism of X with the ordered set of quotients of X.

The subobject m and the quotient p determined by the element $a \in X$ form a *short exact sequence*

$$\downarrow a \rightarrowtail X \twoheadrightarrow \uparrow a, \qquad m = \ker p, \quad p = \operatorname{cok} m. \tag{A.18}$$

Conversely, every short exact sequence (m, p) in **Mlc** with central object X is isomorphic to a unique sequence of this type, with $a = m_\bullet 1 = p^\bullet 0$.

Finally it is easy to see that the functor $\mathrm{Sub}_R \colon R\,\mathbf{Mod} \to \mathbf{Mlc}$ is exact, i.e. preserves kernels and cokernels.

(Let us take an R-homomorphism $h \colon A \to B$ with kernel $m \colon h^{-1}\{0\} \subset A$ and cokernel $p \colon B \to B/h(A)$. Then the embedding $m_* \colon \mathrm{Sub}(h^{-1}\{0\}) \to \mathrm{Sub}A$ is the same as the inclusion $\downarrow h^*(0) \subset \mathrm{Sub}A$, while the embedding $p^* \colon \mathrm{Sub}(B/h(A)) \to \mathrm{Sub}B$ can be identified with the inclusion $\uparrow h_*(1) \subset \mathrm{Sub}B$.)

A1.5 Double categories of lattices

The double category $\mathbb{A}\mathrm{dj}_0\,\mathbf{Ord}$ introduced in 3.1.7 has double subcategories that are of interest here.

(a) First we have $\mathbb{M}\mathrm{lhc} \subset \mathbb{A}\mathrm{dj}_0\mathbf{Ord}$, the *cellwise-full* double subcategory formed of modular lattices, their homomorphisms and modular connections. It amalgamates the categories

$$\mathbf{Mlh} = \mathrm{Hor}_0\mathbb{M}\mathrm{lhc}, \qquad \mathbf{Mlc} = \mathrm{Ver}_0\mathbb{M}\mathrm{lhc}, \qquad (\mathrm{A}.19)$$

of modular lattices and homomorphisms, or modular lattices and modular connections, respectively. The flat double cells are 'bicommutative squares', as in (3.16)

$$
\begin{array}{ccc}
X & \xrightarrow{\ f\ } & X' \\
u \downarrow & = & \downarrow v \\
Y & \xrightarrow[\ g\ ]{} & Y'
\end{array}
\qquad v_\bullet f = g u_\bullet, \quad f u^\bullet = v^\bullet g. \qquad (\mathrm{A}.20)
$$

We have already seen above, in A1.1, that $\mathbb{M}\mathrm{lhc}$ comes out naturally when we want to formalise direct and inverse images of subobjects for modules (or abelian groups). The same will hold for all Puppe-exact categories, including the abelian ones (see Section A2).

We also note that every horizontal isomorphism has a vertical companion, and symmetrically (by Exercise A1.3(b)), so that the double category $\mathbb{M}\mathrm{lhc}$ is horizontally and vertically invariant.

(b) We are also interested in a larger structure: $\mathbb{L}\mathrm{thc} \subset \mathbb{A}\mathrm{dj}_0\mathbf{Ord}$, the cellwise-full double subcategory formed of all lattices, their homomorphisms and adjunctions. It amalgamates the categories

$$\mathbf{Lth} = \mathrm{Hor}_0\mathbb{L}\mathrm{thc}, \qquad \mathbf{Ltc} = \mathrm{Ver}_0\mathbb{L}\mathrm{thc}, \qquad (\mathrm{A}.21)$$

of lattices and homomorphisms, or lattices and Galois connections, respectively. The flat double cells are again 'bicommutative squares'.

The double category $\mathbb{L}$thc plays a role similar to that of $\mathbb{M}$lhc, in the much more general contexts of 'semiexact' and 'homological' categories, studied in [G9].

*(c) One can form similar larger structures, cellwise-full in AdjOrd, so that a cell $(u \, {}^f_g \, v)$ has only to satisfy the relation $v_{\bullet}f \leqslant gu_{\bullet}$ (or, equivalently, $fu^{\bullet} \leqslant v^{\bullet}g$). This seems to be less interesting for the present applications.

A1.6 Double categories of additive relations

The rest of this section is only sketched. The interested reader can find more information in Section A4 and [G8].

The (ultraflat) double category $\mathbb{R}\mathrm{el}(R\,\mathbf{Mod})$ of modules, homomorphisms and relations has been briefly introduced in 3.1.3.

Let $F \colon R\,\mathbf{Mod} \to S\,\mathbf{Mod}$ be an exact functor between categories of modules; it is well known that F preserves finite limits (see A3.8). As a consequence, F can be extended to a (unique) functor

$$\mathrm{Rel}(F) \colon \mathrm{Rel}(R\,\mathbf{Mod}) \to \mathrm{Rel}(S\,\mathbf{Mod}), \qquad u \mapsto F(u),$$

that preserves order and involution, by sending a submodule $u \subset A \times B$ to the submodule $F(u) \subset FA \times FB$. Globally, we have a double functor

$$\mathbb{R}\mathrm{el}(F) \colon \mathbb{R}\mathrm{el}(R\,\mathbf{Mod}) \to \mathbb{R}\mathrm{el}(S\,\mathbf{Mod}), \tag{A.22}$$

acting as F on homomorphisms, as $\mathrm{Rel}(F)$ on relations, and preserving the flat double cells.

If $\varphi \colon F \to G \colon R\,\mathbf{Mod} \to S\,\mathbf{Mod}$ is a natural transformation between exact functors, the *same* components $\varphi A \colon FA \to GA$ (which are homomorphisms) give, for a relation $u \colon A \nrightarrow B$, a 'lax-naturality square' which generally does not commute, but is a double cell in $\mathbb{R}\mathrm{el}(S\,\mathbf{Mod})$

$$\tag{A.23}$$

This is shown by the right commutative diagram above, in $S\,\mathbf{Mod}$, where the slanting arrows are restrictions of cartesian projections: every pair $(x, y) \in F(u)$ gives a pair $(\varphi_A(x), \varphi_B(y)) \in G(u)$.

We have thus a *horizontal transformation of double functors*

$$\mathbb{Rel}(\varphi)\colon \mathbb{Rel}(F) \to \mathbb{Rel}(G)\colon \mathbb{Rel}(R\,\mathbf{Mod}) \to \mathbb{Rel}(S\,\mathbf{Mod}). \qquad (A.24)$$

In Section A4 all this will be extended: $\mathbb{Rel}$ becomes a 2-functor from the 2-category of (small) Puppe-exact categories, exact functors and natural transformations to the 2-category **Dbl** of double categories, double functors and horizontal transformations.

A1.7 Subquotients and regular induction

The prime example of a 'subquotient' is the homology

$$H = \operatorname{Ker} d/\operatorname{Im} d$$

of a differential module (A, d), i.e. an R-module A equipped with an endomorphism $d\colon A \to A$ such that $dd = 0$, or equivalently $\operatorname{Im} d \subset \operatorname{Ker} d$. A consistent part of Homological Algebra, from the homology of chain complexes to the theory of spectral sequences, is about subquotients and induced morphisms between them (see [M2, G8]).

In general, a *subquotient* $S = M/N$ of a module A is a quotient of a subobject (M) of A, or equivalently a subobject of a quotient (A/N) of A. It is determined by a decreasing pair (M, N) of submodules of A, via a commutative square

$$
\begin{array}{ccc}
M & \xrightarrow{\;m\;} & A \\
{\scriptstyle p}\downarrow & & \downarrow{\scriptstyle q} \\
S & \xrightarrow{\;n\;} & A/N
\end{array}
\qquad\qquad N \subset M, \qquad (A.25)
$$

where m and n denote inclusions of submodules, p and q projections on quotients. This square is *bicartesian*, i.e. *pullback* and *pushout* at the same time (as can be easily verified, see A2.5(b)).

This bicartesian square determines *one relation* $s = mp^\sharp = q^\sharp n\colon S \nrightarrow A$, that sends the class $[x] \in M/N$ to all the elements of the lateral $[x] = x + N \subset A$. It is actually a *monorelation*, i.e. a monomorphism in the category of relations, since $s^\sharp s = \operatorname{id}(S)$. (We shall see that, because of the canonical factorisation (A.51) of a relation, every monorelation $s\colon S' \nrightarrow A$ is of this type, up to isomorphism. The *subquotients* of the module A amount thus to the *subobjects of A in* $\mathbb{Rel}(R\,\mathbf{Mod})$.)

Now we consider induced homomorphisms between subquotients; we go on in a diagrammatic way, so that everything can be readily extended to more general categories.

A homomorphism $f\colon A \to B$ is given. If M and H are submodules of

A and B respectively, and $f(M) \subset H$, we have a commutative diagram with short exact rows, displaying the homomorphism f' induced by f on our subobjects, and the homomorphism f'' induced by f on the associated quotients

$$
\begin{array}{ccc}
M & \rightarrowtail & A & \twoheadrightarrow & A/M \\
{\scriptstyle f'}\big\downarrow & & {\scriptstyle f}\big\downarrow & & \big\downarrow{\scriptstyle f''} \\
H & \rightarrowtail & B & \twoheadrightarrow & B/H
\end{array}
\qquad (f_*M \leqslant H) \qquad (\text{A.26})
$$

More generally, given two subquotients M/N of A and H/K of B, we say that f *has a regular induction* from M/N to H/K whenever

$$
f(M) \subset H \quad \text{and} \quad f(N) \subset K. \qquad (\text{A.27})
$$

In this situation we can obtain a homomorphism $g \colon M/N \to H/K$ *regularly induced* by f. We begin by the commutative diagram below, applying (A.26) in two different ways (on M, H and N, K)

$$
\begin{array}{ccc}
M & \rightarrowtail & A & \twoheadrightarrow & A/N \\
{\scriptstyle f'}\big\downarrow & & {\scriptstyle f}\big\downarrow & & \big\downarrow{\scriptstyle f''} \\
H & \rightarrowtail & B & \twoheadrightarrow & B/K
\end{array}
\qquad (f_*M \leqslant H, \; f_*N \leqslant K). \qquad (\text{A.28})
$$

By epi-mono factorisation of these rows we get the diagram below, and the induced homomorphism $g \colon M/N \to H/K$

$$
\begin{array}{ccc}
M & \twoheadrightarrow & M/N & \rightarrowtail & A/N \\
{\scriptstyle f'}\big\downarrow & & \big\downarrow{\scriptstyle g} & & \big\downarrow{\scriptstyle f''} \\
H & \twoheadrightarrow & H/K & \rightarrowtail & B/K
\end{array}
\qquad (\text{A.29})
$$

These data can be combined in a double cell $fs \leqslant tg$ of $\mathbb{R}\mathrm{el}(R\,\mathbf{Mod})$, called a (regular) *inductive square*

$$
\begin{array}{ccc}
M/N & \xrightarrow{\ g\ } & H/K \\
{\scriptstyle s}\big\downarrow & \leqslant & \big\downarrow{\scriptstyle t} \\
A & \xrightarrow[\ f\ ]{} & B
\end{array}
\qquad (\text{A.30})
$$

where the vertical arrows are monorelations. These cells form a double subcategory $\mathbb{I}\mathrm{nd}(R\,\mathbf{Mod}) \subset \mathbb{R}\mathrm{el}(R\,\mathbf{Mod})$: see [G8], Subsection 2.6.9.

A2 Puppe-exact categories

The theory of abelian categories, initiated by Buchsbaum and Grothendieck [Bu1, Bu2, Gt], was an important contribution to Homological Algebra. Yet some aspects of the theory are better viewed in the more general context of Puppe-exact categories [Pu, Mit], or p-exact categories for short.

After reviewing the main definitions, we are particularly interested in the category **Mlc** of modular lattices and modular connections, already introduced in Section A1. We show now that this category, which is p-exact and not abelian, abstracts the behaviour of direct and inverse images of subobjects *for all p-exact categories* (including the abelian ones); p-exact categories form thus the natural setting of this theory and its developments (exposed in [G8]).

A2.1 Three forms of exactness

The term 'exact category' has assumed different meanings in category theory, *and at least three of them are still in use* (listed as (b), (c), (d) below).

(a) It was first used in 1955–56 by Buchsbaum [Bu1, Bu2], essentially meaning what is now called an abelian category (even if the existence of finite biproducts was seemingly deferred to an additional axiom). The name subsisted – in this sense – in various papers of the 1950's and 1960's (by Atiyah, Hilton, Heller, etc.), together with the term 'abelian category' used by Grothendieck in his Tôhoku paper of 1957 [Gt]. Gradually the last term was universally accepted, including the influential books by Freyd [Fr1] (1964) and Mitchell [Mit] (1965), both focused on the embedding of abelian categories in categories of modules.

(b) Meanwhile, in 1962, Puppe [Pu] had introduced a more general notion, called a 'quasi exact category', still selfdual but not additive. This framework was investigated by Tsalenko [T1, T2] (also transliterated as 'Calenko') in 1964 and 1967, for the construction of the category of relations, and by many researchers for diagram lemmas. This notion became an 'exact category' in Mitchell's book [Mit] (1965), where abelian categories are defined as additive exact categories; the new name was also used – in this sense – in subsequent works by Brinkmann and Puppe [Bri, BriP], in 1969, and in the text by Herrlich and Strecker [HeS], in 1973. The books [AHS, FrS] still use in 1990 the name of 'exact category' in the sense of Puppe and Mitchell.

(c) In 1971 Barr [Ba] used the term 'exact category' for a different generalisation of abelian categories, based on regular categories and not selfdual. This setting, which also contains the categories of sets and groups (and, more generally, every variety of algebras in the sense recalled in 1.1.8(e)), became popular in category theory and has been extended in various forms (see [Bou1, JaMT, BoB, Bo4, Bou2]).

(d) In 1973, a paper by Quillen [Qu] on higher K-theory introduced another notion of 'exact category', as a full additive subcategory **E** of an abelian category, closed under extensions; this means that, for every short exact sequence $A \rightarrowtail C \twoheadrightarrow B$ of the abelian category, if A and B are in **E**, so is C. This notion can be defined intrinsically, without reference to an abelian environment, as an additive category equipped with a class of 'short exact sequences' satisfying some axioms.

We must thus distinguish *Puppe-exact*, *Barr-exact* and *Quillen-exact* categories. It is often remarked that the first two notions both satisfy the 'equation': exact + additive = abelian; on the other hand, Quillen-exact categories are additive. A Puppe-exact category is abelian if and only if it has finite products, or equivalently finite sums (see Theorem A3.7). Therefore the intersection of each pair of these three frameworks only contains the abelian categories.

A2.2 Exact and abelian categories

Kernels and cokernels in pointed categories have been defined in Section 1.8.

A *Puppe-exact*, or *p-exact*, category **E** is a category satisfying two selfdual axioms (already considered in 1.8.5):

(pex.1) **E** is a pointed and every morphism $f : A \to B$ has a kernel and a cokernel,

(pex.2) in the *canonical factorisation* of a morphism f through its *coimage* and its *image*

$$\mathrm{Ker}\, f \;\overset{k}{\rightarrowtail}\; A \;\overset{f}{\longrightarrow}\; B \;\overset{c}{\twoheadrightarrow}\; \mathrm{Cok}\, f$$
$$q \downarrow \qquad\qquad \uparrow n$$
$$\mathrm{Coim}\, f \;\underset{g}{\longrightarrow}\; \mathrm{Im}\, f \qquad\qquad\qquad (\mathrm{A.31})$$

$$\mathrm{Coim}\, f = \mathrm{Cok}\,(\mathrm{ker}\, f), \qquad\qquad \mathrm{Im}\, f = \mathrm{Ker}\,(\mathrm{cok}\, f),$$
$$q = \mathrm{coim}\, f = \mathrm{cok}\,(\mathrm{ker}\, f), \qquad n = \mathrm{im}\, f = \mathrm{ker}\,(\mathrm{cok}\, f),$$

the unique morphism g such that $f = ngq$ is an *isomorphism*.

This formulation is redundant, but clearer and often simpler to check than the more concise ones. For instance – as is easy to verify – it is sufficient to say that **E** is pointed and every morphism factorises as a normal epi followed by a normal mono.

As an easy consequence of the axioms, each monomorphism is normal (i.e. a kernel of some arrow) and each epimorphism is normal (i.e. a cokernel). Every morphism has an essentially unique epi-mono factorisation, the canonical one, as in diagram (A.31) – which motivates our names of coimage and image, above. **E** is always a balanced category, i.e. epi and mono implies iso.

Following the notation of 1.1.7, we write $\mathrm{Sub}_{\mathbf{E}}A$ and $\mathrm{Quo}_{\mathbf{E}}A$ for the (possibly large) ordered sets of subobjects and quotients of an object A in **E**, often shortened to $\mathrm{Sub}A$ and $\mathrm{Quo}A$. Cokernels and kernels give two *decreasing* mappings

$$\mathrm{cok} : \mathrm{Sub}_{\mathbf{E}}A \;\rightleftarrows\; \mathrm{Quo}_{\mathbf{E}}A : \mathrm{ker} \qquad (\textit{kernel duality}), \qquad (A.32)$$

that are easily proved to be inverse to each other, using the fact that all monos and epis are normal

$$\mathrm{ker}\,(\mathrm{cok}\,m) = m, \qquad \mathrm{cok}\,(\mathrm{ker}\,p) = p. \qquad (A.33)$$

(This should not be confused with the *categorical* duality between $\mathrm{Sub}_{\mathbf{E}}A$ and $\mathrm{Quo}_{\mathbf{E}^{\mathrm{op}}}A$, which preserves the ordering.)

From now on we assume that a p-exact category is well powered (see 1.1.7), i.e. that all sets $\mathrm{Sub}A$ (and $\mathrm{Quo}A$, as a consequence) are small. We shall see in A2.6 that $\mathrm{Sub}A$ and $\mathrm{Quo}A$ are modular lattices, anti-isomorphic by kernel duality.

A p-exact category is said to be *trivial* if all its objects are zero objects; or, in other words, if it is equivalent to the singleton category **1**.

As partially anticipated in 1.8.5, an *abelian category* can be defined as a p-exact category having finite products, or equivalently finite sums, or finite limits and colimits; this equivalence will be examined in detail in Theorem A3.7, but we shall use from now the notion of abelian category, much better known than that of p-exact category.

A2.3 Examples and exercises

The pointed categories **Gp** and **Rng**$'$ are not p-exact, since there are non-normal monomorphisms; **Set.** is not p-exact, since there are non-normal epimorphisms. The following examples of p-exact categories are taken from [G8], Subsection 1.5.6. The reader can easily fill-in the omitted verifications for many of them.

(a) Every abelian category is p-exact, by definition.

(b) Every non-empty full subcategory of a p-exact category that is closed under subobjects and quotients, is also p-exact. For instance, the categories of:

- cyclic groups; finite cyclic groups,
- vector spaces on a fixed field, of dimension lower than a fixed integer,
- abelian groups of cardinality (or rank) lower than a fixed integer.

(Replacing a 'fixed integer' by a 'fixed infinite cardinal' in these examples would just give an abelian subcategory.)

(c) We have already seen that the category **Mlc** of modular lattices and modular connections is p-exact (see A1.2, A1.4). It abstracts the behaviour of direct and inverse images of subobjects in (abelian or) p-exact categories (as we shall see in A2.6). It is *not* abelian and cannot be exactly embedded in an abelian category (cf. [G8], Lemma 2.3.5).

(d) Its full subcategory **Dlc** of *distributive lattices* (also defined in A1.2) is p-exact, by (b), and plays the same role as **Mlc** for *distributive p-exact categories*, i.e. those p-exact categories whose lattices of subobjects are distributive. The importance of distributive p-exact categories in Homological Algebra, with respect to coherence of induction in complicated systems like those produced by spectral sequences, is studied in [G8]. This theory cannot be effectively developed within abelian categories, which are never distributive except in the trivial case.

(e) The category $\mathcal{I}$ of sets and partial bijections, analysed in [G8] (Section 1.6), has been recalled in 2.1.7: a morphism $f\colon X \to Y$ in $\mathcal{I}$ is a bijection between a subset of X and a subset of Y, or equivalently a single-valued, injective relation; they compose as relations in Rel(**Set**).

$\mathcal{I}$ is selfdual; the zero object is $\emptyset$, with $\mathrm{Ker}\, f = X \setminus \mathrm{Def}\,(f)$ and $\mathrm{Cok}\, f = Y \setminus \mathrm{Val}\,(f)$. This category is distributive p-exact (with $\mathrm{Sub}X = \mathcal{P}X$), and 'universal' in this domain, in the sense that every small category of this kind has an exact embedding in $\mathcal{I}$ (cf. [G8], 4.6.7(b)).

*(f) The *projective category* $\mathrm{Pr}\,\mathbf{E}$ associated to any p-exact category **E** is defined as the quotient of **E** modulo the congruence that correlates two morphisms $f, g\colon A \to B$ when $f_* = g_*$ (or, equivalently, $f^* = g^*$). It is studied in [G8], Section 2.3.

In particular, if $\mathbf{A} = K\,\mathbf{Mod}$ is the abelian category of vector spaces over the commutative field K, $\mathrm{Pr}\,\mathbf{A}$ 'is' the (p-exact) category of projective spaces and projective linear maps over K. $\mathrm{Pr}\,\mathbf{A}$ cannot be exactly embedded in an abelian category ([G8], Lemma 2.3.5), *unless* K is the 2-element field, in which case $\mathrm{Pr}\,\mathbf{A} = \mathbf{A}$ is (isomorphic to) the category of the abelian groups where $2x = 0$ (for every element x).

One can similarly work with the abelian category $K\mathbf{Mod}_f$ of finite dimensional vector spaces; for $\mathbf{A} = \mathbb{R}\mathbf{Mod}_f$ we get the p-exact category $\mathbb{R}\mathbf{Prj}_f$ of finite-dimensional real projective spaces, considered in 2.1.8(c).

*(g) The *distributive expansion* $\mathrm{Dst}\,\mathbf{E}$ of any p-exact category $\mathbf{E}$, defined in [G8], Section 2.8, is a distributive p-exact category. It is never abelian – unless $\mathbf{E}$ (and therefore $\mathrm{Dst}\,\mathbf{E}$) is trivial.

(h) If $\mathbf{E}$ is a p-exact category and $\mathbf{S}$ is a small category, one can easily see that the category of functors $\mathbf{E}^{\mathbf{S}}$ is also p-exact.

A2.4 Exact functors and exact sequences

A functor $F\colon \mathbf{E} \to \mathbf{E}'$ between p-exact categories is said to be *exact* if it preserves kernels and cokernels, in the usual sense of preserving limits, that is – in the present case – up to equivalence of monos and epis, respectively). As a consequence it also preserves the zero object (that is the kernel and cokernel of any identity), canonical factorisations and exact sequences.

The latter are defined in the usual way: in the p-exact category $\mathbf{E}$ the sequence

$$A \xrightarrow{f} B \xrightarrow{g} C \qquad\qquad (\text{A.34})$$

is said to be *exact* (in B) if $\mathrm{im}\,f = \ker g$, or equivalently $\mathrm{cok}\,f = \mathrm{coim}\,g$. A sequence of consecutive morphisms is said to be *exact* if it is in all locations where the condition makes sense.

Short exact sequences are of particular interest. By definition, this is a sequence of the following form, that is exact (in A, B, C)

$$0 \longrightarrow A \xrightarrow{m} B \xrightarrow{p} C \longrightarrow 0. \qquad\qquad (\text{A.35})$$

Plainly, this means that the sequence satisfies the following equivalent conditions:

(i) m is a monomorphism, p is an epimorphism and $\mathrm{im}\,m = \ker p$,

(ii) $m \sim \ker p$ (as a mono to B) and $p \sim \mathrm{cok}\,m$ (as an epi from B).

A functor between p-exact categories is *left exact* (resp. *right exact*) if it preserves kernels (resp. cokernels), or equivalently exact sequences of the form $0 \to A \to B \to C$ (resp. $A \to B \to C \to 0$). A functor is exact if and only if it is left and right exact. A left adjoint functor between p-exact categories is necessarily right exact.

We shall see in A3.8 that any left exact (or right exact) functor between abelian categories preserves finite products and sums, and the additive structure.

A2.5 Lemma (Pullbacks and pushouts in p-exact categories)

A p-exact category **E** *has* pullbacks of monos *and* pushouts of epis *along arbitrary maps. The following points are a more detailed formulation.*

(a) Along a map f, the pullback (or preimage) $f^*(n)$ of a monomorphism n and the pushout (or direct image) $f_\circ(p)$ of an epimorphism p always exist, and are computed as follows

$$f^*(n) = \ker((\operatorname{cok} n).f), \qquad f_\circ(p) = \operatorname{cok}(f.(\ker p)), \qquad (A.36)$$

(b) In the same situation, if in the left diagram f is epi, so is g and the square is also a pushout (a bicartesian square*). Dually, if in the right diagram f is mono, so is g and the square is also a pullback.*

(c) Given the following commutative diagram with (short) exact rows

$$
\begin{array}{ccccc}
A & \overset{m}{\rightarrowtail} & B & \overset{p}{\twoheadrightarrow} & C \\
{\scriptstyle u}\downarrow & & {\scriptstyle v}\downarrow & & {\scriptstyle w}\downarrow \\
A' & \underset{n}{\rightarrowtail} & B' & \underset{q}{\twoheadrightarrow} & C'
\end{array}
\qquad (A.37)
$$

the left square is a pullback if and only if w is mono. The right square is a pushout if and only if u is epi.

Proof The proof of (a) is an easy exercise. For the rest, see [G8], Lemma 2.2.4.   □

A2.6 Theorem and Definition (The transfer functor)

(a) In a p-exact category **E** *all the ordered sets of subobjects* Sub$A$ *are modular lattices. Moreover every morphism $f: A \to B$ gives rise to two increasing mappings between ordered sets, called – respectively – direct and inverse image of subobjects along f*

$$
\begin{aligned}
&f_* : \operatorname{Sub}A \to \operatorname{Sub}B, && f_*(m) = \operatorname{im}(fm) = \ker(\operatorname{cok}(fm)), \\
&f^* : \operatorname{Sub}B \to \operatorname{Sub}A, && f^*(n) = \ker((\operatorname{cok} n).f),
\end{aligned}
\qquad (A.38)
$$

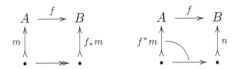

The left square is commutative and the right one is a pullback. The pair (f_*, f^*) is a modular connection, i.e. a morphism of the category **Mlc** studied in Section A1.

There is an exact functor (also because we have assumed that **E** is well-powered, in A2.2)

$$\mathrm{Sub}_{\mathbf{E}} \colon \mathbf{E} \to \mathbf{Mlc},$$

$$A \mapsto \mathrm{Sub}A, \qquad f \mapsto \mathrm{Sub}(f) = (f_*, f^*). \tag{A.39}$$

It also reflects exactness: a sequence of **E** is exact if and only if its image in **Mlc** is exact.

$\mathrm{Sub}_{\mathbf{E}}$ will be called the transfer functor (for subobjects), or also the projective functor of **E**. (The name is motivated by the projective category $\mathrm{Pr}\,\mathbf{E}$ associated to **E**, see A2.3(f).)

(b) The lattice operations and the ordering of $\mathrm{Sub}A$ can be described as follows, for $m, n \in \mathrm{Sub}A$ and $p = \mathrm{cok}\,m$, $q = \mathrm{cok}\,n$ (the corresponding quotients)

$$m \wedge n = m_* m^*(n) = n_* n^*(m), \qquad m \vee n = p^* p_*(n) = q^* q_*(m),$$

$$n \leqslant m \quad \Leftrightarrow \quad n^*(m) = 1. \tag{A.40}$$

(c) An exact functor $F \colon \mathbf{E} \to \mathbf{E}'$ preserves direct and inverse images of subobjects, as well as their finite joins and meets. Therefore it induces a homomorphism of lattices, for every A in **E**

$$\mathrm{Sub}_F A \colon \mathrm{Sub}_{\mathbf{E}} A \to \mathrm{Sub}_{\mathbf{E}'}(FA), \qquad \mathrm{Sub}_F(m) = \mathrm{im}\,(Fm), \tag{A.41}$$

(where $\mathrm{im}\,(Fm)$ is the subobject associated to the monomorphism Fm).

Proof The particular case $\mathbf{E} = R\,\mathbf{Mod}$ has been partially studied in A1.1, A1.4. The general case is proved in [G8], Subsection 2.2.5 and Theorem 2.2.6. □

A2.7 Exercises and complements (Vertical limits for adjoints)

(a) Prove that the kernels and cokernels of $\mathbf{Mlc} = \mathrm{Ver}_0 \mathbb{M}\mathrm{lhc}$ (resp. $\mathbf{Ltc} = \mathrm{Ver}_0 \mathbb{L}\mathrm{thc}$) become vertical double limits and colimits in $\mathbb{M}\mathrm{lhc}$ (resp. $\mathbb{L}\mathrm{thc}$).

(b) The double category $\mathbb{M}\mathrm{lhc}$ is *vertically Puppe-exact*, in the following sense:

(i) the category $\mathrm{Ver}_0\mathbb{M}\mathrm{lhc} = \mathbf{Mlc}$ is p-exact,

(ii) the category $\mathrm{Ver}_1\mathbb{M}\mathrm{lhc}$ is p-exact,

(iii) the faces and degeneracy functors between them are exact.

(c) The category $\mathbf{Ltc} = \mathrm{Ver}_0\mathbb{L}\mathrm{thc}$ has finite biproducts (see 1.8.4), which become vertical double limits and colimits in $\mathbb{L}\mathrm{thc}$. (Moreover they give $\mathbf{Ltc}$ a semiadditive structure, see A3.6.)

*(d) One can prove that the double category $\mathbb{L}\mathrm{thc}$ is *vertically p-homological*, extending as above the notion of a p-homological category, defined in [G9].

A3 Additive and abelian categories

Biproducts have been defined in pointed categories (see 1.8.4). After examining preadditive and additive categories, we characterise abelian categories as p-exact categories satisfying additional conditions. Some proofs are only referred to.

The reader can find further results in [M4, Bo2, Gt]. The embedding of abelian categories in categories of modules is studied in [Fr1, Mit]; it provides a metatheorem according to which many diagram lemmas in abelian categories need only be proved in categories of modules.

A3.1 Preadditive categories

As recalled in 2.4.2(e), a *preadditive*, or $\mathbb{Z}$-*linear*, category is a category $\mathbf{C}$ where every hom-set $\mathbf{C}(A, B)$ is equipped with a structure of abelian group, so that composition is bilinear over $\mathbb{Z}$.

More generally, we are also interested in $\mathbb{N}$-*linear categories*, where every hom-set $\mathbf{C}(A, B)$ is equipped with a structure of abelian monoid, and composition is bilinear over $\mathbb{N}$. The zero element of $\mathbf{C}(A, B)$ is generally written as $0_{AB} \colon A \to B$, or simply 0. If $\mathbf{C}$ has small hom-sets, this means that $\mathbf{C}$ is enriched over $\mathbf{Abm}$, the symmetric monoidal category of abelian monoids (see Exercise 2.2.4(g)).

A3.2 An exercise (Zero object)

Let $\mathbf{C}$ be an $\mathbb{N}$-linear category. Prove that the following conditions on an object Z are equivalent:

(i) Z is terminal,

(i*) Z is initial,

(ii) Z is the zero object,

(iii) $\mathbf{C}(Z, Z)$ is the trivial group,

(iv) $\mathrm{id}Z = 0_{ZZ}$.

A3.3 Lemma (Binary biproducts)

In the $\mathbb{N}$-linear category $\mathbf{C}$, consider a commutative diagram, as in (1.93)

$$
\begin{array}{ccc}
A & & B \\
\| & \searrow^{u} \quad \swarrow^{v} & \| \\
& C & \\
\| & \swarrow_{p} \quad \searrow_{q} & \| \\
A & & B
\end{array}
\qquad (\text{A.42})
$$

Using the notation recalled at the end of 1.8.4, the following conditions on (A.42) *are equivalent:*

(i) $(C; p, q)$ is the product of A, B, and $u = (\mathrm{id}A, 0)$, $v = (0, \mathrm{id}B)$,

(i^*) $(C; u, v)$ is the sum of A, B, and $p = [\mathrm{id}A, 0]$, $q = [0, \mathrm{id}B]$,

(ii) $pu = \mathrm{id}A$, $qv = \mathrm{id}B$, $up + vq = \mathrm{id}C$,

(iii) p, q are jointly mono, $pu = \mathrm{id}A$, $qv = \mathrm{id}B$, $pv = 0$, $qu = 0$,

(iii^*) u, v are jointly epi, $pu = \mathrm{id}A$, $qv = \mathrm{id}B$, $pv = 0$, $qu = 0$.

Therefore in an $\mathbb{N}$-linear category (a fortiori in a preadditive category), the existence of finite products is equivalent to the existence of finite sums, which are then biproducts.

Note. This is the binary counterpart of the previous exercise, and can also be verified as an exercise.

Proof (i) $\Rightarrow$ (iii) Obvious. (iii) $\Rightarrow$ (ii) We have $p(up + vq) = p$ and $q(up + vq) = q$, whence $up + vq = \mathrm{id}C$. (ii) $\Rightarrow$ (i) The maps $f\colon X \to A$ and $g\colon X \to B$ are given. If $h\colon X \to C$ satisfies $ph = f$ and $qh = g$, then $h = (up + vq)h = uf + vg$; conversely, the morphism $h = uf + vg\colon X \to C$ does have f, g as components. $\qquad\square$

A3.4 Theorem and Definition (Semiadditive categories)

Let $\mathbf{C}$ be a category.

(a) $\mathbf{C}$ is said to be semiadditive *if it satisfies the following equivalent conditions:*

(i) $\mathbf{C}$ is an $\mathbb{N}$-linear category with finite products,

(i^*) $\mathbf{C}$ is an $\mathbb{N}$-linear category with finite sums,

(ii) $\mathbf{C}$ has finite biproducts.

When these conditions hold, the sum $f + g$ of two maps $f, g \colon A \to B$ is determined by the categorical structure

$$A \xrightarrow{\ d\ } A \oplus A \xrightarrow{\ f \oplus g\ } B \oplus B \xrightarrow{\ \partial\ } B \qquad (A.43)$$

$$f + g = \partial(f \oplus g)d,$$

where $d = (1,1) \colon A \to A \oplus A$ is the diagonal of the product and $\partial = [1,1] \colon B \oplus B \to B$ is the codiagonal of the coproduct.

(b) A functor $F \colon \mathbf{C} \to \mathbf{C}'$ between semiadditive categories is said to be additive *if it satisfies the following equivalent conditions:*

(i) *F preserves finite products,*

(i*) *F preserves finite sums,*

(ii) *F preserves finite biproducts,*

(iii) *F preserves (finite) sums of parallel maps.*

Proof (a) We already know, from the previous lemma, that (i) and (i*) are equivalent. Furthermore, if they hold, $\mathbf{C}$ has finite biproducts and the binary ones satisfy the conditions of A3.3.

Therefore, for $f, g \colon A \to B$, the composition (A.43) is computed as follows (letting p', q' be the projections of $B \oplus B$):

$$\partial(f \oplus g)d = (p' + q')(u'fp + v'gq)(u + v) = f + g.$$

Now suppose that $\mathbf{C}$ has finite biproducts (as defined in 1.8.4), and define the sum of parallel maps as above, in (A.43). One verifies that this is indeed an enrichment over **Abm** (see [M4], Section VIII.2, exercise 4a); moreover, in diagram (A.42), the map $up + vq \colon C \to C$ must be the identity, because:

$$p(up + vq) = pup + pvq = p, \qquad q(up + vq) = qup + qvq = q.$$

(b) It is a straightforward consequence.   □

A3.5 Definition (Additive categories)

As a consequence of the previous theorem, we can define an *additive category* $\mathbf{C}$ as a preadditive category (i.e. a $\mathbb{Z}$-linear category) with finite products, or equivalently with finite sums.

It follows that $\mathbf{C}$ has finite biproducts; moreover the sum of parallel maps is determined by the categorical structure, as in (A.43).

It is easy to see that a $\mathbb{Z}$-linear category is finitely complete if and only if it is additive and has kernels. In fact, in any category the existence of finite limits is equivalent to the existence of finite products and equalisers (see 1.3.6). Furthermore, if $\mathbf{C}$ is $\mathbb{Z}$-linear, the equaliser of two maps $f, g\colon A \to B$ amounts to the kernel of $f - g$, while the kernel of f amounts to the equaliser of f and 0_{AB}.

A3.6 Exercises and complements (Semiadditive categories)

(a) **Abm** is obviously a semiadditive category, and is not additive.

(b) A selfdual category with finite products is automatically semiadditive, as in the following two cases.

(c) We have seen in (A2.7)(c) that the category **Ltc** of lattices and Galois connections has finite biproducts, realised as cartesian products. It is thus semiadditive; compute the sum $f + g$ of two Galois connections, and show that the structure is not additive.

(d) We have seen in 2.1.4(c) that Rel**Set** has arbitrary biproducts, realised as disjoint unions. Compute the sum of a family of parallel relations $a_i \colon X \rightarrowtail Y$.

*(e) **Ban** is additive while **Ban**$_1$ is not.

A3.7 Theorem and definition (Abelian categories)

Let $\mathbf{E}$ be a Puppe-exact category. We say that $\mathbf{E}$ is abelian *if it satisfies the following equivalent properties:*

(i) $\mathbf{E}$ *has finite products,*

(i*) $\mathbf{E}$ *has finite sums,*

(ii) $\mathbf{E}$ *has finite biproducts (in the sense of 1.8.4),*

(iii) $\mathbf{E}$ *has pullbacks,*

(iii*) $\mathbf{E}$ *has pushouts,*

(iv) $\mathbf{E}$ *has finite limits,*

(iv*) $\mathbf{E}$ *has finite colimits,*

(v) $\mathbf{E}$ *is additive.*

The sum of parallel maps is then determined by the categorical structure, as in (A.43).

Note. A preadditive Puppe-exact category need not be abelian: see the examples of A2.3(b).

Proof See [G8], Theorem 2.1.5. The essential part of this result was stated in [HeS] and proved in [FrS]; in [Mit] there is a weaker result, based on the existence of biproducts.   □

A3.8 Biproducts in abelian categories

We already know that, in any $\mathbb{N}$-linear category $\mathbf{C}$, the biproduct $C = A \oplus B$ can be characterised by a diagram (A.42) satisfying the condition:

(iii) p, q are jointly mono, $pu = \text{id} A$, $qv = \text{id} B$, $pv = 0$, $qu = 0$.

It is now easy to see that, if $\mathbf{C}$ is *preadditive* (i.e. $\mathbb{Z}$-linear) and *p-exact*, the biproduct can also be characterised by a diagram (A.42) satisfying:

(iv) the diagram is commutative, with (short) exact diagonals.

Indeed, if (iv) holds and a morphism $f \colon X \to C$ gives $pf = 0$, $qf = 0$, then it must factorise as $f = vg$, through $v \sim \ker p$. It follows that $g = qvg = qf = 0$ and $f = 0$. For parallel morphisms $f_i \colon X \to C$ on which p, q coincide, take $f = f_1 - f_2$. Conversely, if (iii) is satisfied, let us prove that $v \sim \ker p$. Take $f \colon X \to C$ such that $pf = 0$; since we know that $\text{id} C = up + vq$, it follows that $f = (up + vq)f = vqf$ factorises through v (that is mono).

Let $F \colon \mathbf{E} \to \mathbf{E}'$ be a functor between abelian categories. As an immediate consequence of the previous point, if F is exact (in the previous sense, i.e. it preserves kernels and cokernels), then it preserves biproducts, the additive structure, all finite limits and finite colimits.

Preservation of biproducts and sums of maps also holds if F is just left (or right) exact because, in the diagonals of (A.42), all monos and epis split so that the exactness of the diagonals can be simply expressed by a kernel (or cokernel) condition.

A3.9 Exercises and complements (Exact squares)

Let us start from a commutative square in an *abelian* category $\mathbf{E}$

$$
\begin{array}{ccc}
X & \xrightarrow{\ f\ } & A \\
{\scriptstyle g}\downarrow & & \downarrow{\scriptstyle h} \\
B & \xrightarrow{\ k\ } & Y
\end{array}
\qquad (A.44)
$$

and form the sequence

$$
0 \to X \xrightarrow{\ (f,g)\ } A \oplus B \xrightarrow{\ [h,-k]\ } Y \to 0 \qquad (A.45)
$$

where composed morphisms are null, since $hf - kg = 0$.

(a) Prove that the square (A.44) is a pullback if and only if the sequence (A.45) is exact in X and $A \oplus B$.

(a*) Dually, the square (A.44) is a pushout if and only if the sequence (A.45) is exact in Y and $A \oplus B$.

(b) More generally the square (A.44) is said to be *exact* in the sense of Hilton [Hi] if the sequence (A.45) is exact in the central object $A \oplus B$, which means that

$$\text{Im}\,(f, g) = \text{Ker}\,[h, -k]. \tag{A.46}$$

Prove that this condition is equivalent to each of the following ones:

(i) the span (f, g) and the pullback of (h, k) have the same pushout,

(i*) the cospan (h, k) and the pushout of (f, g) have the same pullback.

*(c) If the square (A.44) is a pullback and h is epi, so is g and the square is also a pushout.

*(d) Every abelian category is regular.

*(e) The square (A.44) is exact (in the present sense) if and only if it is v-exact (as defined in 2.5.6), if and only if it becomes bicommutative in the category Rel**E** constructed in Section 25.

A4 Relations for Puppe-exact and abelian categories

As we have mentioned in Section A1, categories of (additive) *relations* are used in homological algebra to define induced morphisms in a wide sense. This can already be found in Mac Lane's book 'Homology' ([M2], Section II.6), and is well known for abelian categories.

The extension of relations to p-exact categories dates back to Tsalenko [T1, T2] and Brinkmann-Puppe [BriP]; it can also be found in [G8]. Here we briefly recall this construction. For abelian categories, it gives the same result as the construction of Section 2.5, in a form more adapted to the study of subquotients.

Then we show how an adjunction between p-exact categories can always be extended to a colax-lax adjunction between their double categories of relations, coming back to a problem mentioned since the beginning of the general Introduction.

A4.1 Relations of abelian groups

The category Rel**Ab** of relations between abelian groups has been reviewed in 2.1.5; we give now a more detailed analysis to prepare the construction of relations for all p-exact categories.

A relation $u \colon A \to B$ is a subgroup of the direct sum $A \oplus B$. It can be viewed as a 'partially defined, multi-valued homomorphism', that sends an element $x \in A$ to the subset $\{y \in B \mid (x, y) \in u\}$ of B. The composite of u with $v \colon B \to C$ is

$$vu = \{(x, z) \in A \oplus C \mid \exists y \in B \colon (x, y) \in u \text{ and } (y, z) \in v\}.$$

RelAb is an involutive ordered category. For two parallel relations $u, u' \colon A \to B$, $u \leqslant u'$ means that $u \subset u'$ as subgroups of $A \oplus B$. The opposite relation of $u \colon A \to B$ is obtained by reversing pairs, and written as $u^\sharp \colon B \to A$. The involution $u \mapsto u^\sharp$ is regular in the sense of von Neumann, i.e. $uu^\sharp u = u$ for all relations u.

As a consequence, a *monorelation*, i.e. a monomorphism in the category RelAb, is characterised by the condition $u^\sharp u = 1$ (and is a split mono); an epirelation by the dual condition $uu^\sharp = 1$.

The category **Ab** is embedded in the category of relations RelAb, identifying a homomorphism $f \colon A \to B$ with its graph.

A relation $u \colon A \to B$ determines two subgroups of A, called *definition* and *annihilator*

$$\begin{aligned} \mathrm{Def}\, u &= \{x \in A \mid \exists y \in B \colon (x, y) \in u\}, \\ \mathrm{Ann}\, u &= \{x \in A \mid (x, 0) \in u\}, \end{aligned} \tag{A.47}$$

and two subgroups of B, called *values* and *indeterminacy*

$$\begin{aligned} \mathrm{Val}\, u &= \{y \in B \mid \exists x \in A \colon (x, y) \in u\} = \mathrm{Def}\, u^\sharp, \\ \mathrm{Ind}\, u &= \{y \in B \mid (0, y) \in u\} = \mathrm{Ann}\, u^\sharp. \end{aligned} \tag{A.48}$$

A4.2 Exercises and complements

The following properties in RelAb can be easily verified.

(a) A relation $u \colon A \to B$ is a homomorphism if and only if it is *everywhere defined* (i.e. $\mathrm{Def}\, u = A$) and *single-valued* (i.e. $\mathrm{Ind}\, u = 0$), in which case annihilator and values coincide with kernel and image, respectively.

(b) A relation u is a monorelation if and only if $\mathrm{Def}\, u = A$ and $\mathrm{Ann}\, u = 0$.

(b*) A relation u is an epirelation if and only if $\mathrm{Val}\, u = B$ and $\mathrm{Ind}\, u = 0$.

(c) A relation u is invertible in RelAb if and only if it is invertible in **Ab**, i.e. an isomorphism of abelian groups, in which case $u^\sharp$ is the inverse isomorphism.

(d) Every monomorphism (resp. epimorphism) in **Ab** is a monorelation (resp. epirelation).

(e) The relation u has a canonical factorisation $u = (nq^\sharp).i.(pm^\sharp)$, as in the lower part of the following diagram

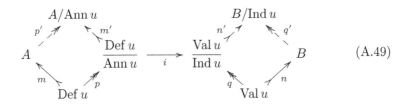

$$(A.49)$$

where m, n are embeddings of subgroups, p, q are projections on quotients and i is an isomorphism; the latter relates a subquotient of A to a subquotient of B. (Equivalently, one can use the factorisation $u = (q'^\sharp n').i.(m'^\sharp p')$, in the upper part of the diagram.)

(f) This factorisation consists of an epirelation $(ipm^\sharp)$ followed by a monorelation $(s = nq^\sharp)$: it is a factorisation epi-mono in $\mathbf{RelAb}$, essentially unique (since all monos and epis split).

Therefore *the image of u in this category should be defined as the subquotient $S =$ Val $u/$Ind u of B*, which is a *subobject of B* in $\mathbf{RelAb}$, via the monorelation $s = nq^\sharp: S \twoheadrightarrow B$. Note that Val $u =$ Val s and Ind $u =$ Ind s.

(g) In the following diagram the monorelation $s: S \twoheadrightarrow B$ is presented, in $\mathbf{Ab}$, as *a quotient of a subobject* $(s = nq^\sharp)$, and equivalently as *a subobject of a quotient* $(s = q'^\sharp n')$

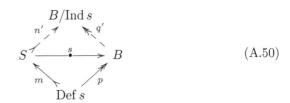

$$(A.50)$$

A4.3 Relations for p-exact categories

Every p-exact category $\mathbf{E}$ has an involutive ordered category of relations Rel$\mathbf{E}$, which we now describe (the proofs can be found in [G8]). Let us recall that $\mathbf{E}$ has pullbacks of monos and pushouts of epis along arbitrary morphisms (see A2.5), and is assumed to be well powered (see A2.2).

A relation $u: A \twoheadrightarrow B$ has a *w-factorisation* $u = nq^\sharp pm^\sharp$, and a *w*-factorisation* $u = q'^\sharp n'm'^\sharp p'$, each of them determined up to three coherent isomorphisms; they form two bicartesian squares (see A2.5(b))

$$A \qquad \overset{p'}{\nearrow} \; \overset{\bullet}{\nwarrow} \; \overset{m'}{} \quad \overset{n'}{} \; \overset{\bullet}{\nearrow} \; \overset{q'}{\nwarrow} \qquad B \qquad (\text{A.51})$$
$$\overset{m}{\nwarrow} \; \overset{\bullet}{\nearrow} \; \overset{p}{} \qquad \overset{q}{} \; \overset{\bullet}{\nwarrow} \; \overset{n}{\nearrow}$$

The composition vu of relations is computed, on *w-factorisations*, by *pullbacks of monos* (along arbitrary arrows) and *pushout of epis along epis*, as in the following diagram; the relation $u\colon A \twoheadrightarrow B$ appears in the upper row, the relation $v\colon B \twoheadrightarrow C$ in the right-hand column and $vu\colon A \twoheadrightarrow C$ along the dashed diagonal

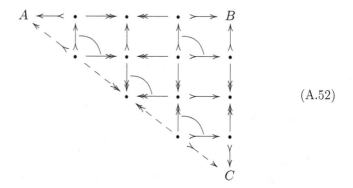

$$(\text{A.52})$$

Dually, it can be computed on w*-factorisations, by means of pushouts of epis along arbitrary arrows and pullbacks of monos along monos.

The involution $u \mapsto u^\sharp$ is obvious; the order relation $u \leqslant v$ is expressed by the existence of a commutative diagram in **E** linking their w-factorisations (or equivalently the w* ones)

$$
\begin{array}{ccccccccc}
A & \leftarrowtail & \bullet & \twoheadrightarrow & \bullet & \leftarrowtail & \bullet & \rightarrowtail & B & \quad (u) \\
\| & & \downarrow & & \downarrow & & \downarrow & & \| & \\
A & \leftarrowtail & \bullet & \twoheadrightarrow & \bullet & \leftarrowtail & \bullet & \rightarrowtail & B & \quad (v)
\end{array}
\qquad (\text{A.53})
$$

The construction of Rel**E** can be obtained by considering equivalence classes of diagrams

$$A \leftarrowtail \bullet \twoheadrightarrow \bullet \leftarrowtail \bullet \rightarrowtail B$$

in **E**, up to three coherent *iso*morphisms (as in diagram (A.53)); the composition is described above; the proof of associativity is not easy (see one of the references [T2, BriP, G8]).

A commutative square of **E** is said to be *exact* if it is *bicommutative* in Rel**E**, which means that it stays commutative when one reverses two parallel sides of the square. To characterise this fact, first note that:

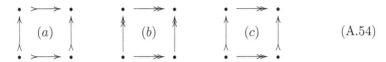

$$(A.54)$$

(i) a commutative square of monos (a) is exact if and only if it is a pullback,

(i*) a commutative square of epis (b) is exact if and only if it is a pushout,

(ii) a commutative *mixed* square (c) is exact if and only if it is a pullback, if and only if it is a pushout.

As a consequence, every pullback of a mono (along any map) produces a bicommutative square, and similarly every pushout of an epi. Finally, a commutative square in **E** becomes bicommutative in Rel**E** if and only if its canonical factorisation is composed of four bicommutative squares of the previous types

$$(A.55)$$

Every p-exact functor $F: \mathbf{E} \to \mathbf{E}'$ has a unique extension to a 2-functor Rel$F$: Rel**E** $\to$ Rel**E**$'$; every natural transformation $\varphi: F \to G: \mathbf{E} \to \mathbf{E}'$ determines a lax transformation of 2-functors

$$\text{Rel}\varphi: \text{Rel}F \to \text{Rel}G: \text{Rel}\mathbf{E} \to \text{Rel}\mathbf{E}',$$

whose components are precisely those of φ (and belong to **E**$'$).

A4.4 Double categories of relations

Every p-exact category **E** produces an ultraflat double category $\mathbb{R}\text{el}\mathbf{E}$, with **E**-morphisms as horizontal arrows and **E**-relations as vertical ones. Also here a cell corresponds to an inequality

$$gu \leqslant vf \quad (\Leftrightarrow\ fu^\sharp \leqslant v^\sharp g) \qquad (A.56)$$

and amounts to a commutative diagram of **E** based on w-factorisations, where the three dashed arrows are uniquely determined, by induction on subobjects and quotients

$$
\begin{array}{ccc}
A & \xrightarrow{\;f\;} & A' \\
\end{array}
$$

(A.57)

Equivalently, one can use a dual diagram for w*-factorisations.

We also have double functors $\mathrm{Rel}(F)\colon \mathrm{Rel}\mathbf{E} \to \mathrm{Rel}\mathbf{E}'$ (for F exact) and horizontal transformations

$$
\mathrm{Rel}(\varphi)\colon \mathrm{Rel}F \to \mathrm{Rel}G\colon \mathrm{Rel}\mathbf{E} \to \mathrm{Rel}\mathbf{E}',
$$

as in A1.6.

A4.5 Left exact functors

Let us take now, more generally, a *left exact* functor $S\colon \mathbf{E} \to \mathbf{E}'$, i.e. a functor which preserves kernels (whence the zero-object, monomorphisms and their pullbacks).

Since S preserves pullbacks of monomorphisms, which become bicommutative squares in $\mathrm{Rel}\mathbf{E}'$, we can extend it to relations by using, equivalently, the w- or w*-factorisation:

$$
\begin{aligned}
S'(nq^{\sharp}pm^{\sharp}) &= S(n)(Sq)^{\sharp}(Sp)(Sm^{\sharp}), \\
S'(q'^{\sharp}n'm'^{\sharp}p') &= (Sq')^{\sharp}(Sn')(Sm')^{\sharp}(Sp'),
\end{aligned}
\tag{A.58}
$$

even if S need *not* preserve epis and the right-hand parts above are *not* w- nor w*-factorisations, generally.

We obtain thus a *lax* double functor $S' = \mathrm{Rel}S\colon \mathrm{Rel}\mathbf{E} \to \mathrm{Rel}\mathbf{E}'$. The fact that S' preserves cells follows trivially by the fact that the cell (A.56) amounts to a commutative diagram (A.57) in **E**, and factorises in four cells of $\mathrm{Rel}\mathbf{E}$, whose images are cells of $\mathbf{E}'$.

Finally, the comparison special cells for vertical composition $S'v.S'u \leqslant$

$S'(vu)$ are an easy consequence of the composition vu as computed in (A.52)

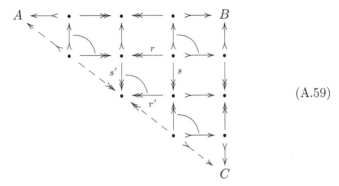

$$(A.59)$$

Indeed the functor S preserves pullbacks of monos (along arbitrary arrows), that are bicommutative in the category of relations; furthermore it carries the pushout $(r, s; r', s')$ to a commutative square $Ss'.Sr = Sr'.Ss$, which gives $(Ss)(Sr)^{\sharp} \leqslant (Sr')^{\sharp}(Ss)$ in RelE'. It follows immediately that $S'v.S'u \leqslant S'(vu)$.

Again, a natural transformation $\varphi \colon S \to T \colon \mathbf{E} \to \mathbf{E}'$ of left exact functors gives a horizontal transformation of lax double functors

$$\mathbb{R}\mathrm{el}\varphi \colon \mathbb{R}\mathrm{el}S \to \mathbb{R}\mathrm{el}T \colon \mathbb{R}\mathrm{el}\mathbf{E} \to \mathbb{R}\mathrm{el}\mathbf{E}',$$

with the same components as φ. (The coherence with the comparison cells of $\mathbb{R}\mathrm{el}S$ and $\mathbb{R}\mathrm{el}T$ is here automatic, since our double categories are flat.)

Dually, a *right exact* functor $S \colon \mathbf{E} \to \mathbf{E}'$ (which preserves cokernels) can be extended to a *colax* double functor $S' = \mathbb{R}\mathrm{el}S \colon \mathbb{R}\mathrm{el}\mathbf{E}' \to \mathbb{R}\mathrm{el}\mathbf{E}$, by the same definition as above, in (A.58).

Of course, if S is exact we get the functor already considered in A4.4.

A4.6 Adjoints between categories of relations

Let us start from an arbitrary adjunction $F \dashv G$ between p-exact categories

$$F \colon \mathbf{E} \to \mathbf{E}', \qquad\qquad G \colon \mathbf{E}' \to \mathbf{E},$$
$$\eta \colon 1 \to GF \colon \mathbf{E} \to \mathbf{E}, \qquad \varepsilon \colon FG \to 1 \colon \mathbf{E}' \to \mathbf{E}', \qquad (A.60)$$
$$\varepsilon F.F\eta = 1_F, \qquad\qquad G\varepsilon.\eta G = 1_G.$$

F preserves the existing colimits and G the existing limits; we have thus a colax and a lax extension, respectively

$$F' = \mathbb{R}\mathrm{el}F \colon \mathbb{R}\mathrm{el}\mathbf{E} \to \mathbb{R}\mathrm{el}\mathbf{E}' \qquad\qquad (colax),$$
$$G' = \mathbb{R}\mathrm{el}G \colon \mathbb{R}\mathrm{el}\mathbf{E}' \to \mathbb{R}\mathrm{el}\mathbf{E} \qquad\qquad (lax), \qquad (A.61)$$

which form a colax-lax adjunction of double categories (Chapter 4). The unit is

$$\text{(A.62)}$$

where the right-hand inequality comes from the colax-property of G' and the definition of F'

$$(GFn)(GFq)^{\sharp}(GFp)(GFm^{\sharp}) \leqslant G'((Fn)(Fq)^{\sharp}(Fp)(Fm^{\sharp})) = G'F'u.$$

The counit is defined dually. The coherence relations hold automatically, by flatness.

A4.7 The abelian case

Working on an abelian category **E**, we get 'the same' relations as constructed in Section 2.5. Essentially, this is a consequence of Exercise A4.8(b).

Going on in the present framework, every relation has now a *binary factorisation* $u = gf^{\sharp}\colon A \to B$. Such a factorisation will be said to be *strong* if the pair (f, g) is jointly mono (corresponding to a monomorphism with values in $A \oplus B$); this can be obtained from a w-factorisation of u, by the pullback of the central epimorphisms (which is also a pushout):

$$\text{(A.63)}$$

Dually, there is a *strong cobinary factorisation* $u = f'^{\sharp}g'$, where (f', g') is jointly epi. Two cobinary factorisations yield the same relation u if and only if they have the same pullback, which is then a strong binary factorisation of u.

A cell in $\mathbb{Rel}\mathbf{E}$ corresponds to a commutative $\mathbf{E}$-diagram for strong binary factorisations (as in the central diagram below), or equivalently, for strong cobinary factorisations (as at the right)

$$
\begin{array}{ccc}
A \xrightarrow{\ f\ } A' \\
\end{array}
\qquad
\begin{array}{ccc}
A \xrightarrow{\ f\ } A' \\
\end{array}
\qquad
\begin{array}{ccc}
A \xrightarrow{\ f\ } A' \\
\end{array}
\qquad (\mathrm{A}.64)
$$

A left exact functor S between abelian categories preserves arbitrary pullbacks; its extension $S' = \mathbb{Rel}S$ can be equivalently computed over w-factorisations, or w*-factorisations, or strong binary factorisations, or arbitrary cobinary factorisations. We get a lax double functor, as we already know from the more general p-exact case considered above. Dually for right exact functors.

A4.8 Exercises and complements

(a) Starting from the following adjunction $F \dashv G$ of abelian groups

$$
F = - \otimes A \colon \mathbf{Ab} \to \mathbf{Ab}, \qquad G = \mathrm{Hom}(A, -) \colon \mathbf{Ab} \to \mathbf{Ab}, \qquad (\mathrm{A}.65)
$$

show that both the extended 'functors' $\mathbb{Rel}\mathbf{Ab} \to \mathbb{Rel}\mathbf{Ab}$ fail to preserve vertical composition of relations, for $A = \mathbb{Z}/2$ (or any non-trivial finite cyclic group).

*(b) Prove that a square of the abelian category $\mathbf{E}$ is bicommutative in the category of relations constructed in Section 2.5 (as characterised in A3.9) if and only if it is bicommutative in the present construction (as characterised in A4.3).

A5 Homotopy quotients of topological spans and cospans

We end this appendix with a few hints of applications in Algebraic Topology. The weak double category $\mathbb{Span}\mathbf{Top}$ *of topological spans* has a quotient $\mathrm{ho}\mathbb{Span}\mathbf{Top}$ where horizontal and vertical arrows are unchanged, but double cells are identified up to homotopy, in a sense specified below (in A5.4). Similarly the weak double category $\mathbb{Cosp}\mathbf{Top}$ *of topological cospan* has a quotient $\mathrm{ho}\mathbb{Cosp}\mathbf{Top}$.

The interested reader is referred to [G2]–[G4], to see how these structures, together with their higher dimensional extensions, can be used to study higher cobordism.

A continuous mapping is also called a *map*. The boolean variable α takes values in $\{0, 1\}$, also written as $\{-, +\}$.

A5.1 Homotopy

As we mentioned in 2.2.4, the standard euclidean interval $\mathbb{I} = [0, 1]$ is exponentiable in **Top**. In other words the *cylinder endofunctor*

$$I: \mathbf{Top} \to \mathbf{Top}, \qquad I(X) = X \times \mathbb{I}, \quad I(f) = f \times \mathrm{id}\mathbb{I}, \qquad (\text{A.66})$$

has a right adjoint, the *cocylinder*, or *path endofunctor*

$$P: \mathbf{Top} \to \mathbf{Top}, \qquad P(Y) = Y^{\mathbb{I}}, \quad P(g) = g^{\mathbb{I}}: c \mapsto gc, \qquad (\text{A.67})$$

where $P(Y)$ is the set $\mathbf{Top}(\mathbb{I}, Y)$ of *paths* $c: \mathbb{I} \to Y$, equipped with the compact-open topology – also called the topology of uniform convergence (since $\mathbb{I}$ is compact). The unit and counit are obvious

$$\begin{aligned}
\eta: 1 &\to PI, & \eta_X(x)(t) &= (x, t) & (x \in X, \, t \in \mathbb{I}), \\
\varepsilon: IP &\to 1, & \varepsilon_Y(c, t) &= c(t) & (c \in PY, \, t \in \mathbb{I}).
\end{aligned} \qquad (\text{A.68})$$

Each of these functors has a basic structure of two *faces* and a *degeneracy*, that are mates under the adjunction, as in 2.3.3 (here $\alpha = 0, 1$)

$$X \underset{e}{\overset{d^\alpha}{\rightrightarrows}} IX \qquad d^\alpha(x) = (x, \alpha), \quad e(x, t) = x, \quad ed^\alpha = 1_X, \qquad (\text{A.69})$$

$$PY \underset{e}{\overset{d^\alpha}{\rightrightarrows}} Y \qquad d^\alpha(c) = c(\alpha), \quad e(y)(t) = y, \quad d^\alpha e = 1_Y. \qquad (\text{A.70})$$

Because of the adjunction, a *homotopy* $\varphi: f^- \simeq f^+$ between two maps $f^\alpha: X \to Y$ can be equivalently defined as

- a map $\hat{\varphi}: IX \to Y$ that coincides with f^-, f^+ on the faces (or bases) of the cylinder, i.e. $\hat{\varphi}.d^\alpha = f^\alpha$,

- a map $\check{\varphi}: X \to PY$ such that $d^\alpha.\check{\varphi} = f^\alpha$.

There is a whisker composition of homotopies and maps

$$(k\varphi h)\hat{} = k.\hat{\varphi}.Ih: IX' \to Y' \qquad (\text{A.71})$$

$$X' \xrightarrow{h} X \underset{g}{\overset{f}{\underset{\Downarrow\varphi}{\rightrightarrows}}} Y \xrightarrow{k} Y'.$$

Homotopies have a defective 'vertical structure' (that would only give a 2-category up to higher homotopies):

- *trivial homotopies* $0_f: f \simeq f$ (represented by $fe: IX \to Y$ or $ef: X \to PY$),

- *reversed homotopies* $\varphi^\sharp: g \to f$ (for $\varphi: f \simeq g$),

- *concatenated homotopies* $\varphi + \psi: f \simeq h$ (for $\varphi: f \simeq g$ and $\psi: g \simeq h$).

All this is well-known and easily described directly – or can be produced by further structure on the (co)cylinder functor, as in [G7].

We shall often write the 'representative map' $\hat{\varphi}$ (or $\check{\varphi}$) of a homotopy as φ, provided this does not lead to ambiguity – as it certainly would above, in formula (A.71).

A5.2 The homotopy category

As already recalled in 1.1.5, the homotopy relation $f \simeq g$ (given by the existence of a homotopy $\varphi: f \simeq g$) is a congruence of categories in **Top**. Its quotient is called the homotopy category $\mathrm{hoTop} = \mathbf{Top}/\simeq$.

We also recall that a map $f: X \to Y$ is a *homotopy equivalence* if there exists a map $g: Y \to X$ such that $gf \simeq \mathrm{id}X$ and $fg \simeq \mathrm{id}Y$, which amounts to saying that its homotopy class $[f]$ is an isomorphism of hoTop.

A functor $F: \mathbf{Top} \to \mathbf{C}$ with values in any category is said to be *homotopy invariant* if it takes homotopic maps $f \simeq g$ to the same morphism $F(f) = F(g)$, so that it induces a functor $\mathrm{hoTop} \to \mathbf{C}$.

Many categories with a notion of homotopy have a structure of 'cylinder-path endofunctors' as sketched above. An interested reader can see [G7] and references therein.

A5.3 Proposition (Homotopies in pullbacks and pushouts)

(a) A commutative solid diagram of maps and homotopies, as in the left diagram below, can be uniquely completed with a homotopy $\omega: w \simeq w': W \to W'$ from the pullback W of (f,g) to the pullback W' of (f',g'), consistent with the pullback-maps, in the sense that $p'\omega = \xi p$ and $q'\omega = \eta q$

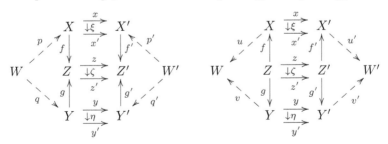

(a*) A commutative solid diagram of maps and homotopies, as in the

right diagram above, can be uniquely completed with a homotopy $\omega\colon w \simeq w'\colon W \to W'$ from the pushout W of (f, g) to the pushout W' of (f', g'), consistent with the pushout-maps, so that $\omega u = u'\xi$ and $\omega v = v'\eta$.

Note. These results hold in any category with a cylinder-path adjunction, as sketched above. They are dual to each other in this selfdual context – not in **Top**.*

Proof The path (resp. cylinder) functor preserves all limits (resp. colimits). It is thus sufficient to transform the diagrams above as follows and apply the universal property of the pullback PW' (resp. the pushout IW) to get the homotopy ω

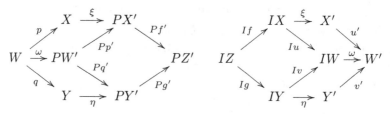

A5.4 Spans and homotopy

In the weak double category $\mathbb{S}\mathrm{pan}\mathbf{Top}$ we introduce an equivalence relation $\sim$ that is trivial (i.e. the identity) on all objects and arrows, and is defined as follows on cells.

Consider two cells $x, x'\colon (u \overset{f}{\underset{g}{}} v)$ *with the same boundary*. By abuse of notation, we write as x, x' also the central maps $U \to U'$ that determine these cells, once their boundary is given. We say that these cells are *h-equivalent*, and write $x \sim x'$, if there exists a homotopy

$$\xi\colon x \simeq x'\colon U \to U'$$

which is consistent with the trivial homotopies 0_f and 0_g, in the sense that:

$$
\begin{array}{ccc}
X & \overset{f}{\longrightarrow} & X' \\
{\scriptstyle u'}\uparrow & & \uparrow{\scriptstyle v'} \\
U & \overset{x}{\underset{\downarrow\xi}{\longrightarrow}} U' & \\
{\scriptstyle u''}\downarrow & \overset{x'}{} & \downarrow{\scriptstyle v''} \\
Y & \underset{g}{\longrightarrow} & Y'
\end{array}
\qquad\qquad\text{(A.72)}
$$

$$v'\xi = 0_f u' = 0_{fu'}, \qquad v''\xi = 0_g u'' = 0_{gu''}.$$

We prove below that this equivalence relation gives a weak double category

$$\text{ho}\mathbb{S}\text{pan}\mathbf{Top} = (\mathbb{S}\text{pan}\mathbf{Top})/\sim, \qquad (\text{A.73})$$

with the same objects and arrows of $\mathbb{S}\text{pan}\mathbf{Top}$, and equivalence classes of cells $[x]$: $(u \overset{f}{\underset{g}{}} v)$. It will be called the *homotopy quotient* of $\mathbb{S}\text{pan}\mathbf{Top}$.

Note that the homotopy category ho$\mathbf{Top}$ has few limits and colimits: for instance, products and sums. In particular, it has no pullbacks nor pushouts: one cannot form a weak double category of spans or cospans on it. *But it is interesting to note that ho$\mathbf{Top}$ has weak pullbacks and weak pushouts, produced – respectively – by the homotopy pullbacks and homotopy pushouts of $\mathbf{Top}$ [Mat].*

A5.5 Theorem (Spans and homotopy)

The relation of h-equivalence is consistent with horizontal and vertical composition of double cells.

Proof Consistency with horizontal composition is shown in the diagram below

$$(\text{A.74})$$

In fact, composing the homotopies ξ, η 'along the diagonal', we get a homotopy

$$\varphi \colon yx \simeq y'x' \colon U \to U''$$

that proves that $(x \,|\, y) \sim (x' \,|\, y')$

$$\varphi(z,t) = \eta(\xi(z,t),t), \qquad \varphi(-,0) = yx, \qquad \varphi(-,1) = y'x',$$
$$w'(\eta(\xi(z,t),t)) = f'v'(\xi(z,t)) = f'fu'(z),$$
$$w''(\eta(\xi(z,t),t)) = g'v''(\xi(z,t)) = g'gu''(z).$$

For vertical composition we apply Proposition A5.3(a) to the central part of the concatenation of x, x': $(u \overset{f}{\underset{g}{}} v)$ with y, y': $(w \overset{g}{\underset{h}{}} t)$

This gives a homotopy $\omega\colon w \simeq w'\colon W \to W'$ between the pullbacks of (f, g) and (f', g'), consistent with the pullback-maps: $p'\omega = \xi p$ and $q'\omega = \eta q$, so that

$$v'p'\omega = v'\xi p = 0_f u'p, \qquad t''q'\omega = t''\eta q = 0_h w''q.$$

Finally we have proved that $x \otimes y \sim x' \otimes y'$. $\qquad\qquad\square$

A5.6 Cospans and homotopy

Similarly we have a weak double category $\mathrm{ho}\mathbb{C}\mathrm{osp}\mathbf{Top} = (\mathbb{C}\mathrm{osp}\mathbf{Top})/\sim$, with the same objects and arrows as $\mathbb{C}\mathrm{osp}\mathbf{Top}$ and equivalence classes $[x]\colon (u \overset{f}{\underset{g}{}} v)$ of cells.

Given two cells x, x': $(u \overset{f}{\underset{g}{}} v)$ of cospans *with the same boundary*, as in the diagram below (again we write as x, x' also the associated maps $U \to U'$), we say that these cells are *h-equivalent*, and write $x \sim x'$, if there exists a homotopy $\xi\colon x \simeq x'\colon U \to U'$ of the associated maps, which is consistent with the trivial homotopies 0_f and 0_g

(A.75)

$$\xi u' = v'0_f = 0_{v'f}, \qquad \xi u'' = v''0_g = 0_{v''g}.$$

Also here one should distinguish the *cell* $[x]$: $(u \overset{f}{\underset{g}{}} v)$ from the *associated central morphism* $[x] \colon U \to U'$ in ho**Top**, even though we use the same notation $[x]$ in both cases.

For instance, the morphism $[x]$ is invertible whenever $x \colon U \to U'$ is a homotopy equivalence (i.e. there is some $y \colon U' \to U$ such that $yx \simeq 1$ and $xy \simeq 1$), but the horizontal invertibility of the cell $[x] \colon (u \,{}^{f}_{g}\, v)$ is a stronger condition: the map y should be coherent with the cospans.

Appendix B
Symmetric cubical sets and cubical categories

Cubical sets are a classical topic of Algebraic Topology, introduced by Kan [K1] and extensively studied by Brown and Higgins [BroH1, BroH2]; see also [GM]. Strict cubical categories and their links with globular categories are dealt with in [AlBS].

Weak cubical categories and their symmetric version were introduced and studied in [G2]–[G6], as a basis for the study of cubical cospans in Algebraic Topology and higher cobordism. We have seen in 6.4.5 that these structures can be seen as weak multiple categories of cubical type. Here they are treated as an independent subject.

As a basic example we use again the weak multiple category $\mathsf{Cosp}(\mathbf{C})$ presented in 6.4.6. Its components, the ordinary categories $\mathrm{Cosp}_n(\mathbf{C}) = \mathbf{Cat}(\wedge^n, \mathbf{C})$, form a cubical object in $\mathbf{Cat}$, with obvious faces and degeneracies. The details of its construction are analysed here, in Section B4.

In fact $\mathsf{Cosp}(\mathbf{C})$ is a weak *symmetric* cubical category, when equipped with the obvious action of the symmetric group S_n on $\mathbf{Cat}(\wedge^n, \mathbf{C})$: namely, the action of permuting the factors of $\wedge^n$, i.e. the directions of n-cubical cospans in $\mathbf{C}$. These symmetries allow to only consider the faces, degeneracies and geometric compositions in a single direction (see B2.2), which can help to simplify the coherence conditions.

The present appendix gives an analysis of the role of symmetries, based on [G5], which goes much beyond the simplification of coherence properties. In fact the category $\mathbf{Cub}$ of ordinary cubical sets has a Kan tensor product, which is *non* symmetric and biclosed, with left and right internal homs based on the right and left path functors (see B1.6, B1.7). On the other hand, *symmetric cubical sets* have one path functor (see B2.3) leading to one internal hom and form a *symmetric monoidal closed* category sCub.

Similar facts hold for cubical and symmetric cubical categories, and have played a relevant role in the study of limits and adjunctions in these higher dimensional categories [G6, G10].

I still denotes the standard interval $[0, 1]$ with euclidean topology, while $\underline{u}$ will be the 'elementary' interval of **Cub**. Also here the boolean variables α, β take values in $\{-, +\}$ or $\{0, 1\}$, as may be convenient.

B1 Cubical sets and internal homs

We start from the category **Cub** of ordinary cubical sets and their (*non* symmetric) monoidal biclosed structure [K1, BroH2]. As in [G5] we stress the role of the 'transposer' S, which reverses the order of faces; in particular, we are interested in the *external symmetry* $s\colon S(X \otimes Y) \to (SY) \otimes (SX)$, which partially surrogates here the 'internal' symmetry of a symmetric tensor product.

B1.1 Cubical sets

A cubical set $X = ((X_n), (\partial_i^\alpha), (e_i))$, in the usual sense [K1, K2, BroH1, BroH2], has *faces* (∂_i^α) and *degeneracies* (e_i)

$$\partial_i^\alpha\colon X_n \rightleftarrows X_{n-1} : e_i \qquad (i = 1, ..., n;\ \alpha = \pm), \qquad (\text{B.1})$$

satisfying the cubical relations:

$$\partial_i^\alpha \partial_j^\beta = \partial_j^\beta \partial_{i+1}^\alpha \ (j \leqslant i), \qquad e_j e_i = e_{i+1} e_j \ (j \leqslant i),$$
$$\partial_i^\alpha e_j = e_j \partial_{i-1}^\alpha \ (j < i), \qquad \partial_i^\alpha e_i = 1, \qquad \partial_i^\alpha e_j = e_{j-1} \partial_i^\alpha \ (j > i). \qquad (\text{B.2})$$

Elements of X_n are called *n-cubes*; or *vertices* for $n = 0$ and *edges* for $n = 1$. Every n-cube $x \in X_n$ has 2^n vertices: $\partial_1^{\alpha_1} ... \partial_n^{\alpha_n}(x)$.

The classical example is the *singular cubical set* $\square(S)$ of a topological space S, whose n-component is the set $\square_n(S) = \mathbf{Top}(\mathbb{I}^n, S)$ of *singular n-cubes* $x\colon \mathbb{I}^n \to S$; faces and degeneracies are obvious.

A *morphism* $f = (f_n)\colon X \to Y$ of cubical sets is a sequence of mappings $f_n\colon X_n \to Y_n$ that commutes with faces and degeneracies.

Cubical sets and their morphisms form a category **Cub**, which has all limits and colimits. In fact, as we already recalled in 6.2.3, it is the presheaf category of functors $X\colon \underline{I}^{\mathrm{op}} \to \mathbf{Set}$, on the cubical site $\underline{I}$. The latter is the subcategory of **Set** consisting of the *elementary cubes* $2^n = \{0, 1\}^n$, together with the maps $\{0, 1\}^m \to \{0, 1\}^n$ which delete some coordinates and insert some 0's and 1's, without modifying the order of the remaining coordinates (see [GM]).

The terminal object $\top$ of **Cub** is freely generated by one vertex $*$ and will also be written as $\{*\}$; but note that each of its components is a singleton, and all faces and degeneracies are identities.

The initial object is empty, i.e. all its components are; the other cubical sets have a non-empty component in each degree.

The category **Cub** has two (covariant) involutive endofunctors, which we call *reversor* (R) and *transposer* (S)

$$R: \mathbf{Cub} \to \mathbf{Cub}, \qquad RX = X^{\mathrm{op}} = ((X_n), (\partial_i^{1-\alpha}), (e_i)), \qquad (\mathrm{B.3})$$

$$S: \mathbf{Cub} \to \mathbf{Cub}, \qquad SX = ((X_n), (\partial_{n+1-i}^{\alpha}), (e_{n+1-i})). \qquad (\mathrm{B.4})$$

If $x \in X_n$, the corresponding element in $(RX)_n = X_n$ will often be written as x^{op}, so that $\partial_i^-(x^{\mathrm{op}}) = (\partial_i^+ x)^{\mathrm{op}}$. We say that a cubical set X is *reversive* if $RX \cong X$ and *permutative* if $SX \cong X$.

B1.2 Tensor product

As any category of presheaves, **Cub** is cartesian closed (see 2.2.4(k)). But we are interested in a different monoidal structure (see [K1, BroH2])

$$(X \otimes Y)_n = (\textstyle\sum_{p+q=n} X_p \times Y_q)/\!\sim_n, \qquad (\mathrm{B.5})$$

where $\sim_n$ is the equivalence relation, in the previous sum of sets, generated by correlating each pair $(e_{r+1}x, y)$ with $(x, e_1 y)$, for all $(x, y) \in X_r \times Y_s$ (where $r + s = n - 1$).

Writing as $x \otimes y$ the equivalence class of (x, y), the faces and degeneracies of $X \otimes Y$ are defined as follows, when x is of degree p and y of degree q

$$\partial_i^{\alpha}(x \otimes y) = \begin{cases} (\partial_i^{\alpha} x) \otimes y, & \text{if } 1 \leqslant i \leqslant p, \\ x \otimes (\partial_{i-p}^{\alpha} y), & \text{if } p < i \leqslant n, \end{cases} \qquad (\mathrm{B.6})$$

$$e_i(x \otimes y) = \begin{cases} (e_i x) \otimes y, & \text{if } 1 \leqslant i \leqslant p+1, \\ x \otimes (e_{i-p} y), & \text{if } p+1 \leqslant i \leqslant n+1. \end{cases} \qquad (\mathrm{B.7})$$

Note that

$$e_{p+1}(x \otimes y) = (e_{p+1} x) \otimes y = x \otimes (e_1 y)$$

is well defined, because of the previous equivalence relation.

The identity of the tensor product is the terminal object $\top = \{*\}$, which is obviously reversive and permutative.

B1.3 The external symmetry

The tensor product is not symmetric, but is related to reversor and transposer by an identity and an *external symmetry* s

$$R(X \otimes Y) = RX \otimes RY,$$

$$s(X,Y): S(X \otimes Y) \to (SY) \otimes (SX), \qquad x \otimes y \mapsto y \otimes x. \qquad (B.8)$$

The latter is a functorial isomorphism of functors $\mathbf{Cub} \times \mathbf{Cub} \to \mathbf{Cub}$, with inverse $S(s(SY, SX))$.

Therefore reversive objects are stable under tensor product while permutative objects are stable under tensor *power*: if $SX \cong X$, then

$$S(X^{\otimes n}) \cong (SX)^{\otimes n} \cong X^{\otimes n}.$$

B1.4 The standard interval

The *standard interval* $\underline{u}$ of $\mathbf{Cub}$ is the cubical set freely generated by one cube u, of degree 1

$$0 \xrightarrow{u} 1 \qquad \partial_1^-(u) = 0, \quad \partial_1^+(u) = 1. \qquad (B.9)$$

This cubical set is reversive and permutative. It is the representable presheaf $Y(2) = \underline{I}(-, 2): \underline{I}^{\mathrm{op}} \to \mathbf{Set}$ (see (1.14)).

The *standard n-cube* is its n-th tensor power $\underline{u}^{\otimes n} = \underline{u} \otimes ... \otimes \underline{u}$ (for $n \geqslant 0$), freely generated by its n-cube $u^{\otimes n}$. It is still reversive and permutative and 'coincides' with the representable presheaf $Y(2^n) = \underline{I}(-, 2^n): \underline{I}^{\mathrm{op}} \to \mathbf{Set}$.

The *standard square* $\underline{u} \otimes \underline{u}$ can be represented as follows, showing the generator $u \otimes u$ and its faces

$$
\begin{array}{ccc}
00 & \xrightarrow{\;u \otimes 0\;} & 10 \\
{\scriptstyle 0 \otimes u}\big\downarrow & u \otimes u & \big\downarrow{\scriptstyle 1 \otimes u} \\
01 & \xrightarrow[\;u \otimes 1\;]{} & 11
\end{array}
\qquad
\begin{array}{c}
\bullet \xrightarrow{\;1\;} \\
{\scriptstyle 2}\big\downarrow
\end{array}
\qquad (B.10)
$$

As usual the face $\partial_1^-(u \otimes u) = 0 \otimes u$ is drawn as orthogonal to direction 1 (which can be done in every dimension). For each cubical object X we have $\mathbf{Cub}(\underline{u}^{\otimes n}, X) = X_n$, by the Yoneda Lemma (in 1.2.8).

Note that the cartesian power $\underline{u} \times \underline{u}$ has *two* non-degenerate 2-cubes, namely $(e_1 u, e_2 u)$ and $(e_2 u, e_1 u)$.

B1.5 Left and right cylinder functors

Let us start from the standard interval $\underline{u}$, and work with the monoidal structure recalled above, with unit $\{*\}$ and reversor R. Recall that u

denotes the 1-dimensional generator of $\underline{u}$, and u^{op} is the corresponding edge of $\underline{u}^{\mathrm{op}}$ (as in B1.1).

The cubical set $\underline{u}$ has a structure consisting of two *faces* (∂^0, ∂^1), a *degeneracy* (e) and a reflection or external reversion (r):

$$\partial^\alpha \colon \{*\} \to \underline{u}, \qquad \partial^\alpha(*) = \alpha,$$

$$e \colon \underline{u} \to \{*\}, \qquad e(\alpha) = *, \qquad e(u) = e_1(*), \qquad (\mathrm{B.11})$$

$$r \colon \underline{u} \to \underline{u}^{\mathrm{op}}, \qquad r(0) = 1^{\mathrm{op}}, \quad r(1) = 0^{\mathrm{op}}, \qquad r(u) = u^{\mathrm{op}}.$$

Since the tensor product is not symmetric, the elementary directed interval yields a *left (elementary) cylinder* $\underline{u} \otimes X$ and a *right cylinder* $X \otimes \underline{u}$. These functors are not isomorphic, but each of them determines the other, using the transposer S, defined in (B.4), and the property $S(\underline{u}) = \underline{u}$

$$I \colon \mathbf{Cub} \to \mathbf{Cub}, \qquad\qquad IX = \underline{u} \otimes X,$$
$$SIS \colon \mathbf{Cub} \to \mathbf{Cub}, \qquad SIS(X) = S(\underline{u} \otimes SX) \cong X \otimes \underline{u}. \qquad (\mathrm{B.12})$$

The last relation is the canonical isomorphism $s(\underline{u}, SX)$ of (B.8).

The left cylinder $IX = \underline{u} \otimes X$ inherits from the structure of $\underline{u}$, in (B.11), two faces, a degeneracy and a reflection, as follows

$$\partial^\alpha = \partial^\alpha \otimes X \colon X \to IX, \qquad \partial^\alpha(x) = \alpha \otimes x,$$

$$e = e \otimes X \colon IX \to X,$$

$$e(u \otimes x) = e_1(*) \otimes x = * \otimes e_1(x) = e_1(x), \qquad (\mathrm{B.13})$$

$$r = r \otimes RX \colon IRX \to RIX,$$

$$r(\alpha \otimes x^{\mathrm{op}}) = ((1 - \alpha) \otimes x)^{\mathrm{op}}, \qquad r(u \otimes x^{\mathrm{op}}) = (u \otimes x)^{\mathrm{op}}.$$

As a right cylinder functor we shall use SIS, rather then the isomorphic functor $- \otimes \underline{u}$.

B1.6 Left and right path functors

The category $\mathbf{Cub}$ has a *left path* functor P, right adjoint to the left cylinder functor $IX = \underline{u} \otimes X$.

The functor P shifts down all components discarding the faces and degeneracies of index 1; the latter are then used to build three natural transformations, the *faces* and *degeneracy* of P

$$P \colon \mathbf{Cub} \to \mathbf{Cub}, \qquad PY = ((Y_{n+1}), (\partial^\alpha_{i+1}), (e_{i+1})),$$
$$\partial^\alpha = \partial^\alpha_1 \colon PY \to Y, \qquad\qquad e = e_1 \colon Y \to PY. \qquad (\mathrm{B.14})$$

The transposer S produces the *right path* functor $P' = SPS$, right adjoint to the right cylinder SIS. Explicitly, SPS shifts down all components and discards the faces and degeneracies of highest index (used again to build faces and degeneracy)

$$SPS: \mathbf{Cub} \to \mathbf{Cub}, \qquad SPS(Y) = ((Y_{n+1}), (\partial_i^\alpha), (e_i)),$$

$$\partial^\alpha: SPS(Y) \to Y, \qquad \partial^\alpha = (\partial_{n+1}^\alpha: Y_{n+1} \to Y_n)_{n \geqslant 0}, \qquad (\text{B.15})$$

$$e: Y \to SPS(Y), \qquad e = (e_{n+1}: Y_n \to Y_{n+1})_{n \geqslant 0}.$$

An elementary (or immediate) *left homotopy* $f: f^- \to_L f^+: X \to Y$ is defined as a map $f: X \to PY$ with $\partial^\alpha f = f^\alpha$. This leads to a simple expression of f as a family of mappings satisfying the following conditions

$$f_n: X_n \to Y_{n+1}, \qquad \partial_{i+1}^\alpha f_n = f_{n-1} \partial_i^\alpha, \qquad (\text{B.16})$$

$$\partial_1^\alpha f_n = f^\alpha, \qquad e_{i+1} f_{n-1} = f_n e_i.$$

Similarly, an elementary *right homotopy* $f: f^- \to_R f^+: X \to Y$ is a map $f: X \to SPS(Y)$ with faces $\partial^\alpha f = f^\alpha$. This amounts to a family (f_n) such that

$$f_n: X_n \to Y_{n+1}, \qquad \partial_i^\alpha f_n = f_{n-1} \partial_i^\alpha, \qquad (\text{B.17})$$

$$\partial_{n+1}^\alpha f_n = f^\alpha, \qquad e_i f_{n-1} = f_n e_i.$$

The transposer can be viewed as an isomorphism $S: \mathbf{Cub}_L \to \mathbf{Cub}_R$ between the left and the right structure. One can define an *external transposition* $s: PP' \to P'P$ (replacing, from a formal point of view, the transposition $s: P^2 \to P^2$ of topological spaces, which permutes the two variables); it is actually an identity $PP' = P'P$

$$s: PSPS \to SPSP, \qquad (sY)_n = \mathrm{id}Y_{n+2}, \qquad (\text{B.18})$$

since both functors PP' and $P'P$ shift down all components of two degrees, discarding the faces and degeneracies of least and greatest index.

B1.7 Internal homs

The category **Cub** has left and right internal homs [BroH2]. The *right* internal hom $\mathrm{CUB}(A, Y)$ can be constructed with the *left* cocylinder functor P and its natural transformations (which produce a cubical object $P^\bullet Y$ in **Cub**)

$$- \otimes A \dashv \mathrm{CUB}(A, -), \qquad \mathrm{CUB}_n(A, Y) = \mathbf{Cub}(A, P^n Y). \qquad (\text{B.19})$$

The natural bijection

$$\varphi(X, Y): \mathbf{Cub}(X \otimes A, Y) \to \mathbf{Cub}(X, \mathrm{CUB}(A, Y)), \qquad (\text{B.20})$$

is constructed as follows, on an arbitrary morphism $f = (f_n): X \otimes A \to Y$. Its n-component f_n can be decomposed into a family of mappings

$$f_{pq}: X_p \times A_q \to Y_{p+q} \qquad (p + q = n), \tag{B.21}$$

consistent with the equivalence relations $\sim_n$ (see (B.5)). By the exponential law in **Set**, these mappings amount to mappings

$$g_{pq}: X_p \to \mathbf{Set}(A_q, Y_{p+q}).$$

Keeping p fixed, we get a mapping

$$g_p = (g_{pq}): X_p \to \mathbf{Cub}(A, P^p Y) \subset \textstyle\prod_q \mathbf{Set}(A_q, Y_{p+q}), \tag{B.22}$$

whose family forms a morphism of cubical sets $g = (g_p): X \to \mathrm{CUB}(A, Y)$.

B1.8 Higher path functors

We have seen that the two path functors $P, P': \mathbf{Cub} \to \mathbf{Cub}$ commute (in (B.18)). Therefore, every composition of n occurrences of them can be written as

$$P_i^n = P^{n-i}.P'^i = P^{n-i}.SP^i S: \mathbf{Cub} \to \mathbf{Cub} \qquad (i = 0, ..., n). \tag{B.23}$$

$P_i^n X$ has p-component X_{p+n}; its faces and degeneracies $X_{p+n} \rightleftarrows X_{p+n-1}$ are those of X with indices $n - i + 1, ..., n - i + p$ (renumbered as $1, ..., p$). In particular, $P_0^1 = P$ and $P_1^1 = P'$.

There are generalised faces linking higher path functors

$$
\begin{aligned}
(P^i \partial^\alpha P^j).(SP^k S): P^{i+j+1}.SP^k S &\to P^{i+j}.SP^k S, \\
P^i.(SP^j \partial^\alpha P^k S): P^i.SP^{j+k+1} S &\to P^{i+j}.SP^k S,
\end{aligned}
\tag{B.24}
$$

and similar generalised degeneracies.

B2 Symmetric cubical sets and their closed structure

We consider now *symmetric* cubical sets, equipped with transpositions. In the singular cubical set of a topological space these mappings correspond to transposing variables in the standard cubes $\mathbb{I}^n$.

Lifting the previous left or right path functors (in B1.6) to the symmetric case, we get isomorphic functors, and essentially *one* path functor. The latter produces one internal hom, and a *symmetric monoidal closed structure*.

The real points of interest are the path functor and the internal hom. On the other hand, the symmetric tensor product (of symmetric cubical sets) and the corresponding cylinder functor are complicated and not really needed, here; they will only be sketched in Section B6.

B2.1 Symmetric cubical sets

As in [G2], a *symmetric cubical* set, or *sc-set*, is a cubical set which is further equipped with mappings, called *transpositions*

$$s_i \colon X_n \to X_n \qquad (i = 1, ..., n - 1; \ n \geqslant 2). \qquad (B.25)$$

These have to satisfy the Moore relations

$$s_i.s_i = 1, \qquad s_i.s_j = s_j.s_i \ \ (i < j - 1),$$
$$s_i.s_j.s_i = s_j.s_i.s_j \ \ (i = j - 1), \qquad (B.26)$$

and the following conditions of coherence with faces and degeneracies:

	$j < i$	$j = i$	$j = i + 1$	$j > i + 1$
$\partial_j^\alpha.s_i =$	$s_{i-1}.\partial_j^\alpha$	∂_{i+1}^α	∂_i^α	$s_i.\partial_j^\alpha,$
$s_i.e_j =$	$e_j.s_{i-1}$	e_{i+1}	e_i	$e_j.s_i.$

(B.27)

Assigning the mappings (B.25) under conditions (B.26) amounts to letting the symmetric group S_n operate on X_n. Indeed, it is well known that S_n is generated, under the Moore relations, by the 'ordinary' transpositions $s_1, ..., s_{n-1}$, where s_i, acting on the set $\{1, ..., n\}$, exchanges i with $i + 1$ (see Coxeter-Moser [CoM], 6.2; or Johnson [Jh], Section 5, Theorem 3).

A *morphism* $f = (f_n) \colon X \to Y$ of sc-sets is a sequence of mappings $f_n \colon X_n \to Y_n$ that commutes with faces, degeneracies and transpositions.

The resulting category s**Cub**, of small symmetric cubical sets and their morphisms, is again a category of presheaves $X \colon \underline{I}_s^{\mathrm{op}} \to$ **Set**, for the *symmetric cubical site* $\underline{I}_s$. The latter can be defined as the subcategory of **Set** consisting of the elementary cubes $2^n = \{0, 1\}^n$ together with the maps $2^m \to 2^n$ which delete some coordinates, permute the remaining ones and insert some 0's and 1's. It is a subcategory of the *extended cubical site* of [GM], which also contains the 'connections' (or higher degeneracies).

The reversor and transposer of cubical sets (in B1.1) have obvious liftings to s**Cub**

$$R \colon s\mathbf{Cub} \to s\mathbf{Cub}, \quad RX = X^{\mathrm{op}} = ((X_n), (\partial_i^{1-\alpha}), (e_i), (s_i)), \qquad (B.28)$$

$$S \colon s\mathbf{Cub} \to s\mathbf{Cub}, \quad SX = ((X_n), (\partial_{n+1-i}^\alpha), (e_{n+1-i}), (s_{n+1-i})). \qquad (B.29)$$

Here the transpositions make S isomorphic to the identity functor and inessential, as we shall see in B2.6.

Similarly one defines the category $s\mathbf{Cub}(\mathbf{C})$ of *symmetric cubical objects* over any category $\mathbf{C}$.

B2.2 Reduced presentation of symmetric cubical sets

In a symmetric cubical set the presence of transpositions makes all faces and degeneracies determined ·by the 1-directed ones, ∂_1^-, ∂_1^+ and e_1. In fact, from $\partial_{i+1}^\alpha = \partial_i^\alpha.s_i$ and $e_{i+1} = s_i.e_i$, we deduce that:

$$\partial_i^\alpha = \partial_1^\alpha.\underline{s}_i', \qquad e_i = \underline{s}_i.e_1 \qquad (i = 2,\dots,n;\ \alpha = \pm), \tag{B.30}$$

where we are using the 'permutations' $\underline{s}_i$ and $\underline{s}_i'$, inverse to each other

$$\underline{s}_i = s_{i-1}\dots s_1, \qquad\qquad \underline{s}_i' = s_1\dots s_{i-1}. \tag{B.31}$$

This leads to a more economical presentation of our structure, as proved in [G4]. Namely, an sc-set can be equivalently defined as a system

$$X = ((X_n), \partial_1^-, \partial_1^+, e_1, (s_i)), \tag{B.32}$$

under the Moore relations for transpositions, in (B.26), and the axioms:

$$\partial_1^\alpha.e_1 = \mathrm{id}, \qquad \partial_1^\alpha.\partial_1^\beta = \partial_1^\beta.\partial_1^\alpha.s_1, \qquad e_1e_1 = s_1.e_1e_1,$$
$$s_i.\partial_1^\alpha = \partial_1^\alpha.s_{i+1}, \qquad e_1.s_i = s_{i+1}.e_1. \tag{B.33}$$

In other words, X can be presented as a system $((X_n), \partial_1^-, \partial_1^+, e_1)$ where each X_n is an S_n-set (i.e. a set equipped with an action of the symmetric group S_n) and the axioms (B.33) are satisfied.

B2.3 The symmetric path functor

We now define the *path functor* P of sc-sets, by lifting the left path functor of ordinary cubical sets (see B1.6): P shifts down all the components, discarding the faces, degeneracy and transpositions *of index* 1

$$P\colon s\mathbf{Cub} \to s\mathbf{Cub},$$
$$PX = ((X_{n+1}), (\partial_{i+1}^\alpha), (e_{i+1}), (s_{i+1})), \qquad (Pf)_n = f_{n+1}. \tag{B.34}$$

Again, the discarded faces and degeneracy are used to build three natural transformations, the faces and degeneracy of P (while the discarded s_1 will give the transposition of P^2, in B2.4)

$$\partial^\alpha\colon PX \to X, \qquad \partial^\alpha = (\partial_1^\alpha\colon X_{n+1} \to X_n)_{n\geqslant 0},$$
$$e\colon X \to PX, \qquad e = (e_1\colon X_n \to X_{n+1})_{n\geqslant 0}. \tag{B.35}$$

The transposer S, in (B.29), produces the *right path* functor $P' = SPS$, which shifts down all components discarding the faces, degeneracies and transpositions *of highest index*. However $S \cong \text{id}$ and $SPS \cong P$ (as we shall see in B2.6), so that one path functor is sufficient.

B2.4 The transposition of the path functor

The 'second order' path functor is computed as:

$$P^2 \colon \mathbf{sCub} \to \mathbf{sCub},$$
$$P^2 X = ((X_{n+2}), (\partial^\alpha_{i+2}), (e_{i+2}), (s_{i+2})), \qquad (P^2 f)_n = f_{n+2}. \tag{B.36}$$

It has two pairs of faces $P\partial^\alpha$, $\partial^\alpha P \colon P^2 \to P$ and two degeneracies $Pe, eP \colon P \to P^2$

$$P(\partial^\alpha X) = (\partial^\alpha_1 \colon X_{n+2} \to X_{n+1})_{n \geqslant 0},$$
$$\partial^\alpha(PX) = (\partial^\alpha_2 \colon X_{n+2} \to X_{n+1})_{n \geqslant 0}, \tag{B.37}$$

$$P(eX) = (e_1 \colon X_{n+1} \to X_{n+2})_{n \geqslant 0},$$
$$e(PX) = (e_2 \colon X_{n+1} \to X_{n+2})_{n \geqslant 0}. \tag{B.38}$$

The important fact is that we have a *transposition* for the path functor P

$$s \colon P^2 \to P^2, \qquad x \mapsto s_1(x). \tag{B.39}$$

First, $s \colon P^2 X \to P^2 X$ is indeed a morphism of sc-sets, as it follows immediately from the symmetric cubical relations (in B2.1)

Second, because of other symmetric cubical relations (not applied above), the involution s interchanges the faces of P^2 of type $P\partial^\alpha$ and $\partial^\alpha P$, and the degeneracies of type Pe and eP

$$\partial^\alpha P.s = P\partial^\alpha, \qquad s.eP = Pe. \tag{B.40}$$

B2.5 Internal homs

We define the internal-hom functor of the category $\mathbf{sCub}$ as

$$\mathrm{sCUB} \colon \mathbf{sCub}^{\mathrm{op}} \times \mathbf{sCub} \to \mathbf{sCub},$$
$$\mathrm{sCUB}_n(A, Y) = \mathbf{sCub}(A, P^n Y). \tag{B.41}$$

In particular

$$sCUB(\{*\}, Y) = Y, \qquad sCUB(\underline{u}, Y) = PY, \qquad \text{(B.42)}$$

where $\underline{u} = Y(2)$ is now the representable *symmetric* cubical set freely generated by one 1-cube u.

This is the same as the cubical set $\underline{u}$ of Section B1, equipped with the unique symmetric structure that permutes its (degenerate) cubes of degree $\geqslant 2$

$$s_1 e_1(u) = e_2(u), \qquad s_1 e_2(u) = e_1(u),$$

$$s_1 e_1(v) = e_2(v), \qquad s_1 e_2(v) = e_1(v), \qquad s_1 e_3(v) = e_3 s_1(v),$$

$$s_2 e_1(v) = e_1 s_1(v), \qquad s_2 e_2(v) = e_3(v), \qquad s_2 e_3(v) = e_2(v), \dots$$

where v is a 2-cube, namely $e_1(u)$ or $e_2(u)$.

B2.6 Exercises and complements

The results of the following exercises make clear how the structure of ordinary cubical sets is simplified by the addition of symmetries in **sCub**.

(a) Construct a natural isomorphism $\sigma \colon 1 \to S \colon \mathbf{sCub} \to \mathbf{sCub}$.

(b) Construct a natural isomorphism $\gamma \colon P \to P' = SPS$ between the left and the right path functors.

(c) Show that the transposition $s \colon P^2 \to P^2$ defined in (B.39) can be obtained combining the isomorphism γ and its inverse.

B3 Weak cubical categories and the symmetric case

Weak cubical categories and their symmetric version were introduced in [G2]. After reformulating the definition in a simpler way, we introduce the path functor of these cubical structures, which plays an important role in their theory – for instance for the study of cubical limits (cf. [G6]).

B3.1 Weak cubical categories

The present definition follows the same pattern of weak multiple categories, in Section 6.4. Besides replacing a multiple object (in **Cat**) with a cubical one, the main difference is that – here – faces and degeneracies (the operators that modify the dimension) carry a shift of indices, as already remarked for cubical sets in 6.2.3.

(wcc.1) A *weak cubical category* A is, first of all, a sequence of categories

$\mathrm{tv}_n(\mathsf{A})$, for $n \geqslant 0$, called the *transversal categories*, or *transversal compo-nents* of A.

In $\mathrm{tv}_n(\mathsf{A})$, the objects are called n-cubes and form a set A_n; the mor-phisms $f \colon x^- \to x^+$ are called (transversal) n-maps and form a set M_n. The categorical structure has faces $\partial_0^\alpha(f) = x^\alpha$ ($\alpha = \pm$), identities $e_0(x) = \mathrm{id}(x) = 1_x$ and a composition law $c_0(f, g) = f +_0 g = gf$ defined on the set $M_n \times_\bullet M_n$ of consecutive pairs (f, g) of n-maps ($\partial_0^+ f = \partial_0^- g$)

$$A_n \underset{e_0}{\overset{\partial_0^\alpha}{\rightleftarrows}} M_n \xleftarrow{c_0} M_n \times_\bullet M_n. \tag{B.43}$$

(wcc.2) These categories form a cubical object in **Cat**. In other words, we have functors of faces and degeneracies

$$\partial_i^\alpha \colon \mathrm{tv}_n(\mathsf{A}) \rightleftarrows \mathrm{tv}_{n-1}(\mathsf{A}) : e_i, \tag{B.44}$$

that satisfy the cubical relations of (B.2).

(wcc.3) Furthermore n-cubes and n-maps have a *geometric i-concatenation* functor

$$+_i \colon \mathrm{tv}_n\mathsf{A} \times_i \mathrm{tv}_n\mathsf{A} \to \mathrm{tv}_n\mathsf{A} \qquad (1 \leqslant i \leqslant n), \tag{B.45}$$

whose action we write in additive notation and diagrammatic order:

$$x +_i y, \qquad f +_i g \qquad (\partial_i^+ x = \partial_i^- y, \ \partial_i^+ f = \partial_i^- g).$$

(As a functor it commutes with the transversal faces, degeneracies and composition of (B.43).)

(wcc.4) The following 'geometrical' interactions with faces and degeneracies are required, for i-consecutive n-cubes x, y:

$$\partial_i^-(x +_i y) = \partial_i^-(x), \quad \partial_i^+(x +_i y) = \partial_i^+(y) \qquad (1 \leqslant i \leqslant n),$$

$$\partial_j^\alpha(x +_i y) = \begin{cases} \partial_j^\alpha(x) +_{i-1} \partial_j^\alpha(y), & \text{for } 1 \leqslant j < i \leqslant n, \\ \partial_j^\alpha(x) +_i \partial_j^\alpha(y), & \text{for } 1 \leqslant i < j \leqslant n, \end{cases}$$

$$e_j(x +_i y) = \begin{cases} e_j(x) +_{i+1} e_j(y), & \text{for } 1 \leqslant j \leqslant i \leqslant n, \\ e_j(x) +_i e_j(y), & \text{for } 1 \leqslant i < j \leqslant n+1. \end{cases}$$

The same holds for i-consecutive n-maps. Writing $\partial_i^\alpha x$, $e_i x$, $x +_i y$ or $\partial_i^\alpha f$, $e_i f$, $f +_i g$ it will be understood that such terms make sense.

An n-map $f \colon x \to y$ is said to be *i-special*, or *special in direction i*, if its two i-faces are transversal identities

$$\partial_i^\alpha f = e_0 \partial_i^\alpha x = e_0 \partial_i^\alpha y. \tag{B.46}$$

This implies that the n-cubes x, y have the same i-faces: $\partial_i^\alpha x = \partial_i^\alpha y$. We say that f is *ij-special* if it is special in both directions i, j.

(wcc.5–8) The geometric compositions are *weakly* categorical and satisfy interchange, up to invertible transversal maps called *comparisons*, defined for n-cubes x, y, z, u and $i, j \leqslant n$

$$\lambda_i x \colon (e_i \partial_i^- x) +_i x \to x \qquad \text{(left i-unitor)},$$

$$\rho_i x \colon x +_i (e_i \partial_i^+ x) \to x \qquad \text{(right i-unitor)},$$

$$\kappa_i(x, y, z) \colon x +_i (y +_i z) \to (x +_i y) +_i z \qquad \text{(i-associator)},$$

$$\chi_{ij}(x, y, z, u) \colon (x +_i y) +_j (z +_i u) \to (x +_j z) +_i (y +_j u)$$
$$\text{(ij-interchanger, for $i < j$)}.$$

These comparisons are natural with respect to transversal maps; λ_i, ρ_i and κ_i are i-special and χ_{ij} is ij-special. The remaining faces ∂_k^α commute with the comparisons up to the usual shifting of indices, which we now express by a term $i \circ k$ whose value is $i - 1$ or i, when $k < i$ or $k > i$, respectively

$$\partial_k^\alpha \lambda_i(x) = \lambda_{i \circ k}(\partial_k^\alpha x), \qquad \partial_k^\alpha \rho_i(x) = \rho_{i \circ k}(\partial_k^\alpha x),$$

$$\partial_k^\alpha \kappa_i(x, y, z) = \kappa_{i \circ k}(\partial_k^\alpha x, \partial_k^\alpha y, \partial_k^\alpha z),$$

$$\partial_k^\alpha \chi_{ij}(x, y, z, u) = \chi_{i \circ k, j \circ k}(\partial_k^\alpha x, \partial_k^\alpha y, \partial_k^\alpha z, \partial_k^\alpha u).$$

(wcc.9) Finally these comparisons must satisfy various conditions of coherence, similar to those of weak multiple categories in 6.4.3, 6.4.4.

If all comparisons are identities, A is a (strict) *cubical category*.

B3.2 Weak symmetric cubical categories

Let us begin from the strict case.

A *symmetric cubical category* is a cubical category A equipped with an action of the symmetric group S_n (non trivial for $n \geqslant 2$) on each category $\mathrm{tv}_n(\mathsf{A})$, generated by the *transposition* functors

$$s_i \colon \mathrm{tv}_n(\mathsf{A}) \to \mathrm{tv}_n(\mathsf{A}), \qquad i = 1, \dots, n-1 \quad (n \geqslant 2), \tag{B.47}$$

under the Moore relations of (B.26).

These actions must give a symmetric cubical object in **Cat**, which means that the transpositions satisfy the relations (B.27), with faces and degeneracies. Moreover they are assumed to be coherent with concatenations: if X and Y are n-cubes *or* n-maps, we have (for $i, j > 0$ and $j \neq i, i+1$)

$$s_i(X +_i Y) = s_i X +_{i+1} s_i Y, \qquad s_i(X +_j Y) = s_i X +_j s_i Y. \tag{B.48}$$

Equivalently, for every permutation $s\colon [n] \to [n]$ that takes i to i' we have:

$$s(X +_i Y) = sX +_{i'} sY. \tag{B.49}$$

Finally, in a *weak symmetric cubical category* A, the symmetries are also assumed to be coherent with the comparisons $\lambda_i, \rho_i, \kappa_i, \chi_{ij}$. Namely, for every permutation $s\colon [n] \to [n]$ that takes i to i' and j to j' we have (introducing $\chi_{ji} = (\chi_{ij})^{-1}$, for $i < j$):

$$s.\lambda_i = \lambda_{i'}.s, \qquad s.\rho_i = \rho_{i'}.s, \qquad s.\kappa_i = \kappa_{i'}.s,$$
$$s.\chi_{ij} = \chi_{i'j'}.s \qquad (i \neq j). \tag{B.50}$$

Truncation works in the usual way, as in Chapter 6. Since the symmetric groups S_0 and S_1 are trivial, a 2-dimensional truncated weak symmetric cubical category has no transpositions and is the same as a weak double category.

A weak symmetric cubical category A can be viewed as a weak multiple category $A^\sharp$ of symmetric cubical type (see 6.4.5) by letting, for every positive multi-index **i** of degree $n \geqslant 0$:

$$(A^\sharp)_\mathbf{i} = A_n, \qquad (A^\sharp)_{0\mathbf{i}} = M_n, \tag{B.51}$$

so that any **i**-cube (or **i**-map) of degree n of $A^\sharp$ is an n-cube (or n-map) of A.

A (strict) *functor* $F\colon$ A $\to$ B *of weak symmetric cubical categories*, or sc-functor, is a morphism of symmetric cubical objects that preserves all compositions and comparisons.

B3.3 Path functors of cubical categories

We write as cb**Cat** the 2-category of (small) cubical categories, their (strict) functors and transversal transformations, and as **Wsc** the 2-category of weak symmetric cubical categories, their functors and transversal transformations.

Cubical categories have a left and a right path 2-functor, which are obvious liftings of the path functors of cubical sets

$$P\colon \mathrm{cb}\mathbf{Cat} \to \mathrm{cb}\mathbf{Cat}, \qquad P' = SPS\colon \mathrm{cb}\mathbf{Cat} \to \mathrm{cb}\mathbf{Cat},$$
$$\mathrm{tv}_n.P = \mathrm{tv}_{n+1}, \qquad\qquad \mathrm{tv}_n.P' = \mathrm{tv}_{n+1}. \tag{B.52}$$

In every degree P discards faces, degeneracies and concatenations in direction 1 while P' discards those in the last direction. Again, P and P' are related to each other by the transposer $S\colon \mathrm{cb}\mathbf{Cat} \to \mathrm{cb}\mathbf{Cat}$, which in every degree reverses the order of faces, degeneracies and concatenations.

Also here, P and P' have isomorphic liftings to the symmetric case (where $S \cong \mathrm{id}$), and we will only use the path 2-functor which discards direction 1, written as

$$P \colon \mathbf{Wsc} \to \mathbf{Wsc}, \qquad (\text{B.53})$$

for the 2-category of weak sc-categories, sc-functors and their transversal transformations.

B3.4 Exponentials

(a) First, for a small *ordinary* category $\mathbf{X}$ and a symmetric cubical category A, we have the *symmetric cubical category* $\mathsf{A}^{\mathbf{X}}$ *of level functors and their natural transformations*

$$\mathrm{tv}_n(\mathsf{A}^{\mathbf{X}}) = \mathbf{Cat}(\mathbf{X}, \mathrm{tv}_n \mathsf{A}). \qquad (\text{B.54})$$

An n-cube is an ordinary functor $F \colon \mathbf{X} \to \mathrm{tv}_n \mathsf{A}$, and will also be called an *n-level functor* with values in A; an n-map is a natural transformation $f \colon F \to G \colon \mathbf{X} \to \mathrm{tv}_n \mathsf{A}$. Their faces, degeneracies, transpositions and concatenations are obtained by post-composition with the structural functors of A (in B3.1)

$$\partial_i^\alpha \colon \mathrm{tv}_n \mathsf{A} \rightleftarrows \mathrm{tv}_{n-1} \mathsf{A} : e_i, \qquad s_i \colon \mathrm{tv}_n \mathsf{A} \rightleftarrows \mathrm{tv}_n \mathsf{A},$$
$$+_i \colon \mathrm{tv}_n \mathsf{A} \times_i \mathrm{tv}_n \mathsf{A} \to \mathrm{tv}_n \mathsf{A}. \qquad (\text{B.55})$$

(b) If A is a weak symmetric cubical category, also $\mathsf{A}^{\mathbf{X}}$ is, with comparisons obtained from those of A.

(c) Now, let X be a small weak sc-category and A a weak sc-category. We define the weak symmetric cubical category A^{X} of *higher sc-functors from* X to A *and their transversal transformations*.

An n-cube is an sc-functor $F \colon \mathsf{X} \to P^n \mathsf{A}$, an n-map is a transversal transformation of these functors. Faces, degeneracies, transpositions are obtained by post-composition with the structure of the path functor P of weak sc-categories, in (B.53)

$$\partial_i^\alpha \colon P^n \mathsf{A} \rightleftarrows P^{n-1} \mathsf{A} : e_i, \qquad s_i \colon P^n \mathsf{A} \to P^n \mathsf{A}. \qquad (\text{B.56})$$

Similarly, one obtains concatenations and the comparisons of the weak structure of A^{X}.

B4 The weak symmetric cubical category of higher cospans

We give now a detailed description of the weak sc-category $\mathsf{Cosp}(\mathbf{C})$ of cubical cospans over a category $\mathbf{C}$ with pushouts, by a construction based on a formal structure on the cospan category $\wedge$, taken from [G2]; a construction that was only sketched in 6.4.6.

B4.1 The models

Let us recall that the construction is based on the formal-cospan category $\wedge$, together with its cartesian powers; we use the notation of 6.4.6.

The category $\wedge$ has a basic structure of 'formal symmetric interval', with respect to the cartesian product in **Cat**, which consists of two (co)faces ∂^α, a (co)degeneracy e and a transposition s

$$\partial^\alpha : 1 \rightrightarrows \wedge, \qquad e : \wedge \to 1, \qquad s : \wedge^2 \to \wedge^2,$$
$$\partial^\alpha(0) = \alpha, \qquad s(t_1, t_2) = (t_2, t_1) \qquad (\alpha = 0, 1). \tag{B.57}$$

The functors

$$(-)_i^n = \wedge^{i-1} \times - \times \wedge^{n-i} : \mathbf{Cat} \to \mathbf{Cat}, \tag{B.58}$$

produce a symmetric cocubical object in **Cat**, with components $\wedge^n$ $(n \geqslant 0)$ and the following structure (for $1 \leqslant i \leqslant n$ and $\alpha = 0, 1$)

$$\partial_i^\alpha : \wedge^{n-1} \to \wedge^n, \qquad \partial_i^\alpha(t_1, ..., t_{n-1}) = (t_1, ..., \alpha, ..., t_{n-1}),$$
$$e_i : \wedge^n \to \wedge^{n-1}, \qquad e_i(t_1, ..., t_n) = (t_1, ..., \hat{t}_i, ..., t_n), \tag{B.59}$$
$$s_i : \wedge^{n+1} \to \wedge^{n+1}, \qquad s_i(t_1, ..., t_{n+1}) = (t_1, ..., t_{i+1}, t_i, ..., t_n).$$

Further structure on the formal interval $\wedge$, that will not be used here, comprises the reversion symmetry $r : \wedge \to \wedge$ and the connections (cf. [GM]).

B4.2 The symmetric cubical object of cospans

Applying the contravariant functor $\mathbf{Cat}(-, \mathbf{C})$ we obtain, in **Cat**, a symmetric cubical object $\mathsf{Cosp}(\mathbf{C})$ with transversal components

$$\mathrm{tv}_n(\mathsf{Cosp}(\mathbf{C})) = \mathsf{Cosp}_n(\mathbf{C}) = \mathbf{Cat}(\wedge^n, \mathbf{C}), \tag{B.60}$$

so that an n-cube, or n-*dimensional object*, or n-*cubical cospan*, is a functor $x : \wedge^n \to \mathbf{C}$, and an n-map is a natural transformation $f : x \to y : \wedge^n \to \mathbf{C}$.

Faces, degeneracies and transpositions are computed as in (B.59), on cubes (and similarly on transversal maps)

$$\partial_i^\alpha(x)(t_1, ..., t_{n-1}) = x(t_1, ..., t_{i-1}, \alpha, ..., t_{n-1}),$$
$$e_i(x)(t_1, ..., t_n) = x(t_1, ..., \hat{t}_i, ..., t_n), \tag{B.61}$$
$$s_i(x)(t_1, ..., t_{n+1}) = x(t_1, ..., t_{i+1}, t_i, ..., t_{n+1}).$$

B4.3 Formal concatenations

As in 6.4.6, the *model of binary concatenation* is the category $\wedge_2$ displayed below at the left, with one marked square

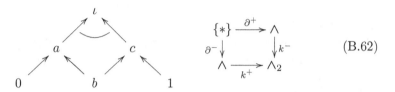

$$(B.62)$$

In the right-hand commutative square above we have:

$$k^-(\wedge) = \{0 \to a \leftarrow b\}, \qquad k^+(\wedge) = \{b \to c \leftarrow 1\}.$$

The pushout in **Cat** of ∂^+ and ∂^- is the subcategory $\wedge_{(2)} \subset \wedge_2$ lying at the bottom of $\wedge_2$

$$0 \to a \leftarrow b \to c \leftarrow 1 \qquad \wedge_{(2)}. \qquad (B.63)$$

The relevant fact – however – is that, for our category **C** with a fixed choice of pushouts, we have the following property of $\wedge_2$

(a) given two consecutive cospans $x, y \colon \wedge \to \mathbf{C}$ (with $\partial_1^+ x = \partial_1^- y$), there is precisely one functor $[x, y] \colon \wedge_2 \to \mathbf{C}$ such that:

- $[x, y].k^- = x$, $[x, y].k^+ = y$,

- $[x, y]$ takes the marked square of (B.63) to a distinguished pushout.

The concatenation $x +_1 y \colon \wedge \to \mathbf{C}$ is obtained by pre-composing $[x, y]$ with the embedding

$$k \colon \wedge \to \wedge_2, \quad k(t) = t \qquad (\textit{concatenation map}), \qquad (B.64)$$

already displayed in (B.62) by the objects $0, \iota, 1$ (named as in $\wedge$).

Acting on $\wedge_2$ and k, the functors $(-)_i^n$ of (B.58) produce the model $\wedge_2^{ni}$ of i-concatenation in dimension n

$$
\begin{array}{ccc}
\wedge^{n-1} & \xrightarrow{\ \partial_i^+\ } & \wedge^n \\
{\scriptstyle \partial_i^-}\downarrow & & \downarrow{\scriptstyle k_i^-} \\
\wedge^n & \xrightarrow[\ k_i^+\ ]{} & \wedge_2^{ni}
\end{array}
\qquad (B.65)
$$

$$
\begin{aligned}
\wedge_2^{ni} &= \wedge^{i-1} \times \wedge_2 \times \wedge^{n-i} && (\textit{i-concatenation model}), \\
k_i &= \wedge^{i-1} \times k \times \wedge^{n-i} \colon \wedge^n \to \wedge_2^{ni} && (\textit{i-concatenation map}).
\end{aligned}
$$

The category $\wedge_2^{ni}$ has 3^{n-1} marked squares, sitting in the following sub-categories (isomorphic to $\wedge_2$)

$$\{(t_1, ..., t_{i-1})\} \times \wedge_2 \times \{(t_{i+1}, ..., t_n)\} \qquad (t_i \in \{0, \iota, 1\}), \qquad (\text{B.66})$$

and has a property extending property (a) above:

(b) given two functors $x, y \colon \wedge^n \to \mathbf{C}$ with $\partial_i^+ x = \partial_i^- y$, there is precisely one functor $[x, y] \colon \wedge_2^{ni} \to \mathbf{C}$ such that:

- $[x, y].k_i^- = x$, $\qquad [x, y].k_i^+ = y$,

- $[x, y]$ takes all the marked squares of $\wedge_2^{ni}$ to distinguished pushouts.

Now the i-concatenation $x +_i y$ of i-consecutive n-cubes is computed by pre-composing $[x, y]$ with the embedding k_i of (B.65)

$$x +_i y = [x, y].k_i \colon \wedge^n \to \wedge_2^{ni} \to \mathbf{C}, \qquad (\text{B.67})$$

and we work in the same way for i-consecutive n-maps.

B4.4 Associativity comparisons

We extend now the basic structure of the formal interval with formal comparisons for associativity (and then for middle-four interchange).

The *model of ternary compositions* $\wedge_3$ is the following preorder-category, with five marked squares (as made explicit below)

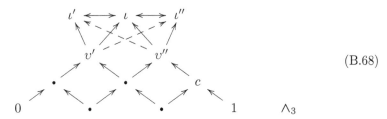

$$(\text{B.68})$$

The role of $\wedge_3$ will come forth from the following commutative diagram, in **Cat**

$$(\text{B.69})$$

Again, the cocone of vertex $\wedge_3$ in this diagram is not a colimit in **Cat**: the corresponding colimit-category only consists of the six (non-trivial) arrows in the bottom row $\wedge_{(3)}$ of the diagram of $\wedge_3$. Besides these arrows, $\wedge_3$ contains:

- a symmetric construction of three *marked squares*, reaching the vertices $\upsilon', \upsilon'', \iota$,

- two more *marked squares*, which attain the vertices ι' and ι'',

- three coherent isomorphisms $i'\colon \iota' \to \iota$, $i''\colon \iota \to \iota''$ and $i = i''i'$.

The embeddings $k', k''\colon \wedge \to \wedge_3$ with the following images

$$k'\colon \quad 0 \to \iota' \leftarrow 1 \qquad\qquad k''\colon \quad 0 \to \iota'' \leftarrow 1 \qquad\qquad (B.70)$$

correspond to two iterated concatenations of the three consecutive cospans $\wedge \to \wedge_3$ which 'cover' the bottom row $\wedge_{(3)}$. They are linked by a functorial isomorphism, the basic *formal associator*, whose components are identities, except the central one

$$\kappa\colon k' \to k''\colon \wedge \to \wedge_3, \qquad\qquad \kappa(\iota) = i\colon \iota' \to \iota''. \qquad (B.71)$$

Applying the functor $(-)_i^n$ we have a natural transformation, for $1 \leqslant i \leqslant n$

$$\kappa_i = (\kappa)_i^n\colon (k')_i^n \to (k'')_i^n\colon \wedge^n \to (\wedge_3)_i^n = \wedge^{i-1} \times \wedge_3 \times \wedge^{n-i}$$
$$\text{(\textit{formal i-associator})}. \qquad (B.72)$$

Given three i-consecutive n-cospans $x, y, z\colon \wedge^n \to \mathbf{C}$, pre-composing with κ_i gives a transversal isomorphism for i-associativity

$$\kappa_i(x,y,z) = [x,y,z].\kappa_i\colon x +_i (y +_i z) \to (x +_i y) +_i z, \qquad (B.73)$$

where the functor $[x,y,z]\colon (\wedge_3)_i^n \to \mathbf{C}$ is defined 'as' in B4.3(b). (Let us note that it also contains a *regular ternary concatenation* $x +_i y +_i z$, corresponding to the vertex ι of $\wedge_3$.)

B4.5 Interchange comparisons

Double cospans $\wedge^2 \to \mathbf{C}$ can be concatenated in two directions.

The model $\wedge_{2 \times 2}$ for the 2-dimensional interchange of concatenations is constructed below, starting from the colimit in **Cat** of the following diagram

$$
\begin{array}{ccccc}
\wedge^2 & \xleftarrow{\ \partial_1^+\ } \wedge \xrightarrow{\ \partial_1^-\ } & \wedge^2 \\[4pt]
{\scriptstyle\partial_2^+}\big\uparrow & & \big\uparrow{\scriptstyle\partial_2^+} \\[4pt]
\wedge & & \wedge \\[4pt]
{\scriptstyle\partial_2^-}\big\downarrow & & \big\downarrow{\scriptstyle\partial_2^-} \\[4pt]
\wedge^2 & \xleftarrow[\ \partial_1^+\ ]{} \wedge \xrightarrow[\ \partial_1^-\ ]{} & \wedge^2
\end{array}
\qquad (B.74)
$$

The colimit is the pasting of four copies of $\wedge^2$, displayed in the solid diagram below, and amounts to the product $\wedge_{(2)} \times \wedge_{(2)}$ (see (B.63))

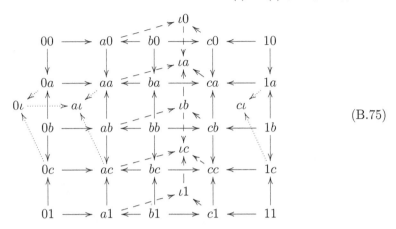

$$(B.75)$$

By definition, the category $\wedge_{2\times2}$ is obtained by adding to $\wedge_{(2)} \times \wedge_{(2)}$ two 'constructions', partially displayed above, which correspond to two symmetric procedures: first composing in direction 1 and then in direction 2, or vice versa. Thus $\wedge_{2\times2}$ contains

(a) a copy of $\wedge_2 \times \wedge_{(2)}$ (adding in the dashed arrows above, pertaining to five marked squares ending in (ι, j), with $j = 0, a, b, c, 1$), together with the marked square ending in ι', displayed below

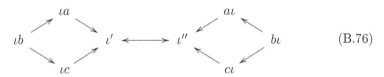

$$(B.76)$$

(b) a symmetric construction, of which we only display in (B.75) some dotted arrows: it is a copy of $\wedge_{(2)} \times \wedge_2$ (with five marked squares ending in (j, ι)), together with the marked square ending in ι'', displayed above,

(c) a coherent isomorphism $i: \iota' \to \iota''$ which links these two objects, so that each of them becomes a colimit of the inclusion $\wedge_{(2)} \times \wedge_{(2)} \to \wedge_{2\times2}$.

The two symmetric procedures correspond to two functors $m', m'': \wedge^2 \to \wedge_{2\times2}$, whose image is displayed below

$$
\begin{array}{ccc}
00 \to \iota0 \leftarrow 10 \\
\downarrow \quad \downarrow \quad \downarrow \\
0\iota \to \iota' \leftarrow 1\iota \\
\uparrow \quad \uparrow \quad \uparrow \\
01 \to \iota1 \leftarrow 11 \quad {}_{m'}
\end{array}
\qquad
\begin{array}{ccc}
00 \to \iota0 \leftarrow 10 \\
\downarrow \quad \downarrow \quad \downarrow \\
0\iota \to \iota'' \leftarrow 1\iota \\
\uparrow \quad \uparrow \quad \uparrow \\
01 \to \iota1 \leftarrow 11 \quad {}_{m''}
\end{array}
\qquad
\begin{array}{c}
\bullet \xrightarrow{1} \\
\downarrow 2 \\
{}
\end{array}
$$

$$(B.77)$$

They are connected by a natural isomorphism, the basic *formal inter-changer*, all of whose components are identities except the central one

$$\chi\colon m' \to m''\colon \wedge^2 \to \wedge_{2\times2}, \qquad \chi(\iota\iota) = i\colon \iota' \to \iota''. \tag{B.78}$$

Now a consistent matrix $\left(\begin{smallmatrix} x & y \\ z & u \end{smallmatrix}\right)$ of double cospans in $\mathbf{C}$ gives a functor

$$\left[\begin{smallmatrix} x & y \\ z & u \end{smallmatrix}\right]\colon \wedge_{2\times2} \to \mathbf{C}$$

(which takes all marked squares to distinguished pushouts) and a transversal isomorphism, the 12-interchange comparison

$$\chi(x, y, z, u) = \left[\begin{smallmatrix} x & y \\ z & u \end{smallmatrix}\right].\chi\colon (x +_1 y) +_2 (z +_1 u) \to (x +_2 z) +_1 (y +_2 u). \tag{B.79}$$

Applying the functor $(-)^n_i$ we get a natural transformation, for $1 \leqslant i < n$

$$\chi_i = (\chi)^n_i\colon (m')^n_i \to (m'')^n_i\colon \wedge^n \to \overset{\bullet}{\wedge}^{i-1} \times \wedge_{2\times2} \times \wedge^{n-i-1} \tag{B.80}$$
$$(\textit{formal i-interchanger}),$$

which gives the interchange of the cubical compositions in directions i and $i + 1$. Each ij-interchanger can be obtained by transpositions.

B4.6 Coherence

Finally the coherence axioms hold, as the terms of each diagram in 6.4.3, 6.4.4 are computed by different systems of pushouts, which end up with various constructions of the same colimit in $\mathbf{C}$, and therefore are linked by coherent isomorphisms.

B5 Symmetries in weak multiple categories

We come back to examining weak multiple categories.

Making precise what we anticipated in Chapter 6, a weak multiple category A is made *symmetric* by assigning an action of the group $\underline{\mathsf{S}}$ of 'finite permutations' of $\mathbb{N}$. If this is possible, A is necessarily *of symmetric cubical type*, up to isomorphism. With respect to weak symmetric cubical categories, the present notion has the advantage that indices need not be normalised and shifted.

The examples $\mathsf{Span}(\mathbf{C})$ and $\mathsf{Cosp}(\mathbf{C})$ are here revisited in this sense. We end by showing that the chiral multiple category $\mathsf{S}_{-\infty}\mathsf{C}_\infty(\mathbf{C})$ can be equipped with a *partial symmetric structure*, which permutes separately spans and cospans.

B5.1 A group of permutations

We write as $\underline{S}$ the group of *finite* permutations of the set $\mathbb{N}$ *pointed at* 0. In other words an element $s \in \underline{S}$ is a bijection $\mathbb{N} \to \mathbb{N}$ which has a finite, positive *support*

$$\mathbf{i}_s = \{i \in \mathbb{N} \mid s(i) \neq i\} \subset \mathbb{N}^*. \tag{B.81}$$

It follows that s induces a permutation $\hat{s} \colon \mathbf{i}_s \to \mathbf{i}_s$ with no fixed points, and is the identity outside. (In fact, by the injectivity of s, if $s(i) = j \neq i$ it follows that $s(j) \neq j$; therefore s has an injective restriction $\mathbf{i}_s \to \mathbf{i}_s$, which must be a bijection of finite sets.)

$\underline{S}$ can also be defined as the subgroup of permutations of $\mathbb{N}$ generated by the elementary transpositions s_i for $i > 0$, where $s_i \in \underline{S}$ is the transposition that (only) interchanges i and $i+1$. (For $s \in \underline{S}$, take a finite interval $\mathbf{j}$ of $\mathbb{N}$ containing $\mathbf{i}_s$; then the restriction $\mathbf{j} \to \mathbf{j}$ of s is generated by the elementary transpositions s_i restricted to $\mathbf{j}$, for $i, i+1 \in \mathbf{j}$.)

B5.2 Definition

A weak multiple category A is made *symmetric* by a left action of the group $\underline{S}$ consistent with faces, degeneracies, compositions and comparisons, according to the following axioms. We assume that $s \in \underline{S}$, $s(i) = j$, $s(i') = j'$ and $x, y, z, u \in A_{\mathbf{i}}$ ($\mathbf{i} \subset \mathbb{N}$); we introduce $\chi_{ji} = (\chi_{ij})^{-1}$ for $i < j$.

(sym.0) The permutation s takes an $\mathbf{i}$-cell $x \in A_{\mathbf{i}}$ to an $s(\mathbf{i})$-cell sx which only depends on x and the restriction of s to $\mathbf{i}$,

(sym.1)	$s(\partial_i^\alpha x) = \partial_j^\alpha(sx)$	$(i \in \mathbf{i})$,
(sym.2)	$s(e_i x) = e_j(sx)$	$(i \notin \mathbf{i})$,
(sym.3)	$s(x +_i y) = sx +_j sy$	$(i \in \mathbf{i})$,
(sym.4)	$s(\lambda_i x) = \lambda_j(sx)$	$(i \in \mathbf{i} \subset \mathbb{N}^*)$,
(sym.5)	$s(\rho_i x) = \rho_j(sx)$	$(i \in \mathbf{i} \subset \mathbb{N}^*)$,
(sym.6)	$s(\kappa_i(x, y, z)) = \kappa_j(sx, sy, sz)$	$(i \in \mathbf{i} \subset \mathbb{N}^*)$,
(sym.7)	$s(\chi_{ii'}(x, y, z, u)) = \chi_{jj'}(sx, sy, sz, su)$	$(i, i' \in \mathbf{i} \subset \mathbb{N}^*)$.

These axioms can be equivalently written using the elementary transpositions s_i. For instance (sym.1) can be replaced with:

(sym.1') $\quad s_i \partial_i^\alpha = \partial_{i+1}^\alpha s_i, \qquad \partial_j^\alpha s_i = s_i \partial_j^\alpha \qquad (j \neq i, i+1).$

B5.3 Cubical spans revisited

All this is better understood from the examples. The weak multiple category $\mathsf{Span}(\mathbf{C})$ can be easily reformulated in the form of a *symmetric weak multiple category* $\mathsf{Span}'(\mathbf{C})$, that will be shown to be isomorphic to the former.

While the construction of $\mathsf{Span}(\mathbf{C})$ is based on the cartesian powers $\vee^n$ of the formal-span category $\vee$, we use now, for every positive multi-index $\mathbf{i}$, the category $\vee^{\mathbf{i}} = \mathbf{Cat}(\mathbf{i}, \vee)$, i.e. the (finite) category of functors $t \colon \mathbf{i} \to \vee$ and their natural transformations, where $\mathbf{i} = \{i_1, ..., i_n\}$ is viewed as a discrete category; of course $\vee^{\mathbf{i}}$ is canonically isomorphic to its normalised form $\vee^n = \vee^{[n]}$.

The weak multiple category $\mathsf{A} = \mathsf{Span}'(\mathbf{C})$ has transversal components

$$\mathrm{tv}_{\mathbf{i}}(\mathsf{Span}'(\mathbf{C})) = \mathbf{Cat}(\vee^{\mathbf{i}}, \mathbf{C}) \qquad (\mathbf{i} \subset \mathbb{N}^*). \tag{B.82}$$

Thus an $\mathbf{i}$-cube is a functor $x \colon \vee^{\mathbf{i}} \to \mathbf{C}$ and an $\mathbf{i}$-map is a natural transformation $f \colon x \to y \colon \vee^{\mathbf{i}} \to \mathbf{C}$ of these functors. There are obvious geometric faces and degeneracies

$$\partial_i^\alpha \colon A_{\mathbf{i}} \rightleftarrows A_{\mathbf{i}|i} \colon e_i \qquad (i \in \mathbf{i}, \alpha = \pm), \tag{B.83}$$

satisfying the multiple relations. (The degeneracy derives from the restriction $\vee^{\mathbf{i}} \to \vee^{\mathbf{i}|i}$, whose two sections $\vee^{\mathbf{i}|i} \to \vee^{\mathbf{i}}$ determined by $\alpha = 0, 1$ give the faces). The remaining structure is defined as in Section 6.4.

Now the weak multiple category $\mathsf{A} = \mathsf{Span}'(\mathbf{C})$ is made symmetric by the following left action of the group $\underline{\mathsf{S}}$. Every permutation $s \in \underline{\mathsf{S}}$ induces, for every positive multi-index $\mathbf{i} = \{i_1, ..., i_n\} \subset \mathbb{N}^*$, the following isomorphisms of categories

$$
\begin{aligned}
s' &\colon \mathbf{i} \to s(\mathbf{i}) = \mathbf{j}, & i &\mapsto s(i), \\
s^* &\colon \vee^{\mathbf{j}} \to \vee^{\mathbf{i}}, & t &\mapsto s^*(t) = t.s', \\
s_* &\colon \mathbf{Cat}(\vee^{\mathbf{i}}, \mathbf{C}) \to \mathbf{Cat}(\vee^{\mathbf{j}}, \mathbf{C}), & x &\mapsto sx = x.s^*.
\end{aligned}
\tag{B.84}
$$

Thus $s \in \underline{\mathsf{S}}$ takes an $\mathbf{i}$-cube $x \in A_{\mathbf{i}}$ to a $\mathbf{j}$-cube sx which only depends on x and the restriction s' of s to $\mathbf{i}$; similarly an $\mathbf{i}$-map $f \in A_{0\mathbf{i}}$ is taken to a $\mathbf{j}$-map. The other axioms (sym.1-7) of B5.2 are also satisfied.

B5.4 Cubical cospans revisited

Similarly the weak multiple category $\mathsf{Cosp}(\mathbf{C})$ can be reformulated in the form of a *symmetric weak multiple category* $\mathsf{Cosp}'(\mathbf{C})$, transversally dual to $\mathsf{Span}'(\mathbf{C}^{\mathrm{op}})$. It has transversal components

$$\mathrm{tv}_{\mathbf{i}}(\mathsf{Cosp}'(\mathbf{C})) = \mathbf{Cat}(\wedge^{\mathbf{i}}, \mathbf{C}) \qquad (\mathbf{i} \subset \mathbb{N}^*), \tag{B.85}$$

where $\wedge = \vee^{\mathrm{op}}$ is the formal-cospan category. An **i**-cube is now a functor $x\colon \wedge^{\mathbf{i}} \to \mathbf{C}$.

B5.5 The associated cubical type

Now, if we start from an arbitrary symmetric weak multiple category A, it is easy to define an associated weak multiple category $\mathsf{A}^{\square}$ of symmetric cubical type, with transversal components 'of normalised form'

$$\mathrm{tv}_{\mathbf{i}}(\mathsf{A}^{\square}) = \mathrm{tv}_{[n]}\mathsf{A} \qquad (\mathbf{i} \subset \mathbb{N}^*). \tag{B.86}$$

If $s \in \underline{\mathrm{S}}$ is any permutation that induces the order-preserving bijection $[n] \to \mathbf{i}$, the action of $\underline{\mathrm{S}}$ induces an isomorphism

$$\mathrm{tv}_{\mathbf{i}}(\mathsf{A}^{\square}) \to \mathrm{tv}_{\mathbf{i}}\mathsf{A}, \qquad (f\colon x \to y) \mapsto (sf\colon sx \to sy), \tag{B.87}$$

which is the transversal **i**-component of an isomorphism $\mathsf{A}^{\square} \to \mathsf{A}$ of weak multiple categories.

This shows that $\mathsf{Span}'(\mathbf{C})$ is canonically isomorphic to $(\mathsf{Span}'(\mathbf{C}))^{\square} = \mathsf{Span}(\mathbf{C})$, and $\mathsf{Cosp}'(\mathbf{C})$ to $\mathsf{Cosp}(\mathbf{C})$.

B5.6 Partial symmetric structures

The 'unbounded' chiral multiple category $\mathsf{A} = \mathsf{S}_{-\infty}\mathsf{C}_{\infty}(\mathbf{C})$ was introduced in 6.6.4. It is indexed by the ordered set $\mathbb{Z}$ of all integers, pointed at 0; its arrows are spans in each negative direction, ordinary morphisms in direction 0 and cospans in positive directions.

We cannot permute negative and positive indices, which are related to different kinds of arrows; but we can equip A with a *partial symmetric structure*, which – loosely speaking – permutes separately spans and cospans. This structure is a left action of the group T of finite permutations of the pointed set $\mathbb{Z}$ that preserve negative numbers and positive numbers; T is generated by the elementary transpositions $s_i\colon \mathbb{Z} \to \mathbb{Z}$ that permute i and $i+1$, for $i < -1$ or $i \geqslant 1$.

A multi-index $\mathbf{ij} = \mathbf{i} \cup \mathbf{j} \subset \mathbb{Z}^*$ has a negative part $\mathbf{i}$ and a positive part $\mathbf{j}$. An $\mathbf{ij}$-cube $x \in A_{\mathbf{ij}}$ can be defined as a functor $x\colon \vee^{\mathbf{i}} \times \wedge^{\mathbf{j}} \to \mathbf{C}$ (by a reformulation of cubes similar to the previous ones, above). A permutation $s \in T$ induces the following isomorphisms of categories

$$s'\colon \mathbf{i} \to s(\mathbf{i}) = \mathbf{i}', \qquad\qquad s''\colon \mathbf{j} \to s(\mathbf{j}) = \mathbf{j}'$$

$$s^*\colon \vee^{\mathbf{i}'} \to \vee^{\mathbf{i}} \ (t \mapsto t.s'), \qquad s^{**}\colon \wedge^{\mathbf{j}'} \to \wedge^{\mathbf{j}} \ (t \mapsto t.s''), \tag{B.88}$$

$$s_*\colon A_{\mathbf{ij}} \to A_{\mathbf{i}'\mathbf{j}'} \qquad\qquad x \mapsto sx = x(s^* \times s^{**}).$$

Thus $s \in T$ takes an **ij**-cube $x \in A_{ij}$ to an s(**ij**)-cube sx which only depends on x and the restriction of s to **ij**; similarly an **ij**-map is taken to an s(**ij**)-map.

In the same way the chiral multiple category $S_p C_q(\mathbf{C})$, whose geometric i-directed arrows are spans of $\mathbf{C}$ for $0 < i \leqslant p$ and cospans of $\mathbf{C}$ for $p < i \leqslant p + q$, has a partial symmetric structure over the subgroup $\underline{S}_{pq}$ of $\underline{S}$ formed of those permutations that preserve $[1, p]$ and $[p + 1, p + q]$. This also makes sense for $q = \infty$.

B6 *Complements on symmetric cubical sets

We end this appendix by defining the symmetric tensor product of symmetric cubical sets (which can also be applied to symmetric cubical categories).

After the internal hom of Section B3, this completes the symmetric monoidal closed structure of s**Cub**; it also yields a cylinder functor, by tensoring with an obvious 'standard interval'. However, these computations are complicated: working with the path functor is more effective.

B6.1 *Remarks*

The *ordinary* tensor product $X \otimes Y$ of two symmetric cubical sets cannot be directly equipped with transpositions. For instance, if x, y are 2-cubes in X and Y, we might define $s_1(x \otimes y) = (s_1 x) \otimes y$ and $s_3(x \otimes y) = x \otimes (s_1 y)$, but $s_2(x \otimes y)$ cannot be obtained from the transposition s_1 of X_2 or Y_2, and has to be formally introduced.

Therefore we shall define a *symmetric tensor product* $X \hat{\otimes} Y$ whose n-component is the free S_n-set on $X \otimes Y$, modulo the identifications exemplified above.

The symmetric group S_n will be viewed as the set of all bijections $u \colon [n] \to [n]$, where $[n] = \{1, ..., n\}$, since this set is here more convenient than the cardinal $\mathbf{n} = \{0, ..., n - 1\}$.

B6.2 *The action of permutations*

We have defined s**Cub** as the category of functors $X \colon \underline{I}_s{}^{\mathrm{op}} \to \mathbf{Set}$, in B2.1. The symmetric cubical site $\underline{I}_s$ is now realised as the subcategory of **Set** consisting of the elementary cubes $2^{[n]}$, together with the maps $2^{[m]} \to 2^{[n]}$ which delete some coordinates, permute the remaining ones and insert some 0's and 1's.

Let $X \colon \underline{I}_s{}^{\mathrm{op}} \to \mathbf{Set}$ be a symmetric cubical set. It will be useful to give an explicit description of the *left* action of the symmetric group S_n

on the component $X_n = X(2^{[n]})$. In fact, the group S_n (of permutations $u \colon [n] \to [n]$) acts contravariantly on the set $2^{[n]} = \mathbf{Set}([n], 2)$ and then covariantly on the set X_n

$$\hat{u} \colon 2^{[n]} \to 2^{[n]}, \qquad \hat{u} \colon t \mapsto t.u \qquad (t \colon [n] \to 2).$$
$$u \colon X_n \to X_n, \qquad ux = (X(\hat{u}))(x) \qquad (x \in X_n). \tag{B.89}$$

The permutation $u \in S_n$ acts as follows, on the face $\partial_1^\alpha \colon X_n \to X_{n-1}$ and the degeneracy $e_1 \colon X_n \to X_{n+1}$

$$\partial_1^\alpha u = u' \partial_i^\alpha, \qquad\qquad e_1 u = u'' e_1, \tag{B.90}$$

where $u'' = \mathrm{id} + u \colon [n+1] \to [n+1]$, while $u' \colon [n-1] \to [n-1]$ is defined below, letting $i = u^{-1}(1)$

$$u'(j) = u(j) - 1 \ \text{ for } j < i, \qquad u'(j) = u(j+1) - 1 \ \text{ for } j \geqslant i.$$

B6.3 Theorem and Definition

The category s**Cub** *of symmetric cubical sets has a symmetric monoidal closed structure, whose internal hom is the functor* sCUB *defined in B2.5.*

The n-th component of the symmetric tensor product $X \hat{\otimes} Y$ of symmetric cubical sets

$$(X \hat{\otimes} Y)_n = S_n((X \otimes Y)_n) / \equiv_n, \tag{B.91}$$

is a quotient of the free S_n-set $S_n((X \otimes Y)_n)$ generated by the n-th component of the ordinary tensor product $X \otimes Y$, and containing all the formal permutations $u(x \otimes y)$, for $u \in S_n$. The quotient is taken modulo the congruence $\equiv_n$ of S_n-sets generated by the following 'identifications'

$$s_i(x \otimes y) = s_i x \otimes y \quad \text{for } i < p = \dim(x),$$
$$s_i(x \otimes y) = x \otimes s_{i-p} y \quad \text{for } i > p. \tag{B.92}$$

Faces and degeneracies in direction 1 are defined as follows, letting $\partial_1^\alpha u = u' \partial_i^\alpha$ and $e_1 u = u'' e_1$ (with u', u'' as at the end of B6.4)

$$\partial_1^\alpha[u.(x \otimes y)] = [u'.\partial_i^\alpha(x \otimes y)] = [u'.((\partial_i^\alpha x) \otimes y))], \quad \text{for } i \leqslant p,$$
$$\partial_1^\alpha[u.(x \otimes y)] = [u'.\partial_i^\alpha(x \otimes y)] = [u'.(x \otimes (\partial_{i-p}^\alpha y))], \quad \text{for } i > p, \tag{B.93}$$
$$e_1[u.(x \otimes y)] = [u''.e_1(x \otimes y)] = [u''.(e_1 x) \otimes y)].$$

This completes the definition of the symmetric cubical set $X \hat{\otimes} Y$, in the reduced form of B2.2

$$X \hat{\otimes} Y = (((X \hat{\otimes} Y)_n), (\partial_1^\alpha), (e_1), (s_i)). \tag{B.94}$$

Proof The verification of the axioms (B.33) is left to the reader. To obtain the exponential law, as a natural bijection

$$\mathbf{sCub}(X \hat{\otimes} A, Y) \to \mathbf{sCub}(X, \mathrm{sCUB}(A, Y)), \qquad (\mathrm{B}.95)$$

let us take a morphism $f = (f_n): X \hat{\otimes} A \to Y$. Its n-component f_n decomposes into a family of mappings

$$f_{pq}: S_n(X_p \times A_q) \to Y_n \qquad (p + q = n),$$

consistent with the action of S_n, the equivalence relation $\sim_n$ of (B.5) and the new equivalence relation $\equiv_n$. Their restrictions

$$f_{pq}: X_p \times A_q \to Y_n,$$

amount to mappings $g_{pq}: X_p \to \mathbf{Set}(A_q, Y_{p+q})$. Keeping p fixed, we get mappings

$$g_p = (g_{pq}): X_p \to \mathbf{sCub}(A, P^p Y) = \mathrm{sCUB}_p(A, Y),$$

which form a morphism of symmetric cubical sets

$$g = (g_p): X \to \mathrm{sCUB}(A, Y).$$

□

B6.4 The cylinder functor

We have already considered the representable symmetric cubical set $\underline{u} = Y(2)$ freely generated by one 1-cube u (see (B.42)), and observed that $\mathrm{sCUB}(\underline{u}, Y) = PY$.

The cylinder functor

$$I: \mathbf{sCub} \to \mathbf{sCub}, \qquad I(X) = X \hat{\otimes} \underline{u}, \qquad (\mathrm{B}.96)$$

is thus left adjoint to the path functor $P: \mathbf{sCub} \to \mathbf{sCub}$

$$\mathbf{sCub}(X \hat{\otimes} \underline{u}, Y) = \mathbf{sCub}(X, \mathrm{sCUB}(\underline{u}, Y)) = \mathbf{sCub}(X, P(Y)). \qquad (\mathrm{B}.97)$$

Appendix C
Solutions and hints

Easy exercises and exercises marked with * are often left to the reader.

C1 Exercises of Chapter 1

C1.1 *Exercises of 1.2.6* (Equivalences and skeletons)

For (a), (b), (f) and (g) one can use the characterisation of equivalences by property 1.2.5(iii).

(c) Use the category $\mathbf{T}$ of 1.1.8(a).

(d) The construction is simpler if the functor U is essentially surjective on objects, as in the previous case: then we construct a category $\mathbf{C}'$ with the same objects as $\mathbf{A}$ and hom-sets $\mathbf{C}'(X,Y) = \mathbf{C}(UX,UY)$. Otherwise one can add to $\mathbf{C}'$ the objects of $\mathbf{C}$, with suitable morphisms.

(h) It is sufficient to prove that an equivalence F between two skeletal categories is an isomorphism. In fact F reflects isomorphisms, because it is full and faithful (see 1.2.3(c)), and is essentially surjective on objects: it follows that it is bijective on objects and morphisms.

(i) The definition of G on the morphisms of $\mathbf{C}$ is determined.

C1.2 *Exercises of 1.3.5* (Pullbacks and pushouts)

(c) We have seen in (a) that the pullback of a symmetric pair (f,f) can only be a symmetric pair (h,h) when f is mono. In particular, if $f\colon X \to \top$ takes values in the terminal object, the pullback will be the pair of projections $X \times X \to X$, which only coincide in very particular situations (and indeed when $X \to \top$ is mono).

(g) In **Top** our property holds in two main, well-known situations: when the subspaces X_i are both open in X, or both closed. The reader will note that the common way of verifying the continuity of a mapping $f: X \to Y$ on the subspaces of a finite closed cover of X is a consequence.

*(i) Proving this fact in **Set** will show how to proceed in the general case.

C1.3 Exercises of 1.5.3

(b) For each forgetful functor $U: \mathbf{A} \to \mathbf{X}$ of the list (1.50), the required universal arrows $(A, \eta: X \to UA)$ are constructed, or sketched, in 1.2.9.

(c) DX (resp. CX) is the set X with the discrete (resp. indiscrete, or coarsest) topology.

(d) Now DX is the set X with the discrete order $x = x'$, while CX is the set X with the indiscrete, or chaotic, preorder: $x, x' \in X$. Here D has a left adjoint $\pi_0: \mathbf{pOrd} \to \mathbf{Set}$, where $\pi_0(X)$ is the quotient of the preordered set X modulo the equivalence relation spanned by the relation: $x \prec x'$.

(e) Similarly, DX is the discrete category on the set X (only having one identity arrow 1_x for each $x \in X$) and CX is the indiscrete category on X (having precisely one arrow $x \to x'$ for each pair of elements of X).

The functor π_0 takes a category $\mathbf{C}$ to its set of connected components, namely the quotient of $\mathrm{Ob}\mathbf{C}$ modulo the equivalence relation generated by the existence of a morphism $x \to x'$. This quotient is the singleton if $\mathbf{C}$ is not empty, and any pair of objects is linked by a chain of morphisms $x_0 \to x_1 \leftarrow x_2 \ldots \to x_n$.

(f) The preorder relation $i \prec j$ in $\mathrm{Ob}\mathbf{S}$ is defined by the existence of an arrow $i \to j$ in $\mathbf{S}$.

(g) The reader can easily give an analytic proof, proving that a universal cocone $(L, (u_i: X_i \to L))$ of the functor $X: \mathbf{I} \to \mathbf{C}$ is taken to a cocone $(FL, (Fu_i: FX_i \to FL))$ of the functor $FX: \mathbf{I} \to \mathbf{D}$, which is also universal.

*As a more formal alternative, one can use the characterisation of colimits as representative objects, in 1.3.8, completing the following outline.

We have an adjunction $F \dashv G: \mathbf{C} \rightharpoonup \mathbf{D}$, and a functor $X: \mathbf{I} \to \mathbf{C}$. By 1.3.8 its colimit L represents the covariant functor

$$H = \mathbf{C}^{\mathbf{I}}(X, \Delta(-)): \mathbf{C} \to \mathbf{Set}$$

via the morphism $u: X \to \Delta L$ in $\mathbf{C}^{\mathbf{I}}$. Then FL represents the covariant functor

$$K = \mathbf{D}^{\mathbf{I}}(FX, \Delta(-)): \mathbf{D} \to \mathbf{Set}$$

via the morphism $Fu: FX \to \Delta FL$, by composing three functorial isomorphisms in the variable Y

$$\mathbf{D}(FL, Y) \to \mathbf{C}(L, GY) \to \mathbf{C}^{\mathbf{I}}(X, \Delta GY)$$
$$= \mathbf{C}^{\mathbf{I}}(X, G^{\mathbf{I}}(\Delta Y)) \to \mathbf{D}^{\mathbf{I}}(F^{\mathbf{I}}(X), \Delta Y) = K(Y).$$

The first comes from the adjunction $F \dashv G$, the second from L representing the functor $H: \mathbf{C} \to \mathbf{Set}$, the third from the extended adjunction $F^{\mathbf{I}} \dashv G^{\mathbf{I}}$ (see 1.5.2(c)).*

(h) The functor $D: \mathbf{Set} \to \mathbf{Top}$ does not preserve infinite products and cannot have a left adjoint, while C does not preserve (even binary) sums and cannot have a right adjoint.

(i) The initial object is not preserved, by any of the functors of the list (1.50).

C1.4 Exercises of 1.5.8

(a) The set of cones $\mathbf{Set}^{\mathbf{I}}(\Delta\{*\}, X)$ can be rewritten as $\mathbf{Set}(\{*\}, \operatorname{Lim} X)$, by the adjunction $\Delta \dashv \operatorname{Lim}$.

(b) By 1.5.4(e), Δ has a left adjoint S (resp. a right adjoint P) if and only if $\mathbf{C}$ has binary sums (resp. products)

$$
\begin{array}{llll}
S: \mathbf{C}^2 \to \mathbf{C}, & S(X, Y) = X + Y & (S \dashv \Delta), \\
P: \mathbf{C}^2 \to \mathbf{C}, & P(X, Y) = X \times Y & (\Delta \dashv P).
\end{array}
\tag{C.1}
$$

(c) If $\mathbf{C}$ is a category of modules, the functors S and P coincide: the direct sum $B(X, Y) = X \oplus Y$ is at the same time

- left adjoint to Δ, with a unit $\eta: (X, Y) \to (X \oplus Y, X \oplus Y)$ giving the injections of the sum,

- right adjoint to Δ, with a counit $\varepsilon: (X \oplus Y, X \oplus Y) \to (X, Y)$ giving the projections of the product.

(d) In any category $\mathbf{C}$ there is a natural isomorphism

$$\mathbf{C}(\Sigma_{i \in I} X, Y) \to \mathbf{C}(X, \Pi_{i \in I} Y),$$

whenever sums and products indexed by the set I exist.

(f) The universal arrow from the embedding $U: \mathbf{Gp} \to \mathbf{Mon}$ to a monoid M is the embedding $\varepsilon: U\operatorname{Inv}(M) \to M$ of the subgroup of invertible elements of M.

*(g) The reflector $\mathbf{Mon} \to \mathbf{Gp}$ can be obtained taking the free group $F|M|$ on the underlying set of a monoid, and then its quotient modulo the

congruence of groups that forces the embedding $|M| \to F|M|$ to become a homomorphism of monoids.

Alternatively, one can apply a theorem saying that any forgetful functor between varieties of algebras (forgetting structure, or properties, or both) has a left adjoint, like Theorem 4.4.5 in [G11].

C1.5 Exercises of 1.6.6

(a) By a zeroary operation, under no axioms; by a binary operation, under one axiom of associativity.

(c) For a functor $H \colon \mathbf{I} \to \mathbf{X}^T$, consider the limit $(L, (u_i \colon L \to G^T Hi)_i)$ of the functor $G^T H \colon \mathbf{I} \to \mathbf{X}$. Prove that there is a unique T-algebra $(L, a \colon TL \to L)$ such that every $u_i \colon L \to G^T Hi$ is a morphism of T-algebras $(L, a) \to Hi$. Prove that this family is the universal cone of H.

(f)–(h) Following the way suggested in (h), we define the functor

$$F \colon \mathbf{Set} \to \mathbf{C},$$
$$F(X) = (\mathcal{P}X, \subset), \qquad F(f) = f_* \colon \mathcal{P}X \to \mathcal{P}Y. \tag{C.2}$$

Now $F \dashv U$, with unit and counit computed as follows, on a set X and a complete lattice L

$$\eta X \colon X \to UFX = \mathcal{P}X, \qquad \eta X(x) = \{x\} \qquad (x \in X),$$
$$\varepsilon L \colon FUL \to L, \qquad \mathcal{A} \mapsto \vee \mathcal{A} \qquad (\mathcal{A} \subset L). \tag{C.3}$$

This adjunction gives the monad $UF = \mathcal{P}$, with the unit above and the required multiplication

$$\mu X(\mathcal{A}) = U\varepsilon FX(\mathcal{A}) = \bigcup_{A \in \mathcal{A}} A.$$

Finally the comparison $K \colon \mathbf{C} \to \mathbf{Set}^T$ sends a complete lattice L to its underlying set $|L|$, equipped with the structure $U\varepsilon L \colon UFUL \to UL$ that acts as in (C.3). Plainly, K is an isomorphism.

C1.6 Exercises of 1.6.8

(a) Take two consecutive morphisms $f^\sharp \colon X \nrightarrow Y$ and $g^\sharp \colon Y \nrightarrow Z$, and a fixed element $x \in X$. With some abuse of notation we can write $f^\sharp(x) = \sum_y \lambda_y y$, as a formal linear combination of elements in Y with quasi-null coefficients in Z (the *weights*). For each of them $g^\sharp(y) = \sum_z \mu_{yz} z$, and we obtain

$$g^\sharp f^\sharp(x) = \sum_{y,z} \lambda_y \mu_{yz} z, \tag{C.4}$$

as a quasi-null linear combination of elements in Z.

(b) We have to verify that $L\colon \mathbf{X}_T \to \mathbf{X}^T$ is full and faithful, which means that the mapping

$$L = L_{X,Y}\colon \mathbf{X}_T(X,Y) \to \mathbf{X}^T(LX, LY), \quad L(f^\sharp) = \mu Y.Tf, \qquad \text{(C.5)}$$

is bijective, for all X, Y in $\mathbf{X}$. In fact there is an inverse mapping

$$M\colon \mathbf{X}^T(LX, LY) \to \mathbf{X}_T(X,Y), \quad M(g) = (g.\eta X)^\sharp \colon X \to Y,$$

$$ML(f^\sharp) = M(\mu Y.Tf) = (\mu Y.Tf.\eta X)^\sharp = (\mu Y.T\eta Y.f)^\sharp = f^\sharp, \qquad \text{(C.6)}$$

$$LM(g) = L(g.\eta X)^\sharp = \mu Y.Tg.T\eta X = g.\mu X.T\eta X = g.$$

C1.7 Exercises of 1.7.6

(a) It is sufficient to recall that a monad on an ordered set is always idempotent, as we have already seen in 1.6.3.

(b) In fact $FG^T K = FG$ is isomorphic to the identity via ε. Similarly $KFG^T = F^T G^T$ is isomorphic to the identity via ε^T, since every component $\varepsilon^T(X,a) = a\colon (TX, \mu X) \to (X,a)$ is invertible.

(c) The embedding $\mathbf{Rng} \subset \mathbf{Rng}'$ has a left adjoint $(-)^+\colon \mathbf{Rng}' \to \mathbf{Rng}$, that universally adds a unit. It is a well-known construction, based on the abelian group $R \oplus \mathbb{Z}$ equipped with a suitable product that makes $(0,1)$ the unit of R^+

$$(\lambda, h).(\mu, k) = (\lambda\mu + h\mu + k\lambda, hk). \qquad \text{(C.7)}$$

An algebra $(R, a\colon R^+ \to R)$ on $\mathbf{Rng}'$ detects a unit in R.

Note that this embedding is not full, and the corresponding adjunction does not yield an idempotent monad: the ring R^{++} will have a new unit added to R^+.

(The reader may know that in Algebra it is often preferred to define R^+ by 'adding a unit when it does not already exist in the ring'; this procedure is idempotent, but is not a functor.)

(d) The adjunction $D \dashv U$ gives a trivial monad $UD = 1$ on **Set**, whose algebras are just sets.

On **Top** it gives an idempotent comonad:

$$S = DU\colon \mathbf{Top} \to \mathbf{Top}, \qquad \varepsilon X\colon DU(X) \to X, \qquad U\varepsilon = 1, \qquad \text{(C.8)}$$

(even strictly idempotent, see 1.7.8). The coalgebraic objects X (for which εX is a homeomorphism) are the discrete spaces. In fact, D embeds **Set** in **Top** as a full, replete, coreflective subcategory.

The adjunction $U \dashv C$ gives a trivial comonad on **Set**, and an idempotent monad on **Top**; the latter corresponds to the embedding $C\colon$ **Set** $\to$ **Top** as the full, replete, reflective subcategory of indiscrete spaces.

This exercise can be reviewed at the light of 1.7.7(c).

(e) We are simply interpreting point (d) of Theorem 1.7.3, for the abelianisation monad. The category $\mathbf{Gp}^T$ can be identified with **Ab**. In $\mathbf{Gp}_T$ an object is a group, but a morphism $f^\sharp\colon X \to Y$ is 'represented by' a homomorphism $f\colon X \to Y^{\mathrm{ab}}$, and amounts to a homomorphism $X^{\mathrm{ab}} \to Y^{\mathrm{ab}}$ of abelian groups. Since the objects X and X^{ab} are isomorphic in $\mathbf{Gp}_T$, the latter is equivalent to $\mathbf{Ab} = \mathbf{Gp}^T$.

C1.8 Exercises of 1.8.3

(a) Kernels in these categories are obvious and well known. In $R\,\mathbf{Mod}$ and **Set.** every subobject is normal, while in **Gp** and $\mathbf{Rng}'$ the normal subobjects correspond to normal subgroups (in the usual sense) and bilateral ideals, respectively.

The natural cokernel of $f\colon A \to B$ in $R\,\mathbf{Mod}$ is the projection $B \to B/f(A)$. In **Gp** (resp. $\mathbf{Rng}'$) one takes the quotient of B modulo the normal subgroup (resp. the bilateral ideal) generated by $f(A)$. In **Set.** and **Top.** one has to collapse the subset $f(A)$ to the base point.

C2 Exercises of Chapter 2

C2.1 Exercises of 2.1.4

(b) It is sufficient to find a suitable endorelation of the set $\{0,1\}$.

(c) One proves that a sum $X = \sum X_i$ in **Set** is still a sum in RelSet, with the same injections $u_i\colon X_i \to X$. Then $(X, (u_i^\sharp\colon X \to X_i))$ is a product.

(d) In fact a relation of sets $u\colon X \to Y$ can be viewed as a mapping $X \to \mathcal{P}Y$. The composition of relations agrees with this interpretation

$$(vu)(x) = (\mu\mathcal{P}v.u)(x) = \bigcup \{v(y)\,|\,y \in u(x)\} \qquad (x \in X).$$

(e) Taking into account the description of $\mathbf{Set}^{\mathcal{P}}$ in 1.6.6, as the category of upper-complete semilattices, the full and faithful embedding

$$L\colon \mathbf{Set}_{\mathcal{P}} \to \mathbf{Set}^{\mathcal{P}}$$

takes a mapping $u\colon X \to \mathcal{P}Y$ to its join-preserving extension $L(u)\colon \mathcal{P}X \to \mathcal{P}Y$. In other words, a relation of sets $u\colon X \to Y$ can also be viewed as a mapping $\mathcal{P}X \to \mathcal{P}Y$ that preserves arbitrary unions.

(f) In the adjunction $F_{\mathcal{P}} \dashv G_{\mathcal{P}}$ associated to the category $\mathbf{Set}_{\mathcal{P}}$ of Kleisli algebras (in (1.70)), the functor $F_{\mathcal{P}}\colon \mathbf{Set} \to \mathrm{RelSet}$ is the embedding.

C2.2 Exercises of 2.1.8 (Partial continuous mappings)

(a) The functor R destroys topological information, namely the filter of the neighbourhoods of the base point. Therefore it does not reflect isomorphisms, and cannot be full and faithful (by a remark in 1.2.3(c)).

In fact, it is easy to see that R is not full. On the other hand, it is obviously faithful and surjective on objects.

C2.3 Exercises of 2.2.4

(a) The unit and counit of this adjunction are:

$$\eta X \colon X \to \mathbf{Set}(A, X \times A), \qquad x \mapsto (a \mapsto (x, a)),$$
$$\varepsilon Y \colon Y^A \times A \to Y, \qquad (h, a) \mapsto h(a). \tag{C.9}$$

(b) The canonical bijection is as in (a), after showing that a mapping $f \colon X \times A \to Y$ preserves orders if and only if it does in each variable.

(c) The proof is similar, if more complex.

(d), (e) If our category has a zero object, then $0 \times A = A$, for every object A. Therefore the product $- \times A$ preserves the initial object if and only if A is itself initial, i.e. a zero object.

(g) One adapts the construction of the tensor product of R-modules, in 2.2.1, replacing the ring R with the semiring $\mathbb{N}$.

(i) The abelian group $R \otimes_{\mathbb{Z}} S$ becomes a ring $R \otimes S$, with multiplication $(r \otimes s)(r' \otimes s') = rr' \otimes ss'$ and unit $1_R \otimes 1_S$. The unit object of the monoidal structure is the ring of integers.

(j) In $\mathbf{CRng}$, $R \otimes S$ is the categorical sum of R and S, with 'injections'

$$R \to R \otimes S, \quad r \mapsto r \otimes 1, \qquad S \to R \otimes S, \quad s \mapsto 1 \otimes s. \tag{C.10}$$

(These need not be mono: for instance, the ring $\mathbb{Z}/2 \otimes \mathbb{Z}/3$ is null.)

We have seen, at the end of 1.2.9, that the polynomial ring $\mathbb{Z}[X]$ is the free commutative ring on a set X 'of variables'. Since left adjoints preserve sums, the sum $\mathbb{Z}[X] \otimes \mathbb{Z}[Y]$ in $\mathbf{CRng}$ is the polynomial ring on the set $X + Y$, a categorical sum of sets.

The categorical sum in $\mathbf{Rng}$, called a 'free product of rings', is more complex: the fact that the elements $r \otimes 1$ and $1 \otimes s$ always commute in $R \otimes S$ readily shows that the morphisms (C.10) cannot give the sum of R and S in the general case.

Using again what we have seen at the end of 1.2.9, the polynomial ring $R = \mathbb{Z}[x]$ in one variable is – at the same time – the free ring and the free commutative ring on the singleton. Therefore the categorical sum $R * R$ in

Rng is the ring of 'non-commutative polynomials' on two variables (with integral coefficients), while the polynomial ring $\mathbb{Z}[x, y]$ on two variables is the categorical sum $R \otimes R$ in **CRng**.

(In a parallel way, we have the sum $\mathbb{Z} * \mathbb{Z}$ in **Gp**, versus the sum $\mathbb{Z} \oplus \mathbb{Z}$ in **Ab**.)

C2.4 Exercises of 2.2.6

(c) The inversion morphism $i\colon G \to G$ has to make the following diagram commutative:

$$
\begin{array}{ccccc}
G & \xrightarrow{(i,1)} & G^2 & \xleftarrow{(1,i)} & G \\
\downarrow & & \downarrow{m} & & \downarrow \\
G^0 & \xrightarrow{e} & G & \xleftarrow{e} & G^0
\end{array}
\tag{C.11}
$$

(d) In an ordered group, the inversion mapping necessarily reverses the order.

C2.5 Exercises of 2.3.7

(a) A 2-cell $\varphi\colon u \to v\colon X \nrightarrow Y$ of the bicategory Span**C** is a natural transformation $\varphi\colon u \to v\colon \vee \to \mathbf{C}$, where φ_0 and φ_1 are identities. It amounts to its central component $\varphi\iota\colon u(\iota) \to v(\iota)$, that has to form two commutative triangles with the given spans u and v

$$
\begin{array}{ccccc}
X & \xleftarrow{u'} & U & \xrightarrow{u''} & Y \\
 & \llap{\scriptstyle u'}\nwarrow & \downarrow{\varphi\iota} & \nearrow\rlap{\scriptstyle u''} & \\
 & & V & &
\end{array}
\tag{C.12}
$$

The vertical composition of cells is obvious (a vertical composition of natural transformations) and categorical. The horizontal composition of arrows and cells is computed by pullbacks; its comparison cells come from the universal property of the latter. A more detailed analysis will be given in 3.4.1.

(c) It is an easy consequence of the triangular equations.

C2.6 Exercises of 2.4.6

(a) The forgetful functor $\mathbb{R}_+(0, -)\colon \mathbb{R}_+ \to \mathbf{Set}$ gives a singleton on 0, and the empty set on any $\lambda > 0$. It corresponds to the functor

$$
q\colon \mathbb{R}_+ \to \mathbf{2}, \qquad q(0) = 1, \quad q(\lambda) = 0 \text{ for } \lambda > 0,
\tag{C.13}
$$

i.e. the truth-value of the condition $0 \geqslant \lambda$.

(b) The forgetful functor $\mathbf{Mtr} \to \mathbf{Cat}$ takes an L-metric space X to the category on the elements of the set $|X|$, with one arrow $x \to y$ when $d_X(x,y) = 0$ and none otherwise. It can be described as a functor with values in preordered sets

$$Q \colon \mathbf{Mtr} \to \mathrm{pOrd}, \qquad x \prec_Q y \iff d_X(x,y) = 0. \qquad \text{(C.14)}$$

C2.7 Exercises of 2.5.5

(a) The diagram below shows that the composed relation $f^\sharp f \colon X \nrightarrow X$ is the kernel pair $K \rightarrowtail X \times X$ of f (i.e. the pullback of the pair (f,f)), which is greater than 1_X

$$\text{(C.15)}$$

The property $f f^\sharp \leqslant 1_Y$ comes from the following diagram

(b) If $f \leqslant g$ in $\mathbf{C}(X,Y)$ then $g \leqslant g f^\sharp f \leqslant g g^\sharp f \leqslant f$.

*(c) This is more difficult to prove and not needed here. One can see [Grt], Proposition 4.12.

C2.8 Exercises of 2.5.8

Point (a) is obvious, and (b) is a consequence, since in all these cases the forgetful functor $\mathrm{Rel}(\mathbf{C}) \to \mathrm{RelSet}$ is faithful.

(c) Take for instance the two constant maps $h, k \colon \{0,1\} \to \{0,1\}$. Their pullback is $\emptyset$; this pullback square is (obviously) v-exact, but is not v*-exact.

On the other hand, there are two endorelations $\{0,1\} \nrightarrow \{0,1\}$ strictly containing the identity and not total; each of them gives a jointly monic span whose pushout is the singleton; this pushout square is v*-exact and not v-exact.

C3 Exercises of Chapter 3

C3.1 Exercises of 3.1.8

(a) The horizontal composition of double cells in the double category $\mathbb{Q}\mathbf{C}$

$$
\begin{array}{ccccc}
X & \xrightarrow{f} & X' & \xrightarrow{f'} & X'' \\
u\downarrow & \alpha & v\downarrow & \beta & \downarrow w \\
Y & \xrightarrow{g} & Y' & \xrightarrow{g'} & Y''
\end{array}
\qquad \varphi = \alpha \,|\, \beta,
\qquad\qquad (\text{C.16})
$$

is represented by the 2-cell $\varphi = g'\alpha.\beta f\colon wf'f \to g'vf \to g'gu$.

(b) We know, from 2.1.4(a), that an adjoint pair in the 2-category RelSet amounts to an ordinary mapping $f\colon X \to Y$, with $f \dashv f^{\sharp}$. Thus a double cell of $\mathbb{A}\mathrm{dj}(\mathrm{RelSet})$ is formed as in the left diagram below: u and v are relations, f and g are mappings, and $gu \leqslant vf$

$$
\begin{array}{ccc}
X & \overset{u}{\dashrightarrow} & X' \\
f\downarrow & \downarrow & \downarrow g \\
Y & \overset{}{\underset{v}{\dashrightarrow}} & Y'
\end{array}
\qquad\qquad
\begin{array}{ccc}
X & \xrightarrow{f} & Y \\
u\downarrow & \leqslant & \downarrow v \\
X' & \xrightarrow{g} & Y'
\end{array}
\qquad (\text{C.17})
$$

By symmetry with respect to the main diagonal (from X to Y'), this cell is transformed into a cell of the double category $\mathbb{R}\mathrm{elSet}$, as in the right diagram above. Since both structures are flat, this proves that $\mathbb{A}\mathrm{dj}(\mathrm{RelSet})$ is the *transpose* of $\mathbb{R}\mathrm{elSet}$, as defined in 3.2.2.

*(c) Everything works as in (b), using Exercise 2.5.5.

*(d) Take a horizontal composite $\varphi = \alpha \,|\, \beta$ in $\mathbb{A}\mathrm{dj}\mathbf{C}$. By definition, its covariant part $\varphi_{\bullet}$ is obtained by pasting the corresponding double cells $\alpha_{\bullet}$ and $\beta_{\bullet}$ in $\mathbb{Q}\mathbf{C}$, as above in (C.16)

$$
\alpha_{\bullet}\colon v_{\bullet}f \to gu_{\bullet}, \qquad \beta_{\bullet}\colon w_{\bullet}h \to kv_{\bullet},
$$
$$
\varphi_{\bullet} = (k\alpha_{\bullet})(\beta_{\bullet}f)\colon w_{\bullet}hf \to kgu_{\bullet}.
\qquad (\text{C.18})
$$

The contravariant parts $\alpha^{\bullet}$ and $\beta^{\bullet}$ are computed by means of the units and counits of the adjoint pairs u, v, w

$$
\alpha^{\bullet} = (fu^{\bullet} \to v^{\bullet}(v_{\bullet}f)u^{\bullet} \to v^{\bullet}(gu_{\bullet})u^{\bullet} \to v^{\bullet}g),
$$
$$
\beta^{\bullet} = (hv^{\bullet} \to w^{\bullet}(w_{\bullet}h)v^{\bullet} \to w^{\bullet}(kv_{\bullet})v^{\bullet} \to w^{\bullet}k).
\qquad (\text{C.19})
$$

Their pasting produces a 2-cell

$$
\psi = \beta^{\bullet}g.h\alpha^{\bullet}\colon hfu^{\bullet} \to hv^{\bullet}g \to w^{\bullet}kg.
\qquad (\text{C.20})
$$

Replacing $\alpha^\bullet$ and $\beta^\bullet$ with their expressions in (C.19), repeated applications of middle-four interchange show that ψ coincides with the contravariant part of φ, mate to $\varphi_\bullet$

$$\varphi^\bullet = (hfu^\bullet \to w^\bullet(w_\bullet hf)u^\bullet \to w^\bullet(kgu_\bullet)u^\bullet \to w^\bullet kg).$$

C3.2 Exercises of 3.2.8 (Exponential)

(a) The proof is written down in 7.7.1, 7.7.2. But its infinite-dimensional extension in 7.7.3 is clearer and simpler, as all the items of a multiple category – of any dimension and direction – can be treated in the same way.

(c) The objects of $\mathbb{A}^{\mathbf{X}}$ are the ordinary functors $F\colon \mathbf{X} \to \mathrm{Hor}_0\mathbb{A}$, and the horizontal morphisms are their natural transformations $h\colon F \to F'$.

A vertical morphism $u\colon F \nrightarrow G$ is a family of vertical arrows and double cells of $\mathbb{A}$, which respects the horizontal composition and identities of the latter

$$ux\colon Fx \nrightarrow Gx, \qquad\qquad uf\colon (ux \, {}^{Ff}_{Gf} \, vy) \qquad\qquad (\text{C.21})$$

(for x and $f\colon x \to y$ in $\mathbf{X}$) and amounts to a functor $u\colon \mathbf{X} \to \mathrm{Hor}_1\mathbb{A}$.

A double cell $\alpha\colon (u \, {}^h_k \, v)$ is a coherent family of double cells of $\mathbb{A}$

$$\alpha x\colon (ux \, {}^{hx}_{kx} \, vx), \qquad\qquad (\alpha x \,|\, vf) = (uf \,|\, ay) \qquad\qquad (\text{C.22})$$

(for x and $f\colon x \to y$ in $\mathbf{X}$) and amounts to a natural transformation $\alpha\colon u \to v\colon \mathbf{X} \to \mathrm{Hor}_1\mathbb{A}$.

C3.3 Exercises of 3.3.6

(a) Follows from the following computations ·

$$x \otimes y = (x \,|\, e) \otimes (e \,|\, y) = (x \otimes e) \,|\, (e \otimes y) = x \,|\, y,$$
$$x \otimes y = (e \,|\, x) \otimes (y \,|\, e) = (e \otimes y) \,|\, (x \otimes e) = y \,|\, x.$$

(b) These double cells can be composed horizontally and vertically; they satisfy the interchange property and have the same identity, $\square A$.

*(c) One completes the construction of 3.2.8(c) defining the comparisons of $\mathbb{A}^{\mathbf{X}}$.

C3.4 Exercise 3.4.8

We obtain now the weak double category of *monoids, homomorphisms and bimodules*. The description is similar to that of $\mathbb{R}$ng, forgetting the additive part of the structure.

Now a *bimodule* $X\colon R \nrightarrow S$ between monoids is a set X with a left action of the monoid R and a right action of the monoid S, consistent as previously. The vertical composite of X with $Y\colon S \nrightarrow T$ is

$$X \, _S{\otimes}_S \, Y = (X \times Y)/{\sim}, \qquad x \otimes y = [(x, y)], \qquad \text{(C.23)}$$

where the equivalence relation is generated by letting $(xs, y) \sim (x, sy)$. The actions of 'scalars' on $X \, _S{\otimes}_S \, Y$ are given again by: $r.(x \otimes y).t = rx \otimes yt$.

C3.5 Exercises of 3.5.7 (Spans of sets as discrete profunctors)

(a) As usual a discrete (small) category is identified with its set of objects, and a functor $f\colon X \to X'$ between such categories with the corresponding mapping of sets.

A profunctor $u\colon X^{\mathrm{op}} \times Y \to \mathbf{Set}$ between discrete categories is the same as a family of sets $u(x, y)$ indexed by $(x, y) \in X \times Y$, and amounts to a span $X \nrightarrow Y$ of sets

$$X \leftarrow \Sigma_{xy} \, u(x, y) \to Y. \qquad \text{(C.24)}$$

Similarly a cell $\alpha\colon u \to v(f, g)$ is a family of mappings

$$\alpha(x, y)\colon u(x, y) \to v(fx, gy)$$

and defines a mapping $m(\alpha)\colon \Sigma \, u(x, y) \to \Sigma \, v(x', y')$ consistent with the associated spans.

C3.6 Exercises of 3.6.2

All these facts for $\mathbb{R}$el$\mathbf{Set}$, $\mathbb{R}$el$\mathbf{Ab}$ and $\mathbb{R}$el$\mathbf{Gp}$ are easily verified.

C3.7 Exercises of 3.6.7

(a) This is already known, since 1.2.7.

*(c) In this case, a 2-cone with vertex A amounts to an ordinary cone with vertex $\mathbf{2} \otimes A$.

C3.8 Exercises of 3.7.2 (On adjoint functors)

(a) In $\mathbb{Adj}\mathbf{Cat}$ the tabulator $\top u$ of an adjunction $u\colon X \nrightarrow Y$ is its graph, defined in 1.5.7: a comma category

$$\top u = (u_{\bullet} \downarrow Y) \cong (X \downarrow u^{\bullet}),$$

$$(x, y, c\colon u_{\bullet}x \to y) \;\leftrightarrow\; (x, y, c'\colon x \to u^{\bullet}y), \tag{C.25}$$

equipped with obvious projections p, q and a universal cell $\pi\colon (e \,{}^{p}_{q}\, u)$

$$p\colon \top u \to X, \qquad q\colon \top u \to Y,$$

$$\pi_{\bullet}\colon u_{\bullet}p \to q\colon \top u \to Y, \quad \pi_{\bullet}(x, y, c\colon u_{\bullet}x \to y) = c\colon u_{\bullet}x \to y.$$

The tabulator of a cell $\alpha\colon (u \,{}^{f}_{g}\, v)$, with components $\alpha_{\bullet}x\colon v_{\bullet}fx \to gu_{\bullet}x$, is the following functor

$$\top\alpha\colon \top u \to \top v,$$

$$(\top\alpha)(x, y, c\colon u_{\bullet}x \to y) = (fx, gy, g(c).\alpha_{\bullet}x\colon v_{\bullet}fx \to gy). \tag{C.26}$$

This proves that $\mathbb{Adj}\mathbf{Cat}$ is span representable: in fact the component $\alpha_{\bullet}x\colon v_{\bullet}fx \to gu_{\bullet}x$ is determined by

$$(\top\alpha)(x, u_{\bullet}x, 1\colon u_{\bullet}x \to u_{\bullet}x) = (fx, gu_{\bullet}x, \alpha_{\bullet}x\colon v_{\bullet}fx \to gu_{\bullet}x).$$

Tabulators (and span-representability) are inherited by the flat double category $\mathbb{Adj}\mathbf{Ord}$ where

$$\top u = \{(x, y) \in X \times Y \mid u_{\bullet}x \leqslant y\} = \{(x, y) \in X \times Y \mid x \leqslant u^{\bullet}y\},$$

$$p\colon \top u \to X, \quad q\colon \top u \to Y, \quad \pi_{\bullet}\colon u_{\bullet}p \leqslant q\colon \top u \to Y. \tag{C.27}$$

Let us note that tabulators are *not* inherited by the double subcategory $\mathbb{Adj}_0\mathbf{Ord}$ (see 3.1.7), which likely does not have them.

(b) $\mathbb{Adj}\mathbf{Cat}$ has cotabulators, where

$$C = \bot u = X +_{u} Y \tag{C.28}$$

is the category consisting of the disjoint union $X + Y$, together with new maps from objects of X to objects of Y

$$\hat{c}_x = (x, y, c\colon u_{\bullet}x \to y)\hat{}\; \in C(x, y), \tag{C.29}$$

that are 'represented' by objects $(x, y, c\colon u_{\bullet}x \to y)$ of $\top u = (u_{\bullet} \downarrow Y)$. The composition of the new maps with old maps $\varphi\colon x' \to x$ and $\psi\colon y \to y'$ (of X and Y) is defined in the obvious way

$$\psi.\hat{c}_x.\varphi = (x', y', \psi.c.u_{\bullet}(\varphi)\colon u_{\bullet}x' \to u_{\bullet}x \to y \to y')\hat{}. \tag{C.30}$$

The universal cell $\iota_\bullet\colon i \to ju_\bullet\colon X \to \bot u$ is given by

$$\iota_\bullet x = (x, u_\bullet x, 1_{u_\bullet x})\hat{} \in C(x, u_\bullet x).$$

The cotabulator of a cell $\alpha\colon (u \overset{f}{\underset{g}{}} v)$, with components $\alpha_\bullet x\colon v_\bullet fx \to gu_\bullet x$, works as f and g on the old objects and arrows, as $\bot\alpha$ on the additional arrows

$$\bot\alpha\colon \bot u \to \bot v,$$

$$(\bot\alpha)(x, y, c\colon u_\bullet x \to y)\hat{} = (fx, gy, g(c).\alpha_\bullet x\colon v_\bullet fx \to gy)\hat{}.$$

(C.31)

This determines $\alpha_\bullet x$ as above: $\mathbb{A}\mathrm{dj}\mathbf{Cat}$ *is also cospan representable.*

Restricting to the flat double category $\mathbb{A}\mathrm{dj}\mathbf{Ord}$, we get the disjoint union $\bot u = X +_u Y$, with the order relations of X and Y supplemented with $x \leqslant y$ when $u_\bullet x \leqslant y$ (or equivalently $x \leqslant u^\bullet y$).

*(c) The tabulators of $\mathbb{A}\mathrm{dj}\mathbf{Cat}$ are inherited by $\mathbb{A}\mathrm{dj}\mathbf{Ac}$, which is also span representable.

In fact, if the categories X and Y (in an adjunction $u\colon X \rightarrowtail Y$) have some type of limits, or colimits, the same holds for $\mathsf{T}u$ (since $u_\bullet$ preserves the existing colimits and $u^\bullet$ the existing limits), and the projections p, q preserve them. Moreover the functor $(p, q)\colon \mathsf{T}u \to X \times Y$ reflects isomorphisms and is faithful. Using these facts one shows that, if X and Y are abelian, so is $\mathsf{T}u$.

Tabulators are also inherited by $\mathbb{A}\mathrm{dj}\mathbf{Tp}^{\vee}$, by Artin gluing (see [Jo], 4.27).

C3.9 Exercises of 3.7.3 (On spans and cospans)

(a) Let $\mathbf{C}$ be a category with pullbacks. We have already remarked that the weak double category $\mathbb{S}\mathrm{pan}(\mathbf{C})$ is (trivially) span representable, with the obvious tabulator of a span $u = (u', u'')$ given by its central object U, equipped with the left cell below

$$
\begin{array}{ccc}
U \xrightarrow{u'} X & \qquad & X \xrightarrow{i} V \\
{\scriptstyle 1}\uparrow \quad \uparrow{\scriptstyle u'} & & {\scriptstyle u'}\uparrow \quad \uparrow{\scriptstyle 1} \\
U \xrightarrow{1} U & & U \longrightarrow V \\
{\scriptstyle 1}\downarrow \quad \downarrow{\scriptstyle u''} & & {\scriptstyle u''}\downarrow \quad \downarrow{\scriptstyle 1} \\
U \xrightarrow{u''} Y & & Y \xrightarrow{j} V
\end{array}
\qquad (\text{C.32})
$$

The cotabulator of u is the pushout (V, i, j) of the span (if it exists in $\mathbf{C}$), with the right cell above. Cospan representability is examined below: it can only happen in very particular situations.

(b) To prove that $\mathbb{S}\mathrm{pan}\mathbf{Set}$ is not cospan representable one can take a morphism of spans $\alpha: u \to u$ represented in the diagram below, where the objects are cardinal sets

$$
\begin{array}{ccc}
1 & \longrightarrow & 1 \\
\uparrow & & \uparrow \\
2 & \xrightarrow{\;m\alpha\;} & 2 \\
\downarrow & & \downarrow \\
1 & \longrightarrow & 1
\end{array}
\tag{C.33}
$$

All the arrows to 1 are determined and the mapping $m\alpha: 2 \to 2$ is arbitrary; the cotabulator pushout of both spans is $\bot u = 1$, and $\bot\alpha$ does not determine α.

To give a similar counterexample in $\mathbb{S}\mathrm{pan}(\mathbf{C})$ it is sufficient to have in $\mathbf{C}$ a terminal object $\top$, an epimorphism $X \twoheadrightarrow \top$ and at least two maps $X \to X$. All this exists in most categories of structured sets, but not – of course – in the category associated to a preorder (see (d)).

(c) One can give a counterexample for span-representability in $\mathbb{C}\mathrm{osp}\mathbf{Set}$, replacing the epimorphism $2 \twoheadrightarrow 1$ of diagram (C.33) with the monomorphism $0 \rightarrowtail 2$ (four times).

(d) Let X be a preordered set with pullbacks and pushouts: this means that every upper-bounded pair of elements has a meet and every lower-bounded pair has a join (see 1.3.4). By (a), $\mathbb{S}\mathrm{pan}(X)$ has cotabulators. Being obviously flat, it is also cospan representable. Dually $\mathbb{C}\mathrm{osp}(X)$ is span representable.

C3.10 Exercises of 3.7.4 (On profunctors)

(a) The cotabulator $\bot u = X +_u Y$ of the profunctor $u: X \nrightarrow Y$ is the gluing, or collage, of X and Y along u, with new maps given by $(\bot u)(x,y) = u(x,y)$ (already considered in 3.4.3). The inclusions $i: X \to \bot u$ and $j: Y \to \bot u$ are obvious, as well as the structural cell ι

$$
\iota: u \to e_{\bot_u}(i^{\mathrm{op}} \times j): X^{\mathrm{op}} \times Y \to \mathbf{Set},
$$
$$
\iota(x,y): u(x,y) = \bot u(x,y).
\tag{C.34}
$$

(b) The tabulator $\top u$ is the *category of elements* of u, or Grothendieck construction. It has objects (x, y, λ) with $x \in \mathrm{Ob}X$, $y \in \mathrm{Ob}Y$, $\lambda \in u(x,y)$ and maps (f, g) of $X \times Y$ which form a commutative square in the collage $X +_u Y$

$$
(f, g): (x, y, \lambda) \to (x', y', \lambda'),
$$
$$
f: x \to x', \qquad g: y \to y', \qquad g\lambda = \lambda' f,
\tag{C.35}
$$

where the last equality means that: $u(1_x, g)(\lambda) = u(f, 1_y)(\lambda')$, in $u(x, y')$.
The functors p, q are obvious, and the structural cell $\pi \colon e_{\mathsf{T}_u} \to u$ is

$$\pi \colon e_{\mathsf{T}_u} \to u(p^{\mathrm{op}} \times q) \colon (\mathsf{T}u)^{\mathrm{op}} \times \mathsf{T}u \to \mathbf{Set},$$

$$\pi(x, y, \lambda; x', y', \lambda') \colon (\mathsf{T}u)(x, y, \lambda; x', y', \lambda') \to u(x, y'), \qquad (\text{C.36})$$

$$(f, g) \mapsto g\lambda = \lambda'f.$$

(c) For a cell $\alpha \colon (u \overset{f}{\underset{g}{\,}} v)$, viewed as a natural transformation

$$\alpha \colon u \to v(f^{\mathrm{op}} \times g) \colon X^{\mathrm{op}} \times Y \to \mathbf{Set},$$

both the functors $\mathsf{T}\alpha$ and $\bot\alpha$ determine every component $\alpha_{xy} \colon u(x, y) \to v(fx, gy)$ of α

$$\mathsf{T}\alpha \colon \mathsf{T}u \to \mathsf{T}v, \qquad (\mathsf{T}\alpha)(x, y, \lambda) = (fx, gy, \alpha_{xy}(\lambda)),$$

$$\bot\alpha \colon \bot u \to \bot v, \qquad (\bot\alpha)(\lambda \colon x \to y) = \alpha_{xy}(\lambda) \colon fx \to gy, \qquad (\text{C.37})$$

where $\lambda \in u(x, y)$.

C3.11 Exercises of 3.7.5 (On extending functors)

(a) (A more analytic description will be given in 7.1.5, taking advantage of the non-truncated framework of multiple categories.)

On the arrows, the 'functor' Span sends a functor $F \colon \mathbf{X} \to \mathbf{A}$ (between categories with pullbacks) to the obvious pointwise extension

$$\mathrm{Span}(F) \colon \mathrm{Span}\mathbf{X} \to \mathrm{Span}\mathbf{A}, \qquad (\text{C.38})$$

that acts by computing F over the spans of $\mathbf{X}$ and their double cells.

$\mathrm{Span}(F)$ is *unitary* (in a trivial way, independent of choices) and *colax*, with a comparison for vertical composition

$$\underline{F}(u, v) \colon F(u \otimes v) \to Fu \otimes Fv, \qquad (\text{C.39})$$

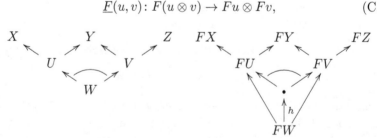

given by the $\mathbf{A}$-morphism h determined by the universal property of the concatenation-pullback in $F(u) \otimes F(v)$. Let us note that $\mathrm{Span}(F)$ *is a pseudo functor if and only if F preserves pullbacks*.

On 2-cells, Span sends a natural transformation $\varphi \colon F \to F' \colon \mathbf{X} \to \mathbf{A}$ to the horizontal transformation $\mathbb{S}\mathrm{pan}(\varphi)\colon \mathbb{S}\mathrm{pan}(F) \to \mathbb{S}\mathrm{pan}(F')$, which acts on a span by computing φ on its three vertices.

(b) Here a functor $F\colon \mathbf{X} \to \mathbf{A}$ (between categories with pushouts) is sent to the pointwise extension

$$\mathbb{C}\mathrm{osp}(F)\colon \mathbb{C}\mathrm{osp}\mathbf{X} \to \mathbb{C}\mathrm{osp}\mathbf{A}, \tag{C.40}$$

that acts by computing F on the cospans of $\mathbf{X}$ and their transversal maps. $\mathbb{C}\mathrm{osp}(F)$ is a unitary lax functor, and is pseudo if and only if F preserves pushouts.

(c) The extension $\mathbb{S}\mathrm{pan}\mathbf{X} \to \mathbb{R}\mathrm{el}\mathbf{A}$ is obtained by composing the functor (C.38) with the obvious projection $R\colon \mathbb{S}\mathrm{pan}\mathbf{A} \to \mathbb{R}\mathrm{el}\mathbf{A}$ (already considered in (3.70) for $\mathbf{A} = \mathbf{Set}$). The extension is a strict functor when the original functor F turns every pullback of $\mathbf{X}$ into a bicommutative square of $\mathbb{R}\mathrm{el}\mathbf{A}$, as defined in 2.5.6.

*(d) The extension $F\colon \mathbb{C}\mathrm{osp}\mathbf{X} \to \mathbb{R}\mathrm{el}\mathbf{Ab}$ is similarly obtained, taking into account that $\mathbb{R}\mathrm{el}\mathbf{Ab}$ can also be constructed by equivalence classes of cospans. In fact the dual of $\mathbf{Ab}$ is an abelian category, and therefore a regular one.

C3.12 Exercises of 3.8.6

(b) The lax functor $F\colon \mathbf{1} \to \mathbb{A}$ amounts to a fourtuple (A, t, η, μ), formed of an object $A = F(0)$, a vertical arrow $t = F(e_0)\colon A \nrightarrow A$ and two special cells $\eta = \underline{F}(0)\colon e_A \to t$ and $\mu = \underline{F}(e_0, e_0)\colon t \otimes t \to t$ that satisfy the monad axioms.

(c) A horizontal arrow $(f, \varphi)\colon (A, t, \eta, \mu) \to (A', t', \eta', \mu')$ of $\mathbb{L}\mathrm{x}(\mathbf{1}, \mathbb{A})$ is an arrow $f\colon A \to A'$ in $\mathbb{A}$ with a cell $\varphi\colon (t \overset{f}{\underset{f}{}} t')$ coherent with units and multiplications. A vertical arrow $(r, \rho)\colon (A, t, \eta, \mu) \to (B, s, \vartheta, \nu)$ is an arrow $r\colon A \nrightarrow B$ in $\mathbb{A}$ with a globular isocell $\rho\colon (t \otimes r \overset{1}{\underset{1}{}} r \otimes s)$ coherent with units and multiplications.

A double cell comes from a double cell $\xi\colon (r \overset{f}{\underset{g}{}} r')$ of $\mathbb{A}$.

C4 Exercises of Chapter 4
C4.1 Exercises of 4.1.3

In the strict case, the double categories of *relations*, in 3.1.2–3.1.3, have (obvious) vertical companions; moreover the vertical adjoint to a horizontal morphism $f\colon X \to Y$ is the opposite relation $f^{\sharp}\colon Y \nrightarrow X$, with double cells $\sigma\colon 1 \leqslant f^{\sharp}f$ and $\tau\colon ff^{\sharp} \leqslant 1$.

The double categories of *adjunctions*, in 3.1.5–3.1.8, lack vertical companions and vertical adjoints.

All the weak double categories of Section 3.4 have vertical companions, given by the obvious embedding of horizontal arrows into the vertical ones. In particular, in $\mathbb{R}$ng (see 3.4.5) the vertical companion to a ring-homomorphism $f \colon R \to S$ is the bimodule $S \colon R \nrightarrow S$, with the left R-module structure induced by f.

C4.2 Exercises of 4.1.8

(c) Plainly the double category $\mathbb{T}$g is not horizontally invariant. Its horizontal isomorphisms and vertical isomorphisms (or vertical equivalences) are obvious.

We can also note that the identity of $\mathbb{T}$g is horizontally isomorphic to the double functor $D \colon \mathbb{T}$g $\to \mathbb{T}$g that takes a topological group G to the discrete topological group DG on the same algebraic structure, and also to the double functor $C \colon \mathbb{T}$g $\to \mathbb{T}$g that produces the coarsest topology.

C4.3 Exercises of 4.2.3

(b) The colax functors U, V, being pseudo, have comparisons $\alpha = \underline{U}^{-1}$ and $\beta = \underline{V}^{-1}$ in the lax direction. The coherence condition (c.2) of 4.2.1 can be rewritten as

$$(\beta(FA) \mid V\underline{F}(A) \mid \pi e_A) = (e_{\pi A} \mid \underline{G}(UA) \mid G\alpha(A)), \qquad (\text{C.41})$$

where

$$(\beta(FA) \mid V\underline{F}(A)) = \underline{VF}(A), \quad (\underline{G}(UA) \mid G\alpha(A)) = \underline{GU}(A)$$

are the comparisons of the lax functors VF and GU on the object A (see (3.63)). Thus condition (C.41) amounts to the coherence condition (ht.2) of π as a horizontal transformation of lax functors $VF \to GU$ (see 3.5.4).

Similarly, (c.3) amounts to (ht.3).

C4.4 *Exercise 4.2.7 (Internal monoids)

The beginning of the statement being obvious, we write down the profunctor $\mathrm{Mon}(U) \colon \mathrm{Mon}(\mathbf{A}) \nrightarrow \mathrm{Mon}(\mathbf{C})$ of point (iii), for a comonoidal functor $U \colon \mathbf{A} \to \mathbf{C}$ between monoidal categories.

Given two monoids $M = (M, m, i)$ in $\mathbf{A}$ and $N = (N, m, i)$ in $\mathbf{C}$, the set

$\mathrm{Mon}(U)(M, N)$ consists of all morphisms $f \colon UM \to N$ in $\mathbf{C}$ which make the following diagrams commute

$$
\begin{array}{ccc}
UI \xrightarrow{\;U\;} I & \qquad U(M \otimes M) \xrightarrow{\;U\;} UM \otimes UM \xrightarrow{\;f \otimes f\;} N \otimes N \\[2pt]
{\scriptstyle Ui}\downarrow \qquad \downarrow{\scriptstyle i} & \quad {\scriptstyle Um}\downarrow \qquad\qquad\qquad\qquad\qquad\qquad \downarrow{\scriptstyle m} \qquad (C.42)\\[2pt]
UM \xrightarrow[f]{} N & \qquad UM \xrightarrow{\qquad\qquad\qquad\qquad\; f \;\qquad\qquad\qquad\qquad} N
\end{array}
$$

Now, given a cell π in $\mathbb{M}\mathrm{nc}$, as in the left diagram below, we define the corresponding cell $\mathrm{Mon}(\pi)$ between functors and profunctors

$$
\begin{array}{ccc}
\mathbf{A} \xrightarrow{\;F\;} \mathbf{B} & \qquad & \mathrm{Mon}(\mathbf{A}) \xrightarrow{\;\mathrm{Mon}(F)\;} \mathrm{Mon}(\mathbf{B}) \\[2pt]
{\scriptstyle U}\downarrow \quad {\scriptstyle \pi} \quad \downarrow{\scriptstyle V} & & {\scriptstyle \mathrm{Mon}(U)}\downarrow \quad {\scriptstyle \mathrm{Mon}(\pi)} \quad \downarrow{\scriptstyle \mathrm{Mon}(V)} \qquad (C.43)\\[2pt]
\mathbf{C} \xrightarrow[G]{} \mathbf{D} & & \mathrm{Mon}(\mathbf{C}) \xrightarrow[\mathrm{Mon}(G)]{} \mathrm{Mon}(\mathbf{D})
\end{array}
$$

by a natural transformation whose general component on M, N is

$$
\mathrm{Mon}(\pi)(M, N) \colon \mathrm{Mon}(U)(M, N) \to \mathrm{Mon}(V)(FM, GN),
$$
$$
(f \colon UM \to N) \mapsto (Gf.\pi M \colon VFM \to GN). \tag{C.44}
$$

Finally, given a comonoidal functor $U' \colon \mathbf{C} \to \mathbf{C}'$ composable with U, one constructs the special cell $\underline{M}(U, U')$ in $\mathbb{C}\mathrm{at}$ (the laxity comparison for vertical composition) as the following natural transformation of profunctors

$$
\underline{M}(U, U') \colon \mathrm{Mon}(U) \otimes \mathrm{Mon}(U') \to \mathrm{Mon}(U'U) \colon \mathrm{Mon}(\mathbf{A}) \to \mathrm{Mon}(\mathbf{C}'),
$$
$$
\underline{M}(U, U')(M, P)[f, g] = g.U'f \colon U'U(M) \to P,
$$

where $f \colon UM \to N$ and $g \colon U'N \to P$ are in $\mathbf{C}$ and $\mathbf{C}'$, while the class $[f, g]$ belongs to the composition-colimit $(\mathrm{Mon}(U) \otimes \mathrm{Mon}(U'))(M, P)$.

C4.5 Exercises of 4.2.8

(a) It will be useful to draw a cell of $\mathbb{D}$ in the following way, adding a *formal* arrow inside the cell

$$
\begin{array}{ccc}
\bullet & \xrightarrow{\;f\;} & \bullet \\[2pt]
{\scriptstyle u}\downarrow & {\nwarrow}{\scriptstyle \pi} & \downarrow{\scriptstyle v} \qquad (C.45)\\[2pt]
\bullet & \xrightarrow[g]{} & \bullet
\end{array}
$$

In $\mathrm{HV}_{\mathbb{D}}(A, B)$ we have two (disjoint) kinds of objects: the *h-objects* are the horizontal arrows $f \colon A \to B$ of $\mathbb{D}$, and the *v-objects* are the vertical arrows $u \colon A \nrightarrow B$ of $\mathbb{D}$.

Then there are four kind of morphisms, given by the double cells of $\mathbb{D}$ whose boundary contains two identities:

- an *hh-map* $\varphi\colon f \dashrightarrow g$ is a cell $\varphi\colon (e_A \begin{smallmatrix} f \\ g \end{smallmatrix} e_B)$,

- a *vv-map* $\omega\colon v \dashrightarrow u$ is a cell $\omega\colon (u \begin{smallmatrix} 1_A \\ 1_B \end{smallmatrix} v)$ with reversed horizontal domains, as in (C.45),

- an *hv-map* $\varepsilon\colon f \dashrightarrow u$ is a cell $\varepsilon\colon (u \begin{smallmatrix} f \\ 1_B \end{smallmatrix} e_B)$,

- a *vh-map* $\eta\colon v \dashrightarrow g$ is a cell $\eta\colon (e_A \begin{smallmatrix} 1_A \\ g \end{smallmatrix} v)$.

These four kinds of morphisms give 8 kinds of composition:

- an *hhh-composition* $\psi\varphi\colon f \dashrightarrow g \dashrightarrow h$ is given by the horizontal composition $\varphi|\psi$ in $\mathbb{D}$,

- an *hhv-composition* $\varepsilon\varphi\colon f \dashrightarrow g \dashrightarrow u$ is given by the vertical composition $\varphi \otimes \varepsilon$, as in the left diagram below

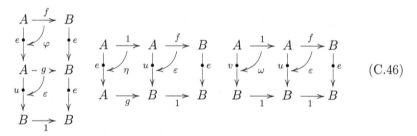

$$(C.46)$$

- an *hvh-composition* $\eta\varepsilon\colon f \dashrightarrow u \dashrightarrow g$ is given by the horizontal composition $\eta|\varepsilon$, as in the central diagram above,

- an *hvv-composition* $\omega\varepsilon\colon f \dashrightarrow u \dashrightarrow v$ is given by the horizontal composition $\omega|\varepsilon$, as in the right diagram above.

The other four cases, namely the vhh-, vhv-, vvh- and vvv-compositions, derive from the previous ones, by transposition.

Unitarity of the composition is rather obvious:

- the unit of the object $f\colon A \to B$ is the cell $e_f\colon (e_A \begin{smallmatrix} f \\ f \end{smallmatrix} e_B)$ as an hh-map $f \dashrightarrow f$,

- the unit of the object $u\colon A \nrightarrow B$ is the cell $1_u\colon (u \begin{smallmatrix} 1_A \\ 1_B \end{smallmatrix} u)$ as a vv-map $u \dashrightarrow u$.

The proof of associativity breaks in many cases, and we write down three of them:

- the hhhh-case $\chi\psi\varphi\colon f \dashrightarrow g \dashrightarrow h \dashrightarrow k$ comes from associativity of horizontal composition in $\mathbb{D}$,

- the hhvv-case $\omega\varepsilon\varphi\colon f \dashrightarrow g \dashrightarrow u \dashrightarrow v$ comes from interchange in $\mathbb{D}$, in the left diagram below

$$w(\varepsilon\varphi) = (w\,|\,\varphi \otimes \varepsilon) = (\square_A \otimes w)\,|\,(\varphi \otimes \varepsilon)$$
$$= (\square_A\,|\,\varphi) \otimes (w\,|\,\varepsilon) = \varphi \otimes (w\,|\,\varepsilon) = (w\varepsilon)\varphi,$$

- the hhvh-case $\sigma\varepsilon\varphi\colon f \dashrightarrow g \dashrightarrow u \dashrightarrow h$ comes again from interchange in $\mathbb{D}$, in the right diagram above.

C4.6 Exercises of 4.3.3

(a) This is already known, from the composition of orthogonal adjoints in $\mathbb{D}$bl: see 4.1.2. The unit and the counit of the composed adjunction are displayed in diagram 0.7 of the general Introduction.

(b) The first equation in (ad.3) says that the adjoint cell of $\underline{F}X$, i.e. $(\underline{F}X)' = (\eta e_X\,|\,G\underline{F}X)$, must be equal to $(e_{\eta X}\,|\,\underline{G}\underline{F}X)$. Similarly for the comparison $\underline{F}(u, u')$, in (ad.4).

(c) A straightforward consequence.

C4.7 Exercises of 4.5.1

The 'functors' M, P, Q are produced by an embedding $m\colon (\mathbf{2}, \leqslant, \times) \to (\mathbb{R}_+, \geqslant, +)$ of strict monoidal categories, with strictly monoidal adjoints $p \dashv m \dashv q$

$$
\begin{aligned}
&m\colon \mathbf{2} \to \mathbb{R}_+, &&0 \mapsto \infty, &&1 \mapsto 0, \\
&p\colon \mathbb{R}_+ \to \mathbf{2}, &&\lambda \mapsto 1, &&\infty \mapsto 0 &&\text{(for } \lambda < \infty\text{)}, \\
&q\colon \mathbb{R}_+ \to \mathbf{2}, &&0 \mapsto 1, &&\lambda \mapsto 0 &&\text{(for } \lambda > 0\text{)}, \\
&pm = qm = 1, &&mq \geqslant 1 \geqslant mp\colon \mathbb{R}_+ \to \mathbb{R}_+.
\end{aligned}
\tag{C.47}
$$

(Note that q amounts to the forgetful functor $\mathbb{R}_+(0, -)\colon \mathbb{R}_+ \to \mathbf{Set}$, as interpreted in (C.13). Similarly, the associated functor Q will correspond to the forgetful functor $\mathbf{Mtr} \to \mathbf{Cat}$, described in C2.6.2.)

The corresponding 'functions' M, P, Q act in the obvious way on objects and similarly on profunctors (whereas they 'do not modify' the horizontal arrows)

$$d_{MX}(x, x') = 0 \quad \text{(or } \infty) \quad \Leftrightarrow \quad x \prec x' \quad \text{(otherwise)},$$

$$y \prec_{PY} y' \quad \Leftrightarrow \quad d(y, y') < \infty, \qquad \text{(C.48)}$$

$$y \prec_{QY} y' \quad \Leftrightarrow \quad d(y, y') = 0.$$

Moreover M and P also preserve the vertical composition and are double functors (since p and m preserve colimits, hence coends). Q is unitary lax; to show that it is not pseudo one can use the following composition of profunctors between euclidean metric spaces

$$v \otimes v' \colon \{0\} \rightarrow (0, \infty) \rightarrow \{0\}, \qquad v(0, \lambda) = \lambda = v'(\lambda, 0),$$

$$(Qv \otimes Qv')(0, 0) = \sup(qv(0, \lambda).qv'(\lambda, 0)) = 0,$$

$$Q(v \otimes v')(0, 0) = q(\inf(v(0, \lambda) + v'(\lambda, 0))) = q(\inf(2\lambda)) = q(0) = 1.$$

M, P and Q form two adjunctions (respectively strict and strict-lax)

$$P \dashv M, \qquad \eta \colon 1 \to MP, \qquad PM = 1, \tag*{(C.49)}$$

$$M \dashv Q, \qquad 1 = MQ, \qquad \varepsilon \colon MQ \to 1,$$

where also the underlying mapping of the unit ηY and the counit εY are identities, so that the four triangle conditions are trivially satisfied

$$\eta Y \colon Y \to MPY, \qquad d(y, y') \geqslant mp(d(y, y')) = d_{MPY}(y, y'), \tag*{(C.50)}$$

$$\varepsilon Y \colon MQY \to Y, \qquad d_{MQY}(y, y') = mq(d(y, y')) \geqslant d(y, y'),$$

$$
\begin{array}{ccccc}
MQY & \xrightarrow{\ \varepsilon Y\ } & Y & \xrightarrow{\ \eta Y\ } & MPY \\
{\scriptstyle MPv}\downarrow & \geqslant & \downarrow{\scriptstyle v} & \geqslant & \downarrow{\scriptstyle MQv} \\
MQY' & \xrightarrow[\ \varepsilon Y'\ ]{} & Y' & \xrightarrow[\ \eta Y'\ ]{} & MPY'
\end{array}
$$

(The diagram above shows the cells εv, ηv for a profunctor $v \colon Y \rightarrow Y'$ of metric spaces.)

C4.8 Exercises of 4.5.2

(a) In the chain $\pi_0 \dashv D \dashv F = \mathrm{Ob}$, we have seen that $D \colon \mathbf{Set} \to \mathbf{Cat}$ associates to a set X its discrete category DX, while $\pi_0 \colon \mathbf{Cat} \to \mathbf{Set}$ takes a category to its set of connected components. Units and counits are obvious

$$\eta \colon 1 \to D\pi_0, \qquad \pi_0 D = 1 \qquad (\pi_0 \eta = 1), \tag*{(C.51)}$$

$$1 = FD, \qquad \varepsilon \colon DF \to 1 \qquad (F\varepsilon = 1).$$

To begin the required extension to double categories, the discrete embedding $D \colon \mathbf{Set} \to \mathbf{Cat}$ is easily turned into a strict double embedding $D \colon \mathbb{S}\mathrm{pan}\mathbf{Set} \to \mathbb{C}\mathrm{at}$, *viewing a span as a set-profunctor*, i.e. a profunctor between discrete categories (see 3.5.7(b)).

D has a left adjoint $\pi_0 \dashv D$ (whose unit and counit are the obvious extensions of the previous ones). To define π_0 on a profunctor $u \colon X \nrightarrow Y$, the set $(\pi_0 u)(\bar{x}, \bar{y})$ is a quotient of $\sum u(x, y)$, under the equivalence relation which identifies a formal arrow $\lambda \colon x \dashrightarrow y$ of u with all composites $\beta \lambda \alpha \colon x' \to x \dashrightarrow y \to y'$.

The functor π_0 is unitary colax, with comparison special cells $\pi_0(u \otimes v) \to \pi_0 u \otimes \pi_0 v$. The latter are not invertible, as shown by the profunctor $u \colon \mathbf{2} \nrightarrow \mathbf{2}$ whose collage $\mathbf{2} +_u \mathbf{2}$ has one added arrow $0 \to 1$. Then $\pi_0 u$ is the vertical identity of the singleton, while $u \otimes u \colon \mathbf{2} \nrightarrow \mathbf{2}$ is the empty profunctor and $\pi_0(u \otimes u)$ is the empty endospan of $\mathbf{2}$.

D has also a lax right adjoint $F \colon \mathbb{C}\mathrm{at} \to \mathbb{S}\mathrm{pan}\mathbf{Set}$ that assigns to a profunctor $u \colon X \nrightarrow Y$ its restricted set-profunctor $Fu \colon \mathrm{Ob}X \nrightarrow \mathrm{Ob}Y$.

(b) Follows from Theorem 4.4.3, since the lax functor F is not pseudo unitary.

(c) First, cotabulators in $\mathbb{S}\mathrm{pan}\mathbf{Set}$ are quotients of the corresponding ones in $\mathbb{C}\mathrm{at}$ (see 3.7.4). Second, the lax functor F is *not unitary*, even in the weak sense, because the unit-profunctor of a category X is taken to the set-profunctor $Fu \colon \mathrm{Ob}X \nrightarrow \mathrm{Ob}X$ having $(Fu)(x, x') = X(x, x')$: the comparison cell $\underline{F}X \colon e_{FX} \to Fe_X$ is invertible if and only if X is discrete.

C4.9 Exercise 4.5.4

Extending the functor po: $\mathbf{Cat} \to \mathbf{pOrd}$ of 1.5.3(f), we construct a double functor po: $\mathbb{C}\mathrm{at} \to \mathbb{p}\mathbb{O}\mathrm{rd}$ that sends the profunctor $u \colon \mathbf{S} \nrightarrow \mathbf{T}$ to the order-profunctor po$(u) \colon \mathrm{po}(\mathbf{S}) \nrightarrow \mathrm{po}(\mathbf{T})$, where

$$\mathrm{po}(u)(x, y) = 1 \iff u(x, y) \neq \emptyset.$$

In other words, we equip the set $\mathrm{Ob}\mathbf{S} + \mathrm{Ob}\mathbf{T}$ with the preorder of po$(\mathbf{S} +_u \mathbf{T})$, associated to the gluing of the profunctor u.

C4.10 Exercises of 4.6.5

(a) We verify the axiom (coh) for the double cell $\pi \,|\, \vartheta$: this means that

$$t(\varphi'.Tf \otimes f'\varphi) \otimes (\vartheta f \otimes g'\pi)a \otimes g'g\rho$$
$$= \tau.T(f'f) \otimes d'.T(\vartheta f \otimes g'\pi) \otimes (\gamma'.Tg \otimes g'\gamma).Tr, \tag{C.52}$$

where $a\colon TA \to A$ and $d'\colon TD' \to D'$ are the structures of the pseudo algebras $\mathbb{A}$ and $\mathbb{D}'$.

The proof is similar to that of Theorem 4.2.2. Writing cells as arrows between morphisms, our property amounts to the commutativity of the outer diagram below

$$
\begin{array}{ccccc}
tf'fa & \xrightarrow{\ \vartheta fa\ } & g'sfa & \xrightarrow{\ g'\pi a\ } & g'gra \\
{\scriptstyle tf'\varphi}\uparrow & & \uparrow{\scriptstyle g's\varphi} & & \downarrow{\scriptstyle g'g\rho} \\
tf'b.Tf & \xrightarrow{\ \vartheta b.Tf\ } & g'sb.Tf & & g'gc.Tr \\
{\scriptstyle t\varphi'Tf}\uparrow & & \downarrow{\scriptstyle g'\sigma.Tf} & & \uparrow{\scriptstyle g'\gamma.Tr} \\
tb'.Tf'.Tf & & g'd.T(sf) & \xrightarrow{\ g'd.T\pi\ } & g'd.T(gr) \\
{\scriptstyle \tau.Tf'.Tf}\downarrow & & \uparrow{\scriptstyle \gamma'T(sf)} & & \uparrow{\scriptstyle \gamma'.T(gr)} \\
d'.T(tf'f) & \xrightarrow[\ d'.T(\vartheta f)\ ]{} & d'.T(g'sf) & \xrightarrow[\ d'.T(g'\pi)\ ]{} & d'.T(g'gr)
\end{array}
\qquad \text{(C.53)}
$$

Here the two hexagons commute by coherence of the double cells π and ϑ, and the two parallelograms by interchange of 2-cells in $\mathbf{C}$.

(b) We have two lax morphisms

$$
\mathbf{f} = (f, \varphi)\colon (A, a, \omega, \kappa) \to (B, b, \omega, \kappa),
$$
$$
\mathbf{g} = (g, \gamma)\colon (B, b, \omega, \kappa) \to (A, a, \omega, \kappa),
$$

which are inverse to each other:

$$
gf = 1, \quad fg = 1, \qquad \gamma.Tf \otimes g\varphi = 1_a, \quad \varphi.Tg \otimes f\gamma = 1_b.
$$

$$
\begin{array}{ccccccc}
TA & \xrightarrow{\ Tf\ } & TB & \xrightarrow{\ Tg\ } & TA & \xrightarrow{\ Tf\ } & TB \\
{\scriptstyle a}\downarrow\ {\scriptstyle\varphi}\nwarrow & & {\scriptstyle b}\downarrow\ {\scriptstyle\gamma}\nwarrow & & {\scriptstyle a}\downarrow\ {\scriptstyle\varphi}\nwarrow & & \downarrow{\scriptstyle b} \\
A & \xrightarrow[\ f\ ]{} & B & \xrightarrow[\ g\ ]{} & A & \xrightarrow[\ f\ ]{} & B
\end{array}
\qquad \text{(C.54)}
$$

Then it is easy to see that the 2-cell $\psi = f.\gamma.Tf\colon fa \to b.Tf$ is vertically inverse to φ.

C5 Exercises of Chapter 5

C5.1 Exercises of 5.1.6

(d) By (tp.1) we have two cells $\eta\colon 1 \to t$ and $\mu\colon t \otimes t \to t$. The monad axioms are necessarily satisfied.

C5.2 Exercises of 5.4.1 (Metric spaces, preordered sets and relations)

(a) We know that a product $\prod X_i$ in $\mathrm{Hor}_0\mathbb{M}\mathrm{tr}$ is the cartesian product of the underlying sets, with the l_∞-metric

$$d((x_i),(x_i')) = \vee_i\, d(x_i, x_i'). \qquad (\mathrm{C.55})$$

To prove that $\mathbb{M}\mathrm{tr}$ has lax functorial products it is sufficient to form the product u of a family of profunctors $u_i\colon X_i \nrightarrow Y_i$

$$u\colon \prod X_i \nrightarrow \prod Y_i, \qquad u((x_i),(y_i)) = \vee_i\, u_i(x_i, y_i), \qquad (\mathrm{C.56})$$

and remark that, when each u_i is the vertical identity of X_i (i.e. $u_i(x_i, x_i') = d(x_i, x_i')$) we come back to (C.55).

The double categories $\mathrm{p}\mathbb{O}\mathrm{rd}$ and $\mathbb{R}\mathrm{el}\mathbf{Set}$ are closed under such constructions. Note that $u\colon \prod X_i \nrightarrow \prod Y_i$ is the obvious cartesian product of a family of relations. Products are easily seen to be functorial in $\mathrm{p}\mathbb{O}\mathrm{rd}$ and $\mathbb{R}\mathrm{el}\mathbf{Set}$, but they are not in $\mathbb{M}\mathrm{tr}$: a counterexample can be found in [GP1], Section 6.2.

(b) For a sum $\sum X_i$ in $\mathbb{M}\mathrm{tr}$ the distance within each component is completed by letting $d(x, y) = \infty$ when x and y are in different summands; a sum of metric profunctors is similarly computed. Again $\mathrm{p}\mathbb{O}\mathrm{rd}$ and $\mathbb{R}\mathrm{el}\mathbf{Set}$ are closed under this construction. In each case sums are functorial.

(c) In $\mathrm{Hor}_0\mathbb{M}\mathrm{tr}$ the equaliser of a pair of weak contractions $f, g\colon X \to Y$ is the equaliser $\mathrm{Eq}\,(f, g)$ in $\mathbf{Set}$ with the restricted metric. The equaliser of a pair of double cells in $\mathrm{Hor}_1\mathbb{M}\mathrm{tr}$, between $u\colon X \nrightarrow X'$ and $v\colon Y \nrightarrow Y'$

$$(f, f')\colon u \to v, \quad (g, g')\colon u \to v, \quad (u \geqslant v(f, f'),\ u \geqslant v(g, g')), \quad (\mathrm{C.57})$$

is the restriction $\mathrm{Eq}\,(f, g) \nrightarrow \mathrm{Eq}\,(f', g')$ of u. Similarly for $\mathrm{p}\mathbb{O}\mathrm{rd}$ and $\mathbb{R}\mathrm{el}\mathbf{Set}$. Equalisers are not functorial [GP1].

(d) In $\mathrm{Hor}_0\mathbb{M}\mathrm{tr}$ the coequaliser of $f, g\colon X \to Y$ is the set-coequaliser Y/R with the induced metric, the largest making the projection $Y \to Y/R$ into a weak contraction

$$d([y], [y']) = \inf \sum_i d(y_i, y_i'), \qquad (\mathrm{C.58})$$

where the inf is taken on all finite sequences $y_1, y_1', y_2, y_2', \dots y_n, y_n'$ with $y_1 = y$, $y_n' = y'$ and $y_i' R y_{i+1}$ for $1 \leqslant i < n$.

In degree 1, the metric profunctor $\overline{v}\colon Y/R \nrightarrow Y'/R'$ is similarly defined. Again, $\mathrm{p}\mathbb{O}\mathrm{rd}$ and $\mathbb{R}\mathrm{el}\mathbf{Set}$ are closed under this construction.

(e) In $\mathbb{M}\mathrm{tr}$ we have described the 1-dimensional tabulator of a metric profunctor $u\colon X \nrightarrow Y$ as a graph of u (in 3.7.1)

$$\begin{aligned} \top u &= \{(x, y) \in X \times Y \mid u(x, y) = 0\}, \\ d((x, y), (x', y')) &= d(x, x') \vee d(y, y'). \end{aligned} \qquad (\mathrm{C.59})$$

Given a commutative square of vertical arrows $(r, s): u \twoheadrightarrow v$, its limit $\top(r, s)$ is the restriction of $r \times s: X \times Y \twoheadrightarrow X' \times Y'$. pOrd and $\mathbb{R}$el**Set** are closed under this solution.

(f) We have also seen that the cotabulator of u is the collage $X +_u Y$, with $d(x, y) = u(x, y)$ and $d(y, x) = \infty$ for $x \in X$, $y \in Y$.

Given a commutative square of vertical arrows $(r, s): u \twoheadrightarrow v$, its colimit $t = \bot(r, s)$ is calculated on pairs (x, y') by means of the diagonal of the square $z = r \otimes v = u \otimes s: X \twoheadrightarrow Y'$ (the other values of t being obvious)

$$\begin{aligned} t(x, y') = z(x, y') &= \wedge_{x'}(r(x, x') + v(x', y')) \\ &= \wedge_y(u(x, y) + s(y, y')), \end{aligned} \tag{C.60}$$

$$t(x, x') = r(x, x'), \qquad t(y, y') = s(y, y'), \qquad t(y, x') = \infty.$$

The double category pOrd is closed in $\mathbb{M}$tr under these constructions, whereas $\mathbb{R}$el**Set** has (generally) different cotabulators: computing the $\mathbb{M}$tr-cotabulator of a relation $u: X \twoheadrightarrow Y$ we may find $d(x, y) = u(x, y) = 0$, in which case the elements x, y must be identified in the cotabulator of $\mathbb{R}$el**Set**.

Cotabulators are functorial in $\mathbb{M}$tr and pOrd.

C5.3 Exercises of 5.4.2 (On adjoint functors)

(a) The prototype $\mathbb{A}$dj**Cat** has (obvious) functorial products and sums. For instance, the product $u = (u_\bullet, u^\bullet, \eta, \varepsilon)$ of a family of adjunctions $(u_i: X_i \twoheadrightarrow Y_i)_{i \in I}$ is computed by ordinary products in **Cat**

$$\begin{aligned} u_\bullet &= \Pi u_{i\bullet}: \Pi X_i \to \Pi Y_i, & u^\bullet &= \Pi u_i^\bullet: \Pi Y_i \to \Pi X_i, \\ \eta &= \Pi \eta_i: 1 \to u^\bullet u_\bullet, & \varepsilon &= \Pi \varepsilon_i: u_\bullet u^\bullet \to 1, \end{aligned} \tag{C.61}$$

and equipped with the obvious cells $\pi_i: (u \;{}^{p_i}_{p_i}\; u_i)$, whose covariant and contravariant part are identities

$$\pi_{i\bullet}: u_{i\bullet} p_i = p_i u_\bullet: \Pi X_j \to Y_i, \qquad \pi_i^\bullet: u_i^\bullet p_i = p_i u^\bullet: \Pi Y_j \to X_i. \tag{C.62}$$

The double subcategories $\mathbb{A}$dj**Ord** and $\mathbb{A}$dj$_0$**Ord** are closed under products.

(b) $\mathbb{A}$dj**Cat** and $\mathbb{A}$dj**Ord** lack equalisers and coequalisers, but $\mathbb{A}$dj$_0$**Ord** has them.

(c) Tabulators and cotabulators of vertical arrows are computed in C3.8. The proof that, in $\mathbb{A}$dj**Cat**, they are, respectively, lax functorial and colax functorial (in fact, functorial) is left to the reader, or can be seen in [GP1], Section 6.5.

C5.4 Exercises of 5.4.3 (On spans and cospans)

(a) As to level limits, we know that, if the category $\mathbf{C}$ is complete (resp. cocomplete), its functor category $\mathbf{C}^\vee$ has all limits (resp. colimits), that are computed pointwise (see 1.5.4(f)). Therefore, in this hypothesis, the weak double category $\mathbb{S}\text{pan}(\mathbf{C})$ has all level limits (resp. colimits), that are computed separately on the three vertices of the spans.

Tabulators are obvious and cotabulators depend on pushouts in $\mathbf{C}$ (see 3.7.3).

C5.5 Exercises of 5.4.4 (On profunctors)

(a) *Products and sums.* In $\mathbb{C}\text{at}$ the product $u\colon X \nrightarrow Y$ of a family of profunctors $u_i\colon X_i \nrightarrow Y_i$ is $u((x_i),(y_i)) = \prod u_i(x_i,y_i)$.

Their sum $\sum u_i\colon \sum X_i \nrightarrow \sum Y_i$ takes an object (x,y) of the category $(\sum X_i)^{\text{op}} \times \sum Y_i$ to the set $u_i(x,y)$ if $(x,y) \in X_i \times Y_i$, and to $\emptyset$ when this does not occur.

(b) *Equalisers and coequalisers.* Equalisers and coequalisers in $\mathbf{Cat}$ have been recalled in 1.3.1 and 1.3.2. In degree 1, a pair of parallel double cells $\sigma,\tau\colon u \to v$ in $\text{Hor}_1\mathbb{C}\text{at} = \mathbf{Cat}/\mathbf{2}$ amounts to a pair of functors

$$\sigma,\tau\colon X +_u Y \to X' +_v Y'$$

over $\mathbf{2}$ (see (3.50)); their ordinary (co)equaliser is a category over $\mathbf{2}$, yielding the (co)limit of the transformation. This agrees with faces and degeneracy.

For instance the equaliser of σ,τ is the profunctor $\overline{u}\colon \overline{X} \to \overline{Y}$ where $\overline{X}$ and $\overline{Y}$ are the equalisers of the faces of σ and τ, while

$$\overline{u}(x,y) = \{\lambda \in u(x,y)\,|\,\sigma(\lambda) = \tau(\lambda)\}.$$

(c) The tabulator and cotabulator of a profunctor $u\colon X \nrightarrow Y$ have been computed in 3.7.4.

For their functoriality, we start from a pseudo vertical transformation (x,y,φ), where φ is a special isocell between composed profunctors, as in (5.17)

$$(x,y,\varphi)\colon u \nrightarrow v\colon \mathbf{2}^{\mathrm{t}} \to \mathbb{C}\text{at},$$

$$\varphi\colon u \otimes y \to x \otimes v\colon A \nrightarrow B'.$$

(C.63)

Its limit $t = \top(x, y, \varphi)\colon \top u \nrightarrow \top v$ is induced by the product profunctor $x \times y\colon A \times B \nrightarrow A' \times B'$

$$t\colon \top u^{\mathrm{op}} \times \top v \to \mathbf{Set}, \quad (a, b, \lambda; c, d, \mu) \mapsto x(a, c) \times y(b, d),$$

$$(f, g; h, k) \mapsto x(f, h) \times y(g, k)\colon \tag{C.64}$$

$$x(a, c) \times y(b, d) \to x(a', c') \times y(b', d'),$$

for $(f, g)\colon (a', b', \lambda') \to (a, b, \lambda)$ in $\top u$ and $(h, k)\colon (c, d, \mu) \to (c', d', \mu')$ in $\top v$.

Its colimit $s = \bot(x, y, \varphi)\colon \bot u \nrightarrow \bot v$ is computed as follows, on the objects

$$s\colon (A +_u B)^{\mathrm{op}} \times (A' +_v B') \to \mathbf{Set},$$

$$s(a, a') = x(a, a'), \qquad s(b, b') = y(b, b'), \tag{C.65}$$

$$s(a, b') = (x \otimes v)(a, b'), \qquad s(b, a') = \emptyset.$$

Note that on an object $(a, b') \in \mathrm{Ob}(A^{\mathrm{op}} \times B')$ it acts as the 'diagonal' $x \otimes v \cong u \otimes y\colon A \nrightarrow B'$ of (C.63).

C5.6 *Exercises of 5.4.6* (Level limits in $\mathbb{Dbl}$)

(a) *Products.* In the double category $\mathbb{Dbl}$ every (small) family $(\mathbb{A}_i)$ of weak double categories has a *transpositive product* $(\mathbb{A}, (P_i))$, the obvious cartesian product $\mathbb{A}$ equipped with the projections $P_i\colon \mathbb{A} \to \mathbb{A}_i$.

In fact, these are strict double functors, and $(\mathbb{A}, (P_i))$ is at the same time the horizontal and vertical product of $(\mathbb{A}_i)$; namely, for every family of *lax* (resp. *colax*) functors $F_i\colon \mathbb{X} \to \mathbb{A}_i$ there is precisely one lax (resp. colax) functor $F\colon \mathbb{X} \to \mathbb{A}$ such that $P_i F = F_i$ (for all indices i)

$$F(x) = (F_i(x))_i, \qquad \underline{F}(x) = (\underline{F}_i(x))_i \tag{C.66}$$

(in the first case x is any item of $\mathbb{X}$, in the second any object or any pair of consecutive vertical arrows).

Furthermore:

- every family $(U_i\colon \mathbb{A}_i \to \mathbb{B}_i)$ of colax functors has an obvious *horizontal product* $U\colon \mathbb{A} \to \mathbb{B}$, with projections $\pi_i\colon (U \overset{P_i}{\underset{Q_i}{\to}} U_i)$ that are consistent with the projections of $\mathbb{A}$ and $\mathbb{B}$,

- every family $(F_i\colon \mathbb{A}_i \to \mathbb{B}_i)$ of lax functors has an obvious *vertical product* $F\colon \mathbb{A} \to \mathbb{B}$, with projections $\pi_i\colon (P_i \overset{F}{\underset{F_i}{\to}} Q_i)$, consistent with the projections of $\mathbb{A}$ and $\mathbb{B}$.

(b) *Equalisers in $\mathbb{Dbl}$ are defective.* A pair $F, G\colon \mathbb{A} \to \mathbb{B}$ of *strict* functors has a horizontal (and vertical) equaliser $E\colon \mathbb{E} \to \mathbb{A}$, given by the strict embedding of the weak double subcategory $\mathbb{E}$ of $\mathbb{A}$ formed by all the items of $\mathbb{A}$ on which F, G coincide. This works because strict functors (of weak double categories) strictly preserve the whole structure (see 3.5.1), but would not work for lax or colax functors.

C5.7 Exercises of 5.4.7 (Tabulators in $\mathbb{Dbl}$)

(a) Every colax functor $U \colon \mathbb{A} \to \mathbb{B}$ has a horizontal tabulator $(\mathbb{T}, P, Q, \tau)$.

The weak double category $\mathbb{T} = U \Downarrow \mathbb{B}$ is a 'one-sided' double comma (see 4.2.4), with strict projections P and Q. We write the vertical companion Q_* as Q; the unit η is simply represented by the horizontal transformation $1_Q \colon Q \to Q$ and the tabulator cell τ is related to the comma-cell π by the unit η and counit ε of the companionship of Q with 'itself'

$$
\begin{array}{ccccc}
\mathbb{T} & \xrightarrow{\;1\;} & U \Downarrow \mathbb{B} & \xrightarrow{\;P\;} & \mathbb{A} \\
{\scriptstyle e}\downarrow & \eta & {\scriptstyle Q}\downarrow & \pi & \downarrow{\scriptstyle U} \\
\mathbb{T} & \xrightarrow{\;Q\;} & \mathbb{B} & \xrightarrow{\;1\;} & \mathbb{B}
\end{array}
\qquad \tau = \eta \,|\, \pi, \quad \pi = \tau \otimes \varepsilon. \tag{C.67}
$$

To be more explicit, the tabulator $\mathbb{T}$ has objects

$$(A, B, b \colon UA \to B), \tag{C.68}$$

with A in $\mathbb{A}$ and b horizontal in $\mathbb{B}$. A horizontal arrow of $\mathbb{T}$

$$(a, b) \colon (A_1, B_1, b_1) \to (A_2, B_2, b_2), \tag{C.69}$$

makes a commutative square in $\mathrm{Hor}_0\mathbb{B}$, as in the upper square of the diagram below (where the slanting direction must be viewed as horizontal). A vertical arrow of $\mathbb{T}$

$$(u, v, \omega) \colon (A_1, B_1, b_1) \to (A_3, B_3, b_3), \tag{C.70}$$

'is' a double cell in $\mathbb{B}$, as in the left square of the diagram below. A double cell (β, β') of $\mathbb{T}$ forms a commutative diagram of double cells of $\mathbb{B}$

$$(\beta, \beta') \colon \left((u, v, \omega) \xrightarrow[(a',b')]{(a,b)} (u', v', \omega') \right), \qquad \beta \,|\, \omega' = \omega \,|\, \beta', \tag{C.71}$$

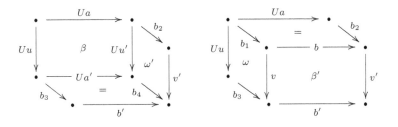

The composition laws of $\mathbb{T}$ are obvious, as well as the (strict) double functors P, Q. The double cell τ has components

$$\tau(A, B, b) = b\colon UA \to B, \qquad \tau(u, v, \omega) = \omega\colon Uu \to v. \qquad \text{(C.72)}$$

Its universal property follows trivially from that of the double comma, in Theorem 4.2.5(a).

(b) We have thus a span representation

$$S\colon \mathbb{D}\mathrm{bl} \to \mathrm{Span}(\mathrm{Lx}\mathbf{Dbl}), \qquad \text{(C.73)}$$

where $\mathrm{Lx}\mathbf{Dbl} = \mathrm{Hor}_0\mathbb{D}\mathrm{bl}$ is the category of weak double categories and lax functors. (Note that, even though the projections P, Q of the double comma are strict double functors, a cell $\varphi\colon U \to V$ in $\mathbb{D}\mathrm{bl}$ gives a *lax* functor $\mathsf{T}\varphi\colon \mathsf{T}U \to \mathsf{T}V$.)

To prove that $\mathbb{D}\mathrm{bl}$ is horizontally span representable, we use the vertical universal property of the double comma $\mathbb{T} = U \Downarrow \mathbb{B}$, in Theorem 4.2.5(b), and deduce the existence of a colax functor $W\colon \mathbb{A} \to \mathbb{T}$ and a cell ξ such that $\xi \otimes \pi = 1_U$ (and $QW = U$)

$$\text{(C.74)}$$

Now a cell $\varphi\colon (U \overset{F}{\underset{G}{\to}} V)$ in $\mathbb{D}\mathrm{bl}$ can be recovered from the lax functor $\mathsf{T}\varphi\colon \mathsf{T}U \to \mathsf{T}V$ as follows

$$\varphi = 1_U \mid \varphi = (\xi \otimes \pi) \mid (e_F \otimes \varphi) = (\xi \otimes t_U \otimes \varepsilon) \mid (e_F \otimes \varphi \otimes e_G)$$
$$= (\xi \mid e_F) \otimes (t_U \mid \varphi) \otimes (\varepsilon \mid e_G) = (\xi \mid e_F) \otimes (\mathsf{T}\varphi \mid t_V) \otimes (\varepsilon \mid e_G).$$

(c) The vertical analogues of the previous topics come from the isomorphism $(-)^{\mathrm{h}}\colon \mathbb{D}\mathrm{bl}^{\mathrm{t}} \to \mathbb{D}\mathrm{bl}$ that sends an object $\mathbb{A}$ to the horizontal opposite $\mathbb{A}^{\mathrm{h}}$, and transposes double cells (in 4.2.3(a)).

Thus, in the *vertical tabulator* $(\mathbb{T}', P, Q, \tau)$ the weak double category $\mathbb{T}' = \mathbb{B} \Downarrow F$ has objects $(A, B, b\colon B \to FA)$, and the cell τ is vertically universal

Dbl is also *vertically span representable*, which means that $\mathbb{D}\mathrm{bl}^t$ is span representable by a lax functor

$$S' \colon \mathbb{D}\mathrm{bl}^t \to \mathbb{S}\mathrm{pan}(\mathrm{Cx}\mathbf{Dbl}).$$

The latter sends a lax functor $F \colon \mathbb{A} \to \mathbb{B}$ to the span

$$S'(F) = (\mathbb{A} \leftarrow \mathbb{T}' \to \mathbb{B})$$

associated to the vertical tabulator.

C5.8 Exercises of 5.4.8 (On rings and bimodules)

(a) Level limits in the weak double category $\mathbb{R}\mathrm{ng}$ of (unitary) rings, homomorphisms and bimodules are computed separately on rings and bimodules. Thus a product of vertical arrows $X_i \colon R_i \to S_i$ is $\Pi X_i \colon \Pi R_i \to \Pi S_i$, with the obvious bimodule structure on ΠX_i. Similarly for the equaliser of two cells $h \colon (X \overset{f}{\underset{g}{\to}} X')$, $h' \colon (X \overset{f'}{\underset{g'}{\to}} X')$.

The 1-dimensional cotabulator of the bimodule $X \colon R \to S$ is the free ring U generated by the set $|R| + |S| + |X|$ modulo the bilateral ideal J that forces the embeddings of R, S, X into U to become two ring homomorphisms and a homomorphism of bimodules, so to give a double cell $X \to e(U/J)$.

A reader familiar with the Adjoint Functor Theorem mentioned in 1.5.4 can deduce the existence of the left adjoint of $e \colon \mathrm{Hor}_0\mathbb{R}\mathrm{ng} \to \mathrm{Hor}_1\mathbb{R}\mathrm{ng}$ from the previous results on level limits, by verifying the Solution Set Condition – which is not difficult.

(b) The 1-dimensional initial object is the ring $\mathbb{Z}$ of integers, but the initial vertical arrow is the zero bimodule $0 \colon \mathbb{Z} \to \mathbb{Z}$ instead of the vertical identity of $\mathbb{Z}$. Thus $(\mathbb{Z}, 0)$ is an *initial pair* (in the sense of 5.1.5) and there is no initial object, in the full 2-dimensional sense. More formally, the double functor $\mathbb{R}\mathrm{ng} \to \mathbf{1}$ has a lax left adjoint $\mathbf{1} \to \mathbb{R}\mathrm{ng}$ which is *not unitary*: a vertical monad $0 \colon \mathbb{Z} \to \mathbb{Z}$, with trivial unit and multiplication.

(c) Therefore the ordinary functor $e \colon \mathrm{Hor}_0\mathbb{R}\mathrm{ng} \to \mathrm{Hor}_1\mathbb{R}\mathrm{ng}$ does not preserve the initial object and cannot have a right adjoint: $\mathbb{R}\mathrm{ng}$ cannot have all 1-dimensional tabulators.

(d) Similarly the weak double category of monoids, homomorphisms and bimodules has level limits and 1-dimensional cotabulators, but few level colimits and tabulators. The singleton monoid $X = \{*\}$ gives an initial pair $(X, \emptyset \colon X \to X)$, when equipped with the initial vertical arrow.

C5.9 Exercises of 5.5.7

(a) This simply depends on the ordinary adjunctions $F_n \dashv G_n$ in degree $n = 0, 1$ (see 4.3.2(c)), which are preserved by faces and degeneracy.

(b) As we have seen in Exercise 3.8.6(b), a lax functor $F: \mathbf{1} \to \mathbb{A}$ amounts to a vertical monad (A, t, η, μ) in $\mathbb{A}$, i.e. a monad in the vertical bicategory VerA (where $\eta = \underline{F}(0)$ and $\mu = \underline{F}(e_0, e_0)$). F is strict if and only if it is unitary, and then it amounts to the object $A = F(0)$, with its trivial vertical monad $(A, e_A, 1, 1)$.

Every object A has a *unary product*, the limit of the strict functor $A: \mathbf{1} \to \mathbb{A}$, which is A.

More generally, every lax functor $F = (A, t, \eta, \mu): \mathbf{1} \to \mathbb{A}$ has a limit cone (A, p)

$$p: A \to F: \mathbf{1} \to \mathbb{A}, \quad p(0) = 1_A, \quad p(e_0) = \eta: e_A \to F(e_0) = t. \quad \text{(C.75)}$$

(c) For a lax functor $S: \mathbb{A} \to \mathbb{B}$, the composite $T = S.A: \mathbf{1} \to \mathbb{A} \to \mathbb{B}$ is a vertical monad in $\mathbb{B}$

$$T(0) = S(A), \qquad T(e_0) = S(e_A) = s: SA \rightarrowtail SA,$$
$$\underline{T}(0) = \underline{S}(A): e_{SA} \to s, \quad \underline{T}(e_0, e_0) = \underline{S}(e_A, e_A): s \otimes s \to s. \quad \text{(C.76)}$$

Its limit, by (b), is the cone $(S(A), q)$, with $q(0) = 1_{SA}$ and $q(e_0) = \underline{S}(A)$.

Therefore $S: \mathbb{A} \to \mathbb{B}$ takes the limit cone $(A, 1)$ of the strict functor $A: \mathbf{1} \to \mathbb{A}$ to the limit cone $(S(A), q)$ of the composed functor $S.A: \mathbf{1} \to \mathbb{B}$ (in the sense made precise above), *even though the latter is no longer a unary product.*

C5.10 Exercises of 5.6.2

(b) The functor $p: \mathbf{S} \to \mathbf{C}(L, X)$ must satisfy:

(i) for every object A in $\mathbf{C}$ and every functor $h: \mathbf{S} \to \mathbf{C}(A, X)$ there is a unique morphism $f: A \to L$ in $\mathbf{C}$ such that $h = \mathbf{C}(L, f).p: \mathbf{S} \to \mathbf{C}(A, X)$,

(ii) for every natural transformation $\xi: h \to k: \mathbf{S} \to \mathbf{C}(A, X)$ there is a unique 2-cell $\alpha: f \to g: A \to L$ in $\mathbf{C}$ such that $\xi = \mathbf{C}(L, \alpha).p: \mathbf{S} \Rightarrow \mathbf{C}(A, X)$.

C5.11 Exercises of 5.6.9

(a) The 1-dimensional universal property of the comma-object is obvious. The double category $\mathbb{I} = \mathbb{El}(W)$ is represented in the left diagram below

$$
\begin{array}{ccc}
0' \longrightarrow 1' & & X' \xrightarrow{\ f\ } Y' \\
\Big\downarrow & & \Big\downarrow{\scriptstyle u} \\
0'' \longrightarrow 1'' & & X'' \xrightarrow[g]{} Y''
\end{array}
\qquad (\text{C.77})
$$

A functor $F\colon \mathbb{I} \to \mathbb{A}$ with values in a weak double category is represented at the right, and its double limit can be constructed by tabulators and pullbacks, provided they exist. (The construction is easy, but one can also follow the proof of Theorem 5.3.6.) A tabulator is a particular case, when f and g are identities.

(b) The inserter of $f, g\colon X \to Y$ is an object H equipped with a morphism $i\colon H \to X$ and a cell $\varphi\colon fi \to gi\colon H \to Y$. The 1-dimensional universal property is:

- for every $i'\colon H' \to X$ and $\varphi'\colon fi' \to gi'\colon H' \to Y$ there is precisely one $h\colon H' \to H$ such that $i' = ih$ and $\varphi' = \varphi h$.

In cases (b) and (c), the double categories $\mathbb{El}(W)$ are, respectively:

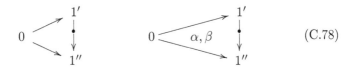

$$(\text{C.78})$$

(d) It is a consequence of Theorem 5.6.6, together with various exercises of this section. First, it is easy to see that 2-products (resp. 2-equalisers) in $\mathbf{C}$ are 'the same' as double products (resp. double equalisers) in $\mathbb{QC}$. Second, we know that the cotensor $\mathbf{2} \pitchfork X$ amounts to the tabulator of the vertical identity of X, from Exercise 5.6.2(c).

Conversely, if the $\mathbf{C}$-morphism $u\colon X \to Y$ is viewed as vertical in $\mathbb{QC}$, its tabulator $(\mathsf{T}u, p, q, \pi)$ can be constructed as the following inserter (see point (b), above)

$$
\mathsf{T}u \xrightarrow{\ i\ } X \times Y \underset{p''}{\overset{up'}{\rightrightarrows}} Y \qquad \pi\colon up'i \to p''i\colon \mathsf{T}u \to Y, \qquad (\text{C.79})
$$

letting $p = p'i\colon \mathsf{T}u \to X$ and $q = p''i\colon \mathsf{T}u \to Y$.

If $\mathbf{C}$ is 2-complete, $\mathbb{QC}$ is span representable because the lax span-representation $S\colon \mathbb{QC} \to \mathbb{Span}(\mathbf{C})$ defined in 3.6.4 operates on a double cell $\alpha\colon vf \to gu$ of $\mathbb{Q}\mathbf{C}$ producing a morphism of spans $S\alpha\colon Su \to Sv$ whose central map $\mathsf{T}\alpha\colon \mathsf{T}u \to \mathsf{T}v$ is defined as follows

$$
\begin{array}{ccccc}
\mathsf{T}u & \xrightarrow{\ i\ } & X\times Y & \underset{p''}{\overset{up'}{\rightrightarrows}} & Y \\
{\scriptstyle \mathsf{T}\alpha}\big\downarrow & & {\scriptstyle f\times g}\big\downarrow & & \big\downarrow{\scriptstyle g} \\
\mathsf{T}v & \xrightarrow[\ j\ ]{} & X'\times Y' & \underset{q''}{\overset{vq'}{\rightrightarrows}} & Y'
\end{array}
\qquad (\text{C.80})
$$

$$\pi_u\colon up'i \to p''i\colon \mathsf{T}u \to Y, \qquad\qquad \pi_v\colon vq'j \to q''j\colon \mathsf{T}v \to Y',$$

$$j.\mathsf{T}\alpha = (f\times g)i, \qquad\qquad \pi_v.\mathsf{T}\alpha = g\pi_u.\alpha p'i\colon vfp'i \to gp''i.$$

Now f and g are determined as the vertical faces of the morphism $S\alpha$. To recover the 2-cell $\alpha\colon vf \to gu$ of $\mathbf{C}$ from the morphism $\mathsf{T}\alpha$, one uses the map $h\colon X \to \mathsf{T}u$ determined by the conditions $ih = (1, u)\colon X \to X\times Y$ and $\pi_u.h = 1_u$, so that

$$\pi_v.\mathsf{T}\alpha.h = (g\pi_u.\alpha p'i)h = g\pi_u h.\alpha p'ih = \alpha p'(1, u) = \alpha.$$

C6 Exercises of Chapter 6

C6.1 *Exercises of 6.1.5* (Higher tabulators in Stc$\mathbb{D}$bl, I)

An object X of the weak double category $\mathbb{T}$ consists of four objects, one in each of $\mathbb{A}, \mathbb{B}, \mathbb{C}, \mathbb{D}$, and four horizontal morphisms of $\mathbb{B}, \mathbb{C}, \mathbb{D}$ (two of them in $\mathbb{D}$)

$$X = (A, B, C, D, b, c, d', d),$$

$$b\colon B \to FA, \quad c\colon UA \to C, \quad d'\colon D \to GC, \quad d\colon VB \to D,$$

$$(\text{C.81})$$

so that the following pentagon of horizontal arrows commutes in $\mathbb{D}$

$$
\begin{array}{ccccc}
VB & \xrightarrow{\ d\ } & D & \xrightarrow{\ d'\ } & GC \\
{\scriptstyle Vb}\searrow & & & \nearrow{\scriptstyle Gc} & \\
& VFA & \xrightarrow[\pi A]{} & GUA &
\end{array}
\qquad (\text{C.82})
$$

The arrows and double cells of $\mathbb{T}$ are essentially as in 5.4.7, if more complicated. The strict double functors P, Q, R, S are obvious projections and the double cells $\varphi, \psi, \omega, \zeta$ have the following components on the object X of (C.81) (and similar components on the vertical arrows of $\mathbb{T}$, which we have not described)

$$\varphi X = b\colon B \to FA, \qquad \omega X = c\colon UA \to C,$$

$$\psi X = d\colon D \to GC, \qquad \zeta X = d'\colon VB \to D.$$

C6.2 Exercises of 6.1.6 (Higher tabulators in StcDbl, II)

(b) $\mathsf{T}_2\pi$ is a lax double functor $\mathsf{T}U \to_1 \mathsf{T}V$ between the tabulators of the two vertical arrows $\partial_1^\alpha \pi$, namely U and V (described in 6.1.3)

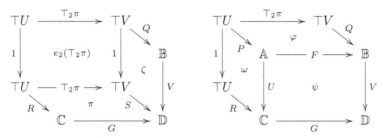

Thus the tabulator $\mathsf{T}U$ has objects $(A, C, c\colon UA \to C)$, the tabulator $\mathsf{T}V$ has objects $(B, D, d\colon VB \to D)$ and

$$(\mathsf{T}_2\pi)(A, C, c\colon UA \to C) = (FA, GC, Gc.\pi A\colon VFA \to GC),$$

$$\varphi(A, C, c\colon UA \to C) = 1_{FA}, \quad \psi(A, C, c\colon UA \to C) = 1_{GC},$$

$$\omega(A, C, c\colon UA \to C) = c, \quad \zeta(B, D, d\colon VB \to D) = d.$$

(d) Computing for instance $\mathsf{T}_1\mathsf{T}_2\pi$, we find that an object is a family

$$((A, C, c), (B, D, d), b, d'),$$

$$c\colon UA \to C, \quad d\colon VB \to D, \quad b\colon B \to FA, \quad d'\colon D \to GC,$$

(with A in $\mathbb{A}$, etc.) such that the following square of horizontal arrows commutes in $\mathbb{D}$, as in the pentagon (C.82)

$$
\begin{array}{ccc}
VB & \xrightarrow{\ d\ } & D \\
{\scriptstyle Vb}\downarrow & & \downarrow{\scriptstyle d'} \\
VFA & \xrightarrow[Gc.\pi A]{} & Gc
\end{array}
$$

C6.3 Exercises of 6.7.3

We compute all tabulators in $\mathsf{SC}(\mathbf{C})$.

(a) The tabulator of a 1-arrow f (i.e. a span) is already known from the theory of weak double categories. It is an object $\mathsf{T}_1 f$ with a universal 1-map $e_1(\mathsf{T}_1 f) \to_0 f$, and the solution is the (trivial) limit of the span f, i.e. its middle object.

(b) The tabulator of a 2-arrow u (i.e. a cospan) is also known. It is an object $\mathsf{T}_2 u$ with a universal 2-map $e_2(\mathsf{T}_2 u) \to_0 u$, and the solution is the pullback of u.

(c) There are then three tabulators of a 12-cube $\pi\colon \vee \times \wedge \to \mathbf{C}$ (a span of cospans). First, the total tabulator $\mathsf{T}_{12}\pi$ is an object with a universal 12-map $e_{12}(\mathsf{T}_{12}\pi) \to_0 \pi$; the solution is the limit of the diagram, i.e. the pullback of its central cospan.

(d) The e_1-tabulator of π is a 2-arrow $\mathsf{T}_1\pi$ (a cospan) with a universal 12-map $e_1(\mathsf{T}_1\pi) \to_0 \pi$; the solution is the middle cospan of π.

(e) The e_2-tabulator of π is a 1-arrow $\mathsf{T}_2\pi$ (a span) with a universal 12-map $e_2(\mathsf{T}_2\pi) \to_0 \pi$; the solution is the obvious span whose objects are the pullbacks of the three cospans of π.

(f) These limits are preserved by faces and degeneracies. For instance:

- $\partial_1^-(\mathsf{T}_2\pi) = \mathsf{T}_2(\partial_1^-\pi)$, which means that the domain of the span $\mathsf{T}_2\pi$ (described above) is the pullback of the cospan $\partial_1^-\pi$,

- $\mathsf{T}_2(e_1 u) = e_1(\mathsf{T}_2 u)$, i.e. the e_2-tabulator of the degenerate 12-cube $e_1 u$ (on the cospan u) is the degenerate span on the pullback of u.

C6.4 Exercises of 6.7.8

(a) Of course, for every category $\mathbf{C}$ with pullbacks, the weak cubical category $\mathsf{Span}(\mathbf{C})$ is span representable, in a strict sense: the functor $T\colon \mathsf{Span}(\mathbf{C}) \to \mathsf{Span}(\mathbf{C})$ is an isomorphism, and even the identity for the natural choice of the total tabulator of a cubical span, namely its central object.

(b) In the same hypothesis on $\mathbf{C}$, it is easy to see that also the cubical categories $\mathsf{Pmap}(\mathbf{C})$ and $\mathsf{Piso}(\mathbf{C})$ of 6.6.5 are span representable, by pseudo functors

$$T\colon \mathsf{Pmap}(\mathbf{C}) \to \mathsf{Span}(\mathbf{C}), \quad T\colon \mathsf{Piso}(\mathbf{C}) \to \mathsf{Span}(\mathbf{C}).$$

(c) The chiral triple category $\mathsf{SC}(\mathbf{Set})$ of spans and cospans of sets is neither span nor cospan representable. In fact $\mathsf{SC}(\mathbf{Set})$ amalgamates the weak double categories $\mathbb{S}\mathrm{pan}\mathbf{Set}$ and $\mathbb{C}\mathrm{osp}\mathbf{Set}$; the conclusion follows easily from the Exercises 3.7.3(b), (c).

C6.5 Exercises of 6.9.3

(a) The tabulator of a 1-arrow $f = 0_{XY}$ (i.e. a null span) is an object $T = \mathsf{T}_1 f$ with a universal 1-map $e_1(T) = 0_{TT} \to_0 f$; the solution is the product $X \times Y$ in $\mathbf{C}$.

(b) The tabulator of a 2-arrow u (a cospan) is an object $\mathsf{T}_2 u$ with a universal 2-map $e_2(\mathsf{T}_2 u) \to_0 u$; the solution is the pullback of u in $\mathbf{C}$.

(c) The total e_1e_2-tabulator $T = \mathsf{T}_{12}\pi$ is an object with a universal 12-map $e_1e_2(T) \to_0 \pi$, where $e_1e_2(T) = ((1_T, 1_T), (1_T, 1_T), (1_T, 1_T))$ (see (6.151)). The solution is the limit in $\mathbf{C}$ of π, viewed as the left diagram below

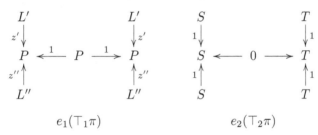

(c') The total e_2e_1-tabulator $T = \mathsf{T}_{21}\pi$ is an object with a universal 12-map $e_2e_1(T) \to_0 \pi$, where $e_2e_1(T) = ((1_T, 1_T), 0_{TT}, (1_T, 1_T))$ (see (6.151)). The solution is the limit in $\mathbf{C}$ of the right diagram above, namely the product $\mathsf{T}u \times \mathsf{T}v$ of two pullbacks.

(d) The e_1-tabulator is a 2-arrow (a cospan) $\mathsf{T}_1\pi$ with a universal 12-map $e_1(\mathsf{T}_1\pi) \to_0 \pi$; the solution, as in the left diagram below, is the cospan $z = (L' \to P \leftarrow L'')$, where $L' = \mathrm{Lim}\,(u', f', f'', v')$ is the limit of the upper part of π and $L'' = \mathrm{Lim}\,(u'', f', f'', v'')$ is the limit of its lower part

$$
\begin{array}{ccc}
L' & \quad & L' \\
{\scriptstyle z'}\big\downarrow & & \big\downarrow{\scriptstyle z'} \\
P \xleftarrow{\ 1\ } P \xrightarrow{\ 1\ } P \\
{\scriptstyle z''}\big\uparrow & & \big\uparrow{\scriptstyle z''} \\
L'' & & L''
\end{array}
\qquad
\begin{array}{ccc}
S & \quad & T \\
{\scriptstyle 1}\big\downarrow & & \big\downarrow{\scriptstyle 1} \\
S \xleftarrow{\ 0\ } T \\
{\scriptstyle 1}\big\uparrow & & \big\uparrow{\scriptstyle 1} \\
S & & T
\end{array}
$$

$$
e_1(\mathsf{T}_1\pi) \qquad\qquad e_2(\mathsf{T}_2\pi)
$$

(e) The e_2-tabulator of π is a 1-arrow $\mathsf{T}_2\pi = 0_{ST}$ (a null span) with a universal 12-map $e_2(\mathsf{T}_2\pi) \to_0 \pi$; the solution, as in the right diagram above, is the null span $(S \leftarrow 0 \to T)$ on the pullbacks of the two cospans of π, namely $S = \mathsf{T}_2u$ and $T = \mathsf{T}_2v$.

(f) These limits are *only partially* preserved by faces and degeneracies, as shown by the following cases.

- $\partial_2^\alpha(\mathsf{T}_1\pi)$ *need not coincide with* $\mathsf{T}_1(\partial_2^\alpha\pi)$. For instance, for $\alpha = 0$, the domain L' of the cospan $\mathsf{T}_1\pi$ (described above) need not be the product $X \times Z$ of the 1-faces of $\partial_2^-\pi = 0_{XZ}$.

- $\mathsf{T}_1(e_2f) = e_2(\mathsf{T}_1f)$, i.e. the e_1-tabulator of the 2-degenerate cell

$$
e_2f = ((1_X, 1_X), 0_{XY}, (1_Y, 1_Y))
$$

on the null span $f = 0_{XY}$, namely

$$
(1_X, 1_X) \times (1_Y, 1_Y) = (X \times Y \leftarrow X \times Y \to X \times Y),
$$

coincides with $e_2(X \times Y)$.

- $\partial_1^\alpha(\mathsf{T}_2\pi) = \mathsf{T}_2(\partial_1^\alpha \pi)$, which means, for $\alpha = 0$, that the domain of the null span $\mathsf{T}_2\pi$ (described above) is the pullback of the cospan $\partial_1^- \pi$.

- $\mathsf{T}_2(e_1 u) = e_1(\mathsf{T}_2 u)$, i.e. the e_2-tabulator of the 1-degenerate cell $e_1 u = (u, 0, u)$ on the cospan u is the degenerate span on the pullback of u.

C7 Exercises of Chapters 7, 8

C7.1 Exercise 7.1.3

The proof is by induction, following the same pattern as in 4.5.6 for ordinary spans and cospans.

In degree zero the adjunction is trivial, and $\eta x \colon x \to GFx$ is the identity, for every object x of $\mathbf{C}$.

Supposing that $F\eta x \colon Fx \to FGFx$ is invertible for every cube of degree $< n$, we take an $\mathbf{i}$-cube x of degree n and the transversal map $\eta_{\mathbf{i}} x \colon x \to G_{\mathbf{i}} F_{\mathbf{i}} x$ of the unit.

The inductive assumption says that all the faces

$$\partial_i^\alpha(F_{\mathbf{i}}\eta_{\mathbf{i}}(x)) = F_{\mathbf{i}|i}(\eta_{\mathbf{i}|i}(\partial_i^\alpha x))$$

are invertible.

Now the transversal map $F\eta x \colon Fx \to FGFx \colon \wedge^{\mathbf{i}} \to \mathbf{C}$ of n-cubical cospans has 3^n components, and we already know that all of them are invertible, excepted the central one on the vertex $\underline{\iota} = (\iota, ..., \iota)$. By definition of F in (7.4), this component is computed as a colimit

$$(F_{\mathbf{i}}\eta_{\mathbf{i}} x)(\underline{\iota}) = \mathrm{Colim}\,(\eta x) \colon \mathrm{Colim}\, x \to \mathrm{Colim}\,(GFx). \tag{C.83}$$

As in 4.5.6 (for $n = 1$), composing with $\eta x \colon x \to GFx$ gives a bijective correspondence between the cocones of the diagram $x \colon \vee^{\mathbf{i}} \to \mathbf{C}$ and those of GFx, whence $(F_{\mathbf{i}}\eta_{\mathbf{i}} x)(\underline{\iota})$ is an isomorphism.

The proof is done, but it may help to draw what happens for a 2-cubical span $x \colon \vee^{\mathbf{i}} \to \mathbf{C}$, represented as the left part of the following diagram, with vertices $x_{00}, ..., x_{\iota\iota}$

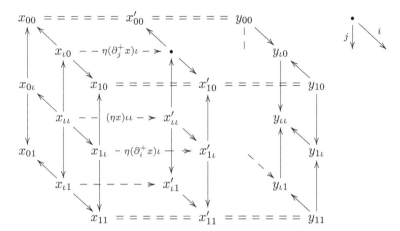

We form the 2-cubical cospan $y = Fx\colon \wedge^{\mathbf{i}} \to \mathbf{C}$ (partially drawn at the right) by four pushouts and $y_{\iota\iota} = \mathrm{Colim}\,x$. Then we take the 2-cubical span $x' = GFx\colon \vee^{\mathbf{i}} \to \mathbf{C}$ (partially drawn in the middle) by four pullbacks and $x'_{\iota\iota} = \mathrm{Lim}\,y$.

On the four vertices $\partial_i^\alpha \partial_j^\beta$ we have $x_{\alpha\beta} = y_{\alpha\beta} = x'_{\alpha\beta}$ (for $\alpha, \beta = 0, 1$). Then we have four other components of $\eta\colon x \to GFx$

$$\eta_j(\partial_i^\alpha x)\colon \partial_i^\alpha x \to GF(\partial_i^\alpha x), \qquad \eta_i(\partial_j^\beta x)\colon \partial_j^\beta x \to GF(\partial_j^\beta x),$$

defined by the universal property of four pullbacks.

Finally the central component $(\eta_{\mathbf{i}} x)_{\iota\iota}\colon x_{\iota\iota} \to x'_{\iota\iota}$ is defined by the universal property of $x'_{\iota\iota}$ as $\mathrm{Lim}\,y$.

C7.2 Exercise 7.2.3

We omit the index i in the composition $+_i$ and the comparisons $\underline{F}_i$, $\underline{R}_i$, κ_i. Let us rewrite the two **i**-cubes of $F \downarrow R$ we are interested in as:

$$(a, x; c) + ((a', x'; c') + (a'', x''; c'')) = (a_1, x_1; c_1),$$
$$((a, x; c) + (a', x'; c')) + (a'', x''; c'') = (a_2, x_2; c_2). \tag{C.84}$$

These are computed by the following formulas and commutative diagrams

$$a_1 = a + (a' + a''), \qquad x_1 = x + (x' + x''),$$
$$a_2 = (a + a') + a'', \qquad x_2 = (x + x') + x'',$$
$$c_1\colon Fa_1 \to Rx_1, \qquad c_2\colon Fa_2 \to Rx_2,$$

$$
\begin{array}{ccc}
Fa_1 & \xrightarrow{\ \ c_1\ \ } & Rx_1 \\
F\downarrow & & \uparrow R \\
Fa + F(a' + a'') & & Rx + R(x' + x'') \\
1+F\downarrow & & \uparrow 1+R \\
Fa + (Fa' + Fa'') & \xrightarrow[\ c+(c'+c'')\ ]{} & Rx + (Rx' + Rx'')
\end{array}
$$

$$
\begin{array}{ccc}
Fa_2 & \xrightarrow{\ \ c_2\ \ } & Rx_2 \\
F\downarrow & & \uparrow R \\
F(a + a') + Fa'' & & R(x + x') + Rx'' \\
F+1\downarrow & & \uparrow R+1 \\
(Fa + Fa') + Fa'' & \xrightarrow[\ (c+c')+c''\ ]{} & (Rx + Rx') + Rx''
\end{array}
$$

Now our claim, i.e. the condition on $(\kappa(\underline{a}), \kappa(\underline{x}))$ expressed in diagram (7.26), amounts to

$$
R\kappa(\underline{x}).c_1 = c_2.F\kappa(\underline{a}) \colon Fa_1 \to Rx_2.
$$

This follows from three facts. First, the naturality of κ on the triple of transversal maps $(c, c', c'') \colon F\underline{a} \to R\underline{x}$ gives the commutative diagram

$$
\begin{array}{ccc}
Fa + (Fa' + Fa'') & \xrightarrow{\ \kappa(F\underline{a})\ } & (Fa + Fa') + Fa'' \\
{\scriptstyle c+(c'+c'')}\downarrow & & \downarrow{\scriptstyle (c+c')+c''} \\
Rx + (Rx' + Rx'') & \xrightarrow[\ \kappa R(\underline{x})\ ]{} & (Rx + Rx') + Rx''
\end{array}
$$

Second and third, we apply the coherence axiom (lmf.4) of 6.5.2, for the lax functor R and the associator κ, and the corresponding axiom (cmf.4), for the colax functor F and κ

$$
R\kappa(\underline{x}).\underline{R}(x, x' + x'').(1 + \underline{R}(x', x''))
$$
$$
= \underline{R}(x + x', x'').(\underline{R}(x, x') + 1).\kappa R(\underline{x}),
$$
$$
(\underline{F}(a, a') + 1).\underline{F}(a + a', a'').F\kappa(\underline{a})
$$
$$
= \kappa(F\underline{a}).(1 + \underline{F}(a', a'')).\underline{F}(a, a' + a'').
$$

C7.3 Exercises of 7.6.5

All points are easy and straightforward. Points (c) and (d) are not used elsewhere, and are entirely left to an interested reader, which will note that point (b) follows from them.

(a) For a positive multi-index $\mathbf{i}$ we let $\mathrm{tv}_{\mathbf{i}}(A^{\mathbf{X}}) = (\mathrm{tv}_{\mathbf{i}}A)^{\mathbf{X}}$, an exponential of ordinary categories: an **i**-cube is an ordinary functor $F: \mathbf{X} \to \mathrm{tv_i A}$, and an **i**-map $\varphi: F \to G$ is a natural transformation of these functors. The multiple structure of $\mathsf{A}^{\mathbf{X}}$ is now defined pointwise, by that of A.

Faces and degeneracies of an **i**-cube F and an **i**-map $\varphi: F \to G$ work in the obvious way on objects and arrows of $\mathbf{X}$ (it is understood that $i \in \mathbf{i}$ for faces and $i \notin \mathbf{i}$ for degeneracies)

$$(\partial_i^\alpha F)(x) = \partial_i^\alpha(Fx), \qquad (\partial_i^\alpha F)(f) = \partial_i^\alpha(Fx),$$
$$(e_i F)(x) = e_i(Fx), \qquad (e_i F)(f) = e_i(Ff), \qquad (\mathrm{C}.85)$$
$$(\partial_i^\alpha \varphi)(x) = \partial_i^\alpha(\varphi x), \qquad (e_i \varphi)(x) = e_i(F\varphi).$$

Similarly, the i-operations are defined pointwise

$$(F +_i G)(x) = Fx +_i Gx, \quad (F +_i G)(f) = Ff +_i Gf,$$
$$(\varphi +_i \psi)(x) = \varphi x +_i \psi x, \qquad (\mathrm{C}.86)$$

and the same works for unitors, associators and interchangers. For instance

$$\kappa_i(F, G, H)(x) = \kappa_i(Fx, Gx, Hx): Fx +_i (Gx +_i Hx) \to (Fx +_i Gx) +_i Hx.$$

The axioms of chiral multiple categories lift pointwise, from A to $\mathsf{A}^{\mathbf{X}}$.

(b) Obviously an **i**-cube a of A is sent to the constant functor $Da: \mathbf{X} \to \mathrm{tv_i A}$ whose value is a. Similarly an **i**-map $h: a \to b$ of A is sent to a constant natural transformation $Dh: Da \to Db$.

C7.4 Exercises of 7.7.4

(a) For $i = \emptyset$, a $*$-cell is a multiple functor $F: \mathbf{X} \to \mathsf{A}$.

(b) A 0-cell is a transversal transformation $F: \partial_0^- F \to_0 \partial_0^+ F: \mathbf{X} \to \mathsf{A}$ of multiple functors. Similarly, for $i > 0$, an i-cell can be called an i-*directed transformation* $F: \partial_i^- F \to_i \partial_i^+ F: \mathbf{X} \to \mathsf{A}$.

(c) An ij-cell can be called an ij-*modification*; its boundary consists of two i-directed and two j-directed transformations, involving four multiple functors.

C7.5 *Exercises of 8.3.3 (Kleisli algebras for lax monads)

(a) Given a lax monad (T, η, μ) on the chiral multiple category X, we want to construct a chiral multiple category X_T of Kleisli T-algebras.

(i) The transversal category $\mathrm{tv_i}(\mathsf{X}_T)$ is the ordinary category of Kleisli algebras $(\mathbf{X_i})_{T_i}$ (see 1.6.7). In other words:

- an **i**-cube x of X gives an **i**-cube of X_T, which we write as $\hat{x}$,

- an **i**-map $\hat{f} \colon \hat{x} \to \hat{x}'$ comes from an **i**-map $f \colon x \to Tx'$ in X,
- the transversal composition of $\hat{f}$ with $\hat{g} \colon \hat{x}' \to \hat{x}''$ is

$$\hat{g}.\hat{f} = (\mu x''.Tg.f \colon x \to Tx'')\hat{\ }.$$

Let us remark that we already have a family of *ordinary functors* $F_T \colon$ $\mathbf{X_i} \to (\mathbf{X_i})_{T_i}$, which we write omitting the positive multi-index $\mathbf{i}$

$$F_T(x) = \hat{x}, \quad F_T(f \colon x \to x') = (\eta x'.f \colon x \to Tx')\hat{\ }. \tag{C.87}$$

(ii) Faces in X_T come from those of X: $\partial_i^\alpha(\hat{x}) = (\partial_i^\alpha x)\hat{\ }$, $\partial_i^\alpha(\hat{f}) = (\partial_i^\alpha f)\hat{\ }$.

(iii) Degeneracies and concatenations in a positive direction i operate on cubes as in X; on transversal maps their action is modified with the laxity comparisons of T

$$e_i(\hat{f} \colon \hat{x} \to \hat{x}') = (\underline{T}_i(x').e_if \colon e_ix \to T(e_ix'))\hat{\ },$$
$$(\hat{f} \colon \hat{x} \to \hat{x}') +_i (\hat{g} \colon \hat{y} \to \hat{y}') = (\underline{T}_i(x',y').(f +_i g))\hat{\ }. \tag{C.88}$$

(iv) Unitors, associators and interchangers of X_T are the F_T-images of those of X

$$\lambda_i(\hat{x}) = F_T(\lambda_i x) = (\eta x.\lambda_i x)\hat{\ }, \quad \rho_i(\hat{x}) = F_T(\rho_i x) = (\eta x.\rho_i x)\hat{\ },$$
$$\kappa_i(\hat{x}, \hat{y}, \hat{z}) = F_T(\kappa_i(x,y,z)),$$
$$\chi_{ij}(\hat{x}, \hat{y}, \hat{z}, \hat{u}) = F_T(\chi_{ij}(x,y,z,u)).$$

We only verify the interchange-property $f'f +_i g'g = (f' +_i g')(f +_i g)$. This follows from the outer part of the following diagram of transversal arrows in X (omitting the index i)

$$
\begin{array}{ccccccc}
x + y & & & & & & \\
{\scriptstyle f+g}\downarrow & & & & & & \\
Tx' + Ty' & \xrightarrow{Tf'+Tg'} & T^2x'' + T^2y'' & = & T^2x'' + T^2y'' & \xrightarrow{\mu+\mu} & Tx'' + Ty'' \\
{\scriptstyle \underline{T}}\downarrow & & \downarrow{\scriptstyle \underline{T}} & & \downarrow{\scriptstyle \underline{TT}} & & \downarrow{\scriptstyle \underline{T}} \\
T(x' + y') & \xrightarrow[T(f'+g')]{} & T(Tx'' + Ty'') & \xrightarrow[T\underline{T}]{} & T^2(x'' + y'') & \xrightarrow[\mu]{} & T(x'' + y'')
\end{array}
$$

In fact the left square commutes by naturality of $\underline{T}$, the central one by definition of the comparison cells of T^2 and the right one by coherence of $\mu \colon T^2 \to T$ with respect to i-composition.

(b) The lax monad $T = (T, \eta, \mu)$ produces a *strict-lax* adjunction of chiral multiple categories

$$F_T \colon X \rightleftarrows X_T \colon G_T, \quad F_T \dashv G_T, \tag{C.89}$$

by extending the usual procedure.

F_T is already defined, and we know that it preserves the transversal structure; moreover it preserves (strictly) the positive degeneracies of transversal maps (and the same holds for concatenations)

$$F_T(e_i f) = (\eta e_i x'.e_i f)^\wedge = (\underline{T}_i(x').e_i \eta x'.e_i f)^\wedge = e_i(\eta x'.f)^\wedge = e_i(F_T f).$$

The lax functor $G_T \colon \mathsf{X}_T \to \mathsf{X}$ is defined as follows

$$G_T(\hat{x}) = Tx, \; G_T(\hat{f}) = \mu x'.Tf,$$

$$(\underline{G_T})_i(\hat{x}) = \underline{T}_i(x) \colon e_i(Tx) \to T(e_i x) = G_T(e_i \hat{x}),$$

$$(\underline{G_T})_i(\hat{x}, \hat{y}) = \underline{T}_i(x, y) \colon Tx +_i Ty \to T(x +_i y) = G_T(\hat{x} +_i \hat{y}).$$

The unit $\eta \colon 1 \to G_T F_T = T$ coincides with the unit of T; the counit $\varepsilon \colon F_T G_T \to 1$ is defined as

$$\varepsilon(\hat{x}) = (1_{Tx})^\wedge \colon T\hat{x} \to \hat{x}.$$

C7.6 Exercises of 8.5.7

(a) Let us take two consecutive relations $u \colon X \nrightarrow Y$ and $v \colon Y \nrightarrow Z$ as in diagram (4.51).

To prove that $G\underline{F}_1(u, v)$ need not be invertible we take

$$X = Y = \{0, 1\}, \qquad\qquad Z = \{0\},$$

$$U = \{(0,0), (1,0), (1,1)\}, \qquad V = \{(1,0)\}.$$

The mappings u', v' work by first projection, u'' by second projection and $v'' \colon Y \to Z$ is determined. The mapping $X \to F(u +_1 v)$ is invertible and $GF(u +_1 v) \cong Z$ has one element; on the other hand $Fu +_1 Fv$ is the singleton and $G(Fu +_1 Fv) \cong X \times Z$ has two elements.

(b) If $\mathbf{C}$ is abelian, it is easy to see that the premultiple endofunctor GF of $\mathbb{S}\mathrm{pan}(\mathbf{C})$ coincides with the lax endofunctor T of the jointly-monic monad of relations. Its algebras are the relations presented as jointly monic spans, while the coalgebras for FG are the same thing, presented as jointly epic cospans. Strict algebras or coalgebras come out, again, of strict factorisations.

C8 Exercises of Appendices A, B

C8.1 Exercises of A1.3

(a) From property (ii) of A1.2 and the modularity of lattices we have:

$$x \vee f^\bullet y = (x \vee f^\bullet y) \vee f^\bullet 0 = f^\bullet f_\bullet (x \vee f^\bullet y) = f^\bullet(f_\bullet x \vee f_\bullet f^\bullet y)$$
$$= f^\bullet(f_\bullet x \vee (y \wedge f_\bullet 1)) = f^\bullet((f_\bullet x \vee y) \wedge f_\bullet 1))$$
$$= f^\bullet f_\bullet f^\bullet(f_\bullet x \vee y)) = f^\bullet(f_\bullet x \vee y)).$$

The second part of (a) is an easy consequence. Points (b) and (c) are also straightforward.

(d) Using distributivity, we get:

$$f_\bullet(x \wedge x') = f_\bullet((x \wedge x') \vee f^\bullet 0) = f_\bullet((x \vee f^\bullet 0) \wedge (x' \vee f^\bullet 0))$$
$$= f_\bullet(f^\bullet f_\bullet x \wedge f^\bullet f_\bullet x') = f_\bullet f^\bullet(f_\bullet x \wedge f_\bullet x')$$
$$= f_\bullet x \wedge f_\bullet x' \wedge f_\bullet 1 = f_\bullet x \wedge f_\bullet x'.$$

C8.2 Exercises of A2.7

(a) The zero-object $\{*\}$ of the category **Mlc** (and **Ltc**) becomes a vertical double limit and colimit in $\mathbb{M}$lhc (and $\mathbb{L}$thc), with uniquely determined bicommutative cells $f \twoheadrightarrow 1_{\{*\}} \twoheadrightarrow f$

$$
\begin{array}{ccc}
X & \xrightarrow{\ f\ } & X' \\
{\scriptstyle t}\downarrow & = & \downarrow{\scriptstyle t} \\
\{*\} & \longrightarrow & \{*\} \\
{\scriptstyle s}\downarrow & = & \downarrow{\scriptstyle s} \\
X & \xrightarrow[\ f\ ]{} & X'
\end{array}
\qquad
t^\bullet(*) = 1, \quad s_\bullet(*) = 0.
\qquad (\text{C.90})
$$

We have also seen that the kernel and cokernel of a covariant connection $u = (u_\bullet, u^\bullet): X \twoheadrightarrow Y$ in the category **Mlc** (and **Ltc**) are computed as follows

$$m: \downarrow(u^\bullet 0) \twoheadrightarrow X, \qquad m_\bullet(x) = x, \qquad m^\bullet(x) = x \wedge u^\bullet 0,$$
$$p: Y \twoheadrightarrow \uparrow(u_\bullet 1), \qquad p_\bullet(y) = y \vee u_\bullet 1, \qquad p^\bullet(y) = y. \qquad (\text{C.91})$$

These become a vertical double kernel and a vertical double cokernel in $\mathbb{M}$lhc (or $\mathbb{L}$thc). Indeed, given the central bicommutative cell $a: (u \overset{f}{\underset{g}{\to}} v)$ below (with $gu_\bullet = v_\bullet f$, $fu^\bullet = v^\bullet g$), the vertical kernels $m = \ker(u)$ and $n = \ker(v)$ determine the upper bicommutative cell of the following diagram, where the homomorphism f' is the restriction of f (and preserves the new maxima because $f(u^{\bullet}0) = v^{\bullet}g(0) = v^{\bullet}(0)$)

$$
\begin{array}{ccc}
\downarrow(u^{\bullet}0) & \xrightarrow{\ f'\ } & \downarrow(v^{\bullet}0) \\
m\downarrow & = & \downarrow n \\
X & \xrightarrow{\ f\ } & X' \\
u\downarrow & = & \downarrow v \\
Y & \xrightarrow{\ g\ } & Y' \\
p\downarrow & = & \downarrow q \\
\uparrow(u_{\bullet}1) & \xrightarrow[g']{} & \uparrow(v_{\bullet}1)
\end{array}
\tag{C.92}
$$

This cell is the kernel of a in $\mathrm{Ver}_1\mathbb{M}\mathrm{lhc}$ (or $\mathrm{Ver}_1\mathbb{L}\mathrm{thc}$). Dually the cokernel of a is the lower cell above, where $p = \mathrm{cok}\,(u)$, $q = \mathrm{cok}\,(v)$ and the homomorphism g' is the restriction of g.

(b) Property (i) is already known, and we have verified above that $\mathrm{Ver}_1\mathbb{M}\mathrm{lhc}$ has kernels and cokernels, preserved by faces and degeneracy. Property (ii) follows now easily: the induced (bicommutative) cell from $\mathrm{Cok}\,(\ker\,(a))$ to $\mathrm{Ker}\,(\mathrm{cok}\,(a))$ is vertically invertible because its vertical arrows are.

(Note that this is not true in $\mathbb{L}\mathrm{thc}$: the category **Ltc** is not p-exact.)

(c) The product and sum of a finite family (X_i) in the category **Ltc** are both constructed with the cartesian product of lattices $X = \prod X_i$, and are biproducts

$$
X \xrightarrow{p_i} X_i \xrightarrow{m_i} X \qquad m_{i\bullet} \dashv m_i{}^{\bullet} = \mathrm{pr}_i = p_{i\bullet} \dashv p_i{}^{\bullet}. \tag{C.93}
$$

(Here pr_i is a cartesian projection, while $m_{i\bullet}$ and $p_i{}^{\bullet}$ are determined by the adjunctions.)

For an infinite family of lattices (X_i) the universal properties of product and sum would fail (unless we restrict to complete lattices).

C8.3 Exercise A3.2 (Zero object)

(i) $\Rightarrow$ (iii) $\Rightarrow$ (iv) Obvious. (iv) $\Rightarrow$ (i) Every object A has a zero morphism $A \to Z$; there cannot be any other morphism. The rest follows by duality.

C8.4 Exercises of A3.6 (Semiadditive categories)

(c) In **Ltc** the sum $f + g$ of two Galois connections $f, g \colon X \nrightarrow Y$ is defined as in (A.43)

$$
X \xrightarrow{\ d\ } X \times X \xrightarrow{\ f \times g\ } Y \times Y \xrightarrow{\ \partial\ } Y \tag{C.94}
$$

$$d_\bullet(x) = (x, x), \quad d^\bullet(x, x') = x \wedge x',$$
$$\partial_\bullet(y, y') = y \vee y', \quad \partial^\bullet(y) = (y, y),$$
$$f + g = \partial.(f \times g).d = (f_\bullet \vee g_\bullet, f^\bullet \wedge g^\bullet).$$

In other words, $f + g$ is the join $f \vee g$ with respect to the order relation of Galois connections, defined in (A.11). The sum is idempotent, and our structure is not additive.

Restricting to the category of *complete* lattices and Galois connections we even have arbitrary biproducts of objects and arbitrary sums of parallel morphisms $\vee f_i$.

(d) The sum of a family of relations $a_i \colon X \rightarrowtail Y$ is again their join $\bigcup a_i$, as subsets of $X \times Y$.

*(e) In $\mathbf{Ban}_1$ a binary product $X \times Y$ is the vector space $X \oplus Y$ with the l_∞-norm $||(x, y)||_\infty = ||x|| \vee ||y||$, while a binary sum $X + Y$ is the same vector space with the l_1-norm $||(x, y)||_1 = ||x|| + ||y||$.

These objects are not isomorphic in $\mathbf{Ban}_1$. But they are in $\mathbf{Ban}$, where the biproduct $X \oplus Y$ can be given any l_p-norm, for $1 \leqslant p \leqslant \infty$ (all of them being Lipschitz-equivalent).

C8.5 Exercises of A3.9 (Exact squares in abelian categories)

We are considering a commutative square in the abelian category $\mathbf{E}$

$$
\begin{array}{ccc}
X & \xrightarrow{f} & A \\
{\scriptstyle g}\downarrow & & \downarrow{\scriptstyle h} \\
B & \xrightarrow{k} & Y
\end{array}
\qquad\qquad (\text{C.95})
$$

(a) We know (by 1.3.4) that the pullback-object of the cospan (h, k) is the equaliser of the morphisms $hp_1, kp_2 \colon A \oplus B \to Y$, that is the following subobject of $A \oplus B$

$$\mathrm{Ker}\,(hp_1 - kp_2) = \mathrm{Ker}\,[h, -k].$$

Therefore the square (C.95) is a pullback if and only if $(f, g) \colon X \to A \oplus B$ is a monomorphism equivalent to the subobject $\mathrm{Ker}\,[h, -k]$ of $A \oplus B$.

(b) We have already noted that the pullback of (h, k) is $P = \mathrm{Ker}\,[h, -k]$; now, by (a*), the pushout of (f, g) is a cospan (h', k') with

$$\mathrm{Ker}\,[h', -k'] = \mathrm{Im}\,(f, g),$$

and its pullback P' is this subobject of $A \oplus B$. Thus our square is exact if and only if the two pullbacks P, P' are the same subobject of $A \oplus B$, which is condition (i*). The rest follows by duality.

(c) Proving that, in an abelian category, epimorphisms are closed under pullback is not trivial: a proof can be seen in [M4], Section VIII.4, Proposition 2. Now, if our square is a pullback, point (a) says that the sequence (A.45) is exact in the first two positions. Moreover the morphism $[h, -k] \colon A \oplus B \to Y$ is epi, because so is assumed to be h, and the sequence is also exact in B: then the square is a pushout, by (a).

*(d) We have seen that an abelian category has finite limits and colimits, and unique epi-mono factorisations, by a regular epi and a regular mono. It follows that every epimorhism is regular; these are closed under pullback, by the previous point.

*(e) Already proved in (b), once we know that **E** is regular, and the construction of Section 2.5 applies.

C8.6 Exercises of A4.8

(a) To show that the lax functor $G' = \mathrm{Rel}(G)$ is not strict, one can take the canonical projection $p \colon \mathbb{Z} \to \mathbb{Z}/2$. Then $pp^\sharp = 1 \colon \mathbb{Z}/2 \to \mathbb{Z}/2$ but $G(p) = 0 \colon 0 \to \mathbb{Z}/2$, whence $(Gp)(Gp)^\sharp \neq G'(pp^\sharp) = 1$. This also shows that we cannot compute G' over the binary factorisation $1 = pp^\sharp$, which is not strong.

Similarly, the colax extension $F' = \mathrm{Rel}(F)$ is not strict. The monomorphism $m = 2.- \colon \mathbb{Z} \to \mathbb{Z}$ gives $m^\sharp m = 1_{\mathbb{Z}}$ but $F(m) = 0 \colon \mathbb{Z}/2 \to \mathbb{Z}/2$, whence $(Fm)^\sharp(Fm) \neq F'(m^\sharp m) = 1$.

*(b) Here we write as $\mathrm{Rel}_{\mathrm{v}}\mathbf{E}$ the category of relations constructed in Section 2.5 (by subobjects of cartesian products), and as $\mathrm{Rel}_{\mathrm{w}}\mathbf{E}$ the present construction (by equivalence classes of w-diagrams).

Now, if a square of **E** is bicommutative in $\mathrm{Rel}_{\mathrm{w}}\mathbf{E}$, it factorises in four squares as in (A.55); all of them are pullbacks or pushouts, whence bicommutative in $\mathrm{Rel}_{\mathrm{v}}\mathbf{E}$ (by 2.5.6), and the original square is also.

Conversely, it is easy to see that a pullback is also bicommutative in $\mathrm{Rel}_{\mathrm{w}}\mathbf{E}$: in fact, if we factorise it in four squares, as previously, we get

- a pullback of monos,
- two 'mixed' bicartesian squares,
- a pullback of epis.

Now, in an abelian category a pullback of epis is also a pushout, by Exercise A3.9(d). Therefore each of these squares is bicommutative in $\mathrm{Rel}_{\mathrm{w}}\mathbf{E}$, and so is their pasting.

C8.7 Exercises of B2.6

(a) Consider the involutive permutation $\sigma_n \in S_n$

$$\sigma_n = (s_1 s_2 ... s_{n-1})...(s_1 s_2 s_3)(s_1 s_2)s_1, \tag{C.96}$$

which, acting on $\{1, ..., n\}$, reverses all indices. It is easy to verify that, for $i = 1, .., n-1$:

$$\partial_n^\alpha . \sigma_n = \sigma_{n-1}. \partial_1^\alpha, \qquad e_n.\sigma_{n-1} = \sigma_n.e_1, \qquad s_i.\sigma_n = \sigma_n.s_{n+1-i}.$$

According to the reduced presentation of symmetric cubical sets, in B2.2, we have thus a natural isomorphism $\sigma: 1 \to S$

$$\sigma X: X \to SX, \qquad x \mapsto \sigma_n.x,$$

$$
\begin{array}{ccc}
X_n & \xrightarrow{\ \sigma_n\ } & X_n \\[2pt]
{\scriptstyle \partial_1^\alpha}\Big\downarrow \Big\uparrow {\scriptstyle e_1} & & {\scriptstyle \partial_n^\alpha}\Big\downarrow \Big\uparrow {\scriptstyle e_n} \\[2pt]
X_{n-1} & \xrightarrow[\ \sigma_{n-1}\ ]{} & X_{n-1}
\end{array}
$$

(b) From $\sigma: 1 \to S$ we get an isomorphism

$$\gamma = \sigma PS.P\sigma: P \to PS \to SPS, \qquad \gamma_n: X_{n+1} \to X_{n+1},$$

$$x \mapsto (\sigma PSX)_n.(P(\sigma X))_n(x) = \sigma_n.\sigma_{n+1}.x$$

$$= s_n...s_1.x = \gamma_n.x \qquad (x \in X_{n+1}),$$

which is computed with the *main cyclic permutation* $\gamma_n = s_n...s_1 \in S_{n+1}$.

Its inverse is computed with the inverse cyclic permutation $\gamma_n' = s_1...s_n \in S_{n+1}$

$$\gamma': SPS \to P, \qquad \gamma_n': X_{n+1} \to X_{n+1}, \qquad x \mapsto \gamma_n'.x.$$

(c) We have seen that $PSPS = SPSP$, in (B.18). Thus the composite

$$\gamma' P.\mathrm{id}.P\gamma: P^2 \to PSPS = SPSP \to P^2,$$

$$x \mapsto (s_2...s_{n+1})(s_{n+1}...s_1)x = s_1 x,$$

gives indeed the transposition $s: P^2 \to P^2$ defined in (B.39).

References

[AHS] J. Adámek, H. Herrlich, G.E. Strecker, Abstract and concrete categories. The joy of cats, John Wiley & Sons, New York 1990.

[AM] M. Aguiar, S. Mahajan, Monoidal functors, species and Hopf algebras, CRM Monograph Series 29. American Mathematical Society, Providence, RI, 2010.

[AlBS] F.A.A. Al-Agl, R. Brown, R. Steiner, Multiple categories: the equivalence of a globular and a cubical approach, Adv. Math. 170 (2002), 71–118.

[AK] M.H. Albert, G.M. Kelly, The closure of a class of colimits, J. Pure Appl. Algebra 51 (1988), 1–17.

[Ba] M. Barr, Exact categories, in Lecture Notes in Math. Vol. 236, Springer-Verlag 1971, pp. 1–120.

[BaB] M. Barr, J. Beck, Homology and standard constructions, in Seminar on Triples and Categorical Homology Theory, Springer-Verlag 1969, pp. 245–335.

[BaE] A. Bastiani, C. Ehresmann, Multiple Functors I. Limits Relative to Double Categories, Cahiers Top. Géom. Diff. 15 (1974), 215–292.

[BBP] M.A. Bednarczyk, A.M. Borzyszkowski, W. Pawlowski, Generalized congruences–epimorphisms in Cat, Theory Appl. Categ. 5 (1999), No. 11, 266–280.

[Bc] J.M. Beck, Triples, algebras and cohomology, Repr. Theory Appl. Categ. 2 (2003).

[Be1] J. Bénabou, Introduction to bicategories, in: Reports of the Midwest Category Seminar, Lecture Notes in Math. Vol. 47, Springer-Verlag 1967, pp. 1–77.

[Be2] J. Bénabou, Les distributeurs, Inst. de Math. Pure et Appliquée, Université Catholique de Louvain, Rapport n. 33, 1973.

[Bi] G. Birkhoff, Lattice theory, 3rd ed., Amer. Math. Soc. Coll. Publ. 25, Providence 1973.

[BlKP] R. Blackwell, G.M. Kelly, A.J. Power, Two-dimensional monad theory, J. Pure Appl. Algebra 59 (1989), 1–41.

[BKPS] G.J. Bird, G.M. Kelly, A.J. Power, R.H. Street, Flexible limits for 2-categories, J. Pure Appl. Algebra 61 (1989), 1–27.

[BhCZ] G. Böhm, Y. Chen, L. Zhang, On Hopf monoids in duoidal categories, J. Algebra 394 (2013), 139–172.

[BkS] T. Booker, R. Street, Tannaka duality and convolution for duoidal categories, Theory Appl. Categ. 28 (2013), No. 6, 166–205.

[Bo1] F. Borceux, Handbook of categorical algebra 1, Cambridge University Press, Cambridge 1994.

509

[Bo2] F. Borceux, Handbook of categorical algebra 2, Cambridge University Press, Cambridge 1994.

[Bo3] F. Borceux, Handbook of categorical algebra 3, Cambridge University Press, Cambridge 1994.

[Bo4] F. Borceux, A survey of semi-abelian categories, in: Galois theory, Hopf algebras, and semiabelian categories, Fields Inst. Commun. Vol. 43, Amer. Math. Soc., Providence 2004, pp. 27–60.

[BoB] F. Borceux, D. Bourn, Mal'cev, protomodular, homological and semi-abelian categories, Kluwer Academic Publishers, Dordrecht 2004.

[Boo] P. Booth, Sequences of adjoint functors, Arch. Math. 23 (1972), 489–493.

[Bou1] D. Bourn, Normalization equivalence, kernel equivalence and affine categories, in: Category theory (Como, 1990), Lecture Notes in Math. Vol. 1488, Springer-Verlag 1991, pp. 43–62.

[Bou2] D. Bourn, Moore normalization and Dold–Kan theorem for semi-abelian categories, in: Categories in algebra, geometry and mathematical physics, 105–124, Amer. Math. Soc., Providence 2007.

[Bra] H. Brandt, Über eine Verallgemeinerung des Gruppenbegriffes, Math. Ann. 96 (1927), 360–366.

[Bri] H.B. Brinkmann, Relations for exact categories, J. Algebra 13 (1969), 465–480.

[BriP] H.B. Brinkmann, D. Puppe, Abelsche und exacte Kategorien, Korrespondenzen, Lecture Notes in Math. Vol. 96, Springer-Verlag 1969.

[Bro1] R. Brown, Elements of modern topology, McGraw–Hill, New York 1968.

[Bro2] R. Brown, Topology and groupoids, Third edition of Elements of modern topology, BookSurge, LLC, Charleston SC 2006.

[BroH1] R. Brown, P.J. Higgins, On the algebra of cubes, J. Pure Appl. Algebra 21 (1981), 233–260.

[BroH2] R. Brown, P.J. Higgins, Tensor products and homotopies for w-groupoids and crossed complexes, J. Pure Appl. Algebra 47 (1987), 1–33.

[BroM] R. Brown, G. Mosa, Double categories, 2-categories, thin structures and connections, Theory Appl. Categ. 5 (1999), 163–175

[BruMM] R. Bruni, J. Meseguer, U. Montanari, Symmetric monoidal and Cartesian double categories as a semantic framework for tile logic, Math. Structures Comput. Sci. 12 (2002), 53–90.

[Bu1] D.A. Buchsbaum, Exact categories and duality, Trans. Amer. Math. Soc. 80 (1955), 1–34.

[Bu2] D.A. Buchsbaum, Appendix, in: H. Cartan, S. Eilenberg, Homological algebra, Princeton University Press, Princeton 1956.

[Bur] A. Burroni, Structures pseudo-algébriques, I, Cah. Topol. Géom. Différ. 16 (1975), 343–393.

[CaKVW] A. Carboni, G.M. Kelly, D. Verity, R. Wood, A 2-categorical approach to change of base and geometric morphisms II, Theory Appl. Categ. 4 (1998), 82–136.

[CaR] A. Carboni, R. Rosebrugh, Lax monads, indexed monoidal monads, J. Pure Appl. Algebra 76 (1991), 13–32.

[ChG] E. Cheng, N. Gurski, Towards an n-category of cobordisms, Theory Appl. Categ. 18 (2007), 274–302.

[ChSW] A. Cherubini, N. Sabadini, R.F.C. Walters, Timing in the Cospan-Span Model, Electronic Notes in Theoretical Computer Science 104C (2004), 81–97.

[Coh] P.M. Cohn, Universal algebra, Harper & Row, New York 1965.

[CoM] H.S.M. Coxeter, W.O.J. Moser, Generators and relations for discrete groups, Springerr-Verlag 1957.

[Da] R. Dawson, A forbidden-suborder characterization of binarily-composable diagrams in double categories, Theory Appl. Categ. 1 (1995), 146–155.

[DaP1] R. Dawson, R. Paré, General associativity and general composition for double categories, Cahiers Top. Géom. Diff. Catég. 34 (1993), 57–79.

[DaP2] R. Dawson, R. Paré, What is a free double category like?, J. Pure Appl. Algebra 168 (2002), 19–34.

[DaPP1] R. Dawson, R. Paré, D.A. Pronk, Free extensions of double categories, Cah. Topol. Géom. Différ. Catég. 45 (2004), 35–80.

[DaPP2] R. Dawson, R. Paré, D.A. Pronk, The span construction, Theory Appl. Categ. 24 (2010), 302–377.

[Du] J. Duskin, Variations on Beck's tripleability criterion, Lecture Notes in Math. 106, Springerr-Verlag 1969, pp. 74–129.

[EE] A. Ehresmann, C. Ehresmann, Multiple functors II. The monoidal closed category of multiple categories, Cahiers Top. Géom. Diff. 19 (1978), 295–333.

[Eh1] C. Ehresmann, Catégories des foncteurs types, Rev. Un. Mat. Argentina 20 (1962), 194–209.

[Eh2] C. Ehresmann, Catégories structurées, Ann. Sci. Ecole Norm. Sup. 80 (1963), 349–425.

[Eh3] C. Ehresmann, Catégorie double des quintettes; applications covariantes, C. R. Acad. Sci. Paris 256 (1963), 1891–1894.

[Eh4] C. Ehresmann, Catégories et structures, Dunod, Paris 1965.

[Eh5] C. Ehresmann, Oeuvres complètes et commentées, Partie II.2, Editée et commentée par Andrée C. Ehresmann, Amiens 1982.

[EiM] S. Eilenberg, S. Mac Lane, General theory of natural equivalences, Trans. Amer. Math. Soc. 58 (1945), 231–294.

[EiK] S. Eilenberg, G.M. Kelly, Closed categories, in Proc. Conf. Categorical Algebra, La Jolla 1965, Springer-Verlag 1966, pp. 421–562.

[Fi] T.M. Fiore, Pseudo algebras and pseudo double categories, J. Homotopy Relat. Struct. 2 (2007), 119–170.

[FGK1] T.M. Fiore, N. Gambino, J. Kock, Monads in double categories, J. Pure Appl. Algebra 215 (2011), 1174–1197.

[FGK2] T.M. Fiore, N. Gambino, J. Kock, Double adjunctions and free monads, Cah. Topol. Géom. Différ. Catég. 53 (2012), 242–306.

[Fr1] P. Freyd, Abelian categories, An introduction to the theory of functors, Harper & Row, New York 1964. Republished in: Reprints Theory Appl. Categ. 3 (2003).

[Fr2] P. Freyd, On the concreteness of certain categories, in: Symposia Mathematica, Vol. IV, INDAM, Roma 1968/69. Academic Press, London 1970, pp. 431–456.

[Fr3] P. Freyd, Homotopy is not concrete, in: The Steenrod algebra and its applications, Battelle Memorial Inst., Columbus, OH, 1970, Lecture Notes in Math. Vol. 168, Springer-Verlag 1970, pp. 25–34.

[FrS] P.J. Freyd, A. Scedrov, Categories, allegories, North-Holland Publ. Co., Amsterdam 1990.

[Ga] R. Garner, Double clubs, Cah. Topol. Géom. Différ. Catég. 47 (2006), 261–317.

[GaG] R. Garner, N. Gurski, The low-dimensional structures formed by tricategories, Math. Proc. Cambridge Philos. Soc. 146 (2009), 551–589.

[GiKS] A. Gianola, S. Kasangian, N. Sabadini, Cospan/Span(Graph): an algebra for open, reconfigurable automata network, in: 7th Conference on Algebra and Coalgebra in Computer Science, Art. No. 2, 17 pp., Leibniz-Zent. Inform., Wadern 2017.

[GPS] R. Gordon, A.J. Power, R. Street, Coherence for tricategories, Mem. Amer. Math. Soc. 117 (1995), n. 558.

[G1] M. Grandis, Cohesive categories and manifolds, Ann. Mat. Pura Appl. 157 (1990), 199–244.

[G2] M. Grandis, Higher cospans and weak cubical categories (Cospans in Algebraic Topology, I), Theory Appl. Categ. 18 (2007), No. 12, 321–347.

[G3] M. Grandis, Collared cospans, cohomotopy and TQFT (Cospans in Algebraic Topology, II), Theory Appl. Categ. 18 (2007), No. 19, 602–630.

[G4] M. Grandis, Cubical cospans and higher cobordisms (Cospans in Algebraic Topology, III), J. Homotopy Relat. Struct. 3 (2008), 273–308.

[G5] M. Grandis, The role of symmetries in cubical sets and cubical categories (On weak cubical categories, I), Cah. Topol. Géom. Différ. Catég. 50 (2009), 102–143.

[G6] M. Grandis, Limits in symmetric cubical categories (On weak cubical categories, II), Cah. Topol. Géom. Différ. Catég. 50 (2009), 242–272.

[G7] M. Grandis, Directed Algebraic Topology, Models of non-reversible worlds, Cambridge University Press, Cambridge 2009. Available at: http://www.dima.unige.it/~grandis/BkDAT_page.html

[G8] M. Grandis, Homological Algebra, The interplay of homology with distributive lattices and orthodox semigroups, World Scientific Publishing Co., Hackensack NJ 2012.

[G9] M. Grandis, Homological Algebra in strongly non-abelian settings, World Scientific Publishing Co., Hackensack NJ 2013.

[G10] M. Grandis, Adjoints for symmetric cubical categories (On weak cubical categories, III), Cah. Topol. Géom. Différ. Catég. 54 (2013), 91–126.

[G11] M. Grandis, Category Theory and Applications: A textbook for beginners, World Scientific Publishing Co., Hackensack NJ 2018.

[GM] M. Grandis, L. Mauri, Cubical sets and their site, Theory Appl. Categ. 11 (2003), No. 8, 185–211.

[GP1] M. Grandis, R. Paré, Limits in double categories, Cah. Topol. Géom. Différ. Catég. 40 (1999), 162–220.

[GP2] M. Grandis, R. Paré, Adjoint for double categories, Cah. Topol. Géom. Différ. Catég. 45 (2004), 193–240.

[GP3] M. Grandis, R. Paré, Kan extensions in double categories (On weak double categories, Part III), Theory Appl. Categ. 20 (2008), No. 8, 152–185.

[GP4] M. Grandis, R. Paré, Lax Kan extensions for double categories (On weak double categories, Part IV), Cah. Topol. Géom. Différ. Catég. 48 (2007), 163–199.

[GP5] M. Grandis, R. Paré, From cubical to globular higher categories, in: Liber Amicorum en l'honneur de M.me A.C. Ehresmann, Diagrammes 67–68 suppl. (2012), 117–148.

[GP6] M. Grandis, R. Paré, Intercategories, Theory Appl. Categ. 30 (2015), No. 38, 1215–1255.

[GP7] M. Grandis, R. Paré, Intercategories: a framework for three dimensional category theory, J. Pure Appl. Algebra 221 (2017), 999–1054.

[GP8] M. Grandis, R. Paré, An introduction to multiple categories (On weak and lax multiple categories, I), Cah. Topol. Géom. Différ. Catég. 57 (2016), 103–159.

[GP9] M. Grandis, R. Paré, Limits in multiple categories (On weak and lax multiple categories, II), Cah. Topol. Géom. Différ. Catég. 57 (2016), 163–202.

[GP10] M. Grandis, R. Paré, Adjoints for multiple categories (On weak and lax multiple categories, III), Cah. Topol. Géom. Différ. Catég. 58 (2017), 3–48.

[GP11] M. Grandis, R. Paré, Span and cospan representations of weak double categories, Categ. Gen. Algebr. Struct. Appl. 6 (2017), 85–105.

[GP12] M. Grandis, R. Paré, A multiple category of lax multiple categories, Cah. Topol. Géom. Différ. Catég. 58 (2017), 195–212.

[GP13] M. Grandis, R. Paré, Persistent double limits, Cah. Topol. Géom. Différ. Catég., to appear.

[GP14] M. Grandis, R. Paré, Persistent double limits and flexible weighted limits, in preparation.

[GT] M. Grandis, W. Tholen, Natural weak factorisation systems, Archivum Mathematicum (Brno) 42 (2006), 397–408.

[Gr1] G. Grätzer, Universal algebra, Van Nostrand Co., Princeton 1968.

[Gr2] G. Grätzer, General lattice theory, Academic Press, New York 1978.

[Gra1] J.W. Gray, Formal category theory: adjointness for 2-categories, Lecture Notes in Mathematics, Vol. 391, Springer-Verlag 1974.

[Gra2] J.W. Gray, The existence and construction of lax limits, Cah. Topol. Géom. Différ. 21 (1980), 277–304.

[Grt] P.A. Grillet, Exact categories and categories of sheaves, in Lecture Notes in Math. Vol. 236, Springer-Verlag 1971, pp. 121–222.

[Gt] A. Grothendieck, Sur quelques points d'algèbre homologique, Tôhoku Math. J. 9 (1957), 119–221.

[HeS] H. Herrlich, G.E. Strecker, Category theory, an introduction, Allyn and Bacon, Boston 1973.

[Hi] P.J. Hilton, Correspondences and exact squares, in: Proc. of the Conf. on Categorical Algebra, La Jolla 1965, Springer-Verlag 1966, 255–271.

[Hr] T. Hirschowitz, Full abstraction for fair testing in CCS (expanded version), Log. Methods Comput. Sci. 10 (2014), no. 4, 4:2, 80 pp.

[JaMT] G. Janelidze, L. Márki, W. Tholen, Semi-abelian categories, in: Category theory 1999 (Coimbra), J. Pure Appl. Algebra 168 (2002), 367–386.

[Jh] D.L. Johnson, Topics in the theory of presentation of groups, Cambridge Univ. Press, Cambridge 1980.

[Jo] P.T. Johnstone, Topos theory, Academic Press, London 1977.

[K1] D.M. Kan, Abstract homotopy I, Proc. Nat. Acad. Sci. U.S.A. 41 (1955), 1092–1096.

[K2] D.M. Kan, Abstract homotopy II, Proc. Nat. Acad. Sci. U.S.A. 42 (1956), 255–258.

[K3] D.M. Kan, Adjoint functors, Trans. Amer. Math. Soc. 87 (1958), 294–329.

[KaRT] C. Kassel, M. Rosso, V. Turaev, Quantum groups and knot invariants, Panoramas et Synthèses, 5. Soc. Math. de France, Paris 1997.

[KaSW] P. Katis, N. Sabadini, R.F.C. Walters, Span(Graph): A categorical algebra of transition systems, in: Proc. AMAST 97, SLNCS 1349, pp. 307–321, Springerr-Verlag 1997.

[Ke1] G.M. Kelly, On Mac Lane's conditions for coherence of natural associativities, commutativities, etc., J. Algebra 1 (1964), 397–402.

[Ke2] G.M. Kelly, Basic concepts of enriched category theory, Cambridge University Press, Cambridge 1982.

[Ke3] G.M. Kelly, Elementary observations on 2-categorical limits, Bull. Austral. Math. Soc. 39 (1989), 301–317.

[KeS] G.M. Kelly, R. Street, Review of the elements of 2-categories, in: Category Seminar, Sydney 1972/73, Lecture Notes in Math. Vol. 420, Springer-Verlag 1974, pp. 75–103.

[Kk] J. Kock, Frobenius algebras and 2D topological quantum field theories, Cambridge University Press, Cambridge 2004.

[Ko1] S.R. Koudenburg, On pointwise Kan extensions in double categories, Theory Appl. Categ. 29 (2014), 781–818.

[Ko2] S.R. Koudenburg, Algebraic Kan extensions in double categories, Theory Appl. Categ. 30 (2015), 86–146.

[La] M.L. Laplaza, Coherence for categories with group structure: an alternative approach, J. Alg. 84 (1983), 305–323.

[Law] F.W. Lawvere, Metric spaces, generalized logic and closed categories, Rend. Sem. Mat. Fis. Univ. Milano 43 (1974), 135–166. Republished in: Reprints Theory Appl. Categ. 1 (2002).

[Le] T. Leinster, Higher operads, higher categories, Cambridge University Press, Cambridge 2004.

[M1] S. Mac Lane, An algebra of additive relations, Proc. Nat. Acad. Sci. USA 47 (1961), 1043–1051.

[M2] S. Mac Lane, Homology, Springer-Verlag 1963.

[M3] S. Mac Lane, Natural associativity and commutativity, Rice University Studies 49 (1963), 28–46.

[M4] S. Mac Lane, Categories for the working mathematician, Springer-Verlag 1971.

[MaM] S. Mac Lane, I. Moerdijk, Sheaves in geometry and logic. A first introduction to topos theory, Springer-Verlag 1994.

[MaP] S. Mac Lane, R. Paré, Coherence for bicategories and indexed categories, J. Pure Appl. Algebra 37 (1985), 59–80.

[Mar] N. Martins-Ferreira, Pseudo-Categories, J. Homotopy Relat. Struct. 1 (2006), 47–78.

[Mat] M. Mather, Pull-backs in homotopy theory, Can. J. Math. 28 (1976), 225–263.

[Me] J. Meisen, Relations in regular categories, in: Localization in group theory and homotopy theory and related topics, Lecture Notes in Math. Vol. 418, Springer-Verlag 1974, pp. 96–102.

[Mic] E. Michael, Local compactness and Cartesian products of quotient maps and k-spaces, Ann. Inst. Fourier (Grenoble) 18 (1968), 281–286.

[Mit] B. Mitchell, Theory of categories, Academic Press, New York 1965.

[Mo] J.C. Morton, Double bicategories and double cospans, J. Homotopy Relat. Struct. 4 (2009), 389–428.

[Ni] S. Niefield, Span, cospan, and other double categories, Theory Appl. Categ. 26 (2012), 729–742.

[Pa1] R. Paré, Double limits, International Category Theory Meeting, Bangor 1989, unpublished talk.

[Pa2] R. Paré, Simply connected limits, Canad. J. Math. 42 (1990), 731–746.

[Pa3] R. Paré, Yoneda theory for double categories, Theory Appl. Categ. 25 (2011), 436–489.

[Pa4] R. Paré, Composition of modules for lax functors, Theory Appl. Categ. 27 (2012), 393–444.

[Pa5] R. Paré, Wobbly double functors, Tbilisi Mathematical Journal 8 (2015), 63–84.

[Pa6] R. Paré, Morphisms of rings, in preparation.

[PaS] R. Paré, D. Schumacher, Indexed categories and their applications, in: Lecture Notes in Math. 661, Springer-Verlag 1978, pp. 1–125.

[Pu] D. Puppe, Korrespondenzen in abelschen Kategorien, Math. Ann. 148 (1962), 1–30.

[Qu] D. Quillen, Higher algebraic K-theory, I, in: Lecture Notes in Math. Vol. 341, Springer-Verlag 1973, pp. 85–147.

[Sh] K. Shimakawa, Multiple categories and algebraic K-theory, J. Pure Appl. Algebra 41 (1986), 285–304.

[Shu] M.A. Shulman, Constructing symmetric monoidal bicategories, Available at: arXiv:1004.0993 [math.CT].

[St1] R. Street, The formal theory of monads, J. Pure Appl. Algebra 2 (1972), 149–168.

[St2] R. Street, Limits indexed by category-valued 2-functors, J. Pure Appl. Alg. 8 (1976), 149–181.

[St3] R. Street, Cosmoi of internal categories, Trans. Amer. Math. Soc. 258 (1980), 271–318.

[St4] R. Street, Fibrations in bicategories, Cah. Topol. Géom. Différ. 21 (1980), 111–159.

[St5] R. Street, Monoidal categories in, and linking, geometry and algebra, Bull. Belg. Math. Soc. Simon Stevin 19 (2012), 769–821.

[T1] M.S. Tsalenko, Correspondences over a quasi exact category, Dokl. Akad. Nauk SSSR 155 (1964), 292–294.

[T2] M.S. Tsalenko, Correspondences over a quasi exact category, Mat. Sbornik 73 (1967), 564–584.

[Ve] D. Verity, Enriched categories, internal categories and change of base, Ph.D. Dissertation, University of Cambridge, 1992. Published in: Repr. Theory Appl. Categ. No. 20 (2011), 1–266.

[Ye] D.N. Yetter, Functorial knot theory, Categories of tangles, coherence, categorical deformations, and topological invariants, World Scientific Publishing Co., Inc., River Edge, NJ, 2001.

Index